Bio-Inspired Artificial Intelligence

Intelligent Robotics and Autonomous Agents
Edited by Ronald C. Arkin

Dorigo, Marco, and Marco Colombetti, Robot Shaping: *An Experiment in Behavior Engineering*

Arkin, Ronald C., *Behavior-Based Robotics*

Stone, Peter, Layered Learning in Multiagent Systems: *A Winning Approach to Robotic Soccer*

Wooldridge, Michael, *Reasoning about Rational Agents*

Murphy, Robin R., *An Introduction to AI Robotics*

Mason, Matthew T., *Mechanics of Robotic Manipulation*

Kraus, Sarit, *Strategic Negotiation in Multiagent Environments*

Nolfi, Stefano, and Dario Floreano, *Evolutionary Robotics: The Biology, Intelligence, and Technology of Self-Organizing Machines*

Siegwart, Roland, and Illah R. Nourbakhsh, *Introduction to Autonomous Mobile Robots*

Breazeal, Cynthia L., *Designing Sociable Robots*

Bekey, George A., *Autonomous Robots: From Biological Inspiration to Implementation and Control*

Choset, Howie, Kevin M. Lynch, Seth Hutchinson, George Kantor, Wolfram Burgard, Lydia E. Kavraki, and Sebastian Thrun, *Principles of Robot Motion: Theory, Algorithms, and Implementations*

Thrun, Sebastian, Wolfram Burgard, and Dieter Fox, *Probabilistic Robotics*

Mataric, Maja J., *The Robotics Primer*

Wellman, Michael P., Amy Greenwald, and Peter Stone, *Autonomous Bidding Agents: Strategies and Lessons from the Trading Agent Competition*

Floreano, Dario, and Claudio Mattiussi, *Bio-Inspired Artificial Intelligence: Theories, Methods, and Technologies*

Bio-Inspired Artificial Intelligence
Theories, Methods, and Technologies

Dario Floreano
Claudio Mattiussi

The MIT Press
Cambridge, Massachusetts
London, England

MIT Press books may be purchased at special quantity discounts for business or sales promotional use. For information, please email special_sales@ mitpress.mit.edu or write to Special Sales Department, The MIT Press, 55 Hayward Street, Cambridge, MA 02142.

This book was set by the authors using LyX and LaTeX.
Printed and bound in the United States of America.

Library of Congress Cataloging-in-Publication Data

Floreano, Dario.
   Bio-inspired artificial intelligence : theories, methods, and technologies / Dario Floreano and Claudio Mattiussi.
      p. cm. – (Intelligent robotics and autonomous agents series)
   Includes bibliographical references and index.
   ISBN 978-0-262-06271-8 (hardcover : alk. paper)
   1. Artificial intelligence–Data processing. 2. Biologically-inspired computing. 3. Self-organizing systems. 4. Autonomous robots. I. Mattiussi, Claudio. II. Title.
Q336.F66 2008
006.3–dc22

                                                              2008008739

10   9   8   7   6   5   4   3   2

In memory of Annamaria Collovini Floreano – D. F.

To Betty – C. M.

Contents

Preface   **xi**

Acknowledgments   **xiii**

1 *Evolutionary Systems*   **1**

1.1   Pillars of Evolutionary Theory   2
1.2   The Genotype   5
1.3   Artificial Evolution   13
1.4   Genetic Representations   16
1.5   Initial Population   21
1.6   Fitness Functions   22
1.7   Selection and Reproduction   23
1.8   Genetic Operators   26
1.9   Evolutionary Measures   29
1.10 Types of Evolutionary Algorithms   33
1.11 Schema Theory   37
1.12 Human-Competitive Evolution   39
1.13 Evolutionary Electronics   42
1.14 Lessons from Evolutionary Electronics   43
1.15 The Role of Abstraction   45
1.16 Analog and Digital Circuits   49
1.17 Extrinsic and Intrinsic Evolution   53
1.18 Digital Design   58
1.19 Evolutionary Digital Design   62
1.20 Analog Design   77
1.21 Evolutionary Analog Design   79
1.22 Multiple Objectives and Constraints   85

1.23 Design Verification   90

1.24 Closing Remarks   92

1.25 Suggested Readings   97

2 *Cellular Systems*   101

2.1 The Basic Ingredients   101

2.2 Cellular Automata   107

2.3 Modeling with Cellular Systems   110

2.4 Some Classic Cellular Automata   118

2.5 Other Cellular Systems   124

2.6 Computation   134

2.7 Artificial Life   138

2.8 Complex Systems   145

2.9 Analysis and Synthesis of Cellular Systems   153

2.10 Closing Remarks   159

2.11 Suggested Readings   160

3 *Neural Systems*   163

3.1 Biological Nervous Systems   167

3.2 Artificial Neural Networks   175

3.3 Neuron Models   177

3.4 Architecture   189

3.5 Signal Encoding   191

3.6 Synaptic Plasticity   196

3.7 Unsupervised Learning   198

3.8 Supervised Learning   219

3.9 Reinforcement Learning   235

3.10 Evolution of Neural Networks   238

3.11 Neural Hardware   250

3.12 Hybrid Neural Systems   256

3.13 Closing Remarks   261

3.14 Suggested Readings   265

4 *Developmental Systems*   269

4.1 Potential Advantages of a Developmental Representation   270

4.2 Rewriting Systems   272

4.3 Synthesis of Developmental Systems   296

4.4 Evolution and Development   298

4.5 Defining Artificial Evolutionary Developmental Systems   299

4.6   Evolutionary Rewriting Systems   301
4.7   Evolutionary Developmental Programs   310
4.8   Evolutionary Developmental Processes   315
4.9   Closing Remarks   332
4.10   Suggested Readings   334

5 *Immune Systems*   **335**
5.1   How Biological Immune Systems Work   337
5.2   The Constituents of Biological Immune Systems   353
5.3   Lessons for Artificial Immune Systems   366
5.4   Algorithms and Applications   373
5.5   Shape Space   375
5.6   Negative Selection Algorithm   384
5.7   Clonal Selection Algorithm   388
5.8   Examples   390
5.9   Closing Remarks   395
5.10   Suggested Readings   396

6 *Behavioral Systems*   **399**
6.1   Behavior in Cognitive Science   400
6.2   Behavior in Artificial Intelligence   403
6.3   Behavior-Based Robotics   407
6.4   Biological Inspiration for Robots   419
6.5   Robots as Biological Models   437
6.6   Robot Learning   449
6.7   Evolution of Behavioral Systems   460
6.8   Evolution and Learning in Behavioral Systems   482
6.9   Evolution and Neural Development in Behavioral Systems   494
6.10   Coevolution of Body and Control   499
6.11   Toward Self-Reproduction   504
6.12   Simulation and Reality   507
6.13   Closing Remarks   511
6.14   Suggested Readings   513

7 *Collective Systems*   **515**
7.1   Biological Self-Organization   516
7.2   Particle Swarm Optimization   524
7.3   Ant Colony Optimization   527
7.4   Swarm Robotics   531

7.5   Coevolutionary Dynamics: Biological Models   547
7.6   Artificial Evolution of Competing Systems   554
7.7   Artificial Evolution of Cooperation   572
7.8   Closing Remarks   581
7.9   Suggested Readings   583

Conclusion   **585**

References   **587**

Index   **651**

Preface

This book is an introduction to the bio-inspired artificial intelligence that is emerging in the twenty-first century. For almost fifty years, mainstream artificial intelligence focused on creating computers and algorithms that displayed human cognitive abilities. Over time, it gradually departed from its original source of inspiration – biological intelligence – and became increasingly concerned with efficient signal processing, optimal control, and data mining.

Mainstream artificial intelligence has been very successful at designing algorithms and devices that solve problems that most humans are not very good at, such as playing chess, controlling aircraft dynamics, or finding the three-dimensional structure of proteins. But, in doing so, it ended up neglecting fundamental aspects of biological intelligence, such as physical embodiment, behavioral autonomy, self-healing, social interaction, evolution and learning, that make biological organisms prone to errors and sometimes difficult to predict, but also so successful to survive in unknown and changing environments.

The mid-1980s witnessed a renaissance of diverse approaches to the understanding and engineering of intelligent systems. A range of newly born fields, such as embodied cognitive science, neuromorphic engineering, artificial life, behavior-based robotics, evolutionary robotics, and swarm intelligence, to mention only a few, questioned the validity of the assumptions and methods of mainstream artificial intelligence for creating artifacts that could approximate the operational characteristics and performance of biological intelligence.

The *new artificial intelligence*, as it is sometimes called, that emerged at the turn of the millennium expanded its focus of attention from human brains and cognitive reasoning to a wider range of organisms, processes, and phenomena that occur at multiple spatial and temporal scales. This change reflected not only a philosophical revolution where humans are no longer at the center of the biological universe, but also a technological revolution where desktop computers are dissolving into a swarm of virtual and physical artifacts (Internet agents, virtual personae, personal digital assistants, communication devices, mobile robots, intelligent prostheses, etc.) in need of real-time and embedded intelligence, autonomous behavior, self-adaptation, and social awareness to interact, merge, and substitute with us.

This book aims at providing a systematic introduction to the theories and methods of bio-inspired artificial intelligence, and at providing a toolbox of design principles for engineers. Theories and methods are accompanied by sample software and hardware technologies to illustrate the application bio-inspired artificial intelligence. The material is organized into seven chapters that gradually guide the reader through biological and artificial systems that operate at different temporal and spatial scales. On the temporal scale, it progresses from systems that change at evolutionary pace to systems that develop and learn during their lifetime, all the way to systems that interact in real time with the environment and with other individuals. On the spatial scale, it progresses from cells and neurons to multicellular organisms, all the way to societies of individuals. Most systems are introduced by a presentation of biological theories and observations followed by a description of engineering methods and technologies. Each chapter concludes with pointers to new avenues and suggestions for further readings.

Approaches that combine methods and technologies operating at multiple temporal and spatial scales are introduced only at a stage when their constituent elements have already been introduced. For example, evolutionary robotics is described in the chapter on behavioral systems, and not in the earlier chapter on evolutionary systems, because its full appreciation requires also an understanding of neural systems, of developmental systems, and of behavioral systems. The keywords on the text side are designed to allow a rapid browsing through the book. Whether you are a student, an advanced researcher, or simply curious about the future of artificial intelligence, we hope that you will find this book instructive and useful.

Acknowledgments

This idea of writing this book developed from a master's course taught at EPFL by the authors for almost ten years. The course quickly became so successful that we had to restrict access in order to maintain a high standard of teaching. Our main problem however was the lack of a suitable textbook that would accompany lectures and exercises.

Although there are many books that describe specific aspects of the field, such as neural networks, evolutionary algorithms, behavior-based robotics, or cellular automata, there was no book that spanned the entire field, covering both theories, algorithms, and hardware technologies. We felt the need for a book that would not only provide a systematic organization of this material for students, but also serve as a basis for the intellectual consolidation of the emerging field of bio-inspired artificial intelligence for advanced scientists and engineers.

Bob Prior at MIT Press liked the book proposal from the very beginning and, after a quick round of peer reviews, offered us a contract. We were much slower than expected in delivering the manuscript. It took us almost four years to digest the literature cited in this book, extract salient aspects, and organize it in a coherent story that would make sense (in our opinion) for both students and advanced scholars. Bob continuously encouraged us to persist, and accepted to extend and modify the contract a few times as the book was slowly taking shape. When the manuscript was finally ready, Ada Brunstein took over the project as Bob moved to another division within MIT Press. Ada and her team maintained the same level of energy, friendly encouragement, and effective project management during the final production stages.

Many people helped with various aspects of this book. Daniel Marbach designed the icons on the cover illustration and drew several figures that appear in the text. Mototaka Suzuki collected and organized on a website many software tools that may be used to experiment with the approaches described in the book. Along with them, Silvano Chialina, Elisabetta Dosso, Peter Dürr, Sabine Hauert, Julien Hubert, Sara Mitri, and Markus Waibel proofread and commented earlier drafts of some chapters. The feedback of the hundreds of students who took the master's course over the years was extremely valuable to improve the organization of the material and the selection of examples used to illustrate theories and methods.

Many colleagues kindly provided us with originals of the illustrations from their own work, often reworking the digital format to comply with the publisher's guidelines. Their names are individually acknowledged in the figure legends. Similarly, several publishers allowed us to reproduce copyrighted material for free or for a symbolic fee. The copyright holders are acknowledged in the figure legends too.

EPFL provided a unique environment to carry out this project, giving us both state-of-the-art infrastructure and freedom to pursue challenging and innovative projects. The strengthening of life science and the encouragement of interdisciplinary research between engineers and biologists actively pursued by EPFL created a stimulating and dynamic atmosphere that was ideal for the writing of this book. Along with EPFL, the Swiss National Science Foundation and the Information Society and Technology division of the European Commission sponsored several research projects that allowed us to explore and actively contribute to the development of bio-inspired artificial intelligence. We especially appreciate the future-oriented and risk-taking perspectives of these sponsoring organizations that are so important for pursuing creative and innovative research.

We are also indebted to our biologist colleagues who shared with us their knowledge and insights in various collaborative projects. In particular, we would like to mention Laurent Keller, who guided us with great clarity through the evolutionary theories of cooperation and competition; Jean-Luis Deneubourg and Jose Halloy, who introduced us to the fascinating world of self-organizing behavior in social animals; and Mandyam Srinivasan and Nicolas Franceschini, who showed us the elegant mechanisms and behavioral strategies that insects use for vision-based flight.

It is not exaggerated to say that this book would have never been completed without the support of our partners. Krisztina and Betty made sure that we never gave up with the project when tempted to do so and patiently contemplated the prospect of its completion as we spent numerous evenings, weekends, and holidays reading and writing.

1   *Evolutionary Systems*

All biological systems result from an evolutionary process. The sophistication, robustness, and adaptability of biological systems represent a powerful motivation for replicating the mechanisms of natural evolution in the attempt to generate software and hardware systems with characteristics comparable to those of biological systems. More than 40 years ago, computer scientists and engineers began developing algorithms inspired by natural evolution (Rechenberg 1965; Fogel et al. 1966; Holland 1975) to generate solutions to problems that were too difficult to tackle with other analytical methods. Evolutionary computation rapidly became a major field of machine learning and system optimization and, more recently, it spread into the area of hardware design by exploiting new technologies in reconfigurable electronic circuits, computer-assisted manufacturing, material production technologies, and robotics.

Before delving into the features of natural and artificial evolution, we wish to emphasize that there is a major, and often neglected, difference between these two processes. Whereas natural evolution does not have a predefined goal and is essentially an open-ended adaptation process, artificial evolution is an optimization process that attempts to find solutions to predefined problems. Therefore, while in natural evolution the fitness of an individual is defined by its reproductive success (number of offspring), in artificial evolution the fitness of an individual is a function that measures how well that individual solves a predefined problem.

The consequence of this difference is that artificial evolution, as it is formulated today, cannot possibly hope to match the diversity and creativity generated by natural evolution because, by definition, artificially evolved systems will all tend to satisfy the predefined problem.

Even in the context of problem solving, artificial evolution is sometimes criticized by engineers because it contains elements of randomness and lacks formal proofs of convergence proper of other, model-based, optimization techniques. Indeed, artificial evolution is better employed in situations where conventional optimization techniques cannot find a satisfactory solution, for example when the function to be optimized is discontinuous, nondifferentiable, and/or presents too many nonlinearly related parameters.

In this chapter we will start by reviewing key features of natural evolution and molecular genetics, which represent a source of inspiration for the models of artificial evolution described in this and later chapters. We will then proceed to explain the basic steps of an artificial evolutionary system, describe the most common algorithms and genetic representations, and give examples of human-competitive results. The second part of the chapter is devoted to the detailed analysis of a particular application of artificial evolution, namely, evolutionary electronics. We will use this subject as a way to illustrate some issues arising in the application of the evolutionary approach to real-world design problems. In the closing remarks we will point out some ideas to achieve open-ended evolution. The combination of evolution with other bioinspired techniques will be described in later chapters.

1.1   Pillars of Evolutionary Theory

Biology is making continuous progress in the description of the components that make up living organisms and of the ways in which those components work together. However, the ultimate explanation is to be found in the theory of natural evolution. As Dobzhansky (1973) put it, "nothing in biology makes sense except in the light of evolution." A bewildering number of books and articles have been written on the theory of natural evolution, but its foundations are rather simple and elegant.

The theory of natural evolution rests on four pillars: population, diversity, heredity, and selection. The premise for evolution is the existence of a

POPULATION   *population*, which here we will loosely define as a pool of two or more individuals. In other words, we cannot speak of evolution of a single organism.

DIVERSITY   *Diversity* means that the individuals of the population vary from one another to some extent. Individual diversity, both within and between species, has

HEREDITY   been observed and described for thousands of years. *Heredity* indicates that individual characters can be transmitted to offspring through reproduction. The notion that individual characters are hereditary was suggested in the

SELECTION eighteenth century by Maupertuis (1753). *Selection* indicates that only part of the population is capable of reproducing and transmitting its characters to future generations. Natural selection, put forward by Darwin (1859) and Wallace (1870) in the nineteenth century, is based on the premise that individuals tend to make several offspring and that not all of them may reproduce. The selection of individuals that can reproduce is not completely random, but regulated by environmental constraints. For example, if an environment contains too many individuals for the available food , those individuals that are better or faster at gathering food will have a higher chance of survival and reproduction.

Natural selection is the most debated, often misunderstood, and abused pillar of natural evolution. In the engineering community, it is commonly described as selection of the fittest; "fittest" is often associated with "best"; and selective reproduction of the best is often associated with progress. However, organisms are not always selected for how well they score individually. For example, some animal societies maintain a number of altruistic individuals that pay a cost in terms of reproduction for the good of their society. Furthermore, selective reproduction of the fittest does not necessarily imply progress

PROGRESS in the two common meanings of the word. One meaning of progress is that new individuals are better than previous ones. However, natural selection has no comparative memory. The only way in which selection operates is here and now. Individuals are selected against the environment and/or their peers at a specific point in space and time. For example, prey at a given point in evolutionary time may be very good at escaping the current generation of coevolving predators they are confronted with, but may not be better than prey of previous generations when predators were different. In general, any change in the environment over time creates different selection conditions and therefore does not guarantee that recent generations are comparatively better than older generations selected in different environmental conditions. The other meaning of progress is that individuals tend to become better in the future. This notion of progress implies a final goal or optimal state of matter. However, natural evolution has no goal, no plan, and no end. In the best case, the combination of variety, heredity, and selection can increase *today* the rate of individuals whose parents had more suitable characteristics *yesterday*.

Where does population diversity come from? From an evolutionary perspective, generation of diversity takes place during reproduction. Offspring are copies of selected parents with small variations. This error-prone copy process can generate individuals with new or modified characteristics. Some of these characteristics will have an effect on the ability of the organism to survive and reproduce. Those new or modified features that give the organism a better ability to cope with the environment with respect to its peers and therefore to reproduce, have a higher probability of being transmitted to future generations. However, also those new or modified features that do not negatively affect the reproduction rate of the organism can be transmitted to future generations (although not at a higher rate). In this latter case, we speak of *neutral evolution* to indicate that the population is changing over generations in ways that do not affect its reproduction rate (Huynen et al. 1996).

NEUTRAL EVOLUTION

The generation of diversity provides adaptation power to evolving populations. Without continuous generation of diversity and given a constant environment, evolution would simply result in the growth of the number of individuals with suitable characteristics for that environment. The appearance of new characteristics allows individuals to sample new functionalities, behaviors, morphologies, and environmental niches. Although error-prone copy is a random process, natural selection makes sure that characteristics that affect the organism negatively have less probability of being transmitted to the next generations. Other new characteristics instead propagate through generations and, if beneficial to the survival of the species, spread at a higher rate through the population.

Again, evolutionary adaptation does not necessarily imply progress in the two meanings of the word described earlier. Natural evolution may simply increase diversity by continuously generating new organisms that occupy new environmental niches. Or, it may increase complexity by incrementally adding new features to previous ones, provided that previous features do not represent a cost for the organism, do not interact negatively with new features, or simply have a higher probability to be preserved than to be replaced by the error-prone copy mechanism.

Considering the enormous explanatory power and relative simplicity of the basic tenets of evolutionary theory, we might expect to find in the literature a compact and universal model that formally describes the evolution of populations, something akin to the laws of thermodynamics or to Newton's laws of physics. In practice, the complexity of the factors that affect the mechanisms and dynamics of evolution has not yet been sufficiently understood to allow the development of a universal formalism. Nonetheless, several formal models have been developed to address specific issues, mainly in the field of population genetics. It is worth pointing out that the great majority of these formal models describe evolutionary phenomena in terms of their ef-

fect on the variation rate of the population size or of a given character of the
evolving individuals. In other words, formal measures of evolution, if we
may liberally call them so, describe frequencies of the occurrence of given
characters, or of given types of organism, over generations. For example,
these models predict that in a relatively stable environment the percentage
of individuals with fitter characteristics will gradually grow until they dominate the population (Fisher 1930). These models do not address the notion
of performance and progress in evolving populations, but only the change in
proportion of organisms of a certain type.

1.2   The Genotype

So far, we have not yet explained how individual characters can be inherited
and modified. In 1865 Mendel arrived at the conclusion that individuals
reproduce by transmitting specific particles, now known as *genetic material*,
to their own offspring. Recent progress in *genetics* (the discipline studying
the structure and behavior of genes) and in *functional genomics* (the discipline
studying the role of genes in organisms) has provided several clues to the
molecular mechanisms and processes that support inheritance and variation.
Although Darwin was probably not aware of Mendel's conclusions when he
formulated the theory of evolution, genetics has become an integral part of
modern evolutionary theories.

GENOTYPE   The genetic material of an individual is known as the *genotype*, whereas its
PHENOTYPE manifestation as an organism is known as the *phenotype*. Natural selection
operates solely on the phenotype, but the genotype is the ultimate vehicle of
inheritance. The extent to which we are determined by our genotype or phenotype and the relationship between these two aspects of our individuality
is a complex and much debated issue (S.J. Gould 1977; West-Eberhard 2003).

In what follows, we will introduce genes, adopting the rather conventional
framework described in most textbooks. We will then point to recent results
that, at the time of writing, are changing our perspectives on the role of genes
in the development and evolution of organisms.

The conventional story involves three types of molecules and goes as fol-
PROTEINS lows. Cells contain a class of molecules, known as *proteins*, whose shape,
concentration, and behavior determine the properties of the cell. For example hair cells and muscle cells are different because they are composed of
different proteins. The definition of specific proteins depends on another
DNA   molecule, known as *DNA* (deoxyribonucleic acid), which in turn relies on

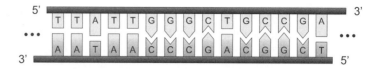

Figure 1.1   Structure of a piece of DNA molecule showing the two strands with matching nucleotides. The numbers 5 and 3 refer to the atomic structure of the molecule and affect the way in which the molecule sequence is translated into a protein. The order of translation always proceeds in the direction from 5' to 3'.

RNA

proteins to become operative and on the mediation of a third type of molecule, known as *RNA* (ribonucleic acid), which is structurally similar to the DNA molecule.

The DNA is the genetic material that is transmitted over generations. It is often enclosed within the nucleus of the cell and all cells in the organism have the same genetic material. DNA molecules (figure 1.1) are long chains of complementary strands composed of four types of chemical units (*nucleotides* or bases): adenine (A), cytosine (C), guanine (G), and thymine (T). The two strands stick together because nucleotides can lock to each other: Adenine binds to thymine and cytosine binds to guanine. This specific binding means that the two DNA strands are perfectly complementary. If we find the sequence ACA on one strand, we know that the corresponding part of the complementary strand will display the sequence TGT (although some mismatch may occur very rarely).

NUCLEOTIDES

The genetic material is organized in several separated DNA molecules, called *chromosomes*. Furthermore, in several organisms chromosomes occur in pairs (also known as diploid organisms in contrast to haploid organisms). The two chromosomes in a pair are approximately homologous in the sense that corresponding areas produce proteins with a similar functionality in similar cells. The number of chromosome pairs and total length of the DNA molecules vary from species to species. For example, humans have 23 pairs of chromosomes totaling several hundreds of millions of nucleotides (International Human Genome Sequencing Consortium 2001). The redundant structure of the genetic material (two chromosomes, two strands) allows replication of DNA molecules during cell replication.

CHROMOSOMES

MITOSIS

There are two types of cell replication: mitosis and meiosis (figure 1.2). *Mitosis* occurs during growth of the organism when a cell divides by producing a copy with the same number of chromosomes (23 times 2 in humans). During mitosis, the two strands of the 46 DNA molecules are separated and each

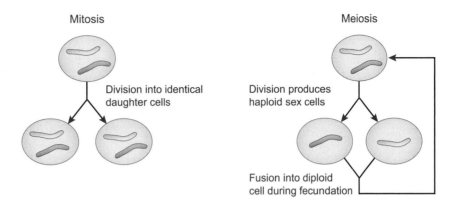

Figure 1.2 Cell replication during mitosis and meiosis. For the sake of simplicity, only a pair of homologous chromosomes are shown.

MEIOSIS

strand goes to one cell. Each strand then rebuilds the missing strand by recruiting the complementary nucleotides. The process ends with two exact copies of the double-stranded DNA molecule, one for each cell. *Meiosis* occurs during the production of sex cells (sperm and eggs). Sex cells receive only one chromosome for each pair. In diploid organisms the pairs of chromosomes are recombined during fecundation of the egg cell (containing the set of chromosomes from the mother) by the sperm cell (containing the set of chromosomes from the father). Although the chromosomes from the mother and father sex cells are homologous, their sequences may be slightly different and produce different proteins for the same functionality. This may result in the expression of features that belong either to the mother or to the father.

1.2.1   Gene Expression

GENES

The sequence of four nucleotides along the DNA chain determines the properties of the cells and the development of the organism. The four nucleotides are effectively the letters of the genetic alphabet. *Genes* are functionally relevant subsequences of nucleotides in the DNA chain (just like words in a sentence), which can produce proteins.

AMINO ACIDS

Proteins are long molecular chains (figure 1.3) composed of hundreds of submolecules, known as *amino acids*. There are 20 types of amino acids that can be combined in various ways and numbers to build up a very large number of different proteins. When amino acids are chained together, the chain bends and twists in the three-dimensional space. The properties of a protein

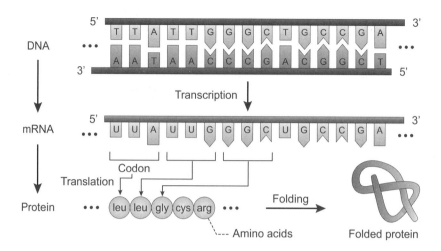

Figure 1.3 Creation of a protein molecule. A sequence from one strand of DNA is transcribed into a single-stranded RNA molecule where uracil (U) is used in place of thymine (T). Triplets of nucleotides (codons) are then translated into corresponding amino acid molecules (a given amino acid can be generated by one or more types of codons). Amino acids are then linked together to form the protein molecule that folds into different shapes according to the specific sequence of amino acids.

are determined mainly by its shape. Each amino acid corresponds to one or

CODON   more specific sequences of three nucleotides (*codon*) in the DNA chain.

The production of proteins (figure 1.3) from DNA is mediated by RNA. RNA is a long molecule similar to DNA, but it consists of only one strand of nucleotides, is much shorter (typically a few thousand nucleotides), and features uracil (U) in place of DNA thymine (T). During protein production, the two strands of DNA are separated and an RNA molecule is assembled along a small part of the DNA strand so as to match the corresponding nucleotides.

TRANSCRIPTION   This process is known as *transcription*. The resulting RNA molecule is used to create a protein by assembling a chain of amino acids that correspond to the sequence of nucleotides. Some proteins regulate cell division and genetic expression of proteins. It is important to notice that while the sequence of DNA nucleotides cannot be modified by proteins, the sequence of protein amino acids is instead determined by DNA. In other words, information flows in one direction only, from genes to proteins. This is the reason why modifications of the phenotype that occur during the life of the individual and are caused by environmental phenomena cannot directly modify the genotype and be inherited by offspring (with the exception of exposure to radiation,

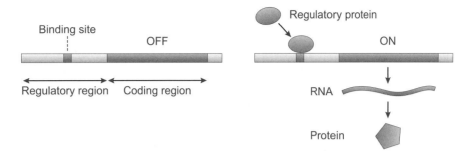

Figure 1.4   Genes are composed of a regulatory region and of a coding region. The activity of the coding region (that is, the production of corresponding proteins) is determined by the activation of the regulatory region following the binding of a specific protein. Since DNA-binding proteins are produced by genes, the activation pattern of a DNA molecule is a complex network of interactions between parts of the genes and its protein products.

which can directly affect the DNA sequence). This is also the reason why the genotype is sometimes considered the blueprint of the organism, that is, the list of instructions to build a fully fledged living system.

The genotype includes both genic and nongenic DNA. Genic DNA is the part of the molecule that can produce proteins, whereas nongenic DNA is the part that does not produce proteins. We will come back later to nongenic DNA, how it emerged, and what role it could play. For the moment, let's focus on genic DNA. At any point in time, a given gene can be active, inactive, or moderately active. The activity level of a gene is used to indicate the rate at which the corresponding protein is produced by means of RNA. Genes are structured into two regions along the DNA molecule (figure 1.4): a coding re-

CODING REGION   gion and a regulatory region. The *coding region* is composed of a sequence of nucleotides that is translated into an RNA molecule and ultimately into the corresponding protein. The translation process and its speed are controlled by the presence of special proteins that bind to the regulatory region of the

REGULATORY REGION   gene. The *regulatory region* is a sequence of nucleotides that do not produce proteins. The shape of binding proteins is such that they can bind only to specific sequences of nucleotides on the regulatory region. If such a binding happens on the regulatory region of a gene, in some cases that gene *expresses* itself by initiating the translation of its coding region into an RNA molecule. In other cases binding proteins can inhibit the expression of the gene or interfere in various ways with other proteins that are binding to different areas of the regulatory region, so as to speed up or slow down the rate at which the coding region is translated.

The proteins that regulate gene expression are themselves produced by genes, which in turn are regulated by other proteins, which are produced by other genes, and so on. Furthermore, now we know that chemical signals from other cells, or induced by the environment, can affect gene expression. The emerging picture is a complex network of interdependences, also known GENE REGULATORY   as the *gene regulatory network*, among genes whose activity can promote or in-
NETWORK   hibit other genes. A single DNA molecule can include several genetic regu-
latory networks, each corresponding to a complex dynamical pattern of gene expression. Therefore, the role of genes cannot be fully understood by simply looking at the DNA sequence or by taking a snapshot of the pattern of gene expression at a given time with today's technology of DNA microarray chips. Understanding gene functionality requires unveiling the gene regula-tory networks in which a gene participates and its level of expression over time. The study of gene regulatory networks and their relation to the way in FUNCTIONAL   which organisms grow and adapt is known as *functional genomics*.
GENOMICS
The interpretation of DNA as a set of instructions to build the organism, which is often used in conventional descriptions of genetic material and in evolutionary computation, does not do justice to the complex networks of mutual interactions between genes and proteins, proteins and cells, cells and organisms, and organisms and environment. Within this larger perspective, DNA molecules should rather be seen as the most constant part of a com-plex dynamical system that unfolds into a full behaving organism. In other words, DNA molecules are the relatively immutable structure that is trans-mitted from parents to offspring.

1.2.2   Genetic Mutations

MUTATIONS   DNA molecules can change by means of *mutations*, or errors, that occur dur-ing replication. Mutations that occur in sex cells can directly affect the evolu-tion of the species. Here we will briefly review only some types of mutations (figure 1.5) that are also used in evolutionary computation (some more often than others).

SUBSTITUTION   *Substitution mutations* change one nucleotide into another (for example, from A to G or from A to C). If the change does not correspond to the produc-tion of an amino acid different from that expressed before the mutation (the same amino acid can be produced by several triplets of nucleotides), the mu-

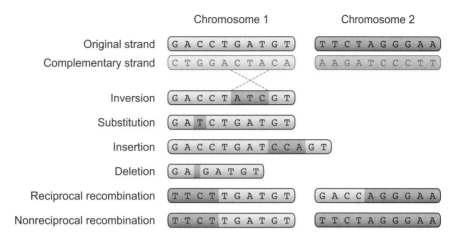

Figure 1.5   Types of mutations, some involving exchange of material from two homologous chromosomes.

tation is called synonymous or silent. Nonsynonymous mutations instead can change a codon so that it produces a different amino acid.

INVERSION   In *inversion mutations* a long sequence of the double-stranded DNA molecule is rotated by 180 degrees.

RECOMBINATION   *Recombination mutations* affect segments of nucleotides between homologous sequences of homologous chromosomes. In the case of reciprocal recombination, chromosomes are *crossed over* so that they exchange homologous sequences. In nonreciprocal recombination, the sequence of one chromosome is replaced by the homologous sequence of the other chromosome, but the replaced sequence is lost. Recombination mutations in sexual individuals effectively correspond to mixing characteristics of the parents because the homologous chromosomes under recombination were separately provided by the father and the mother of the individual. In some evolutionary algorithms, crossover assumes an important role and is considered separately from mutations.

INSERTION AND   *Insertion* and *deletion* are two types of mutation where long sequences of
DELETION   nucleotides are inserted or deleted, respectively, in a DNA molecule. This can happen either during recombination between misaligned sequences of two chromosomes or during replication with slippage of one strand within the same chromosome. The mutation rate of DNA in mammals has been

estimated to be 4^{-10} nucleotides substitution per nucleotide site per year (Li et al. 1985; Kondrashov and Crow 1993).

1.2.3   Nongenic DNA

C-VALUE

The size of genomes varies enormously among species. The size of a genome, also known as the *C-value* to indicate that it is relatively constant within a species, is given by the number of base pairs in DNA. For example, the human genome is approximately 3.6 million kilobases long, quite similar to that of the tobacco plant, but a hundred times shorter than that of an ameba. It seems that there is no relationship between genome size and number of proteins or the morphological and behavioral complexity of an organism (Cavalier-Smith 1978). This can be partly explained by the fact that large parts of the genome are composed of *nongenic DNA*, which is sequences of nucleotides that do not generate a protein. Nongenic DNA arises from several processes, such as insertion and deletion mutations, that add genetic material without function or disable functional genes. Another such process

GENE DUPLICATION

is *gene duplication* whereby adjacent sequences of DNA, which often include functional genes, are copied to other parts of the DNA molecule. Duplication can also occur in the case of entire chromosomes. (In addition, some genes, known as *transposons* or jumping genes, contain all the machinery necessary for their own excision, duplication, and insertion in other parts of the DNA.) Duplicated genes gradually accumulate mutations that make them useless. These defective copies, also known as *pseudogenes*, accumulate and disappear at various rates across species and across evolutionary time within the same species.

SELECTIONIST HYPOTHESIS

There are several hypotheses to explain the role of nongenic DNA. The *selectionist hypothesis* argues that nongenic DNA may play various important roles in gene expression and is therefore actively maintained by evolution (Zuckerkandl 1976). There is indeed experimental evidence for the existence of pseudogenes that regulate the expression of genes they derive from (Zuckerkandl 1976; Hirotsune et al. 2003; Korneev et al. 1999).

NEUTRALIST HYPOTHESIS

The *neutralist hypothesis* instead suggests that nongenic DNA is inert material and is carried along genic DNA during evolution as long as it does not impair the fitness of the organism. (For example, a possible drawback of long genomes is that replication takes a long time and may therefore affect the ability of the species to adapt to the environment.) On similar lines, the

SELFISH DNA HYPOTHESIS

selfish DNA hypothesis argues that nongenic DNA serves no other function than perpetuating itself by incorporating mechanisms for easier replication (Orgel and Crick 1980).

NUCLEOTYPIC
HYPOTHESIS
    The *nucleotypic hypothesis* instead suggests that nongenic DNA plays structural roles in the nucleus of the cell, such as maintaining the volume of the nucleus, that are not related to protein expression (Cavalier-Smith 1978). For example, amebas, which have large cells, have very large genomes (Cavalier-Smith 1978).

Nongenic DNA is not only accumulated but also lost during evolution because the continuous accumulation would create too much of a metabolic and structural burden on the organism. It has been argued that spontaneous deletion of nongenic DNA may be a major factor to account for different genome sizes (Hartl 2000). For example, the fruit fly *Drosophila* loses DNA 60 times faster than mammals and has a comparatively shorter genome (Petrov et al. 1996). Incidentally, 50 years ago researchers begun to search for the shortest genome which contains the minimal set of genes necessary for cellular life and self-replication (Morowitz 1984). Recent analysis, based on counting the set of genes that are common to a group of organisms, points to approximately 240 genes (Koonin and Mushegian 1996). Other results, based on gene knockout, give a similar result of approximately 250 genes (International Human Genome Sequencing Consortium 1995).

Nongenic DNA may have a role in the adaptability of a species because it could eventually result in the appearance of new genes. Indeed, it has been argued that gene duplication and diversification could play an adaptive role in coping with environmental challenges and may account for the rapid evolution of complexity of invertebrates (Ohno 1970). For example, it has been shown that a duplicated gene can mutate into a new type of functional gene, as in the case of olfactory-receptor genes (Glusman et al. 2001).

1.3   Artificial Evolution

Artificial evolution includes a wide set of algorithms that take inspiration from the principles of natural evolution and molecular genetics in order to automatically find solutions to hard optimization problems, improve object shapes, discover novel computer programs, design electronic circuits, and explore several other areas that are usually addressed by human design. Most artificial evolution is based on the very same four pillars of natural evolution: (1) maintenance of a population; (2) creation of diversity; (3) a selection mechanism; and (4) a process of genetic inheritance.

In artificial evolution, the phenotype of an individual is the solution to a problem and undergoes a selection process. The genotype instead is a genetic representation of that solution and is transmitted through generations and manipulated by genetic operators. The mapping between the genetic representation (genotype) and the problem description (phenotype) can take various degrees of complexity ranging from a direct, one-to-one correspondence all the way to sophisticated models of gene expression.

As we mentioned in the introduction to this chapter, the most remarkable difference between artificial and natural evolution is that the former is often formulated and used as a problem-solving technique. The problem-solving feature of artificial evolution is built into the selection process, which consists of two steps: (1) an evaluation of the phenotype that provides a quantitative score, also known as the fitness value; and (2) a reproduction operator that makes a large number of copies of genotypes corresponding to phenotypes with high fitness values. Although this utilitarian and goal-oriented twist of evolution has been successfully applied to living organisms by breeders of plants and animals for hundreds of years, it is not the way in which natural evolution operates. At the end of this chapter we will discuss some consequences of goal-oriented artificial evolution versus open-ended natural evolution.

The structure of an evolutionary algorithm consists of a simple iterative procedure on a population of genetically different individuals (figure 1.6). The phenotypes are evaluated according to a predefined fitness function, the genotypes of the best individuals are copied several times and modified by genetic operators, and the newly obtained genotypes are inserted in the population in place of the old ones. This procedure is continued until a "good enough" solution is found.

Evolutionary algorithms are often used on hard problems where other optimization methods fail or are trapped in suboptimal solutions. Those problems typically include cases that have several free parameters with complex and nonlinear interactions, are characterized by noncontinuous functions, have missing or corrupted data, or display several local optima.

Evolutionary algorithms are applicable to a large number of domains as long as a coherent genetic representation can be formulated. Evolutionary algorithms can also be coupled to other complementary search methods to increase the quality of the solutions. For example, a local gradient ascent technique could be applied to the phenotypes before fitness evaluation so that the selection process could reproduce individuals that are located in better areas of the search space. Evolutionary algorithms also allow inter-

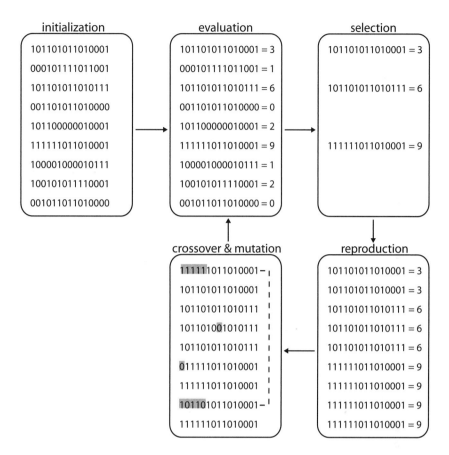

Figure 1.6 A simple evolutionary algorithm. The bit strings represent the genotypes of the individuals. The generational cycle illustrated in the four boxes is continued until a satisfactory genotype is found.

action and collaboration with human designers, for example by letting humans override the fitness function and manually select certain individuals for reproduction or insert in the evolving population genotypes of individuals with desired features.

There are several types of evolutionary algorithms, which are often labeled differently for historical reasons. These algorithms put emphasis on different components, such as the type of mutation operator, or are tailored for specific types of problem, such as the evolution of computer programs. Instead of delving into the details of each type of evolutionary technique, we will provide an overview of the main steps necessary to assemble a "custom-made" evolutionary algorithm. These steps are: (1) choose a genetic representation; (2) build a population; (3) design a fitness function; (4) choose a selection operator; (5) choose a recombination operator; (6) choose a mutation operator; (7) devise a data analysis procedure.

1.4   Genetic Representations

GENETIC ENCODING   A genetic representation, also known as genetic encoding, describes the elements of the genotype and how these elements are mapped into a phenotype. A suitable genetic representation should be devised so that (a) the recombination and mutation operators have a high likelihood of generating increasingly better individuals, and (b) the set of all possible genotypes have a high likelihood of covering the space of optimal solutions for the problem at hand.

Therefore, the choice of a genetic representation can benefit from knowing some properties of the search space. For example, if one intends to evolve a digital electronic circuit, the genetic representation may use a discrete alphabet that has some correspondence to the components of the circuit so that mutations of the genotype are more likely to map into meaningful phenotypes. Instead, if one intends to evolve an analog circuit, the genetic representation may include some real-valued elements to describe the parameters of components such as resistors and capacitors, or at least allocate more characters to genes that describe analog components at a sufficiently fine granularity.

Since artificial evolution is often used for problems that are ill-defined or poorly understood, the choice of a suitable genetic representation is not a simple affair. In this section we describe some common representations that rely on a one-to-one correspondence between the genotype and phenotype space. These representations do not include gene regulation dynamics.

1.4.1   Discrete Representations

The individual is described by a sequence of l discrete values drawn from an alphabet with cardinality k. For example, genetic algorithms (Holland 1975; Goldberg 1989), which are a particular class of evolutionary algorithms, often resort to a binary alphabet $0, 1$ with cardinality $k = 2$.

BINARY
REPRESENTATION   In a few cases, this binary representation can be directly interpreted as a phenotype, such as in the description of the configuration string of field-programmable gate arrays (a specific type of reconfigurable digital electronic

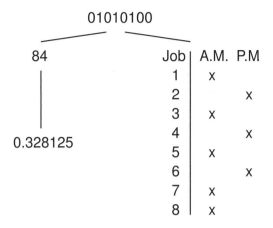

Figure 1.7 Binary representation of an eight-bit-long genome. *Left*: Integer number and real number in the interval $[0, 1]$. *Right*: Job schedule for morning and afternoon shifts.

circuits) that will be described later. However, most often it is necessary to transform a binary representation into a different phenotype representation. Binary genotypes can be mapped into a large number of phenotypes, such as integer numbers, real numbers, schedules, circuit configurations, etc.

Let's consider, for example, the genotype <01010100> (figure 1.7). This genotype can be transformed into an integer phenotype i (figure 1.7, left) using the binary mapping $0 \times 2^7 + 1 \times 2^6 + 0 \times 2^5 + 1 \times 2^4 + 0 \times 2^3 + 1 \times 2^2 + 0 \times 2^1 + 0 \times 2^0 = 84$.

It can also be decoded into a real number r in the range $[min, max]$ by first decoding it into the integer i and then applying the formula $r = min + (i/255)(max - min)$, where 255 is the maximum integer value represented by an eight-bit string.

The same binary genotype can also be used to describe job schedules (figure 1.7, right). Imagine, for example, describing the allocation of eight different jobs in a factory between morning and afternoon shifts. Each position of the genotype corresponds to a different job while its bit value corresponds to the morning (0) or afternoon (1). Our sample genotype will therefore describe a schedule where jobs 1, 3, 5, 7, and 8 are carried out in the morning and the remaining three jobs are carried out in the afternoon. If there are more than two time slots (for example, in the case of a teaching schedule for a school), one should choose a genetic alphabet whose cardinality matches

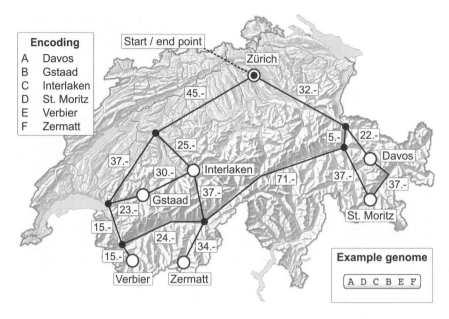

Figure 1.8 Planning a holiday across several resorts with minimal transportation costs (numbers indicate ticket price). *Top*: Some Swiss ski resorts and railway connections (copyright 2005 SwissTopo). *Bottom*: Genetic representation of a holiday plan.

the number of available time slots and whose length matches the number of taught subjects.

Discrete representations can also be used for describing sequences. Consider for, example, the case of planning your two-week winter holiday to visit six skiing resorts in Switzerland with a minimum cost for train tickets (figure 1.8). Each location will be represented by a symbol of the alphabet (in this case, $k = 6$) and the visiting order will be represented by the position TRAVELING SALESMAN of the symbols in the genotype. This example is an instance of the traveling PROBLEM (TSP) salesman problem, a class of problems that consist in visiting all nodes of a graph under multiple constraints (limit in the number of repeated visits, path length, time, etc.).

1.4.2   Real-Valued Representations

The genotype consists of a set of n numbers belonging to the domain of real numbers, typically represented as floating-point values. This representation

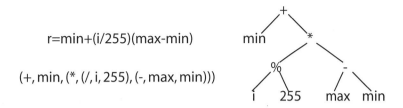

r=min+(i/255)(max-min)

(+, min, (*, (/, i, 255), (-, max, min)))

Figure 1.9   Tree representation of the expression to map integers into real numbers. *Left*: Expression form and nested list. *Right*: Tree.

is suitable for solutions that require high-precision parameter optimization, such as in the case of the description of a wing profile. In this case each number will represent the value of a parameter describing the wing curvature. If we know that the problem does not require a high resolution of the parameter space, we may well use a binary representation and allocate a suitable number of bits for each parameter value, as we have seen above. However, the two types of representations require quite different genetic operators, as we will see later in this chapter.

1.4.3   Tree-Based Representations

These representations are suitable for describing hierarchical structures with branching points and conditions. They are notably used in genetic programming, a particular class of evolutionary algorithms used for evolving computer programs, but are also applied to the description of electronic circuits, construction procedures, and experiment planning, to mention a few.

In genetic programming, each individual is a computer program (or, more generally, an expression) represented as a nested list, which can be directly mapped into a tree (figure 1.9). Consider, for example, the computer program to map an integer value i into a real number r in the range $[min, max]$ that was described above for discrete representations. The program consists of the operators +, -, *, /; the constants min, max, 255; and the variable i. The program is described by the nested list (+, min, (*, (/, i, 255), (-, max, min))), which can be visualized as a tree where a branching point is defined by the opening of a new bracket and the depth of the tree is given by the number of open brackets.

FUNCTIONS   A tree-based representation is composed of a finite set of *functions* and of a
TERMINALS   finite set of *terminals*. The choice of these two sets depends on the problem to

be solved and on some prior knowledge of the solution space. In our simple example, the function set includes the four arithmetic operators. Terminals are the endpoints, or leaves, of the tree. The terminal set may consist of variables, constants, sensor readings from a robot, etc. In the example, the terminal set includes the variable i and the constants 255, min, max.

CLOSURE

SUFFICIENCY

The function set and the terminal set should satisfy the principles of closure and sufficiency. *Closure* means that all the functions should accept any element of the terminal set and any value returned by the functions in the function set. In order to satisfy the principle of closure, the arithmetic division / is replaced by "protected division %" that checks whether the divider is zero in order to avoid premature termination of the program. *Sufficiency* means that the choice of functions and terminals should allow the generation of programs that represent the solution to the problem. This second principle is difficult to satisfy if the user has little prior knowledge of the solution space.

1.4.4   Evolvability

Genetic representations can dramatically affect the *evolvability* of the system, i.e., the probability of generating improvements through the application of genetic operators (Wagner and Altenberg 1996). For example, direct representations, such as those that we have described in this section, use genotypes whose lengths are proportional to the number of free parameters of the problem. Unless the genetic representation is well tailored for the problem to be solved, longer genotypes tend to correspond to larger search spaces and most likely lower the probability of producing improvement through random mutations of few genes.

Another problem is that the dimension of the search space and the number of possible solutions is predefined and constant, which limits the potential of generating more complex solutions over generations. Despite all these caveats, evolutionary algorithms in computer science and engineering often resort to variations of the three genetic representations that have been described in this section because they work reasonably well for a large number of function optimization problems.

The issue of evolvability and of more suitable genotype-phenotype mappings typically arises when one is interested in evolving phenotypes made of several different components, such as complex electronic devices or autonomous robots, or when one is interested in open-ended evolution. Recently, researchers have attempted to devise more efficient genetic representations and mapping processes inspired by the principles of gene expression described earlier in this chapter.

However, it is not always clear to what extent higher biological realism in genetic encoding is useful for artificial evolution. The most promising examples capture a subset of biological features that bring a specific advantage to the problem at hand. Therefore, different problem domains are often tackled with different genotype-to-phenotype mappings. We will see several examples of this approach when we will describe the evolution of electronic circuits, neural networks, plants, and robots in later parts of this book.

1.5 Initial Population

POPULATION SIZE

The initial population should be sufficiently large and diverse to ensure that individuals display different fitness values because, if all individuals have the same fitness, selection cannot operate properly. How large a population should be depends on (a) the properties of the search space at hand and (b) the computational cost of evaluating all the individuals for several generations. Problems where most genotypes have the same fitness or where random mutations have very little probability of generating a fitness improvement require larger populations. In most cases, the initial population size is determined by rule of thumb or computational costs. In the literature we often find populations ranging from a hundred to a few thousand individuals. If the evaluation of an individual requires real-world experiments, as in the case of robot evolution, the population size is often smaller than a hundred individuals.

In the case of binary representations, each genotype is created by generating random sequences of 0s and 1s. A similar process is used for real-valued representations by sampling uniformly within a predefined interval. However, for real-valued representations this works well only if the representation is bounded within that interval, but less well if that is not the case. In this latter situation, a good strategy consists of using a binary representation of the real numbers and a dynamic mapping that zooms in and out of parts of the interval depending on the degree of convergence in the population (Schraudolph and Belew 1992). For example, the eight bits encoding a variable may be initially mapped in the range $[0, 1]$ and later in the range $[0.4, 0.5]$, thus providing a better resolution of the variable in the region that corresponds to higher fitness.

Tree-based genotypes are constructed through a recursive process that expands each node into randomly sampled branches. One starts with the root node by randomly selecting a function from the function set. For each argument of the selected function, one randomly selects among all functions in the function set and all terminals in the terminal set. If a terminal is selected, it becomes a leaf of the tree. If a function is selected, it becomes a new node that is further expanded by randomly selecting a new function or terminal. A maximum depth of branching points is usually imposed in order to prevent the generation of very big trees.

Another way of initializing the population consists in seeding it with mutated copies of one or more genotypes that are known to correspond to good, or promising, phenotypes. However, this strategy carries the risk that the population may not be sufficiently diverse and that evolution may be biased to search in the neighborhood of a suboptimal solution.

1.6   Fitness Functions

The fitness function associates a numerical score to each phenotype in the population. When the phenotype is the result of a growth process or can be modified by other nongenetic processes, such as lifelong fluctuations or learning, the fitness function indirectly evaluates also the quality of the developmental or learning processes. There are two important aspects involved in the design of a fitness function: (a) the choice and combination of fitness components, and (b) the way in which the function is evaluated.

MULTIPLE OBJECTIVES   Fitness functions often attempt to optimize multiple objectives of the problem at hand. For example, one may wish to evolve the design of an airplane wing that maximizes lift, minimizes drag, and is composed of the smallest number of pieces. Although multiobjective optimization in the context of evolutionary computation has been largely discussed (e.g., Michalewicz and Fogel 2004), there is no standard way of combining and weighting the various objectives. Unless one has some knowledge of the properties of the search space that shed light on the relationships among components, the choice is often arbitrary, based on previous experience, or the result of a trial-and-error procedure. We will address again multiobjective optimization in the specific context of evolutionary electronics in the second part of this chapter.

FITNESS EVALUATION   Evaluating the fitness of individuals is often the most time-consuming part of an evolutionary algorithm. The quality of the evolved solutions depends on how exhaustive the evaluation of individuals has been. Later in this book, we will describe examples where artificial evolution finds solutions that capitalize on specific conditions of the fitness evaluation, such as the temperature of the room where an electronic circuit has been evolved. Although this is an instance of evolutionary adaptation to a particular environment, it implies that evolved solutions may not operate properly in situations that are different from the evolutionary conditions (for example, room temperature).

In several situations, especially in the evolution of physical devices or complex nonlinear systems, it is impractical to evaluate an individual in a large number of conditions. This is, for example, the case of synthetic drugs where the evaluation requires computationally expensive modeling of protein folding and protein interaction. In those cases, one may temporarily use methods to approximate the fitness computation by extrapolating fitness data from similar individuals that have already been evaluated, by relaxing the precision of the simulator, or by using adaptive estimation methods, such as artificial neural networks, that gradually build a model of the correspondence between genotypes and fitness values. However, fitness approximation must be used with extreme care to prevent evolution of solutions that do not work in real settings.

SUBJECTIVE FITNESS   *Subjective fitness* is the name used when human observers rate the performance of evolving individuals by visual inspection. An early instance of subjective fitness was the biomorphs software described by Dawkins (1986) where an observer was presented with a screen filled by insect-like creatures whose genes defined their morphologies. The user could select individuals for reproduction by clicking on their shapes. Subjective fitness is often used in artistic fields, such as the evolution of figurative art, architectural structures, and music, where it is difficult to formalize aesthetic qualities into objective fitness functions. Subjective fitness can also be combined with objective fitness (Takagi 2001).

1.7 Selection and Reproduction

The role of selection is to allocate a larger number of offspring to the best individuals of the population. The selection pressure indicates the percentage of individuals that will create offspring for the next generation. High selection pressure means that only a small percentage of individuals will be selected for reproduction. Although this strategy may result in rapid fitness increment, it brings the risk that the population will rapidly lose diversity

SELECTION PRESSURE

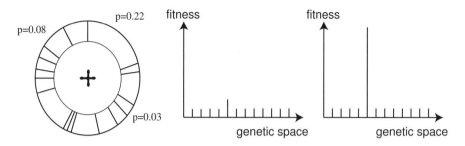

Figure 1.10   Roulette wheel selection. *Left*: Roulette wheel representation of proportionate selection. Each slot corresponds to an individual and the size of the slot is proportional to the reproduction probability of the individual. *Center*: All individuals in the population obtain similar fitness and therefore there is little chance that the best individual will make more offspring than the other individuals. *Right*: One individual obtains much higher fitness than all other individuals and therefore almost all individuals in the next generations will be copies of itself.

and will converge to a local minimum of the search space. It is therefore important to maintain a good balance between selection pressure and other factors that can generate diversity, such as genetic mutations.

PROPORTIONATE
SELECTION

In *proportionate selection*, assuming that the fitness is non-negative, the probability $p(i)$ that an individual i makes a copy of its own genome is given by the ratio between its own fitness value $f(i)$ and the sum of the fitness values of all individuals in the population $\sum_i f(i)$

$$p(i) = \frac{f(i)}{\sum_i^N f(i)}$$

Consequently, the expected number of offspring for an individual i is $Np(i)$ where N is the size of the population. One way of visualizing this process

ROULETTE WHEEL

is to think of a roulette wheel where each slot corresponds to one individual in the population and the size of a slot is directly proportional to the reproduction probability $p(i)$ of that individual (figure 1.10, left). Each offspring is generated by spinning the wheel and making a copy of the individual corresponding to the slot where the ball ends up. Therefore, in order to re-create a population of N individuals, the roulette wheel is spun N times. However, this selection method does not work well in two situations: when all individuals have similar fitness values (figure 1.10, center) and when one individual obtains much higher fitness than the rest of the population (figure 1.10, right). In the first case, all individuals will have almost equal probability of making an offspring and evolution will amount to random search. Furthermore, because of stochastic effects, some individuals won't be reproduced and the population will gradually converge toward a region of the genotype space that may correspond to a local minimum. A similar phenomenon in biology is known as genetic drift (Kimura 1983). In the second case, instead, almost all offspring will be copies of the individual with high fitness, whose genotype will soon dominate the population and cause premature convergence. In order to alleviate this problem, one may scale the fitness values prior to selection so as to emphasize or reduce differences, but the scaling procedure requires additional parameters.

RANK-BASED
SELECTION

Rank-based selection does not suffer from the above-mentioned problems because it is based on the rank of the individual in the population instead of its absolute fitness. It consists of ranking all individuals from best to worst and allocating reproduction probabilities proportional to the rank of the individual. Therefore, no matter how small the difference between any two individuals is, the better one of the two will always have a higher probability of making offspring.

TRUNCATED
RANK-BASED
SELECTION

Truncated rank-based selection is a variation that consists of taking only the top n individuals in the ranked list and making the same number of offspring for each selected individual. For example, one may take the best 20 individuals out of 100 individuals that compose the population and make five copies of each of them in order to create the new population. Provided that n is not too small (which would cause premature convergence), this method ensures that also individuals that have obtained relatively low fitness scores, but still higher than the worst ones, are given the same number of offspring as the best individuals. This method is quite useful when individuals cannot be exhaustively evaluated and thus their fitness scores may not reflect their true fitness, because both high-scoring and low-scoring individuals within the top n individuals of the population are given the same chance to reproduce.

TOURNAMENT
SELECTION

Tournament selection consists of organizing a tournament among a small subset of individuals in the population for every offspring to be generated. The procedure starts by randomly picking k individuals from the population, where k is known as the tournament size. The individual with the best fitness among the k individuals generates an offspring. All k individuals are then put back into the population and are eligible to participate in further tournaments. A new tournament is organized for every offspring to be generated. Tournament selection achieves a good compromise in maintaining both selection pressure and genetic diversity in the population.

GENERATIONAL
REPLACEMENT

In *generational replacement*, by far the most frequently used, the newly produced offspring replace the entire old population of individuals. However, if the search space is very complex, the fitness evaluation is very noisy, or genetic mutations affect very strongly the phenotype, a good individual may be lost in future generations. In this case, a popular replacement strategy, ELITISM known as *elitism*, consists of maintaining the n best individuals from the previous population. It is also possible to relax full generational replacement by inserting only a few offspring into the population in place of individuals that have obtained the worst fitness. In this case, we have a gradual generational rollover.

1.8   Genetic Operators

Genetic operators capture the effects of biological mutations on the genotype. In this section we will describe only a subset of genetic operators that are applicable to the frequently used genetic representations described earlier. These operators are designed to modify genotypes of fixed length that include only coding regions. Other types of genetic operators that can modify the length of genetic strings, such as deletion, insertion, and duplication, are used for more advanced representations tailored to evolve specific problems. We will see an example of those representations when we describe the evolution of electronic analog circuits.

Genetic operators introduce diversity in the population and allow the exploration of novel solutions. The combination, or crossover, of genetic material from two parents may, under some conditions, exploit useful genetic building blocks in the two parents. Since genetic crossover is emphasized as an important source of evolution in genetic algorithms, we will describe it separately from other types of mutations.

1.8.1   Crossover

Crossover makes sure that offspring inherit characteristics from parents by creating pairwise recombinations of the genomes of selected individuals. RECOMBINATION This operator is also known as *recombination*. The newly created offspring are randomly paired and parts of their genotypes are swapped by the crossover operator with a probability p_c. Crossover operators come in different forms, which are tailored to the genetic representations. The idea behind genetic recombination is that some of the resulting offspring may benefit from the synergistic effect that results from the combination of subsolutions found by the two parents.

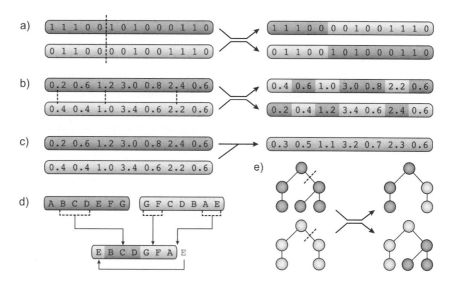

Figure 1.11   Examples of crossover operators. *a)* one-point; *b)* uniform; *c)* arithmetic; *d)* for sequences; *e)* for trees.

However, it is not straightforward to tell whether it is effective in isolating and recombining chunks of genomes that correspond to a subsolution of the phenotype. In that case, genetic recombination may amount to a large random mutation and have a deleterious effect on the fitness of the individual. To check if that is the case, one may compare at every generation the average and best fitness values of genetically recombined individuals with those of other individuals in the population. If recombined individuals report consistently lower fitness values, crossover operates as a large random mutation and consequently p_c should be set to zero.

ONE-POINT   *One-point crossover* can be applied to discrete and real-valued representations. It consists of randomly selecting a crossover point on each of the two strings and swapping genetic material between the individuals around this point (figure 1.11, a)). *Multipoint crossover* consists of randomly selecting n crossover points on the two strings and exchanging genetic material that falls between these points.

For real-valued representations one may also choose between uniform and UNIFORM   arithmetic crossover. *Uniform crossover* consists of exchanging the genetic ARITHMETIC   content at n randomly chosen positions (figure 1.11, b)). *Arithmetic crossover* instead creates a single genotype by taking the average of n randomly chosen positions of the two genetic strings (figure 1.11, c)).

When the genotype represents a sequence, the crossover operator must respect more constraints. For example, in the case of the traveling salesman problem described earlier, one wishes to change the order in which each city is visited, while ensuring that all cities are visited and that no city is visited twice. Therefore, each genotype must contain all the symbols corresponding to all cities and must not contain multiple instances of the same symbol. The crossover operator creates a genotype by taking a randomly selected part of one string and filling the remaining slots with the remaining cities arranged in the order that appears on the other string with wraparound (figure 1.11, d)). For tree-based representations, crossover randomly selects a node on each parent and swaps the two corresponding subtrees (figure 1.11, e)).

1.8.2 Mutation

Mutation operates at the level of the individual. Mutations are small random modifications of the genotype that allow evolution to explore variations of existing solutions. The mutation operator should be designed so that every point in the space of the genetic representation could be potentially reached. Mutations are useful to escape local minima and to achieve further progress in highly converged populations where genetic recombination has little effect. However, the number and size of mutations should be relatively low to prevent loss of previously discovered solutions.

Typically, a mutation consists of changing the contents of each position of the genotype with probability p_m. In the literature we often find mutation probabilities on the order of 0.01 per position (which is much higher than in biology), but the actual values should be chosen considering the effects of mutations on the fitness of the phenotype, which depend on the type of mapping from genotype to phenotype and on the properties of the problem to be solved.

In binary representations, mutation consists of toggling the selected bit values (figure 1.12, a)). In real-value representations, a selected position is modified by adding a random value drawn from a Gaussian distribution $N(0, \sigma)$, where 0 is the mean and σ is the variance, in order to produce few large mutations (figure 1.12, b)). In representations that describe sequences, as in the example of the traveling salesman problem, mutation consists in swapping the contents of two randomly chosen positions on the genotype of the individual (figure 1.12, c)). In this latter case, the probability of mutation

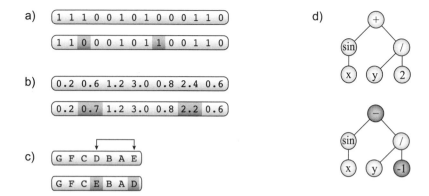

Figure 1.12   Example of mutations. *a*) Toggling a binary position; *b*) Adding a random value to a position in real-valued representations; *c*) Swapping the contents of two positions in a sequence representation; *d*) for trees.

refers to individuals, not to positions in the genotype. In tree-based representations, mutation consists in changing the content of a selected node with another element from the same set (figure 1.12, d)). If the selected node is a terminal, it will be replaced by another element randomly chosen from the terminal set. If the node is a function, it will be replaced by another element randomly chosen from the subset of functions in the function set that have the same number of terminals.

1.9   Evolutionary Measures

FITNESS LANDSCAPE   The evolutionary search space is often described as a *fitness landscape*, which can be visualized as a multidimensional surface obtained by associating a fitness value to all possible individuals that can be obtained from the genetic representation (figure 1.13). Since it is impossible to sample all possible individuals for any realistic problem, in practice one considers a few hundred or thousand genotypes generated by randomly sampling the genetic representation. For the sake of visualization, often fitness landscapes are collapsed to a two-dimensional graph where all sampled individuals are lined on the same axis according to some ordering criterion (for example, string distance or sampling order). If most of the fitness values are equal or zero, the evolvability of the system is very low. In this case, it is advisable to use large populations to change the genetic representation. However, the landscape

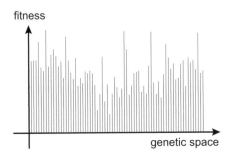

fitness

genetic space

Figure 1.13   The fitness landscape gives an indication of the fitness distribution for a random sample of genotypes. Each point on the x-axis represents a different individual. Neighborhood on this graph does not imply neighborhood on the fitness landscape where two individuals are true neighbors only if the application of the genetic operators allows the direct transition between them.

metaphor can be quite misleading because it implies a notion of neighborhood, which often does not match the way in which genetic operators move on the genetic space.

To better explore the evolutionary search space from the perspective of the evolutionary operators, one may sample the surroundings of individuals by applying multiple times in sequence a genetic operator (for example, mutation) for a given number of steps. The size of the improvement (the larger the better) gives a rough indication of how easy it is for evolution to move over the landscape. Yet another method consists in retaining only operations (for example, mutations) that produce an improvement of the fitness and count the number of steps necessary to obtain a fitness improvement of a given magnitude. Other methods for assessing the properties of fitness landscapes are available (for a critical review, see T.M.C. Smith et al. 2001), although all of them can provide only a partial picture unless the sampling size covers almost the entire genotype space.

FITNESS GRAPH   The *fitness graph* instead is a visualization of performance of an evolutionary algorithm across generations (figure 1.14). Researchers typically plot the average fitness of the population and the fitness of the best individual at every generation. Since artificial evolution builds upon a restricted pool of individuals, multiple runs with different initialization of the population are necessary to draw any solid conclusion about the quality of the final solution. Therefore, fitness graphs often display averages across several runs and include the standard error for each data point. Fitness graphs are meaningful

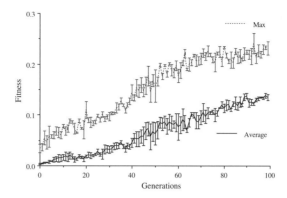

Figure 1.14   A fitness graph shows the average and best fitness of the population across generations (each data point shows the average and standard error over multiple runs from different initial conditions).

only if the problem to be solved is stationary, that is, its search space does not change properties while the evolutionary algorithm is sampling it. If the fitness function, or the problem-defining features change over generations, the fitness values obtained at different generations are no longer comparable unless a suitable scaling of the previously obtained fitness values can be performed (e.g., Floreano 1992).

An evolutionary run lasts as long as a satisfactory individual is generated or when the average and best fitness values do not grow any further. However, in the latter case it is not straightforward to tell whether the individuals have reached the maximum attainable fitness value or are only temporarily stuck in a local minimum. Further information can be gained from the analysis of the diversity of the population. If the population has lost most of its diversity, it is very likely that crossover won't have significant effects and that the population moves in the search space by means of mutations only. That implies that the fitness values may not grow further or may take a lot of generations to display some improvement. However, if the genotype-to-phenotype mapping is redundant or allows for mutations that do not affect the fitness of the phenotype, the population may move along "fitness neutral" paths that could eventually lead to a novel solution with higher fitness. In this case, the population average and best fitness may display long periods of stasis interrupted by rapid increments when a new solution is found.

POPULATION
DIVERSITY

NEUTRAL PATHS

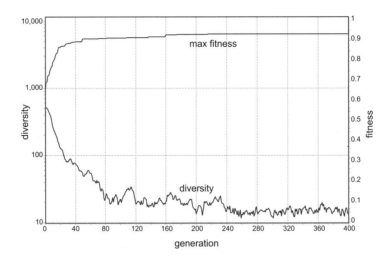

Figure 1.15   Diversity of the population across generations of a function optimization experiment measured with the all-possible-pairs diversity measure described in the text.

There are several ways to measure the diversity of an evolving population that depend on the genetic representation and genotype-to-phenotype mapping. In the case of direct encoding of phenotype parameters with real-value and binary representations, the *all-possible-pairs diversity* (Wineberg and Oppacher 2003) may be an appropriate measure of population diversity:

ALL-POSSIBLE-PAIRS
DIVERSITY

$$D_a(P) = \sum_{i,j \in P} d(g_i, g_j)$$

where the $D_a(P)$ is given by the sum of the Euclidean or Hamming distances, respectively, between the genotypes of all the pairs of individuals in the population. Figure 1.15 shows an example of application of this measure to estimate the population diversity during an evolutionary run.

An alternative measure of diversity that is especially useful for nonbinary genetic alphabets is the *entropic diversity*:

ENTROPIC DIVERSITY

$$D_e(P) = \sum_{k=1}^{l} \sum_{\alpha \in A} f_k(\alpha) \log f_k(\alpha)$$

where $f_k(\alpha)$ is the frequency of the character α of the genetic alphabet A at the position k in the population genome, and l is the length of the genomes. However, when individuals in the population can have genotypes with a different length, these two measures are no longer applicable. This happens, for example, with the tree-based genetic representations used in genetic programming and with other variable-length genotypes. In these cases one must resort to more sophisticated measures of population diversity. Typically, these are based on an estimation of the variety of substructures that exist in the population genomes, for example, the variety of subtrees (Burke et al. 2002) or the variety of substrings (Mattiussi et al. 2004).

1.10 Types of Evolutionary Algorithms

As we mentioned earlier, there are several types of evolutionary algorithms that differ mainly in the choice of genetic representation and operators (Eiben and Smith 2003). The optimal choice of algorithm, or the assembly of a custom-made evolutionary algorithm, depends on the properties of the problem to be solved; in other words, there is no single algorithm that performs better on a majority of problems (Michalewicz 1996). In this section we mention some common types of algorithms that make use of the elements described in the previous sections. We then describe additional variations that have idiosyncratic features.

GENETIC ALGORITHMS
GENETIC PROGRAMMING
EVOLUTIONARY PROGRAMMING

Genetic algorithms (Holland 1975) operate on binary representations of the individuals and emphasize the role of building blocks and crossover. *Genetic programming* (Koza 1992) operates on tree-based representations of computer programs and circuits. *Evolutionary programming* (Fogel et al. 1966) operates directly on the parameters that define the phenotype by applying perturbations drawn from a zero-mean Gaussian distribution (small perturbations are more likely than large perturbations). Evolutionary programming often relies on tournament-based selection with gradual population replacement and does not use crossover. *Evolutionary strategies* (Rechenberg 1973) are similar to evolutionary programming, but the variance of the distribution used for mutation of the individual is genetically encoded and evolved along with the parameters that define the phenotype.

EVOLUTIONARY STRATEGIES

ISLAND MODELS

Island models maintain a set of diverse populations that evolve in parallel and exchange genetic strings every generation. The idea is to let evolution pursue different courses in different islands while maintaining diversity and exploiting potential synergies by exchanging genetic strings between islands. Island models, which are also suitable for parallel implementations of evolutionary algorithms, display better performance than single-population evolutionary algorithms for problems that are linearly separable where good subsolutions discovered in different islands can be recombined (Whitley et al. 1998). The main parameters of island models are the number of different populations, the number and frequency of individuals migrating across populations, and the paths used by migrants.

STEADY-STATE
EVOLUTION

In *steady-state evolution* the individuals that obtained the worst fitness values are replaced by offspring of the individuals that obtained the best fitness values. The idea is to maintain in the population the best solutions found so far while adding better solutions as they become available. It has been experimentally shown that steady-state evolution is suitable when small populations are used Whitley and Kauth (1988); Syswerda (1989). Floreano et al. (2002) devised a steady-state algorithm for very small populations (less than ten individuals) where an individual is randomly chosen from the population, mutated, and evaluated. If its fitness is equal or larger than the fitness of the worst individual in the population, this mutated individual replaces the worst individual; otherwise it is discarded. A mutated individual replaces the worst individual even if it has the same fitness, in order to allow for "neutral walks" (Kimura 1983) on the genetic landscape, which may be a useful strategy for evolution of small converged populations (Harvey and Thompson 1996).

SIMULATED
ANNEALING

Simulated annealing (Kirkpatrick et al. 1983) is a function optimization procedure based on random perturbations of a candidate solution and a probabilistic decision to retain the mutated solution. Simulated annealing takes inspiration from the process of shaping hot metals into stable forms through a gradual cooling process whereby the material transits from a disordered, unstable, high-energy state to an ordered, stable, low-energy state. In simulated annealing, the material is a candidate solution (equivalent to the individual phenotype of an evolutionary algorithm) whose parameters are randomly initialized. The solution undergoes a mutation and, if its energy (equivalent to the inverse of the fitness) is lower than that at the previous stage, the mutated solution replaces the old one. Instead, if the energy is higher, the mutated solution replaces the old one with a probability proportional to the energy difference and the current temperature. Initially, when the temperature of the system is high, mutated solutions with relatively high energy (low fitness) have some probability of being retained. The temperature of the system is lowered every n evaluations, effectively reducing the probability of retaining mutated solutions with higher energy states. The procedure stops when the annealing temperature approaches the zero value. The major differences between simulated annealing and evolutionary algorithms is that the former operates on a single individual whereas the latter operate on a population of individuals that compete for reproduction and in most cases exchange genetic material. For a more extensive comparison, we refer interested readers to (Davis 1989).

POPULATION-BASED
INCREMENTAL
LEARNING (PBIL)

Population-based incremental learning (PBIL) operates on a single genetic string that represents an entire population of individuals (Baluja and Caruana 1995). The algorithm assumes that the genotypes of all individuals have the same length and use a binary encoding. The population string is a real-valued vector P storing at each position P_i the probability of finding a 1 at that position in all the n individuals of the population. At each generation g, n binary genotypes are created by sampling the population string and their fitness evaluated.The population string, which is initialized so that $P_i = 0.5$ for all i, is updated by moving the probability values P_i in the direction of the average values $\hat{I}_i$ observed at location i in the best s individuals of the sampled population:

$$P_i^g = (1 - \eta)\, P_i^{g-1} + \eta \hat{I}_i^g$$

where $0 \leq \eta \leq 1$ is an update constant. At any given time PBIL requires only storage of the population string P and of the s best individuals used for the update. PBIL is therefore suitable for embedded applications of evolutionary algorithms where memory size is an issue. It has been shown experimentally that PBIL achieves solutions of comparable quality to genetic algorithms in a number of problems (Baluja 1996; Urzelai and Floreano 1999), although it is slower where crossover could provide an advantage.

However, PBIL can stagnate in local minima when the problem domain is dynamic. To compensate for that problem, Baluja (1997) suggested adding a small mutation to the population string values. Urzelai and Floreano (1999)

ADAPTIVE PBIL

proposed instead a variation of the algorithm, named *adaptive PBIL* (A-PBIL), that improves both convergence speed and robustness to dynamic environments. In A-PBIL the update constant is proportional to the fitness gain obtained by the s best individuals with respect to the average fitness of the population in the previous generation. After each update of the population string, the values P_i tend to return to their initial value 0.5. Experimental tests of A-PBIL in dynamic environments and with changing fitness functions showed that the algorithm converged as fast as a standard genetic algorithm and, in addition, it always readapted to dynamically changing fitness landscapes where genetic algorithms and the original PBIL failed.

Box 1.1: DNA computing

DNA computing is an approach to parallel problem solving based on DNA replication. Adleman (1994) showed how DNA molecules could be used to find the shortest path of the traveling salesman problem (figure 1.8). The process was articulated in the following three stages.

(1) Encode each of the N cities as a unique sequence of nucleotides (for example, Davos could be ACCTTA and St. Moritz could be TATCTA) and each possible route between two cities as the sequence of nucleotides corresponding to the complement of the second half of the departure city and of the first half of the arrival city (for example, the complementary strands of Davos and St. Moritz are TGGAAT and ATAGAT and the route between Davos and St. Moritz is the concatenation of AAT and ATA). Once all cities and possible routes have been created, make many copies of them (for example, 10^{13}), and mix them together. During mixing, route strands will bind to the complementary strands of two cities, creating long concatenations of cities.

(2) At the end of the mixing process, one will find several strands of different length. In order to select only the strands that represent paths between N cities, suitable DNA strands can be selected according to the number of nucleotides (for example, if the problem involves 10 cities each encoded with 6 nucleotides, suitable strands should have 60 nucleotides.

(3) Many of the DNA strands selected so far may contain several instances of the same city and lack other cities. In order to select only strands that contain all cities, one starts by selecting all DNA strands that contain the first city (for example, Davos). This is achieved by creating a complementary DNA strand attached to a magnetic bead and "fishing out" the subset of DNA strands that bind to it. The operation is repeated on the resulting subset by fishing out the sub-subset of DNA strands that contain also the second city (for example, St. Moritz). This process is repeated until one is left with only the DNA strands that connect all N cities and corresponds to the solution of the traveling salesman problem.

Since a DNA strand can reproduce at a rate of up to 500 bases per second and several strands replicate in parallel, after a few iterations the number of processed data is larger than that of the fastest supercomputer available today. However, the number of DNA strands necessary to solve a computationally hard traveling salesman problem becomes so large that they could not possibly fit in a single building.   *(cont.)*

Box 1.1 (continued)

Furthermore, it is not yet possible to automate the three stages required by the process. It took Adleman seven days to solve the equivalent of a seven-city problem. Also, it is not clear how many problems could be advantageously solved by such a machinery.

1.11   Schema Theory

John Holland (Holland 1975) formulated *schema theory* to formally show how genetic algorithms efficiently explore the search space for increasingly better solutions. Schema theory postulates that a genetic algorithm explores a

SCHEMAS   larger number of potential solutions, also known as *schemas*, than the number of individuals in the population. A schema is a pattern-matching device that encompasses many possible genotypes. It is defined over the same alphabet used for the genetic representation, but includes an additional symbol $*$ interpreted as "don't care." For example, given a binary genotype with length $l = 5$, the schema set is defined over $\{0, 1, *\}$ and schema $<*,1,1,1,1>$ will match genotypes $<0,1,1,1,1>$ and $<1,1,1,1,1>$. For a genetic representation of cardinality k and length l, there are $(k + 1)^l$ schemas, that is, many more than the number of different genotypes that the genetic algorithm evaluates over generations.

The observed fitness of a particular schema is given by the average fitness of all the genotypes represented by that schema. John Holland mathematically showed that selective reproduction allocates an exponentially increasing number of samples to schemas with above-average fitness and an exponentially decreasing number of samples to schemas with below-average fitness. Consequently, the average population fitness will increase because there will be an increasing number of individuals with higher fitness. Schema theory was later generalized to encompass other types of genetic representations (Vose 1991).

However, selective reproduction alone does not explain how evolution can generate individuals with higher fitness over generations. To address this is-

BUILDING BLOCKS   sue, John Holland proposed the so-called *building blocks* hypothesis. A building block is a substructure of a schema that gives a positive contribution to the fitness of the individual and that can be combined with other building blocks to produce better schemas. John Holland argued that schemas with higher fitness are generated through crossover by exploring various combinations of building blocks from different individuals. By incorporating one-point crossover into schema theory, he also showed that evolution will favor schemas with a shorter distance between the first and last fixed position (e.g., $<*,1,1,*,*>$ has a shorter distance than $<1,*,*,*,1>$). This tells us that the genetic representation should be designed to allow for the presence of short building blocks.

The combination of schema theory and of the building blocks hypothesis suggests that the role of crossover consists of exploiting the genetic material present in the evolving population by gradually combining building blocks into better structures. Accordingly, mutation would play a complementary role by exploring parts of the search space that are beyond the region spanned by the current population. An effective genetic algorithm must therefore strike a good balance between exploitation of existing solutions and exploration of new solutions. Practitioners of genetic algorithms that endorse schema theory use a rather high crossover probability (say, more than 0.8 per pair) and a low mutation probability (say, less than 0.01 per position). However, the actual values depend on the properties of the search space and are usually empirically tuned by trying a few combinations.

Over the last years, schema theory has been extensively discussed, expanded, and criticized (several criticisms are summarized in (Mitchell 1996)). One of the main criticisms concerns the evaluation of schemas. Schema theory holds as long as a schema is uniformly sampled to provide a good estimate of its real fitness, which is given by the average of all possible individuals encompassed by that schema. In practice, a genetic algorithm evaluates the fitness of only a subset of all possible individuals encompassed by a schema. Observed fitness is thus only an estimation of the real fitness of a schema. Although the number of observations for good schemas is increased along generations, selective reproduction introduces a strong observation bias toward an increasingly smaller subset of the genetic strings encompassed by that schema. In other words, as the population converges toward a few genetic strings, it no longer represents a uniform sample of the schemas.

Another issue of concern is to what extent building blocks really exist for a given genetic representation. And even if they do exist, the crossover operator may not be properly designed to excise and recombine them in genotypes of higher fitness. In the absence of building blocks or of a suitable crossover operator, the genetic recombination of individuals is analogous to a very large random mutation that may disrupt evolutionary progress.

1.12 Human-Competitive Evolution

Evolutionary algorithms are nowadays used in several engineering fields, from design of electronic boards and wing profiles to biochemical synthesis, all the way to scheduling and process optimization, in order to find solutions for problems that cannot be easily solved by other techniques (for a review of applications, see L. Davis 1991; Dasgupta and Michalewicz 1997; Michalewicz and Fogel 2004).

Furthermore, evolutionary computation has the potential to discover novel solutions that are both innovative and perform better than human-designed solutions. The first Human-Competitive Award in Genetic and Evolutionary Computation was launched in June 2004 at the Conference on Genetic and Evolutionary Computation in Seattle. Entries were solicited for human-competitive results that were produced by any form of genetic and evolutionary computation and that had been published in the open literature (conference proceedings, articles, technical reports, theses, books, etc.) within the 12 months preceding the conference. The selecting committee defined an evolved result as human-competitive if it satisfies at least one of the following eight criteria (`http://isgec.org/gecco-2004/`):

1. The result was patented as an invention in the past, is an improvement over a patented invention, or would qualify today as a patentable new invention.

2. The result is equal to or better than a result that was accepted as a new scientific result at the time it was published in a peer-reviewed scientific journal.

3. The result is equal to or better than a result that was placed in a scientific database or archive of results maintained by an internationally recognized panel of scientific experts.

4. The result is publishable in its own right as a new scientific result independent of the fact that the result was mechanically created.

5. The result is equal to or better than the most recent human-created solution to a long-standing problem for which there has been a succession of increasingly better human-created solutions.

6. The result is equal to or better than a result that was considered an achievement in its field at the time it was first discovered.

7. The result solves a problem of indisputable difficulty in its field.

8. The result holds its own or wins a regulated competition involving human contestants (in the form of either live human players or human-written computer programs).

1.12.1   Example: Evolution of an Antenna

One of the winning entries at the 2004 conference described an evolved X-band antenna design and flight-ready prototype to be deployed on NASA's Space Technology 5 (ST5) spacecraft (Lohn et al. 2004). The NASA ST5 mission consisted of developing three nanosatellites (figure 1.16, top) of 50 cm in diameter and less than 25 kg that fly in the Earth magnetosphere and communicate with the ground station using the same technology of cell phones. In this context, the antenna design plays a crucial role and is a notoriously hard problem because of the difficulty of taking into account complex electromagnetic interactions at design time. The NASA ST5 mission managers provided a specification list to an outside contractor who produced a quadrifilar helical antenna (figure 1.16, bottom left).

The authors of the winning entry instead used an evolutionary algorithm where the fitness function measured the properties of evolving antennas against the specifications given by the mission managers. The genotype used a tree-based representation where each node was an antenna-construction command from the following list:

- **Forward(length, radius)**: Add a wire with the specified length and radius from the current location and then change the current location to the end of the wire.

- **Rotate-x(angle)**: Change the current orientation around the x-axis by the specified amount.

- **Rotate-y(angle)**: As above, for the y-axis.

- **Rotate-z(angle)**: As above, for the z-axis.

The antenna was generated by executing the commands at each node in the tree, starting from the root node.

The authors evaluated each antenna design several times with small random perturbations added to the radius of the segments and to the joints in order to take into account manufacturing imprecisions. Only the worst fitness

Figure 1.16   *Top*: Prototype of nanosatellite. *Bottom left*: Human-designed antenna (quadrifilar helical). *Bottom right*: Evolved antenna. Images courtesy of G. Hornby, NASA Ames Research Center.

score of all evaluations was retained in order to maintain a conservative estimate of the antenna performance. These noisy and worst-fitness conditions produced antennas that worked over a broader range of frequencies than antennas evolved without noise and worst-fitness strategy. In all evolutionary conditions, antenna designs were evaluated using a scaled-down version of computationally intensive simulation software that simplified some aspects of the electromagnetic modeling. The best designs were then tested using the full version of the simulation software and eventually built and measured in an anechoic chamber at NASA Goddard Space Flight Center.

The results of the hardware measurements showed that the best evolved branching antenna (figure 1.16, bottom right) had significantly better gain on a broader range of orientations and lower power requirements than the human-designed antenna. Furthermore, it took only three months to produce the final result versus five months for the human-designed antenna. This result was therefore considered human-competitive against criteria 5 and 7. Although evolved antennas are more difficult to manufacture than those produced by human design, they can be quickly re-evolved if the mission constraints change. Engineers at NASA took advantage of this rapid redesign feature when it was decided that the planned nanosatellites would orbit at a different orientation and altitude and thus require a new antenna design.

1.13 Evolutionary Electronics

We devote the remaining part of this chapter to *evolutionary electronics*, that is, to the application of the evolutionary approach to the design of electronic circuits. By electronic circuit we mean a collection of interconnected electronic devices such as transistors, resistors, and logic gates. The designer must specify the kind of devices that compose an electronic circuit and their

CIRCUIT TOPOLOGY connectivity, which constitute what is called the *topology* of the circuit. Since each kind of device is typically endowed with a certain number of parameters (for example, the value of resistance of a resistor, or the value of a geometric parameter of a transistor), the designer is required to define the

CIRCUIT SIZING values of those parameters, an operation known as the *sizing* of the circuit. Finally, to physically build an electronic circuit it is necessary to specify not only which device is connected with which but also what is the physical layout of the devices and connections. This corresponds to the *placement* of the devices and *routing* of the connections between them.

The process of electronic design starts with the specification of a desired functionality and of a range of operating conditions in which the functionality must be guaranteed. When successful, the design process produces the parameter values, the topology, and, possibly, the physical layout of an electronic circuit that realizes the required functionality in the specified range of operating conditions. In some cases the physical layout influences substantially the behavior of the circuit and in this case placement and routing constitute an integral part of the circuit design problem. In other cases the influence of the physical layout on the functionality can be expected to be negligible and the electronic design problem proper can be considered concluded with the definition of the circuit topology and parameter values.

In the simplest design scenario, the specification of the desired functionality is followed by the choice of a predefined circuit topology which is known to produce that kind of functionality. For example, a designer asked to produce a circuit that amplifies the difference of two signals might know that the circuit shown in figure 1.17 has a topology which, for a suitable choice of the device parameters, can provide the desired functionality and meet all the required specifications. Thus, the design problem in this case is reduced to the determination of a set of parameter values. In an evolutionary perspective this electronic design problem corresponds to a case of evolutionary parameter optimization. The examples reported in the literature (for example, Alpaydin et al. 2003; Nam et al. 2001) testify to the practical relevance of using artificial evolution even in this simple design scenario. Nonetheless, the real nature of evolutionary electronics can be fully appreciated only when considering the possibility of evolving simultaneously both the parameter values and the topology of the circuit. In this case, the definition of a genetic representation for the circuit becomes a much more challenging problem, but the creative potential of the evolutionary process is correspondingly unleashed. The contextual evolution of the physical layout of the electronic circuits along with the parameter values and topology obviously constitutes a further step in this direction. To keep things simple, this subject will not be pursued here. The reader is referred to (Koza et al. 2003) for examples of circuit evolution that include routing and placement.

We start the exploration of evolutionary electronics with an overview of some general aspects of evolutionary electronics and some comments about its place in the general context of evolutionary methods. Then, we will consider separately the cases of digital and analog evolutionary design, describing briefly the conventional approaches to each problem before describing the specificities of their evolutionary counterpart. The methodological discussion is followed by the analysis of some actual examples of circuit evolution. The significance of the current results is finally discussed in the crucial perspective of circuit robustness and verification.

1.14   Lessons from Evolutionary Electronics

There are several reasons for singling out evolutionary electronics among the many existing applications of evolutionary methods and for examining it in detail. A first reason is that electronics is a mature technology which can provide a wealth of tools and devices permitting the exploitation of the potential of evolutionary methodologies. For example, there exist many kinds of reconfigurable devices and simulators which, as explained below, permit the

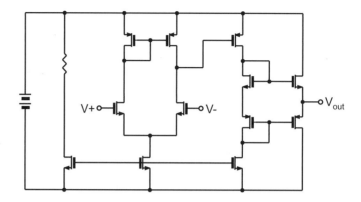

Figure 1.17 The schematic representation of an electronic circuit. When a circuit topology that can realize the required circuit performance is known, it can be used as a starting point for the evolutionary design, which reduces to an evolutionary optimization of a collection of numeric parameters.

automation of the process of evaluation of the fitness of the circuits generated by artificial evolution. A second reason for the choice of electronics is the fact that it provides a unified context for the illustration of many aspects that are relevant to the real-world application of evolutionary methodologies that do not find a place in elementary descriptions of evolutionary algorithms. In this sense, this material can also be considered a concrete illustration and extension of some points that were briefly sketched in the previous pages, and many of the comments of the following sections that refer to electronic circuits can be seen to apply to other domains of evolutionary design. Thus, in what follows, "evolved circuit" can be often read to mean "evolved system." A third and pragmatically more compelling reason for focusing on evolutionary electronics is that electronic circuits can be found in an enormous variety of present-day systems, with an ever-increasing demand for wider applicability, better performances, more complex functionalities, and lower cost. Conventional design practices based on human circuit design expertise are only partially able to satisfy this demand. Thus, alternative design approaches are called for to complement those established practices. Evolutionary electronics has the potential for being one of those alternative approaches because it appears ideally suited to contexts where the limitations of the conventional approach are most apparent, such as

- design problems where the circuit specifications are given in terms that are difficult to formalize in the way required by conventional design practices (for example, a global measure of performance) but are naturally expressed as an evolutionary problem;

- problems where systematic design techniques are scarce or missing altogether and the progress in the generation of satisfying solutions is trusted to the expensive and possibly uncertain insight of the human designer;

- design problems where existing systematic design techniques manage the complexity of circuit design by imposing a number of constraints that result in a waste of devices or performance.

Before entering into the details of these contexts with actual examples of evolutionary experiments and evolved circuits, it is useful to consider in general the characteristics of the conventional design process and compare them with those of the evolutionary approach. In particular, it is useful to consider the crucial role played in any design process by the concept of *abstraction*.

1.15   The Role of Abstraction

Etymologically, "to abstract" means "to draw away." This refers to the fact that the process of abstraction consists in drawing away conceptually, that is, in disregarding, some aspect or property of a system, focusing on a subset of its original properties. We have already met one widely used example of abstraction, namely, the circuit theory abstraction that is implicit in the representation of an electronic circuit in the form exemplified by figure 1.17. This abstraction consists in considering the circuit as composed of electronic devices that interact only through the connecting wires, disregarding the fact that in reality they interact also via electromagnetic fields that are not "guided" by the wires. The cognitive advantage of the process of abstraction consists in the possibility of thinking about the resulting system and its interactions only in terms of the retained properties. Obviously, in this respect an abstraction is meaningful as long as the aspects that are disregarded are actually irrelevant to the system behavior of interest. In other words, the system must possess some properties that make the separate consideration of the retained properties meaningful and useful. For example, the circuit theory abstraction is meaningful and useful only if the interaction via nonguided electromagnetic fields is actually small.

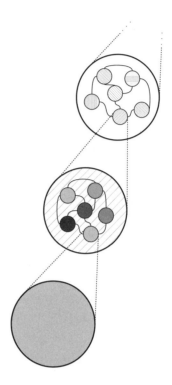

Figure 1.18 Complex systems produced by natural evolution and by human design are typically organized into a hierarchy of levels of abstraction. In this hierarchy, a system at one level becomes a building block for the system defined at the next higher level.

HIERARCHICAL
STRUCTURE

Considering the limits of the human capacity for processing information (G.A. Miller 1956), the use of abstraction appears essential for human design of complex systems, since it decreases the number of effects and interactions that the designer must take into account simultaneously. In fact, the conventional approach to design consists typically in defining and working with a succession of hierarchical layers of abstraction (figure 1.18). This is particularly apparent in the bottom-up design of computer programs, where the design consists in building what can be seen as a succession of more and more specific programming languages with those at lower levels constituting general purpose programming languages, and those at higher levels defined ad hoc by the programmer in view of the particular application at hand (Graham 1993). For example, the assembly language can be used to build a compiler for the C programing language, which can be used to build a compiler for the C++ programming language, which in turn can be used to build a spreadsheet application and, within it, an interpreter for a macrolanguage. This hierarchical decomposition can be also observed in virtually every field of human design and particularly in the field of electronic circuit design, where a hierarchical succession of building blocks (also called *cells* or *modules*) of increasing complexity and specialization is defined, and where at each level of the hierarchy the designer thinks almost exclusively in terms of the "language" of the cells defining that level.

The cognitive advantages provided by the use of abstraction, however, come at a cost. This cost concerns the efficiency in the use of the collection of resources that constitute the system produced by the design process. In other words, we can say that there exists an *abstraction-efficiency tradeoff* or a *cost of abstraction*. The tradeoff is due to the fact that the process of abstraction leads to the disregarding of a certain number of physical effects and interactions. By actively exploiting those effects and interactions there exist typically ways to build a system that implements the same functionalities using fewer resources. Alternatively, there exist ways to use the same resources to build a system with better performances and functionalities. This phenomenon is exemplified by the history of the P2, one of the most successful commercial electronic amplifiers of all time. As reported in (Pease 1991), the exceptional performance of the P2 was due to a interaction between different parts of the circuit which was not contemplated by the circuit theory abstraction on which the design of the circuit was apparently based (it is not clear if this interaction was purposely built in the circuit by the designer, or if it was just a lucky accident). As a consequence, none of the competitors who analyzed and tried to reverse-engineer the circuit on the basis of this abstraction was able to replicate the P2 performance. The original manufacturer thus remained for many years (and very profitably) the sole source of this kind of amplifier, until the development of more performing elementary devices enabled the attainment of the P2 performances using conventional design techniques.

The need for abstraction is not limited to the case of conventional design, but extends also to evolutionary design. This can be understood by considering that a complex system, be it evolved or designed by hand, is a combination of interacting parts that contribute to the workings of the whole system. As discussed in the first part of this chapter, evolution works by producing and testing random changes in the structure of the system. If every part of the system interacts with almost every other part, any change in a part of the sys-

COST OF ABSTRACTION

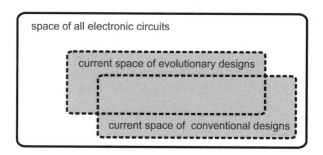

Figure 1.19   An idealized representation of the spaces of electronic circuits and electronic designs (adapted from J.F. Miller et al. 2000). The discovery of new design methodologies could enlarge the space of conventionally designed circuits, whereas the discovery of new evolutionary techniques, as well as the increase of the available computational power, could enlarge the space of circuit designs accessible to evolutionary methods.

tem results in the alteration of numerous functionalities of the system. Since a random change has a greater probability of disrupting an evolved functionality than of improving it, beyond a certain degree of complexity of the system there would be virtually no possibility that the application of a random change producing effects on many parts of the system would produce a global improvement of its functionality. Thus, in the presence of an unlimited global connectivity the evolutionary process would soon come to a halt. The solution to this *evolvability* problem lies in the limitation of the interactions of one part of the evolved system with all the other parts (for example, by imposing a modular or hierarchical structure, which is observed in living organisms and in complex technological systems), so that the evolutionary process can produce changes with local effects. This limitation of the number of interactions permits disregarding a lot of potential interactions, that is, it corresponds to the enforcing of an abstraction. Ideally, it is the evolutionary process itself that should be let free to choose the right kinds of abstraction for the problem at hand. This would correspond to the implementation of a process of *unconstrained evolution*. Unfortunately, for technical and conceptual reasons, the full endowment of an artificial evolutionary process with this kind of flexibility is still an open issue. It is instead usually the case that certain abstractions are imposed from the outside on the evolutionary process, typically in the form of constraints on the kinds of building blocks that

UNCONSTRAINED
EVOLUTION

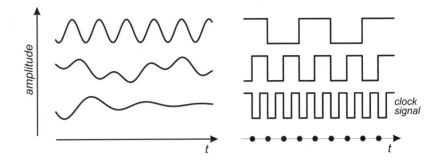

Figure 1.20   Analog signals are continuous in both time and amplitude (*left*). Digital signals (*right*) are discrete in the amplitude and, typically, their amplitude can change only at discrete time instants identified by a periodic clock signal.

are available for the design and on the structure of the genetic representation (Gordon and Bentley 2002).

One might at this point wonder what advantage there could be in using an evolutionary design process rather than a conventional approach, given that both must submit to the abstraction-efficiency tradeoff. The answer is that the limitations of an artificial evolutionary process can be different from those of a human designer. For example, the number of interactions between the elements that compose a module that the evolutionary process can successfully manage might be much greater than the number of those that can be handled by a human designer. Consequently, even if the requirement of abstraction prevents both the conventional and the evolutionary design process from exploring the space of all the possible electronic circuits, the space of designs accessible to the evolutionary process can be different from the space of designs accessible to the conventional design techniques (figure 1.19) and the circuits belonging to the former can thus usefully complement those of the latter. Moreover, it is possible that new design principles can be discovered by analyzing the workings of the evolved circuits, and added to the collection of tools available for conventional design (J.F. Miller et al. 2000).

1.16   Analog and Digital Circuits

The connections between the devices that compose an electronic circuit convey time-varying electric signals. A signal is considered as analog if its amplitude takes its values in a continuous range, and digital if only a discrete

ANALOG
DIGITAL

Box 1.2: Analog vs. digital

Let us consider the differences between analog and digital signals and their consequences in terms of information content, resistance to noise, and power consumption of the corresponding circuits.

The number of distinct signal amplitudes is finite in the digital case and infinite in the analog case. Moreover, analog signals are typically continuous in time, whereas digital signals are discretized also in time. At first sight, the information conveyed by an analog signal is thus potentially infinite, even if the signal has finite duration, whereas that conveyed by a digital signal is necessarily finite. Let us first consider the issue in the time domain. The sampling theorem (Shannon 1949) tells us that a band-limited signal, that is, a signal whose frequency content is confined to a finite frequency interval of width W, can be reconstructed with the information represented by a sequence of signal samples that is discrete in time. Since all actual signals are band limited, we can conclude that in practice a continuous-time signal has the same information content of a discrete sequence of samples. When we consider also the amplitudes, Shannon's theorem for a noisy channel (Shannon 1949) tells us that the information that can be transmitted through it depends on the average power P of the signal, on the power N of the noise, and the bandwidth W of the channel. For example, for a channel with white noise, the maximum rate of information transmission C is

$$C = W \, \log_2 \frac{P + N}{N}$$

Since the bandwidth and power of a physical signal are always limited and the noise is never absent, the information content C of an actual analog signal is always finite. Summing up, the amount of information conveyed by actual signals – be they analog or digital – is always finite.

A fundamental difference between analog and digital signals is the possibility to fight noise. Below a certain noise level, when noise corrupts the amplitude of a digital signal, there is the possibility of restoring it to its original level. In fact, in actual digital circuits each device, in addition to its particular functionality, performs this operation of signal restoration (Sarpeshkar 1998). Thanks to this possibility of easily restoring their level, digital signals are characterized by a good tolerance of noise, and very long sequences of operations can be performed on them without worrying about the accumulation of the effects of noise. On the contrary, analog signals do not permit this simple kind of signal restoration and tend to accumulate progressively the effects of   *(cont.)*

Box 1.2 (continued)

noise. Consequently, long chains of analog processing stages cannot be used, lest the signals be completely swamped by noise.

Tolerance of noise in digital systems comes at some cost. Digital signals use only a few of the amplitude levels that could be ideally distinguished according to Shannon's formula for the channel capacity C. In other words, noise tolerance is obtained at the expense of the (finite) amount of information that is carried by each signal. The reduced information content of each signal forces digital systems to adopt a *distributed representation* of quantities. This can be appreciated considering the structure of a typical analog-to-digital converter (ADC), where a single input line carrying an analog signal becomes many output lines carrying binary digital signals (see adjacent figure). There is also an energetic cost of the strategy used to achieve noise tolerance in digital signals, which follows from the necessity of having signals switch between amplitude levels that are far apart. The relevance of this problem is testified to by the magnitude of the existing trend toward lower operating voltages in digital circuits and by the fact that biological systems apparently evolved to achieve noise (and fault) tolerance with a very different approach, based on low-power (unreliable) devices, a combination of distributed and *redundant* representations (Eliasmith and Anderson 2003), and an extensive use of specialized analog devices (Sarpeshkar 2006). Finally, the use of a distributed representation in digital electronics implies higher wiring costs and complicates the circuitry required for the implementation of transformations and combinations of the represented quantities, even simple ones like the sum of two signal amplitudes. On the other hand, the use of a distributed representation in digital systems has the definite advantage of permitting the increase of the resolution of the representation with the addition of a few more signal lines, a result that cannot be obtained with analog signals (von Neumann 1961).

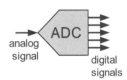

collection of values is admitted (figure 1.20). In relation to the characteristics of those signals, electronic circuits are classified into *analog circuits* and *digital circuits*, with the expression *mixed-signal circuits* used to denote the presence of both. In digital circuits it is the voltage amplitude that is typ-

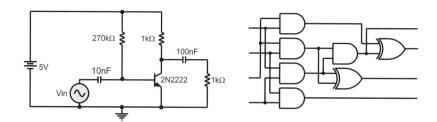

Figure 1.21 Analog circuits (*left*) are characterized by several parameters on which evolution can act to slightly change the behavior of the circuit. These parameters are typically absent from digital circuits (*right*) where evolution can operate only on the topology of the circuit. The digital fitness landscape is therefore more discontinuous and more difficult to explore than its analog counterpart.

ically discretized and assumed to take only two distinct levels. Note that this assumption is actually an abstraction that disregards the values taken by the signals during the transitions between these two levels. The result is the familiar binary circuits that are at the heart of almost all present-day computers. In most digital circuits time is also discretized by allowing changes in the signals level only in correspondence with the changes of level of a special synchronization signal called a clock signal.

The different nature of analog and digital signals has a number of consequences in terms of, for example, the way in which they can convey information, their tolerance of noise, and the power consumption of analog and digital circuits (see box 1.2). From the point of view of evolution, what matters most is the difference between the typical structure of analog and digital circuits (figure 1.21). Analog circuits are characterized by several parameters such as resistance and capacitance values. In general, small changes in these values result in small changes in the behavior of the circuit. This gives to the analog fitness landscape a certain degree of continuity. Adjustable parameters are instead typically absent from digital circuits because most digital design takes place at a level of abstraction above that of elementary devices such as transistors, for example, at the level of the logic gate or higher. The only aspect of the circuit that can be changed while working at this level is the topology of the circuit, and this results in general in major changes of the circuit behavior. Combined with the discreteness of the signal space, this results in more discontinuous fitness landscapes with respect to the analog case and in a greater evolutionary potential of analog circuits relative to digital circuits, provided one adopts a genetic representation that can exploit their continuous transformability (Conrad 1990).

1.17 Extrinsic and Intrinsic Evolution

Evolutionary design is based on the generation and evaluation of a large number of candidate solutions. This means that a substantial number of different circuits must be typically tested during an evolutionary electronics run. Since it is practically impossible to build and test all these circuits by hand, the evolutionary synthesis of electronic circuits requires the availability of a way to perform this task automatically.

EXTRINSIC EVOLUTION

A first solution, which does not require the physical implementation of the circuits, is based on the use of *circuit simulators* and results in what is called *extrinsic* evolution of electronic circuits (de Garis 1993). Circuit simulators (figure 1.22) are computer programs that take as input the formal description of a circuit and the description of the desired operational conditions, and give as output a numerical representation of the behavior of the circuit in these conditions. For example, in a typical extrinsic evolution scenario the simulator could be provided with the description of the circuit, the description of the signals applied to its input terminals, the specification of the circuit operating conditions such as the circuit temperature, and the description of the kinds of analysis that are required. The simulator would then compute the corresponding signals at the output terminals of the circuit, which could be finally compared with the desired response in order to determine the value of fitness of the circuit.

INTRINSIC EVOLUTION

The alternative to the use of a simulator is the physical implementation of the circuits that must be tested in the context of the evolutionary process. As mentioned above, this is practically feasible only if the circuits can be implemented automatically, rather than built by hand one by one. This can be obtained thanks to the availability of *reconfigurable devices* (box 1.3) and results in what is known as *intrinsic* evolution of electronic circuits (de Garis 1993). In practice, the intrinsic evolution of circuits proceeds as follows. The evolutionary algorithm is run on a computer, which is connected to an external system. This system contains the reconfigurable device and some auxiliary programmable circuitry and instrumentation. The auxiliary circuitry permits the generation of the required operating conditions and input signals and the measurement of the circuit response. When the evolutionary algorithm requires the evaluation of a circuit fitness, it instructs the computer to down-

Box 1.3: Reconfigurable devices

Reconfigurable devices are boards or integrated circuits composed of a collection of analog or digital *cells* embedded in a weave of *connections* (see adjacent figure) that can be programmed so as to implement a large number of different circuits. In the simplest kind of reconfigurable devices, each cell implements a fixed circuit functionality and the reconfigurability is restricted to the connections between the cells and between the

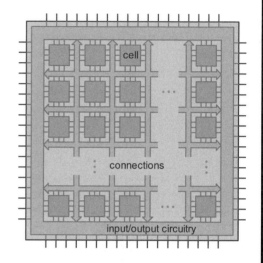

cells and the input/output pins of the device. Typically, to limit the complexity of the wiring and of the interconnection circuitry, not all conceivable connections between the cells are permitted. More flexible devices let the user choose also the functionality of the cells within a predefined set of options.

A reconfigurable device is called *fine-grained* if each cell can provide only simple functionalities such as an elementary logic gate in the digital case or a transistor in the analog case, and *coarse-grained* if each cell can realize more complex circuits and functionalities such as a generic Boolean function of many variables or an operational amplifier. The configuration of the connections and the choice of the functionality of the cells (when available) are done by downloading into the device a string of *configuration bits*. To be suitable for evolutionary experiments a reconfigurable device must permit the execution of this operation a virtually unlimited number of times, and the whole downloading and reconfiguration process must take a very short time. *Dynamic* reconfigurable devices admit the downloading of the configuration bits even during the normal operation of the previously configured circuit, and are thus ideally suited for an evolved circuit displaying online adaptation. On the contrary, in *static* reconfigurable devices the configuration can be done only when the circuit is not operating.

The programming of commercial reconfigurable devices is typically entrusted to a specific configuration tool provided by the   *(cont.)*

Box 1.3 (continued)

manufacturer of the device. The configuration tool transforms a high-level description of the desired circuit in the corresponding sequence of configuration bits, which can be safely downloaded into the chip. Among other things, the mediation of the configuration tool excludes the possibility of incorrect configurations of the device that could damage its cells and connections. For this reason, the structure and meaning of the configuration bits of commercial reconfigurable devices are often not even officially documented by the manufacturer. To obtain a better control of the reconfiguration process and also to have the possibility to tailor the cell functionality and connectivity to the requirements of the evolutionary process, several research groups have developed custom reconfigurable devices expressly conceived for the needs of intrinsic evolutionary electronics (for example, the evolvable motherboard (Layzell 1998), the programmable transistor array (Stoica et al. 2001b), the POEtic circuit (Tempesti et al. 2002), and several other integrated circuits and boards (Zebulum et al. 2002)).

load on the reconfigurable device the configuration information necessary to obtain the desired circuit, and to manipulate the auxiliary circuitry and instrumentation to perform the necessary analyses on the circuit under evaluation. The results are finally sent back to the computer for the computation of the circuit fitness.

Let us consider briefly the strengths and weaknesses of the extrinsic and intrinsic approaches to evolution. Once again, many of these considerations apply to the extrinsic and intrinsic evolutionary design of other kinds of systems. A first aspect that must be considered is the kind of circuits that can be generated and the degree of testing of their functionality that can be achieved. In the case of intrinsic evolution, the limitations of the connectivity, the necessity of avoiding circuit configurations that could damage the physical devices, and the predefined number of cells available on each reconfigurable device tend to put many constraints on the kinds of circuits that can be realized and tested. Moreover, limitations on the kinds of signals that can be generated and measured, and the cost and complexity of the hardware necessary to test more than one physical device and change the environmental conditions of the tests make it difficult to probe extensively the range of conditions in which the system is required to operate correctly. On the other

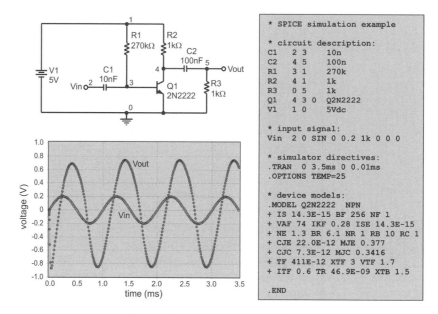

Figure 1.22 An example of circuit description and analysis performed with the circuit simulator SPICE (Vladimirescu 1994). Using a special circuit description language, the circuit is represented as a text file for the simulator (*right*) which contains the description of the circuit, of the operating conditions, and of the kinds of analyses required. The simulator reads the description file and computes and outputs the results of those analyses (*left, bottom*), which can be used to compute the circuit fitness.

hand, in the intrinsic case there is the possibility of testing the behavior of the circuit in operating conditions whose effects on the circuit are difficult to simulate, such as extreme temperatures and exposure to high radiation levels (Stoica et al. 2001a, 2004). In the case of extrinsic evolution, instead, the virtual nature of the circuits permits the evaluation of an almost unlimited variety of circuits and that of a large variety of operating conditions. The only exceptions are a few pathological topologies, some extreme values of device parameters and environmental conditions, and the above-mentioned operating conditions whose effects are difficult to model and simulate.

In general, in the extrinsic case it is simple to probe extensively the range of conditions in which the circuit must operate, because additional probing entails only the execution of additional simulations. On the other hand, the significance of this probing is reduced by the limitations in the accuracy of the models used in extrinsic evolution. This fact could well result in extrinsically evolved circuits whose performance is seriously degraded when they are implemented in hardware. For example, the simulator might accept operating conditions such as overvoltages and overcurrents that cannot be tolerated by actual physical devices, or might model badly or not at all some important real-world phenomena such as noise.

It is clear that intrinsic evolution does not suffer from any limitation in this respect, since the real physical devices are used here. The use of real hardware ensures that the outcome of intrinsic evolution is an actual circuit that certainly works. However, this is guaranteed only in the operating conditions actually tested. Moreover, one must consider that the fabrication processes are not perfectly replicable. Therefore, devices that are nominally identical can differ slightly in their characteristics. This means that the working of a circuit evolved in real hardware is guaranteed only when the circuit is constituted by the physical devices used for the evolution. This last limitation can have serious consequences, since the circuit could be found to be intolerant of changes of the physical device used for the implementation, which could undermine the real-world applicability of a circuit synthesized in this way. This can be especially true if evolution has found a way to exploit some unconventional interactions that exploit the physical properties of the circuit to achieve the required functionality. Although directly linked to the intriguing possibility of transgressing the conventional design abstractions, this behavior can be a mixed blessing, since it can tend to tune the functionality to peculiarities of the hardware used for evolution, eventually interfering with the transfer of the circuit on different physical platforms. An equally mixed blessing is, of course, the fact that the use of a simulator entails the risk of not modeling some physical interactions that could be exploited by evolution.

Let us consider now the relative speed with which the circuits can be tested and evolution can proceed in the extrinsic and intrinsic cases. Thanks to the fast reconfigurability of modern reconfigurable devices, the intrinsic approach can often outperform in speed what can be obtained in simulation. However, it is difficult to make definitive assertions on this issue. If a very detailed model of the circuit and of the environment is required, simulation on a programmable computer typically proceeds at a slower pace with respect to the real hardware, due to the necessity to pay what has been called "the price of programmability" (Conrad 1988). However, if the models can be simplified by the adoption of some abstraction, a *minimalistic simulator* can be defined which can proceed at a much faster pace than the real hardware.

In any case the magnitude of the speed difference seldom results in dramatic differences in the degree of circuit complexity that can be tackled with the two approaches, since the speed advantage is typically rapidly frustrated by the increase of the size of the design space with the complexity of the circuit (Yao and Higuchi 1999).

Finally, a unique property of intrinsic evolution is the possibility of letting the process of evolution continue during the operation of the circuit, achieving *online adaptation* and *fault tolerance* (Keymeulen et al. 1998; Mange et al. 2000). In this case, what is evolved is not the defining features of the circuit but the rules that enable continuous reconfiguration of the circuit. The

EVOLVABLE HARDWARE   term *evolvable hardware* (EHW) is typically used to refer to systems achieving online adaptation, although the term is also used to refer to intrinsic evolutionary electronics in general.

Summing up, there is no absolute winner in the intrinsic vs. extrinsic contest. The choice between the two depends on the kind of application and the possibility to significantly exploit the advantages of one of the two approaches. As a general precept, it is advisable not to forget at the end of the evolution to probe the weak points of the adopted approach. For example, after evolving a circuit in simulation it is advisable to implement in hardware the circuit for a final battery of tests before using it in a real-world application, and after evolving in hardware it is advisable to probe extensively the resulting circuit, which includes testing the circuit in a variety of environmental conditions and using physical devices different from those used in evolution. As discussed at the end of this chapter in the context of design verification, the number of tests required can be dramatically reduced if the working principles of the circuit intrinsically or extrinsically evolved can be understood and analyzed formally.

1.18   Digital Design

After the general remarks of the previous sections, we proceed now to consider the actual problem of circuit design. We start by examining the issue of digital design, first in its conventional and then in its evolutionary form, presenting some actual examples of evolution of digital circuits. The same kind of analysis will be performed later for the analog case.

COMBINATIONAL AND   There are two types of digital circuits: *combinational* and *sequential*. A cir-
SEQUENTIAL CIRCUITS   cuit is called combinational if its outputs depend only on the current inputs and is called sequential if the outputs depend also on the past values of the inputs (Wakerly 2001; Katz and Borriello 2004). In other words, combina-

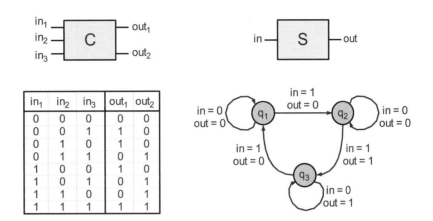

Figure 1.23   The functionality of a combinational circuit C can be represented in terms of its truth table (*left, bottom*), which specifies the configuration of the outputs that are associated with each configuration of the inputs. The functionality of a sequential circuit S can be represented, for example, in terms of a state diagram (*right, bottom*) specifying, for each state q_i and each configuration of the inputs, the new state of the circuit and the configuration of the outputs.

tional circuits are not endowed with memory whereas sequential circuits are. The functionality of a combinational circuit can be represented in terms of its *truth table* whereas a sequential circuit requires a more complicated representation such as a *state/output table* or a *state diagram* (figure 1.23). The greater complexity of the specification and verification of sequential circuits is probably one of the reasons for the prevalence in the literature of examples of evolution of combinational circuits over sequential ones.

We can identify various building blocks for digital circuits, at different levels of abstraction (figure 1.24). The lowest level that is generally considered is that of the *transistors* and of other elementary active and passive devices. These are used to build *logic gates* which are combinational circuits that implement elementary Boolean functions such as the logic negation (NOT), the logic product (AND), the logic sum (OR), the exclusive OR (XOR), and so on. The logic gates are used to build a variety of more complex combinational and sequential building blocks such as encoders, decoders, multiplexers, adders, flip-flops, shift registers, counters, and many others. Those elements of medium complexity can be further used to assemble digital circuits of higher complexity.

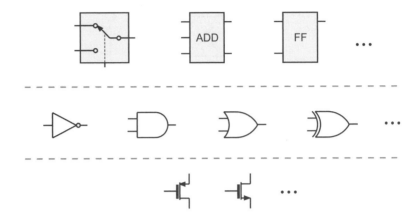

Figure 1.24 The design of digital circuits occurs at various levels of abstraction. The figure shows the level of the discrete components such as transistors (*bottom row*), which are used to synthesize logic gates implementing elementary logic functions such as the NOT, AND, OR, and XOR functions (*center row, from left to right*). The logic gates are used to obtain building blocks of medium complexity such as multiplexers, adders, and flip-flops (*top row, from left to right*), which are used to assemble even more complex circuits.

PROGRAMMABLE
LOGIC ARRAYS (PLAs)

There exist many kinds of commercial reconfigurable digital devices, some of which can be used for the intrinsic evolution of digital circuits. The simplest digital reconfigurable devices are generically called programmable logic devices (PLDs) and permit the implementation of a well-defined class of combinational circuits of medium complexity. For example, programmable logic arrays (PLAs) (figure 1.25) are reconfigurable devices whose cells are constituted by nonreconfigurable elementary logic gates and whose reconfigurable connections are structured so as to permit the implementation of logic expressions in the form of sums of products of the directed and negated inputs. Thanks to some theoretical results on the representation of logic functions (Wakerly 2001; Katz and Borriello 2004), a reconfigurable device based on this structure permits the implementation of all possible logic functions of a given number of inputs. In general, to reduce the complexity of the devices, only a subset of all the possible logic functions of the input signals can be actually implemented on a given PLA. Several PLDs can be put together on a single chip and linked by reconfigurable connections to form devices called complex programmable logic devices (CPLDs). Alternatively, an array of cells that are simpler than PLDs but more complex than elementary

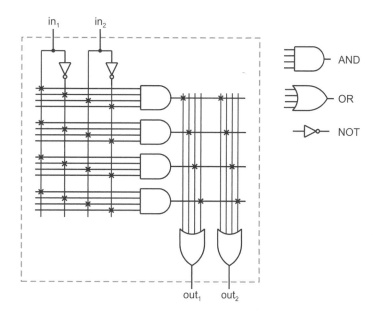

Figure 1.25   An example of a programmable logic array (PLA) with two inputs, two outputs, and the possibility of generating at each output the logic sum of up to four logic products of the directed and negated input variables. The choice of the implemented function is done by programming the reconfigurable connections represented in the schematic by the X symbols. Typical commercial PLAs provide some tens of inputs and outputs each realizing the logic sum of up to several tens of products of the input variables.

FIELD-PROGRAMMABLE
GATE ARRAY (FPGA)

logic gates can be assembled on a chip to form a field-programmable gate array (FPGA). The cell of a typical FPGA can be programmed to implement a certain number of logic functions of a few input variables (see figure 1.31). In FPGAs (and in some CPLDs) there is also the possibly to include in the signal path of each cell a few memory elements. The reconfigurability of the connections between the cells permits typically an almost arbitrary connection with neighboring cells of the array and provides a limited direct access to more distant cells and input/output pins of the device. As discussed in box 1.3 these devices are configured by loading a sequence of configuration bits into the chip.

In general, both the human and the evolutionary design of digital circuits proceed by using the building blocks at one level of abstraction to synthesize circuits at the next higher level of abstraction. We have thus evolutionary experiments where transistors are used to design gates (Koza et al. 2003; Zebulum et al. 2002), experiments in which gates are used to synthesize building blocks of medium complexity (J.F. Miller et al. 2000), and experiments that use function blocks of higher complexity (Damiani et al. 1999). An important aspect of digital design is the existence of systematic design techniques working at the level of the logic gates and elementary memory elements for both combinational and sequential circuits (Wakerly 2001; Katz and Borriello 2004). These powerful techniques ensure the synthesis of a system that realizes the required functionalities. However, these techniques require the specification of the functionality in terms, for example, of a truth table or of a state diagram, and can produce systems that are not optimal in terms of resource usage. As will be explained later, the case of analog design contrasts significantly with that of analog design, where no such systematic techniques exist.

1.19   Evolutionary Digital Design

The setup of an evolutionary electronics experiment requires the specification of a genetic representation for the circuits and that of a fitness function representing the goal of the experiment. Many genetic representations can be devised for digital circuits and, in fact, most of the creativity of the evolutionary experimenter is typically expressed here. In a first class of popular representations the genome is a list of blocks of fixed and predefined structure, each defining the nature of a circuit component and its connections to the other components (see, for example, the genetic representation described in figure 1.28). Another popular representation is used in the case of intrinsic evolution with reconfigurable devices, and is based on the direct use of the string of configuration bits as the genome. Program-based genetic representations such as genetic programming are another popular kind of representation and will be considered in more detail below, in the discussion of analog evolutionary design.

Considering now the definition of the fitness function, if the design goals include a conventional specification of the required combinational input/output function or that of the sequential input/state/output behavior, there is a natural priority for the realization of those requirements. In this case the fitness function is typically defined in terms of a count of the items of the required behavior that are correctly realized by the candidate solution. For example, for the realization of a combinational circuit realizing a given truth table, the fitness could be the number of output entries of the truth table that are correctly realized. In some cases it is advisable to include a mechanism of penalization of the circuit structures that discourages the matching of almost all entries of the truth table with trivial functions that are difficult to improve to an exact solution (Zebulum et al. 2002, p. 174). Once the priority objective of functionality has been realized, other objectives such as the number of devices in the circuit, the input/output delay, the maximum admissible clocking speed, the power consumption, and so on can be also taken into account as objectives of the evolutionary process.

We consider now three examples of evolutionary digital design which illustrate some of the most common circumstances in which the evolutionary approach represents a valid alternative to the conventional digital design approach. In the first example the recourse to evolution is motivated by the fact that the description is not given in a form equivalent to a truth table or to a state diagram required by conventional design. In the second example the objective is the synthesis of circuits that realize logic functions often used as digital building blocks using fewer resources than with conventional synthesis techniques. In the third example, the goal is to explore the potential of evolution by letting it go beyond the limits of the conventional abstractions of digital design.

1.19.1   Example 1: Evolution of a Robot Controller

An example of evolutionary application of a PLA reconfigurable device is the robotic application described in (Keymeulen et al. 1998, 1997). The purpose of the experiments described in these papers is the evolution of a controller for robot navigation. The robot is put in a square arena delimited by walls and populated by a few obstacles and by one special target object. The goal of the experiment is the evolution of a controller driving the robot to the target without getting stuck in the obstacles or bumping into the walls, starting from any position within the arena. The obstacles are low relative to the height of the robot, so that the target is always visible despite the presence of the obstacles.

The robot (figure 1.26) is equipped with a series of obstacle detectors, with a system that can detect the direction of the target, and with two motors that drive independently the two wheels of the robot. The information collected by the sensors is encoded into eight binary signals, six derived from the outputs of the obstacle detectors and two from the system that detects the direction of the target. These eight signals are given as input to a PLA, whose

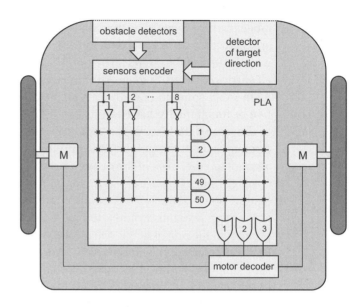

Figure 1.26 The structure of the controller for the experiment of robot evolution reported in (Keymeulen et al. 1997). The information coming from a series of infrared obstacle detectors and from a detector of the direction of the target are encoded into eight binary lines. The corresponding eight signals are given as input to a PLA with 50 AND lines and three output lines. The three PLA output signals are decoded into the signals driving the motors that drive the two wheels of the robot. The goal of the experiment is the evolution of a configuration for the PLA connections (denoted by the X symbols and encoded in the genome), which make the robot reach the target object while avoiding obstacles. Note that to avoid overcrowding the diagram with the input lines of the logic gates, we use here a simplified representation for the PLA structure, where single lines entering the AND and OR gates stand for multiple input lines.

duty is to control the robot by transforming the encoded sensory inputs into encoded motor commands. The encoded motor commands are constituted by three output binary signals which are decoded into actual signals for the motors that drive robot wheels. This means that there are $2^8 = 256$ different input configurations and $2^3 = 8$ different output configurations for the PLA. Thus, the number of different Boolean functions that could be used for the robot control in the given configuration is $8^{256} \approx 10^{231}$. This is the size of the phenotype space. The PLA has 50 AND lines, each of which can be configured to use as input any of the eight input signals, either in its directed or negated form. This is done by setting the state of the $8 \times 2 \times 50 = 800$

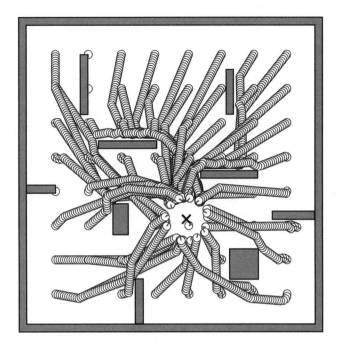

Figure 1.27 The trajectories of a robot successfully evolved in simulation (adapted from Keymeulen et al. 1997). The robot manages to reach the target object (whose position is denoted by the X) without hitting the obstacles (represented by the gray boxes) or touching the walls of the arena from all of the predefined starting positions (except those that are located within an obstacle).

configuration bits (8 input lines, multiplied by two by the generation of the inverted signal, each with the possibility of connecting to the 50 AND gates) that determine the connectivity of the AND connection array. The output of the 50 AND gates can be connected to any of the three output OR gates using the $3 \times 50 = 150$ configuration bits of the OR connection array. This gives 950 bits, which represent univocally the configuration of the PLA and constitute the genetic representation for the individuals of the population used in the evolutionary experiment. Thus, the size of the genotype space is $2^{950} \approx 10^{286}$. Note that the size of the genotype space is greater than the size of the phenotype space. This is an effect of the redundancy that exists in the representation of a logic function in terms of the configuration bits of the PLA, since a given Boolean function can be obtained with several different sets of configuration bits of the PLA (Keymeulen et al. 1997).

The evaluation of the behavior of each robot controller belonging to the evolving population is done by starting the robot from a series of 64 pre-defined positions regularly spaced in the environment (figure 1.27). Each starting position is probed in sequence in a series of trials, each of which lasts until the robot reaches the target object or hits an obstacle or a wall or uses all the allotted time without reaching the target object. The fitness is obtained by counting the number of starting positions from which the robot was able to reach the target object, plus a contribution from the unsuccessful trials that takes into account the final distance from the target object and the number of steps used to reach the final position. The initial population of controllers is obtained generating random sequences of configuration bits corresponding to a connection probability of 0.3 for the AND array and a connection probability of 0.5 for the OR array. Figure 1.27 shows an example of behavior of a robot successfully evolved in simulation in less than 200 generations using a population size of 20. The figure shows that the robot is able to reach the target object avoiding the obstacles (represented by gray boxes) from all the starting positions that do not coincide with the position of an obstacle.

This experiment exemplifies well some of the issues of evolutionary electronics that have been discussed above. A first observation is that the specifications for the PLA are given here in terms of the resulting behavior of the robot rather than in terms of a truth table or of an equivalent input/output description. Thus, we are in one of the cases where the conventional methodologies of digital design are not applicable. We see also at work one of the kinds of genetic representation mentioned above, namely the representation in terms of the string of configuration bits of the reconfigurable device. From the point of view of the verification of the result, note that during evolution the space of initial configurations is sampled at a finite number of discrete locations and with a small set of environmental conditions. Moreover, when evolution is done in simulation the kind of sensory signals experienced by the robot is limited to what is implemented in the simulator. This excludes, for example, the effects of sensor noise and the actual variability of the real environmental conditions. In order to assess the generality of the evolved solution, the evolved robot should be tested extensively in other operational conditions, starting from other initial positions and in other environments. Finally, note that the use of intrinsic evolution in this case does not result in any substantial speed advantage, since the device reconfiguration time is negligible with respect to the time taken by the robot trials.

1.19.2   Example 2: Evolution of Arithmetic Circuits

A common application of combinational digital circuits is the implementation of arithmetic functions such as addition and multiplication of binary numbers. There are well-established techniques for the synthesis of arithmetic circuits, which are based on the decomposition of operations that involve several bits into elementary arithmetic or logic operations. For example, the addition of two n-bit binary numbers can be decomposed into a series of elementary operations involving two-bit additions with carry, and the multiplication of two binary numbers can be decomposed into elementary two-bit additions and two-bit logic operations according to the familiar algorithm of multiplication (Katz and Borriello 2004). It is thus possible to compare the outcome of an evolutionary process with the products of these conventional design techniques. We describe here some of the experiments of arithmetic circuit evolution reported in (J.F. Miller et al. 2000).

CARTESIAN GENETIC PROGRAMMING (CGP)

Figure 1.28 illustrates the genetic representation – called Cartesian genetic programming – used in these experiments. The elements used to compose the circuit are a set of *cells* arranged in an array of n rows and m columns. The genome is a list of blocks of predefined structure, each specifying the functionality of the cells of the array and their connections to the other cells and to the global inputs and outputs of the circuit. To avoid loops and ensure the generation of combinational circuits only, the input of a cell in the array is permitted to connect only to the outputs of cells of the preceding columns or to global inputs. An additional parameter called *levels-back* permits limiting the number of columns to the left of a cell that can connect their input to that cell, although in the experiments reported below this limit is not enforced. The functionality of each cell can be chosen independently from a list of available cell functionalities. In the experiments described below each cell has three inputs and one output and its functionality is chosen from a list of 20 functions including the generation of a constant output, the directed and inverted connection of one input to the output, the logical product and sum of two inputs either directed or negated, and the selection (multiplexing) of one of two inputs directed or negated from the part of the third input. All these functions can be easily implemented in one cell of a typical commercial FPGA.

The experiments are based on an evolutionary strategy which takes as a starting point a population of genomes randomly initialized in terms of connectivity and functionality of the cells. The target of evolution is defined by the truth table of the desired arithmetic circuit, and the fitness of an evolved

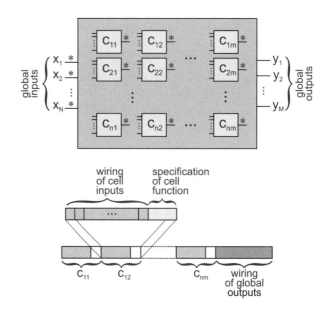

Figure 1.28   The genetic representation of digital circuits adopted in (J.F. Miller et al. 2000). A set of global inputs $\{x_1, \ldots, x_N\}$, a set of global outputs $\{y_1, \ldots, y_M\}$, and an array of cells are given (*top*). Each cell implements a logic function that is selected from a small set of predefined cell functions identified by an integer. The global input lines and the cell output lines (denoted by an asterisk in the figure) are also numbered. The genome (*bottom*) is constituted by a sequence of integers that are subdivided into blocks, one block for each cell of the array, specifying the cell function and to what global input or cell output the input of the cell must be connected. A final block of integers in the genome specifies to what global input or cell output the global outputs must be connected.

circuit corresponds to the number of correct output bits that it generates. We consider first an experiment aimed at the evolution of a two-bit multiplier, which is a circuit that takes as inputs a pair of two-bit binary numbers and produces at its outputs the four-bit binary number corresponding to the product of the two input numbers. The evolution was carried out on an array of $m = 7$ columns and $n = 1$ rows using runs of 100,000 generations with a population size of 5. Figure 1.29 compares the structure of a two-bit multiplier synthesized with the conventional design approach and the circuit structure of an evolved multiplier. The number of two-input gates used by the two circuits is the same and thus the evolved circuit has no advantage in terms of resource usage relative to the conventional design. Still, the re-

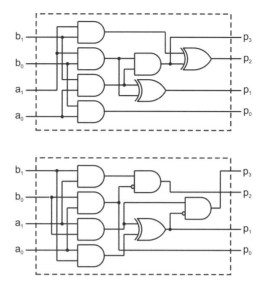

Figure 1.29   The structure of a two-bit multiplier synthesized with conventional design techniques (*top*) and an example of an evolved circuit performing the same function (*bottom*) (adapted from J.F. Miller et al. 2000). The evolved circuit, although composed of the same number of two-input gates of the conventional design, is interesting because it is based on an unconventional reuse of the lowest-order bit p_0 of the product.

sult is interesting because the evolved circuit has an original topology which is based on an unconventional reuse of the lowest-order bit of the product p_0, which permits the use of a single XOR gate in place of the two used by the conventional multiplier (J.F. Miller et al. 2000). Another experiment was aimed at the evolution of a three-bit multiplier. Artificial evolution was carried out on an array of 25 columns and one row using runs of 10 million generations with a population size of 5. Figure 1.30 compares the structure of a three-bit multiplier synthesized with the conventional design approach and the structure of one of the evolved multipliers. The conventional design approach produces a circuit with 24 two-input gates and two multiplexers, whereas evolution produced a circuit using 24 two-input gates and no multiplexers, and is thus more efficient in its use of resources. The evolved circuit does not appear to generate all the terms appearing in the conventional multiplication algorithm and its working is of difficult interpretation from the point of view of the conventional decomposition. Further evolutionary ex-

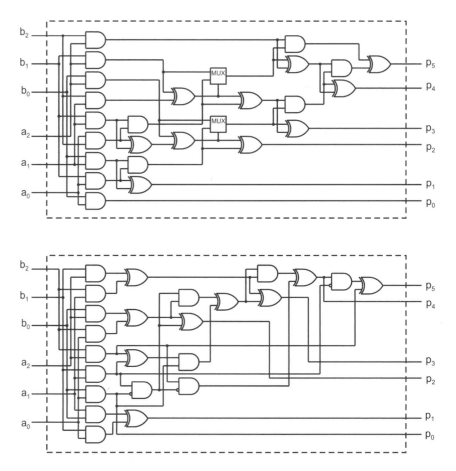

Figure 1.30 The structure of a three-bit multiplier synthesized with conventional design techniques (*top*) and an example of an evolved circuit performing the same function (*bottom*) (adapted from J.F. Miller et al. 2000). The conventionally designed circuit uses 24 two-input gates and two multiplexers, whereas the evolved circuit uses only 24 two-input gates and no multiplexers.

periments reported in (Vassilev et al. 2000), where the evolution was started from the conventional circuit rather than from randomly generated circuits resulted in an even more compact implementation of the three-bit multiplier, using only 23 two-input gates. Although the differences between the conventional and evolutionary results can seem minor, note that even the saving of even a few logic gates can result in substantial savings when the corresponding circuit is instantiated a large number of times, as can be the case for reconfigurable digital devices with hundreds of thousands of cells.

Summing up, we see here at work a genetic representation that permits the specification of the kind of devices composing the circuit and of their connectivity. Compared with the case of the genetic representation in terms of the string of PLA configuration bits considered in the previous example, this representation permits a much more extensive exploration of the space of circuit topologies, including topologies that are not considered by the conventional digital design methodologies. We have seen here two actual examples of structures of arithmetic circuits that differ from those obtained with conventional design techniques. In the three-bit multiplier example illustrated in figure 1.30, the evolved circuit achieves greater efficiency in the use of the resources with respect to the most efficient conventional circuit. As anticipated in the discussion about abstraction, this superior performance follows from the fact that evolution ignores the conventional abstractions based on the decompositions into modules dictated by the algorithm of long multiplication. On the other hand, the workings of the evolved circuits can be of difficult interpretation. In this case, however, the functionality of the evolved circuits can be tested exhaustively since it corresponds to the correct generation of the finite number of entries of the truth tables for the arithmetic functions.

1.19.3   Example 3: Transgressing the Abstractions in FPGA Evolution

In the previous example there was no substantial difference – from the point of view of the kind of functionality that can be potentially evolved – between running the evolution in simulation or running it in real hardware. The reason is that no interactions and operating conditions beyond those contemplated by the digital abstraction were admitted in that experiment. We turn now to an evolutionary experiment that takes a complementary approach. This experiment, despite using building blocks that were explicitly conceived for the implementation of digital circuits, tries to impose as few constraints as possible on their interactions and operating conditions.

The objective of the experiment, described in (A. Thompson et al. 1999), is the evolution of a circuit with one input and one output, capable of discriminating a 1 kHz square wave from a 10 kHz square wave applied at its input and to produce in correspondence of each input signal a distinct predefined logic value at the output. To avoid the limitations on the kind of interactions exploitable by evolution that would be imposed by the use of a simulator, the evolution is performed on the real hardware. The device used is a commercial FPGA constituted by an array of 64×64 cells, of which only the 100 cells of a 10×10 subarray are actually used (figure 1.31, top left). The input and output lines of the circuit are assigned to two predefined cells located on the boundary of this subarray. A particularity of the FPGA used in this experiment is that any sequence of configuration bits is admissible and that it is possible to define quite unconventional circuit topologies. Moreover, the meaning of the configuration bits is fully documented by the manufacturer and this permits the reconstruction of the topology of the evolved circuits from the string of configuration bits. The device is mounted on a board that is connected to a computer for the downloading of its configuration bits, for the definition of the input signals, and for the recording of the circuit behavior. In the FPGA used in this experiment the *function unit* (FU) of each cell permits the implementation of any Boolean function of two inputs and any multiplexing function of two inputs from the part of a third input. As shown in figure 1.31 (top right), each cell has four inputs (N, E, S, W) coming from the adjacent cells of the array and which can be used as arguments of the cell function. Each cell has four outputs going to the adjacent cells, to each of which can be routed any of the inputs coming from the other sides of the cell, or the output of the function unit.

The genetic representation adopted in this experiment recalls the one used in the previous example. The genome corresponds to the 1800 configuration bits that are required to specify the connectivity and functionality of the 100 FPGA cells available to evolution. It is composed of a list of blocks of predefined structure, specifying the functionality of the cells of the FPGA and their connections to the other cells (figure 1.31, bottom). Contrary to the case of the previous example, however, no constraints are imposed on the connectivity in order to avoid loops or to ensure the generation of combinational circuits only. Since no synchronization signal is used for the FPGA and no other precautions are taken to ensure the stabilization of the signals at their digital values, the circuit is not even constrained to operate as a conventional synchronous or asynchronous sequential circuit (Katz and Borriello 2004). The cells are thus free to operate as analog circuits and correspond in fact to a collection of simple and very fast high-gain amplifiers.

In a first series of experiments (Thompson et al. 1999), the starting circuits for the evolution were obtained by randomly generating an initial population of genomes. Each genome was downloaded into the newly initialized FPGA and the corresponding circuit was tested with an uninterrupted sequence of 10 randomly shuffled 500 ms bursts of the two kinds of input signals that the circuit is asked to discriminate, five bursts at 1 kHz and five at 10

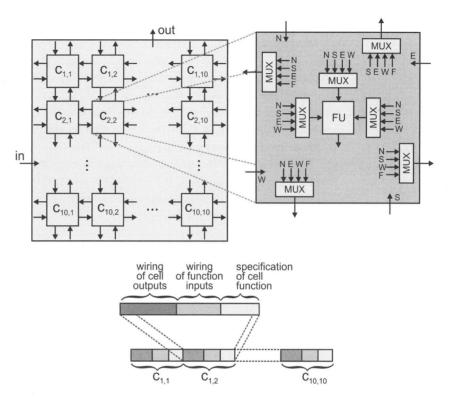

Figure 1.31 The structure of the FPGA and the genetic representation adopted in the experiments described in (A. Thompson et al. 1999). The material available to evolution corresponds to a 10×10 array of cells of an FPGA (*top left*). Each FPGA cell (*top right*) has a function unit (FU) to which can be routed the cell input signals coming from the four adjacent cells. The output of the function unit and the cell inputs can be routed to the cell outputs. The choice of the connectivity and of the functionality of the function unit is done via a set multiplexers (MUX) that are driven by the cell configuration bits (not represented). The genome (*bottom*) corresponds to a sequence of blocks, one for each cell, specifying the value of the configuration bits.

kHz. The output voltage of the circuit was integrated during each burst. For a perfectly discriminating circuit the output should correspond to a constant +5 V amplitude during the 1 kHz bursts and a constant 0 V amplitude during the 10 kHz bursts. To assess the discrepancy of the behavior of the tested circuit from the ideal behavior, the fitness was defined as the weighted difference between the average output amplitude value during the 1 kHz bursts and the average output value during the 10kHz bursts. The goal of evolution was the maximization of the fitness of the evolved circuits.

Figure 1.32 shows the cells composing the circuit with maximum fitness obtained after 5000 generations using a standard genetic algorithm and a population size of 50. The figure shows in gray shading the 21 cells of the FPGA array that contribute to the functionality of the circuit. The output of all the other cells of the array can be clamped to a constant logic value without affecting the circuit performance. The five cells shaded in dark gray in figure 1.32 have the peculiarity of not having their function unit connected in any way to the rest of the circuit and still influence the circuit behavior, as proved by the fact that clamping the output of the function unit of these cells results in a significant degradation of the circuit performance. The evolved circuit exhibits a perfect signal discrimination, producing the required output shortly after the corresponding 1 kHz or 10 kHz input signal is applied.

The successful evolution of a discriminating circuit for the two input signals using the resources represented in figure 1.32 is somewhat surprising, because the characteristic time scales of the signals that must be discriminated is on the order of milliseconds, whereas the characteristic time scales of operation of the devices constituted by the FPGA cells is on the order of nanoseconds. Furthermore, the structure of the evolved circuit considered as a digital circuit (figure 1.33) does not correspond to that of a counter or any other digital accumulator that could bridge the several orders of magnitude separating the two time scales. A series of analyses were therefore performed on the evolved circuit, in an effort to understand its workings (A. Thompson and Layzell 1999). A first outcome of these analyses is that the circuit does not operate according to the conventional digital abstraction. This conclusion follows from the observation that neither a digital simulation nor a hardware implementation in terms of separate logic gates of the circuit shown in figure 1.33 exhibit the frequency discrimination functionality of the evolved circuit. Further tests suggested that the transient behaviors of the evolved circuit following the start of the input pulses play a crucial role in the circuit functionality. Despite extensive scrutiny, however, no satisfying explanation of the circuit behavior could be eventually obtained and the discriminating functionality was attributed by the authors of the experiment to "a subtle property of the [FPGA] medium" (A. Thompson et al. 1999, p. 187). This conclusion, although not unexpected, is somewhat disturbing in the light of the issue of correctness verification, and opens the question of the robustness of the evolved circuit and the sensitivity of its functionality to changes in the environmental parameters and to variations in the characteristics of the

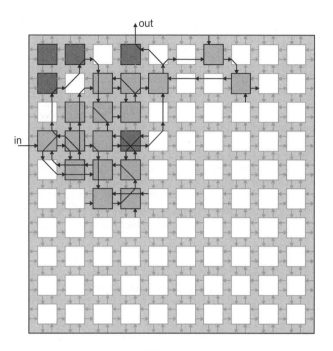

Figure 1.32 In gray shading, the 21 FPGA cells that contribute to the functionality of the evolved signal signal discriminator circuit, and their connectivity (adapted from A. Thompson et al. 1999). In dark gray shading, five cells whose function unit is not connected to the rest of the circuit but whose function unit output cannot be clamped without compromising the circuit performance. The thick black lines linking the shaded cells illustrate the input/output connectivity of the evolved circuit.

physical devices. In fact, a series of tests confirmed that the functionality of the evolved discriminator was seriously degraded if the evolved circuit was required to operate outside the small range of temperatures experienced by the circuits during evolution, or using a logically equivalent but physically different FPGA chip for the implementation of the circuit, and even when the circuit was implemented on a different 10×10 subarray of the original FPGA.

To counter the observed sensitivity of the functionality of the evolved circuit to environmental parameters and chip characteristics, a new series of evolutionary experiments was conducted with the explicit aim of evolving circuit robustness in conjunction with circuit functionality (A. Thompson and Layzell 2000). To this end the experimental setup was changed in or-

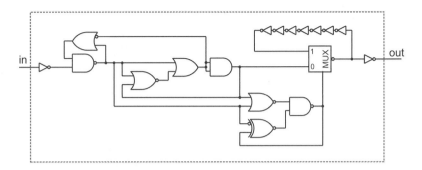

Figure 1.33   The best evolved discriminator circuit seen as a digital circuit, that is, in terms of the logic gates and multiplexers of the light gray-shaded cells represented in figure 1.32 that are connected to the circuit output (adapted from A. Thompson et al. 1999).

der to permit the testing of the evolving circuits on different FPGA chips, on different subarrays of each FPGA, and in different temperature, power supply, and output load conditions. Another major difference of the new setup was that a 6 MHz clock signal was supplied to all the FPGA cells, although no constraint was imposed to force the output of the function unit of the cells to synchronize with the clock. Using this setup it was possible to evolve robust discrimination circuits using a number of FPGA resources comparable to those shown in figure 1.32. The evolved circuits used in an essential way the newly available clock signal and their functioning was no longer dependent on subtle, poorly reproducible effects. In fact, the newly evolved circuit behavior could be easily replicated both in simulation and in circuits composed of separate logic gates.

Summing up, the experiments described above show how unconstrained evolution is able to transgress the conventional design abstractions – in this case the digital circuit abstraction and, possibly, also the circuit theory abstraction – even when these abstractions are strongly built by design into the available material (in this case, the digital cells of the FPGA chips). It is probably this aspect that explains the fascination that the results of these experiments exert, making them one of the most commented on of the whole evolvable electronics literature. On the other hand, we have seen how these transgressions can be paid in terms of robustness of the evolved systems, due to the exploitation of subtle properties and the interactions of the evolving components. We will discuss below, in the section devoted to the issue of design verification, how robustness can be built into the evolved system by a more extensive sampling of the circuit operational conditions at the price of a greater complexity of the evolutionary experiments, of increased constraints imposed on the "creativity" of the evolutionary process, and of reduced performance of the synthesized circuit: in short, by paying the *cost of robustness* (Gilbert 2002).

1.20   Analog Design

We said in the previous section that conventional digital design takes as its starting point a formal description of the circuit functionality (figure 1.23) and proceeds with formal methods to the generation of a circuit that realizes that functionality. On the contrary, there exist no systematic analog synthesis techniques capable of transforming in an almost algorithmic way a set of nontrivial specifications into a circuit complying with them. Consequently, analog design is a much less formalized activity than digital design, and its outcome depends much more on the experience, creativity, and skill of the human designer. This fact makes analog design a very promising domain for the application of evolutionary methods.

One could argue that the actual economical impact of improvements in analog design cannot be too great, since the analog share in electronic circuits is waning due to the steady trend toward the substitution of analog processing with its digital counterpart. This reasoning is disproved by the fact that analog circuits continue necessarily to constitute in most applications the interface between the digital circuitry and the real world, with typically a dramatic impact on the overall system performance. The behavior of the analog subcircuitry of a complex electronic circuit is often the bottleneck and the qualifying aspect of the circuit performance, despite its absorbing a marginal portion of the circuit resources in terms of devices. For example, the analog signal conditioning circuitry of a digital oscilloscope is crucial in determining the performance of the instrument, no matter how complex, sophisticated, and expensive the digital signal processing and displaying hardware and software that follows (Roach 1995). Similar observations can be made about the role of the analog circuitry of many other bestselling devices such as cellular phones and portable music players.

Like digital design, analog design can take place at various levels of abstraction (figure 1.34). Typically, the simplest level considered by circuit designers is that of the transistors and other components such as resistors, capacitors, and the like. There exists then a level of abstraction constituted

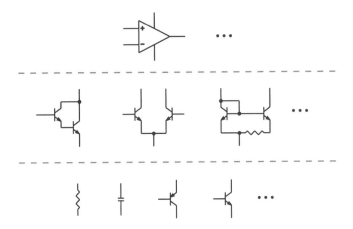

Figure 1.34 Discrete components such as resistors, capacitors, and transistors (*bottom row, from left to right*) are used to synthesize small circuits (*center row*). Discrete components and small circuits are used to obtain building blocks of medium complexity such as operational amplifiers (*top row*), which are used to realize more complex circuits.

by small circuit topologies composed typically of a handful of transistors and passive components. These "little circuits" (Gilbert 1991) can be identified as recurrent building blocks of analog circuits of greater complexity. The importance of these small circuits is witnessed by their omnipresence. The difficulty of their discovery is attested by the fact that they are typically named after their inventor. The most popular analog abstraction above the level of small circuits is that of the *operational amplifier* or *op-amp* (Franco 2001). Existing examples of evolutionary analog design typically take place at the level of abstraction of discrete devices such as transistors. There are also examples of evolutionary experiments that try to capitalize on existing analog design knowledge by also using libraries of standard small circuit topologies and operational amplifiers (Kruiskamp and Leenaerts 1995; Dastidar et al. 2005).

A few manufacturers of integrated circuits have introduced in the recent past models of analog reconfigurable devices called field-programmable analog arrays (FPAAs), in the hope of replicating the commercial success of their digital counterparts constituted by CPLDs and FPGAs. The structure of these devices is typically based on a cell comprising one operational amplifier and a few auxiliary resistors, capacitors, and transistors. By configuring the connections within the op-amp and the auxiliary devices these devices permit the choice for each cell between a handful of different analog functions such as amplifiers with various gains, filters, comparators and the like. None of these examples of commercial analog reconfigurable devices has attained so far a consistent success, and the fabrication of many of them has been discontinued. Probably the cause of this commercial failure must be attributed to the necessity of using in analog design a large variety of op-amp types, characterized by different specifications, in order to fine-tune the different parts of an analog circuit and attain the required global performance. This contrasts with the practical necessity of using only one or at most a few different op-amp types as cell elements of commercial reconfigurable devices, in order to keep at a reasonable level the manufacturing and storage costs. The scarcity of commercial analog reconfigurable devices and the relative inflexibility of the circuit topologies whose production they permit have prompted some research groups to develop custom analog reconfigurable devices, for example, the evolvable motherboard (Layzell 1998) and the programmable transistor array (Stoica et al. 2001b). A common problem in the evolutionary application of analog reconfigurable devices stems from the necessity of dealing also with the connections of the devices with the power supply lines, rather than only with signal lines as in the case of gate-level digital design. This entails the risk of producing circuit topologies that would result in the destruction of the devices due to excessive current, voltage, or power dissipation. To avoid a time-consuming checking of the admissibility of the evolved circuit topologies, some of these custom devices employ for the reconfiguration of their connections switches characterized by a relatively high internal resistance, which ensures that even in the presence of unconventional connectivity the device stress does not reach dangerous levels.

1.21 Evolutionary Analog Design

Like evolutionary digital design, evolutionary analog design experimentation requires the specification of a genetic representation and fitness function. Some of the genetic representations for digital circuits presented in the previous sections can be used also for the evolution of analog circuits. This is obviously the case, for example, for the representation used in intrinsic evolution where the genome corresponds to the sequence of configuration bits of the reconfigurable device. A difference in analog circuits with respect to digital circuits is that the nature of the external sources that produce the input signals, the nature of the loads across which the output signals are delivered,

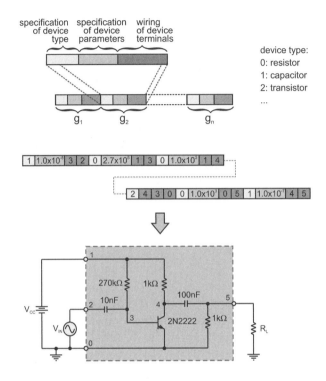

Figure 1.35 The schematic-based genetic representation used for analog circuits has as its starting point the specification of a set of available device types (*top right*), each associated with a unique numeric identifier. Each device type can have some evolvable parameters and has a fixed number of terminals. The variable-length genome (*top right, center*) is composed of a sequence of genes, each comprising a few fields specifying the device type, the value of the evolvable parameters (if any), and the numbers that identify the nodes associated with the device terminals. This genome can be directly decoded into a circuit (shaded box) and connected to the devices of the preassigned external circuit.

and that of the power supplies – in short, the nature of the predefined *external circuit* – can influence significantly the behavior of the evolved circuit. Consequently, it is necessary to define in advance the structure of the external circuit and specify which of its nodes can be connected to the evolving circuit.

A first kind of genetic representation that is frequently used for the evolution of analog circuits is the *schematic-based representation* (Grimbleby 2000;

SCHEMATIC-BASED
GENETIC ENCODING

Zebulum et al. 2002). This representation is based on a sequence of genes that specify the nature of the devices and assign explicitly their connectivity. Moreover, the genome contains typically the representation of many circuit parameters. The genes that form the genome are composed of a number of fields, for example an initial specifier of the device type, followed by the specifiers of the evolvable parameter values and by the specifiers of the circuit nodes to which the device terminals are connected. The genome corresponds thus to a sequence of genes that can be directly decoded into a circuit comprising the preassigned external circuit and the evolved circuit. The genetic operators can alter the values of the gene subfields, insert new genes, or remove existing ones. The length of the genome and the complexity of the circuit can therefore vary during the evolutionary runs. The definition of the evolutionary experiment starts with the specification of the external circuit (figure 1.35). A set of available electronic device types, such as resistors, capacitors, and transistors, is also specified. Each device is associated with a unique numeric identifier and has a predefined number of evolvable parameters and terminals.

An alternative approach to the representation of analog electronic circuits is based on a genome that does not specify directly the devices and their connections but represents instead a sequence of instructions – a program – that can be used to build the circuit. The experimenter must define an *embryonic circuit* or *embryo* that is used as the starting point for the construction of the evolved part of the circuit. The instructions corresponding to the genome are applied to the embryo following a predefined syntax and result in a final circuit comprising the external and the evolved circuit. An example

GENETIC PROGRAMMING (GP)

of this kind of approach is represented by *genetic programming* (GP) (figure 1.36) (Koza 1992; Koza et al. 1999, 2003). Another example is the *circuit-constructing robot* described in (Lohn and Colombano 1999). Program-based approaches can evidently be used not only for the representation of analog circuits but also for the representation of digital circuits and of many other structures. For example, the genetic programming approach has been used to represent and evolve structures such as antennas, control systems, neural networks, and electronic circuits with routing and placement of the devices. Moreover, the possibility to represent parameterized functions permits the evolution of structures where the values of the components are defined by a set of symbolic parameters. This possibility can be used, for example, to design electronic filters with variable passband and stopband frequencies (Koza et al. 2003).

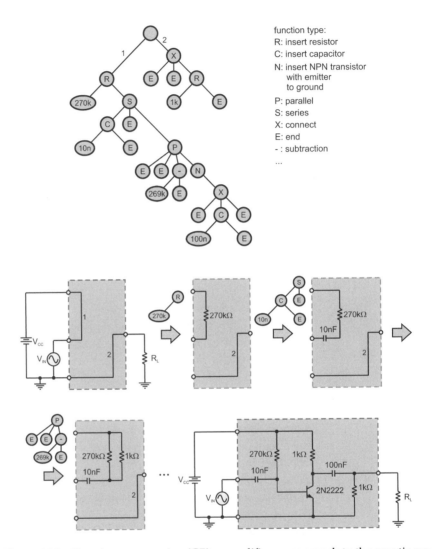

Figure 1.36   Genetic programming (GP) exemplifies an approach to the genetic representation of electronic circuits that is based on the specification of a sequence of instructions – a program – for the construction of the final circuit starting from an embryo (shaded box in the *center row, left*) connected to the predefined external circuit. The program is given in the form of a tree (*top left*), which corresponds to the genome. The nodes of the tree contain either the specification of a function belonging to a predefined list (*top right*) or numeric values used to assign the device parameter values. The instructions of the program specified by the tree are applied sequentially to the embryo and unfold progressively the circuit structure. The final result is a circuit comprising the external and the evolved circuit (*bottom right*).

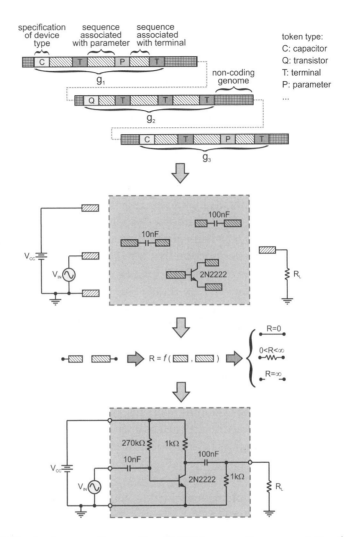

Figure 1.37 Analog genetic encoding (AGE) is a genetic representation for analog circuits based on an implicit definition of the interaction between devices that form the circuit. The genome is constituted of a sequence of genes, possibly separated by fragments of a noncoding genome (*top*). The genes correspond to the devices and specify the sequences associated with their terminals and the parameters. In a first decoding step the devices are extracted from the genome and the corresponding sequences are associated with their terminals (*second row from top*). The value of the parameters is also evaluated during this step. By using a function that maps pairs of sequences to values of resistance (*third row from top*), the connections between the devices can be established and the complete circuit composed of the external and evolved circuit results (*bottom*).

ANALOG GENETIC
ENCODING (AGE)

Another kind of genetic representation tailored to analog circuits is represented by *analog genetic encoding* (AGE) (Mattiussi and Floreano 2004, 2007; Mattiussi 2005; Mattiussi et al. 2008). To set up an evolutionary experiment based on AGE (figure 1.37), the experimenter starts by defining the types of devices that can appear in the circuit. The AGE genome is a sequence of characters where special sequences defined by the experimenter identify subsequences that correspond to genes. Each gene corresponds to a device that will appear in the decoded circuit. However, neither the resistor nor the connections of the device terminals are explicitly represented in the genome. Rather, the connectivity of the devices represented in the genome is specified implicitly by the sequences of characters that form a gene. The decoding process starts with the extraction of these sequences, which are associated with the terminals of the device represented by the gene. The experimenter defines a function that transforms pairs of sequences into a value of resistance. This function is applied to each pair of sequences attached to the terminals of the devices. If the value of resistance thus obtained is finite, a corresponding resistor is inserted between the two device terminals; otherwise the two terminals are left unconnected. The final result is a complete circuit.

IMPLICIT GENETIC
ENCODING

The fundamental characteristic of AGE is that of being based on an *implicit genetic encoding* of the connections between the devices. The use of an implicit encoding has several advantages with respect to a direct encoding of the circuit. First, it reduces the number of elements that must be encoded in the genome with respect to direct encodings, since the resistors do not explicitly appear as devices in an AGE encoding of a circuit. Another advantage of an implicit encoding is that a single mutation can have several effects on the network structure. This is useful in terms of evolvability since the evolutionary process can probe the effect of varying many interactions simultaneously with a single mutation. Moreover, the implicit genetic encoding of AGE permits the use of genetic operators such as the duplication of structures, which are known to be a crucial mechanism for the increase of complexity in biological evolution (Shapiro 2005).

1.21.1   Example: Evolution of a Gaussian Function Generator

We consider an experiment of extrinsic analog circuit evolution aimed at the synthesis of a Gaussian function generator using AGE (Mattiussi and Floreano 2007). This problem was first proposed in (Stoica 1999) and was also considered in (Koza et al. 1999). Figure 1.38 shows the devices of the external circuit: it is composed of a fixed voltage source V_p, and a variable

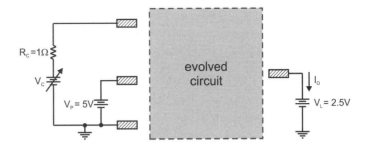

Figure 1.38 The devices of the external circuit in the experiment of evolution of a Gaussian function generator circuit using AGE. The external circuit is composed of a variable voltage source V_c connected to a series resistance R_c, a fixed voltage source V_p, and another fixed voltage source V_L through which the output current I_o of the evolved circuit is measured.

voltage source V_c connected in series to the resistor R_c, and by a "load" voltage source V_L. The goal of the evolutionary experiment is the synthesis of a circuit producing through the load voltage source an output current I_o that is a Gaussian function $g(V_c)$ of the variable input voltage V_c in the range $2\text{V} \leq V_c \leq 3\text{V}$, with a peak value $I_{o_{max}} = 80\,\text{nA}$ in correspondence to $V_c = 2.5\text{V}$, and $\sigma = 0.1\text{V}$.

To assess the performance of the evolved circuits their output current I_o is evaluated in simulation in correspondence to a discrete set of 101 equispaced values of the variable input voltage in the range of interest. The fitness function is defined as the sum of the squared differences $(I_o - I_{o_i}^*)^2$ between the output current I_{o_i} produced by the evolved circuit and the desired output current $I_o^* = g(V_c)$. The device set for this experiment contains a p-channel metal-oxide-semiconductor field-effect transistor (PMOS) and an n-channel metal-oxide-semiconductor field-effect transistor (NMOS) with the bulk terminal of the PMOS transistors connected by default to the positive terminal of V_p, and the bulk terminal of the NMOS transistors connected by default to ground. Figure 1.39 shows an example of an evolved Gaussian function generator circuit obtained after 30,000 generations using a population size of 100. The performance of this circuit is illustrated in figure 1.40.

1.22   Multiple Objectives and Constraints

Figure 1.40 shows that the evolutionary approach can produce a result that complies well with the design objective stated in the example problem de-

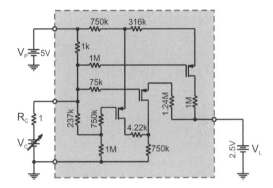

Figure 1.39 An example of Gaussian function generator circuit evolved using AGE as genetic representation. The devices of the preassigned external circuit are drawn outside of the dotted box.

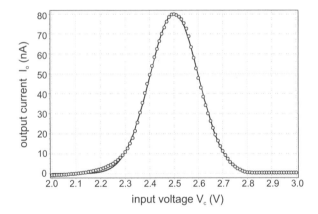

Figure 1.40 The output current I_0 of the evolved Gaussian function generator circuit shown in figure 1.39 plotted as a function of the input voltage V_c. The markers are drawn in correspondence to the values of input voltage used to evaluate the circuit fitness. The background line represents the ideal Gaussian relationship between V_c and I_o.

OBJECTIVE

scribed in the previous section. However, the example is somewhat artificial in its imposing a single objective for the design process. The design of an electronic circuit has typically many conflicting and heterogeneous objectives. For example, an electronic amplifier could be required to provide the maximum possible gain and bandwidth with the minimum possible power consumption, the minimum possible noise, and similar requirements for some other ten or twenty characteristics of the amplifier. The *multiobjective* nature of the design problem means that the problem does not provide a single function whose value represents the quality of the circuit and which could be used as a basis for the definition of a fitness function. To define our evolutionary algorithm we are thus forced to specify how the performance of the circuits with respect to the different objectives must be linked to the relative reproductive success of the circuits in the algorithm. This problem is at the core of the definition of multiobjective evolutionary algorithms, which are one of the most rapidly developing areas in the field of evolutionary computation (Deb 2001; Fonseca and Fleming 2002; Tan et al. 2002). In many cases, to be acceptable the solutions must also comply with a series of

CONSTRAINTS

constraints. For example, in addition to demanding a minimal input-output delay, the design of a digital circuit might dictate that this delay must be kept below a predefined value for the circuit to be acceptable at all. Solutions that do not comply with the constraints are declared *infeasible* and should not appear in the result of the evolutionary process. To this end, the evolutionary algorithms must be endowed with some technique of constraint handling that eventually ensures the production only of *feasible* solutions (Bäck et al. 2000).

Summing up, evolutionary electronics (and many other evolutionary engineering applications) requires in general the use of multiobjective evolutionary algorithms with constraint handling, a detailed discussion of which would take us too far afield. We will thus limit the subsequent discussion to an overview of some possible approaches to the definition of the fitness of the candidate solutions in multiobjective problems with constraints. In the following, unless otherwise stated, we assume that the generic ith objective has been expressed in terms of a function $f_i(\mathbf{x})$, where $\mathbf{x}$ represents a candidate solution in the design space.

PRIORITY-RANKED
OBJECTIVES

In a first design scenario the multiple objectives can be ranked in order of decreasing priority. In this case, the design can be initially focused on the most important objective and then the objectives of lower priority can be progressively taken into account. This corresponds to considering a sequence of single-objective design problems with fitness function $f_i(\mathbf{x})$. Sometimes, after one or a few high-priority objectives are dealt with in this way there remains a collection of objectives that have the same priority and to which one of the approaches described below can be applied.

OBJECTIVES WITH
TARGETS
In a second scenario, for each objective there is a value g_i that represents the *target* of that objective. In this case the multiple objectives can be converted into a single scalar fitness function via a suitably defined distance of the vector $(f_1(\mathbf{x}), \ldots, f_n(\mathbf{x}))$ of the objectives from the vector $(g_1, \ldots, g_n)$ of the goals. Then, a conventional single-objective evolutionary algorithm based on this scalar-valued fitness function can be applied.

A third possibility is to assume that we know how to convert the different objectives in a common utility currency. For example, we might know how much of the amplifier gain can be sacrificed for a given increment of bandwidth, that is, we might know the *tradeoff* between gain and bandwidth. Of

TRADEOFFS BETWEEN
OBJECTIVES
course, similar tradeoffs must be defined for all the pairs of objectives if we want arrive at a single numerical estimate of the performance. In its simplest form the idea of tradeoff corresponds to the definition of an aggregated fitness function $f(\mathbf{x})$ which is a weighted sum of the objective functions

$$f(\mathbf{x}) = \sum_i w_i \, f_i(\mathbf{x})$$

with positive weights w_i. We have at this point a conventional fitness function that must be maximized, and we can use a single-objective evolutionary algorithm, possibly complemented by some constraint-handling technique. The approach just described is much used by virtue of its simplicity. However, it rests on the assumption of the knowledge of the tradeoffs between all the pairs of objectives, which is a condition rarely verified in practice. Moreover, it is seldom the case that the different objectives can be combined linearly, as implied by the above formula. For example, we could be ready to exchange a large amount of bandwidth for some gain to improve an electronic amplifier with a very small gain and plenty of bandwidth, but we should be willing to sacrifice a much smaller amount of bandwidth if the gain is high and the bandwidth is less abundant.

PARETO DOMINANCE
A fourth approach to the management of multiple objectives, and which does not require the specifications of tradeoffs, is based on the concept of *Pareto dominance*. Given two candidate solutions $\mathbf{x}_1$ and $\mathbf{x}_2$, one is said to dominate the other if it is at least as good with respect to all objective functions, and strictly better relative to at least one objective. Thus, the dominating solution can be considered fitter than the dominated one. Of course, it is possible that given two candidate solutions neither dominates the other

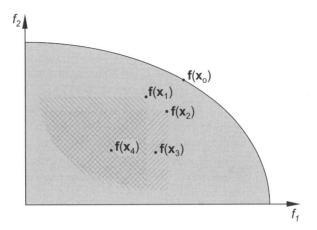

Figure 1.41   An illustration of the concept of Pareto dominance. The axes values correspond to two objective functions f_1 and f_2 with larger values corresponding to better performance. We denote with $\mathbf{x}_i$ the candidate solutions in the design space and represent the objectives as a vector-valued function $\mathbf{f}(\mathbf{x}) = (f_1(\mathbf{x}), f_2(\mathbf{x}))$. In the figure, the solutions $\mathbf{x}_1$ and $\mathbf{x}_2$ both dominate $\mathbf{x}_4$, but $\mathbf{x}_1$ does not dominate $\mathbf{x}_3$, whereas $\mathbf{x}_2$ does. The curve represents the Pareto front, and is composed of the values $\mathbf{f}(\mathbf{x})$ of the solutions that, like $\mathbf{x}_o$, are not dominated by any other element in the space of feasible solutions.

(figure 1.41). In fact, in the space of feasible solutions there will be typically a whole set of solutions that are not dominated by any other solutions. This collection of solutions constitutes the *Pareto-optimal set* and its image in the space of the objectives is called the *Pareto front*. In general the outcome of an evolutionary process based on the concept of Pareto dominance is a set of individuals rather than a single optimal individual. Ideally, this set of individuals should be representative of the whole Pareto-optimal set, and it is assumed that there will be a further phase of the project where a human designer will select one of the solutions produced by the algorithm, possibly applying the concept of tradeoff to this restricted set of solutions.

   The approaches to multiobjective problems considered so far are based on the concept of optimization, that is, of the idea of specifying a criterion to define (and search for) the best solutions in the design space. It has been argued (Simon 1996; Eilon 1972) that in many design contexts the designer aims at SATISFICING   *satisficing* rather than optimizing. In satisficing, the designer sets for each objective a minimal required level of performance and then tries to find a solution that satisfies all these performance criteria. Note that the data sheet of a commercial product can be seen as just the list of performance criteria that the designer has satisfied. The satisficing approach seems appropriate also in the light of our current understanding of the workings of natural evolution. As remarked by Ernst Mayr (2001), natural selection is best viewed as a process of elimination of the individuals that are nonviable or only marginally viable, rather than as a process of selection. The difference is that a process of selection – which corresponds to an optimization – tends to lead to the reproduction of only a few "best" individuals of the population. On the contrary, a process of elimination removes individuals that are nonviable relative to the prevailing viability criteria, that is, it eliminates the individuals that are not "satisficing" the viability constraints. The consequence is a reduction in the variety of individuals present in a population that is subject to selection relative to the variety of those surviving in a population subject to elimination (see also the closing remarks of the end of the chapter). A possible formulation of an evolutionary multiobjective design problem in terms of satisficing can be based on the idea of the distance of an objective from a target described above, with the difference that a component of the distance is considered zero as soon as the required level of performance for that objective is attained.

1.23   Design Verification

A recurrent theme in this chapter was the question of the actual applicability of the results of the evolutionary design process as engineering products. An evolutionary process can produce nonconventional designs whose working is based on mechanisms that go beyond the human cognitive limitations. This entails that the workings of evolved systems can be difficult or even impossible to understand by a human designer. This, in turn, can undermine the possibility of *verifying* the correctness of the system considered as a solution of the original design problem. It is important to realize that a design problem requires the implementation of a set of functionalities in a whole range of operational conditions and values of the device parameters. For example, an electronic amplifier is typically required to amplify correctly all the inputs signals in a certain range of frequencies and voltage ranges. Moreover, these performances must be guaranteed when the power supply voltage, the circuit temperature, and many other variables – including the values of the parameters of the devices that compose the circuit – vary in a predefined range. In short, we can say that the designed circuit must be *ro-*

OPERATIONAL
ENVELOPE

bust and perform correctly within a given *operational envelope* defined by the combined range of all these operational conditions and parameter values (A. Thompson and Layzell 2000; Gilbert 2002; Koza et al. 2005).

Ideally, a design technique should be accompanied by suitable verification techniques ensuring the correct functioning of the circuit within the whole operational envelope. In practice, apart from elementary cases where the operational envelope is composed of a small set of conditions that can be tested exhaustively, the verification of the correctness of a system is seldom attainable. The conventional approach is to choose design abstractions that permit the *generalization* of the verification of the circuit correctness from a few points of the operational envelope to (ideally) the whole of it (Gordon and Bentley 2002; Yao and Higuchi 1999). This means that the behavior of the designed circuit is conceptually or experimentally tested in a few operational conditions, with the assumption that the structure of the system has been chosen so as to ensure that the validity of the tests extends to the operation of the system in all other points of the operational envelope. Since evolution typically works with abstractions that are different from those used in conventional design, the verification of the correctness of evolved circuits can be a major problem. A possible solution is the enforcement on the evolutionary process of the use of the same abstractions that are adopted in conventional design. This can facilitate the understanding of the workings of the evolved circuits and permit the verification of their correctness. However, from our previous discussion it follows that the imposition on the evolutionary process of the same set of abstractions used in the conventional design threatens to undermine the very reasons for using an evolutionary approach in the first place, since it tends to cancel the difference between the conventional and the evolutionary design spaces.

On the other hand, in the absence of some precautionary measure, the very nature of the evolutionary process would tend to work against the satisfaction of the specifications in the whole operational envelope. We must bear in mind that evolution is able to subtly exploit the peculiarities of the context in which the individuals subject to evolution are evaluated. Thus, it can happen that the functionality of an evolved system depends on the presence of conditions that are proper to the way in which the circuits were evaluated during evolution. For example, a circuit whose evolution has been based on fitness evaluations conducted with a constant circuit temperature, or with a constant value of a certain device parameter, could be found to operate improperly when the temperature or the device parameter values are changed, precluding the use of the evolved circuit in real applications. At first sight a solution to this problem could consist in testing each individual in a large variety of operational conditions, so as to sample extensively the operational envelope (A. Thompson and Layzell 2000). Since this approach entails a high evaluation cost for each evolving individual, a diluted version could consist in testing each individual at a different point of the operational envelope, randomly chosen according to a suitable probability distribution. It is clear, however, that neither of these tactics can guarantee the correctness of an evolved system, since unexpected behaviors of the evolved circuit can manifest themselves in points of the operational envelope that have not been experienced during evolution, although the probability of this happening might be rendered small by extensive testing.

To avoid closing this discussion on too pessimistic a note, in considering these difficulties of the evolutionary approach we must not forget that the formal verification of system correctness is a difficult problem for both the conventional and the evolutionary synthesis of complex systems. In the evolutionary case the ultimate answer to the verification riddle lies perhaps in getting back to the real nature of the evolutionary design process, refraining from forcing on it the categories of the conventional design process. This means considering the result of the evolutionary process as a system that is suited to operate in the environment where it has been evolved, rather than as a system satisfying some specifications in a predefined operational envelope (Yao and Higuchi 1999). Moreover, this means entrusting the generalization properties of the evolved systems to the *adaptive* capacities of the system rather than to a set of predefined abstractions derived from the conventional design approach. In other words, the system should be evolved in conditions that favor the use of adaptation as a strategy to attain robustness in a wide range of operating conditions (Hammerstein et al. 2006).

1.24 Closing Remarks

This overview of evolutionary electronics has been an opportunity to illustrate the challenges that one can expect to encounter when applying artificial evolution to real-world problems. Through it we have distilled a series of lessons that include the importance of the concept of abstraction and modularity in the context of synthesizing complex systems. We have also emphasized the centrality and the difficulty of the verification of the correctness of the systems generated by the evolutionary process when they correspond to complex structures. The problem of verification can indeed be considered one of the most important open problems for complex real-world applications of evolutionary methods. The existing examples of evolved circuits witness the possibility of solving interesting engineering problems and achieving a better exploitation of resources relative to conventional approaches. They also show that it is possible to discover new system structures and tackle design areas where no systematic design techniques exist. The experience of electronic circuit evolution suggests that it is necessary to increase the extent of probing of the operational envelope during the evolutionary process. It suggests also that it is important to use multiobjective algorithms that do not force an excessively short-sighted optimization on the result.

The success of artificial evolution in addressing complex engineering problems is somehow at odds with the questions raised in the introduction to this chapter on whether natural evolution is equivalent to a problem-solving process. Paleontologist Richard Lewontin noted that the metaphor of biological evolution as an adaptation process in which problems set by the external world are solved by organisms through evolution originates from the process by which humans modify their environment to meet their own needs. Later, this metaphor was forgotten, he argued, and engineers started to believe that problems can be solved by mimicking natural evolution because organisms solved their problems this way (Lewontin 1996).

The notion of evolutionary adaptation implies a preexisting set of challenges that the environment, or ecological niche, presents to the organisms. In nature, this assumption is contradictory because the ecological niche is defined in function of the relation with the organism, which is by definition already adapted to its own niche. Van Valen (1973) attempted to defend the adaptive role of natural evolution by arguing that the environment is constantly decaying with respect to its organisms and therefore evolution operates so as to maintain organisms fit to their changing niche rather than to improve them over time. But even this elegant explanation is not entirely tenable because it does not predict or explain the diversification of species that move out of existing niches, which brings back the contradictory notion of preexisting niches and challenges available for organisms to be colonized (Lewontin 1978).

As a matter of fact, the four pillars of natural evolution can explain and predict only the differential reproduction and variability of individuals in a diversified population, but not progress or adaptation (see also S.J. Gould 1997). Darwin (1859), however, argued that adaptation, in the sense of amelioration of the current state of affairs, may occur in situations of particular

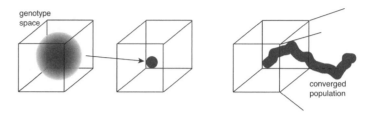

Figure 1.42 *Left*: Evolution in genetic algorithms amounts to convergence of a population (shaded area) toward a region of genotype space with higher fitness. *Right*: Evolution in SAGA corresponds to the displacement of a fairly converged population in a slowly varying genotype space. Adapted from Harvey 1992.

environmental stress, such as in the presence of a competitor or of environmental changes that undermine the survival of the species in specific ways.

Explaining the adaptive power of artificial evolution is easier because the fitness function effectively amounts to a set of preexisting environmental challenges and to a measure of progress along predefined directions. However, the consequence is that artificial evolution is a short-term and limited –albeit powerful– optimization process, whereas natural evolution is an incremental, open-ended, and creative process (Banzhaf et al. 2006). Current attempts to modify artificial evolution so as to allow for emergence of increasingly more complex and creative artifacts focus mainly on two aspects: genetic representations and evolutionary conditions.

Fixed-length, one-to-one, genetic encodings often used for parameter optimization do not allow scalability and evolution of increasingly more complex structures. More suitable genetic codes and mapping from phenotype to genotype should have a number of properties, such as (a) an indirect mapping that allows phenotype variability without necessarily requiring genotype variability, such as growth and learning; (b) redundant mapping that allows the expression of phenotype characteristics in many possible ways from the same set of genes, so that genetic mutations have lower probability of precluding incremental complexity; (c) variable-length genotypes that can grow, shrink, and/or reorganize during evolution; (d) genetic operators that have a significantly higher likelihood of producing small, rather than big, variations in fitness; and (e) evolution of the genetic representation and mapping.

After a few generations of artificial evolution, the population converges around a genotype that encodes the best, or locally best, solution to the problem at hand. Genetic algorithms and schema theory are indeed designed to display and explain, respectively, this behavior. If a new problem needs to be tackled, it is common practice to start a new evolutionary experiment from scratch with a new fitness function. Therefore, open-ended evolution is ruled out by definition.

SPECIES ADAPTATION GENETIC ALGORITHM (SAGA)

Harvey (1992) proposed SAGA (species adaptation genetic algorithm) to address open-ended incremental evolution (figure 1.42). SAGA focuses on the incremental evolution of fairly converged populations, which not only is the final state of most evolutionary algorithms but may also capture the state of affairs in evolving natural species. SAGA is based on redundant and variable-length genotypes where small mutations (instead of crossover) are the main factor that moves the population across genotype space as the environment or fitness function changes. Genetic redundancy, which in SAGA includes nongenic DNA, creates opportunities for "neutral walks" in the fitness landscape, that is the accumulation of mutations that do not affect the fitness of the individual, but may eventually lead to the expression of a much fitter phenotype. In other words, SAGA relies on a genetic code that allows many-to-one mappings from genotype to phenotype where mappings of equivalent fitness can be reached by a single mutation in genotype space. The collection of paths of equivalent fitness in the fitness landscape can be visualized as a "neutral network" (Huynen et al. 1996).

VIABILITY EVOLUTION

Another strategy toward open-ended evolution may consist in reproducing individuals according to "viability constraints," rather than performance against a specific problem (Mattiussi and Floreano 2005). Biological organisms and artifacts must satisfy many constraints in order to stay alive (viable). For example, bacteria and electronic circuits can operate only within a well-defined temperature range; they also require a certain amount of energy per unit of time in order to function properly; etc. In the case of electronic circuits, for example, several viability constraints are readily obtained by the specification sheet of the manufacturer. Each viability constraint defines a range within a specific dimension (temperature, energy, etc.) and the intersection of these ranges defines the viability space of the organism or artifact (figure 1.43, left).

Consider now a population of randomly generated individuals scattered across the entire space. In the absence of selection pressure by competitors or environmental change, all individuals within the viability space can reproduce, but offspring that fall outside the viability space by the effect of mutations are eliminated. This type of "viability evolution" has some benefits: (a) It allows individuals to explore more areas of the fitness landscape

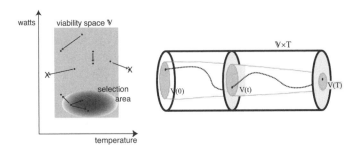

Figure 1.43 Viability evolution. *Left*: The gray area shows a simple viability space for electronic circuits where the individuals (dots) can reproduce with equal proba-bility. Individuals that fall outside the viability space by the effect of mutations are eliminated. Individuals that fall within a selection area (here shown as an oval whose gray level is proportional to fitness) are subjected to selective reproduction. The selec-tion area is not necessarily present and can extend over more (or fewer) dimensions than those spanned by the viability space. *Right*: The viability space can change over time. An individual can reproduce as long as it stays within the viability tube.

than if they were selected against a predefined criterion; (b) the combination of multiple constraints is easier to consider than in conventional multiobjec-tive optimization; (c) the modification of existing constraints, or inclusion of new constraints, at run-time is also easily done by modifying the borders and dimensions of the viability space (figure 1.43, right).

Furthermore, selection pressure toward a specific goal can be introduced in viability evolution by including a higher reproduction rate of individu-als that satisfy a fitness criterion (dark shaded area in the left panel of fig-ure 1.43). The evolutionary process is thus positioned somewhere between reproduction of the viable and selection of the fittest. This position can be dynamically shifted toward either extreme according to experimental needs and results. Viability evolution can lead to growth (in some cases also to ex-tinction) of the population unless a death criterion is introduced to kill the excess of individuals. Individuals can be randomly selected for elimination at periodic intervals or die of age or other factors.

SAGA and viability evolution are only two proposals toward open-ended and incremental evolution that still require further experimental validation. Whether or not they will prove useful models for science and engineering of complex adaptive systems, we believe that emphasis on evolution as a continuous process of adaptation in dynamic and unpredictable environments remains one of the most promising areas of research in evolutionary systems.

1.25   Suggested Readings

On the Origin of Species (Darwin 1859) remains a very readable and inspiring book, especially for its treatment of selection forces, which is a major issue in artificial evolution. However, Darwin was not aware of genes as the unit of hereditary transmission. The role of genes in natural evolution is described by Dawkins (1976) in the book *The Selfish Gene* where he argues that evolution is driven by genes that attempt to replicate in their own interest, even if that does not coincide with the interest of the phenotype. Although some aspects of that theory have been criticized, the book is worth reading for the clarity and strength of the explanations. We also recommend *The Blind Watchmaker* by the same author (Dawkins 1986), not only for its appealing and clear description of the mechanisms of evolution but also for its emphasis that evolution is not a random process, a criticism that is still raised by some engineers and mathematicians against evolutionary algorithms. After reading Dawkins, which may induce you into thinking that evolution is some sort of optimal adaptive system, we recommend *Full House* by Gould (S.J. 1997), which provides compelling evidence that evolution is mainly spread of diversity in a physical world constrained by the laws of physics. That book is also an excellent antidote against the commonly held belief that humans are at the top of the evolutionary ladder.

There are several other excellent books on evolution, such as the textbook *Evolution* by Ridley (2004). *The Major Transitions in Evolution* by Maynard-Smith and Szathmáry (1995) is unique in that it explains key facts in molecular biology and evolution within the perspective of how information is stored and transmitted through generations. According to the authors, there have been eight major transitions in evolutionary history starting from molecular replicators all the way to human societies and language. Although the book requires good knowledge of some biological and chemical aspects, it is particularly relevant for readers interested in the emergence of complexity. A significantly scaled-down version for a wider readership was published later under the title *The Origins of Life* (Maynard-Smith and Szathmáry 1999). We found that starting with this latter version before attacking the original book is a good strategy to fully appreciate all the insights provided by the authors.

For what concerns the molecular biology involved in evolution, we recommend the handbook *Fundamentals of Molecular Evolution* by Graur and Wen-Hsiung (1999), which is well-organized, clearly written, and richly illustrated. *The Art of Genes* by Coen (1999) is an extremely clear and nicely illustrated description of gene expression and interaction networks in the formation of the phenotype.

In the area of artificial evolution, we recommend the original book on genetic algorithms by Holland (1975) where the author also introduces schema theory in formal terms. We also suggest *Genetic Algorithms in Search, Optimization and Machine Learning* by Goldberg (1989), which is an excellent introduction aimed mainly at computer scientists and engineers. A nice aspect of that book is the inclusion of simple programming functions to illustrate the different steps necessary to set up a genetic algorithm. Another good introduction to genetic algorithms and their applications is given by Mitchell (1996), where readers will also find a good description of schema theory with its major criticisms.

For readers interested in genetic programming, we recommend the series of four books by Koza and colleagues (Koza 1992, 1994; Koza et al. 1999, 2003). Although the basic elements of genetic programming are covered in all books, each volume provides additional techniques and examples of applications. A nice feature of that series is that each book is accompanied by a tutorial video that could be looked at before reading the book or used in classes as support material.

For a more comprehensive view of the field of artificial evolution, we suggest the recent *Introduction to Evolutionary Computation* edited by Eiben and Smith (2003), which spans several approaches and variations of evolutionary algorithms. We also recommend *Evolutionary Computation: The Fossil Record* by Fogel (1998), which includes richly annotated reprints of 30 historical and influential papers in various aspects of evolutionary computation. Readers interested in understanding how to design an evolutionary algorithm tailored to specific applications may read the excellent book *How to Solve It: Modern Heuristics* by Michalewicz and Fogel (2004).

Coming now to the second part of this chapter, (Lohn and Hornby 2006) is a simple and short introduction to evolvable hardware and evolutionary electronics. A good critical overview of evolutionary electronics with comments on the importance of verification is presented in (Yao and Higuchi 1999). In the same vein proceed A. Thompson et al. 1999, but with an increased focus on the analysis of a few selected results and a discussion of the concept of the operational envelope. An analysis more centered on digital circuits but with a general discussion of the scope of evolutionary circuit design can be found in (J.F. Miller et al. 2000). Higuchi et al. (1999) present several examples of circuit evolution from different fields. Book-length coverages of evolutionary electronics are (Zebulum et al. 2002), (Sekanina 2004), and (Greenwood and Tyrrell 2007).

The classic paper by von Neumann (1961) contains interesting comments on the analog vs. digital issue in general and in relation to the working of the brain. Eliasmith and Anderson (2003) provide an original approach to the issue of neuronal encoding in the brain, which is relevant also to the general issue of the digital vs. analog dilemma. Sarpeshkar (1998) presents a good discussion of the advantages and limits of the analog representation and describes a mixed approach based on the regeneration of signals and the use of a distributed representation. Sarpeshkar (2006) describes examples of analog computation in biological systems and discusses the challenge of designing prosthetic sensory devices such as artificial cochleas, equaling the performance of their biological counterpart. (Shannon 1949) is a classic reference describing the sampling theorem and the analysis of the information content of analog signals.

There exist many good textbooks on conventional digital design techniques, for example (Katz and Borriello 2004 and Wakerly 2001). Analog reconfigurable devices, especially custom chips and circuits, are discussed in the already mentioned (Zebulum et al. 2002). Good classic analog design textbooks are (Gray et al. 2001) and (Allen and Holberg 2002), which contain also extensive analyses of the "little circuits" traditionally used as analog building blocks. In (Gilbert 2002) one can find a broad description and analysis of tradeoffs in analog design. The real flavor of analog design can be savored from many of the contributions in the two volumes edited by Jim Williams (Williams 1991, 1995) and from Pease's book on analog troubleshooting (Pease 1991). The subsection title "Transgressing the Abstractions" is admittedly inspired by the title of Sokal's hoax paper (Sokal 1996).

An extensive description of evolutionary multiobjective methods is given in (Deb 2001). A more compact overview is given in (Fonseca and Fleming 2002). Several chapters of (Bäck et al. 2000) consider the various facets of the issue of constraints in an evolutionary context. The original reference on satisficing is (Simon 1996), where the relevant comments are scattered in various chapters of the book. A more compact discussion of satisficing and, more generally, of multiple objective problems is given in (Eilon 1972). Interesting comments on natural evolution seen from the point of view of elimination rather than selection can be found in (Mayr 2001).

2 *Cellular Systems*

The simplest kind of systems that we can reasonably consider as living are biological cells (Harold 2001). Although cells are already quite complex systems, the most complex forms of life are multicellular organisms, that is, structured assemblies of cells. In a multicellular organism almost all cells contain the same genetic material and yet the morphology and function of two cells can be strikingly different. This difference is explained by the fact that the state of each cell depends not only on its genetic material but also on the state of the cell when it was generated and on the influences that acted on the cell from that moment onward. A multicellular system is thus an example of a system that is composed of many copies of a fundamental unit – the cell – whose interaction produces a global behavior that is not merely a scaled-up version of the behavior of an isolated unit. In this chapter we will explain how to define models that capture the essence of this property. We are interested in particular in models whose fundamental units are very simple – much simpler than biological cells. We will start by drawing a list of the basic constituents that are required to define such models. Then we will put together these constituents to build various models of cellular systems. We will analyze the properties of these models and discuss some of their applications in computation, artificial life, physics, and complex system modeling and simulation.

2.1 The Basic Ingredients

We use the inspiration provided by biological cellular tissues to define the elements of an abstract cellular system. This abstract system will be later specialized in order to obtain models that are applicable to a variety of phenomena. In abstracting a cellular tissue, the collection of cells becomes a discrete cellular space. The complex internal state of a biological cell is reduced to a numerical or symbolic state variable taking its values into a reasonably simple state set. The complex rules and interactions that govern the temporal dynamics of biological cells are abstracted by a mathematical function or rule that specifies how the state variable must be updated in time, taking into account the interactions of a cell with its neighbors, starting from a given initial configuration of the cellular space.

In more precise and formal terms, an abstract cellular system is composed of the following elements:

Cellular space. The collection of cells in the system is called the *cellular space*. In general, it is a regular d-dimensional lattice of cells. When dealing with cellular systems at an abstract level, the lattice is typically considered infinite. In practice, however, any actual implementation deals necessarily with a finite space. Figure 2.1 shows some common kinds of cellular spaces. Cellular spaces of more than three dimensions are seldom considered in actual implementations because the total number of lattice cells for a given size along each dimension grows exponentially with the number of dimensions.

Time variable. The dynamics of the cellular system unfolds along a time axis that can be discrete or continuous.

State and state set. The *state* of a cell represents the information specifying the current condition of the cell. It is the memory of what happened to the cell in the past. Thus, it is the only way in which the history of the cell can influence the future of the cellular system. The *state set S* is the set of acceptable values for the state of a cell. Often, a special *quiescent state* s_o is specified, which represents the resting or inactive condition of the cell. In most models of cellular systems the state of a cell is represented by the value of a numerical variable. Sometimes an n-tuple of numerical variables rather than a single variable is used to represent the state, even if in principle this n-tuple could be re-encoded as a single variable. The advantage of using an n-tuple is that each variable can represent in a more meaningful way the different aspects of the cell condition that must be modeled and can thus simplify the definition of the transition function described below.

QUIESCENT STATE

Neighborhood. The *neighborhood* of a cell is the set of cells (including the cell itself) whose state can *directly* influence the future state of the cell. In other

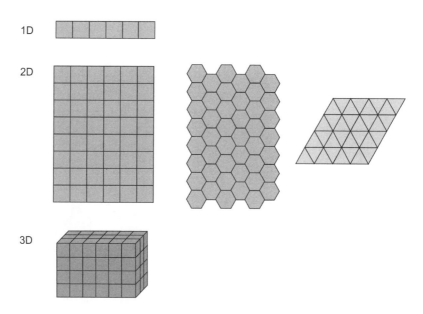

Figure 2.1 Some examples of one-, two-, and three-dimensional cellular spaces.

words, each cell can be thought of as directly connected to and sensing the state of the cells that belong to its neighborhood. In principle, the shape and size of the neighborhood can be arbitrary. However, the neighborhood is typically composed of a small number of adjacent cells because cellular systems are assumed as models of systems that exchange information only locally. A simple way to assign the neighborhood is to define a distance in the cellular space and then specify that the neighborhood of a cell is composed by all cells within a certain *radius* (or *range*) *r* from the cell. Figure 2.2 illustrates this concept and shows some of the most common neighborhoods for low-dimensional cellular systems. If all the cells in the system have the same kind of neighborhood, the cellular system is said to have a *homogeneous* or *uniform* neighborhood. Homogeneity can be further specified as pertaining to space, time, or both.

NEIGHBORHOOD RADIUS

State transition function. The state transition function, often called simply the *transition function*, is the function that specifies how the state of a cell unfolds in time. It depends only on the state of the cells belonging to the cell's neighborhood and, possibly, on the position of the cell and on time. If the transition function is the same for all the cells or does not

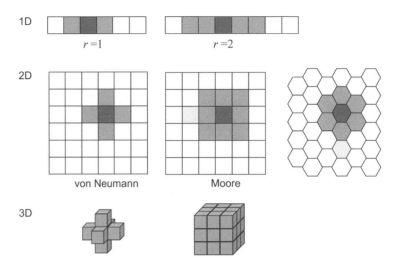

Figure 2.2 Some examples of neighborhoods in cellular spaces. For the one-dimensional case the figure shows how the concept of radius r can be used to define the neighborhood. The most common two-dimensional neighborhoods are the von Neumann neighborhood and the Moore neighborhood.

depend on time, the cellular system is said to be *homogeneous* relative to the transition function, in space and time, respectively. In general, when a cellular system is called homogeneous without further qualifications, it is assumed to be homogeneous relative to both the transition function and the neighborhood, in both space and time. For discrete-time cellular systems the actual implementation of the transition function on a computer can be done by programming a routine that evaluates the function at each time step. For small finite state sets and small neighborhoods the transition function can be implemented with a lookup table that stores all the entries of the transition function.

Boundary conditions. If the cellular space has a boundary, boundary cells may lack some of the cells required to form the prescribed neighborhood. This problem can be solved by specifying suitable *boundary conditions*. The most common kinds of boundary conditions (figure 2.3) are:

- **Periodic:** The simplest solution to the presence of boundaries is to eliminate them by transforming the cellular space from a space with boundaries to a space without boundaries. Typically, a rectangular d-

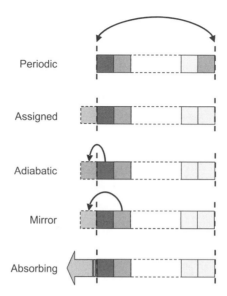

Figure 2.3 Some examples of boundary conditions, illustrated for the one-dimensional case.

dimensional cellular space is transformed into a d-dimensional toroidal cellular space by gluing the opposite sides of the rectangular space. This strategy is known as the assignment of *periodic boundary conditions.*

- **Assigned:** Another strategy to cope with the presence of boundary cells is to define a virtual neighborhood for them. The virtual cells required to complete the neighborhood can also be assigned a state that does not depend on the state of the actual cellular system. In most cases the assigned state is fixed (*fixed boundary conditions*), but it can be generated by a more complex process, for example, it can be generated by a random process (*random boundary conditions*), or it can correspond to a source of one of the quantities modeled in the cellular system (e.g., particles, or vehicles).

- **Copying:** Alternatively, the cells of a virtual neighborhood can be assigned a state that is a copy of the state of the cells of the cellular system. *Adiabatic boundary conditions* are specified by copying the state of the boundary cells. The name derives from systems used to model diffusive thermal phenomena which have the temperature as the state of the cell. For these systems this strategy defines a null temperature gradient and thus a null exchange of heat across the boundary. *Mirror boundary conditions* are specified by copying to the virtual cell the state of the next cell from the boundary. Note that periodic boundary conditions can be interpreted as copying to the virtual cell the state of the cell at the opposite boundary.

- **Reflecting:** *Reflecting boundary conditions* (also called *closed* boundary conditions) correspond to the definition of a process that reflects some of the phenomena that are modeled within the cellular system (for example, a particle colliding with the boundary, or a wave impinging on it). The definition of the reflecting process depends on the details of what is modeled by the cellular system.

- **Absorbing:** *Absorbing boundary conditions* (also called *open* boundary conditions) are a special class of boundary conditions that permit simulating with a finite space the behavior of an infinite cellular space which has all but a finite number of cells in its quiescent state and whose transition function is null state quiescent (see below). The idea is to define at the boundary a process that does not perturb the activity in the finite region that is modeled. The definition of the absorbing process depends on the details of the transition function and can be quite tricky. An alternative solution is to define *moving boundaries*, that is, to keep increasing the size of the cellular space in order to prevent boundary effects from being felt in the finite region of interest.

For some cellular models some of the boundary conditions that are listed above as distinct in fact coalesce. For example, for the cellular systems used to model the motion of a particle, boundary conditions that reflect and absorb particles can be implemented simply as fixed boundary conditions corresponding, respectively, to the fixed presence or absence of particles in the virtual cells beyond the boundary.

Initial conditions. In order to start the updating of the state of the cells of the system according to the transition function it is necessary to specify the initial state of all the cells. This is known as the assignment of the *initial condition* or *seed* of the cellular system.

Stopping condition. The stopping condition specifies when the update of the state of the cellular space must be stopped. Typical stopping conditions are the attainment of a preassigned simulation time and the observation that the state of the cellular system is cycling in a loop.

2.2 Cellular Automata

The simplest and most popular kind of cellular system containing all the above ingredients is the *cellular automaton* (CA). A CA has a discrete time variable, a finite neighborhood, a finite state set, and a synchronous update of all the cells in the cellular space. The integer sequence $S = \{0, \ldots, k-1\}$ is often used as the CA state set, with $s_o = 0$ representing the quiescent state.

TRANSITION RULE   The transition function φ of a CA (also called the *transition rule* or *CA rule*) is a deterministic function that gives the state $s_i(t+1)$ of the ith cell at the time step $t+1$ as a function of the state of the cells in the cell's neighborhood N_i at time t, that is,

$$s_i(t+1) = \varphi(s_j(t) \, : \, j \in N_i) \, .$$

The name CA derives from the mathematical concept of *automaton*[1], a discrete-time system with a finite set of inputs I, a finite set of states S, a finite set of outputs O, a state transition function φ which gives the state at the next time step as a function of the current state and inputs, and an output function η which gives the current output as a function of the current state. In a CA each cell is thus an automaton which issues its state as output and takes as inputs the outputs of the cells in the cell's neighborhood.

TRANSITION TABLE   In principle, the transition rule of a CA can be represented as a *transition table*, that is, a table which specifies the next state of a cell for each possible configuration of the states of the cells in its neighborhood (figure 2.4). If the state set contains k elements and the neighborhood is composed of n cells, the number of possible configurations of the neighborhood (and, thus, the number of entries in the transition table) is k^n. Thus, the representation of the transition rule as a transition table becomes rapidly impractical as k and n increase. The number of possible transition rules grows even more rapidly, and becomes astronomical even for small values of k and n. Since for each configuration of the neighborhood we have k ways to specify the next state, there are k^{k^n} different transition rules for a CA with k possible states and a neighborhood of size n. For a CA with two possible states (a *binary CA*) and a neighborhood of size three, this gives $2^{2^3} = 256$ different transition rules. But already for a CA with three possible states (a *ternary CA*) and a neighborhood of size three there are $3^{3^3} = 7,625,597,484,987$ different transition rules.

---

1. The plural of automaton is *automata*.

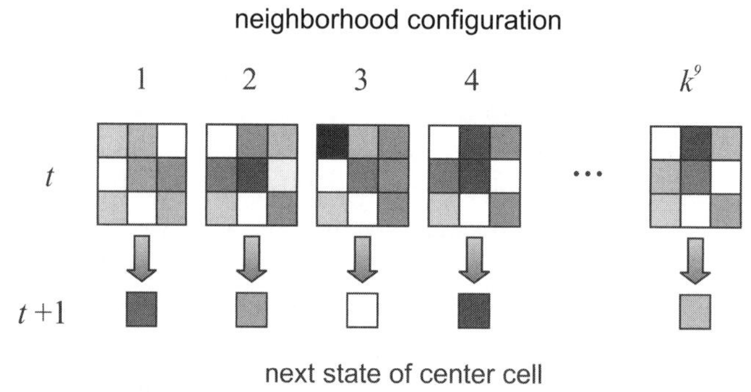

neighborhood configuration

$$1 \qquad 2 \qquad 3 \qquad 4 \qquad\qquad k^9$$

t

$\cdots$

$t+1$

next state of center cell

Figure 2.4 An example of a transition table for a two-dimensional CA with the Moore neighborhood. The cell states are represented as gray levels. The table contains one entry for each configuration of states of the nine cells that form the neighborhood. With k possible states, the table contains k^9 entries.

2.2.1   Special CA Rules

Since the universe of transition rules is in general so vast, it is useful to single out some rules that comply with some additional constraints that make them simpler to specify or ensure to the CA the possession of some special property. Here is a list of the most common special CA rules:

- **Totalistic:** Assuming that the states are represented as numbers, a CA rule is called *totalistic* if it depends only on the sum of the values of the states in the neighborhood. A totalistic transition rule can be written as

$$s_i(t+1) = \varphi(\sum_{j \in N_i} s_j(t)) .$$

With k states and a neighborhood of size n the sum can take only $n(k-1)+1$ different values and thus there are $k^{n(k-1)+1}$ possible totalistic rules. For example, only 16 of the 256 rules of a binary CA with neighborhood size of three, and only 2187 of the more than 10^{12} rules of a ternary CA with neighborhood size of three are totalistic rules.

- **Outer totalistic:** A CA rule is called *outer totalistic* if it depends only on the value of the state of the updated cell (the "center" cell) and the sum of the values of the states of the other cells in the neighborhood (the "outer neighborhood"). An outer totalistic transition rule can be written as

$$s_i(t+1) = \varphi(s_i(t), \sum_{\substack{j \in N_i \\ j \neq i}} s_j(t)) \ .$$

- **Symmetric:** A transition rule is *symmetric* with respect to a permutation of the states of the cells in the neighborhood if it is not affected by the permutation. Since a totalistic rule depends only on the sum of the neighborhood states, it is symmetric with respect to any permutation of the states of the cells of the neighborhood. The same is true of an outer totalistic rule with respect to the permutations of the states of all the cells of the outer neighborhood.

- **Null state quiescent:** A CA rule is called null state quiescent if it maps a quiescent neighborhood to the quiescent state.

2.2.2   Space-Time Diagrams

The most fascinating way to observe the activity of a CA is an animation on a computer screen. When the only medium available is paper the activity of one- and two-dimensional CA with a reasonably small state set can be appreciated using a *space-time diagram* (or *space-time plot*). The left side of figure 2.5 shows an example of a space-time diagram for a one-dimensional CAs. The cellular space at each time step is represented as a horizontal line of squares and the vertical direction is used to show the unfolding in time of the configuration of states of the cellular space. Each state of the state set is represented by a different shade of gray (or color, when available). The right side of figure 2.5 shows the direct generalization of the one-dimensional diagram to the two-dimensional case (the white cells of the cellular space are not shown for clarity). Since this kind of representation hides most of the activity of the two-dimensional CA, the alternative representation of figure 2.6 is often used for a two-dimensional CA. The unfolding in time of the CA is now illustrated by a vertical stack of rectangular regions that represent the state of the cellular space at different time steps.

Figure 2.5 and figure 2.6 both represent a binary CA with the Moore neighborhood implementing the so-called *outer parity rule*. Black cells correspond to the state $s = 1$ and white cells correspond to $s = 0$. The transition rule specifies that the next state of a cell is 1 if the number of 1s in its outer neighborhood is odd, and is 0 otherwise.

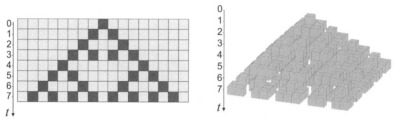

Figure 2.5   Space-time diagrams for (*left*) one- and (*right*) two-dimensional CAs.

2.3   Modeling with Cellular Systems

So far we have described in general terms the elements of cellular system models. In this section we show how to actually build a cellular model of a real-world phenomenon and use the model to investigate the properties of the phenomenon. To define and run a cellular model we proceed according to the following steps:

1. Assign the cellular space.

2. Assign the time variable.

3. Assign the neighborhood.

4. Assign the state set.

5. Assign the transition rule.

6. Assign the boundary conditions.

7. Assign the initial condition.

8. Assign a stopping condition.

9. Proceed to update the state of the cells until the stopping condition is met.

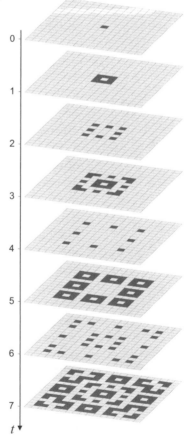

Figure 2.6   Another kind of space-time diagram for a two-dimensional CA.

2.3.1 Example: The Elementary Traffic CA

To illustrate the application of the modeling steps listed above we will now show how to define a simple CA model of traffic. We want to model a finite stretch of a unidirectional, single-lane road (figure 2.7a)). This is done by discretizing the stretch of road into cells of finite length. The resulting cellular space is a one-dimensional finite lattice of cells (figure 2.7b)). In a CA, the time variable is discrete and vehicles are thus modeled as moving at discrete time instants. Therefore, realistic traffic quantities will be measurable only by averaging over several time steps or over several cells. We assume that the state of each cell is influenced only by its adjacent cells, that is, the CA neighborhood is composed of three cells (figure 2.7c)). We assume that each cell contains either a single vehicle or is empty, which means that the state set contains only two states (figure 2.7d)). This assumption is of course linked to the length that we must attribute to each cell in discretizing the road, which must be large enough to contain one vehicle but not so large as to make it implausible that at most one vehicle is contained in it. The transition rule is intended to model vehicles moving from left to right and prescribes that a vehicle can advance and must advance only if the destination cell is free. This rule can be represented with the transition table shown in figure 2.7e). We assign periodic boundary conditions, so that vehicles leaving the cellular space from the right reenter it from the left (figure 2.7f)). To run the simulations we must now assign the initial conditions. We use as the initial condition a random distribution of cars of density ρ that can vary from 0 (empty road) to 1 (each cell is occupied by a vehicle). Note that the transition rule ensures that vehicles are neither created nor destroyed, that is, the number of vehicles is conserved and thus the assigned density is maintained during the whole CA evolution.

Running the CA we observe that after an initial transient the flow of vehicles stabilizes into a configuration that repeats itself periodically. Figure 2.8 shows the space-time diagrams for two different car densities. Black cells correspond to cells occupied by cars and white cells correspond to empty cells. Note that there is a qualitative difference in the aspect of the two diagrams. For the lower density $\rho = 0.3$ once the initial transient is finished, there remain in the space-time diagram only diagonal stripes of white cells, corresponding in the model to empty stretches moving along the road in the same direction of the traffic. For the higher density $\rho = 0.7$ after the transient, there remain only diagonal stripes of black cells, corresponding to traffic jams moving along the road in the direction opposite to that of the traffic.

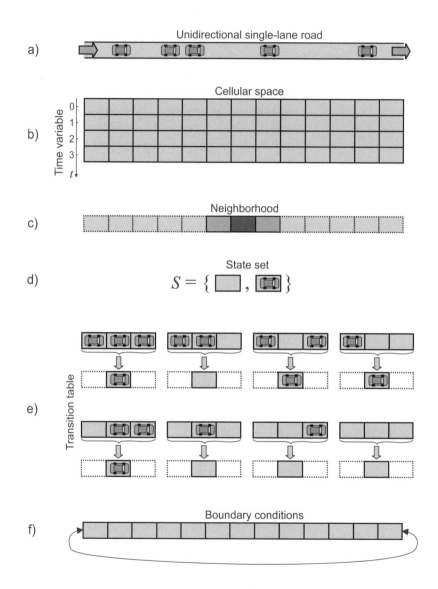

Figure 2.7   The elements of the traffic CA model. *a*) The kind of traffic flow modeled. *b*) The cellular space-time. *c*) The neighborhood. *d*) The state set. *e*) The transition table. *f*) The boundary conditions.

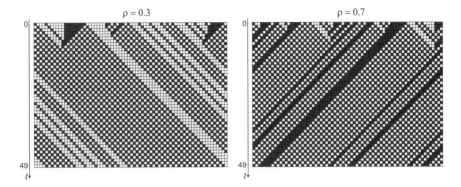

Figure 2.8 Examples of traffic flow in the elementary traffic CA with two different densities of vehicles. The space-time diagram on the left corresponds to a road with 30% occupation by vehicles. The diagram on the right corresponds to a 70% occupation by vehicles.

To analyze and understand the difference between the two plots of figure 2.8 we must repeat the simulation with different values of traffic density. The result can be better appreciated by considering a global property of the traffic flow, for example, the mean speed of all the vehicles measured in number of cells traveled per time step. Figure 2.9 shows the plot of the mean speed of the vehicles as a function of the vehicle density ρ. The plot was obtained by running the CA simulation and averaging over the part of the space-time diagram where the traffic flow has stabilized. The plot reveals that the qualitative change of behavior occurs for $\rho = 0.5$. Below this threshold the traffic is moving freely and above it it is congested. In the language PHASE TRANSITION of physics there is a *phase transition* between the two regimes at the *critical density* $\rho = 0.5$ (Fuks 1997; Maerivoet and De Moor 2005).

2.3.2 Remarks on Cellular Models

We could go on analyzing the properties of the elementary traffic CA but what we want to stress at this point is the general properties of cellular models rather than the particular properties of this example. The first observation is that by defining just the local properties of the model we have revealed interesting global behaviors such as moving traffic "holes" and jams, and the existence of two distinct regimes of traffic. This is one of the fundamental characterizing properties of cellular modeling which makes them an ideal tool to study how simple local rules can produce complex global behaviors.

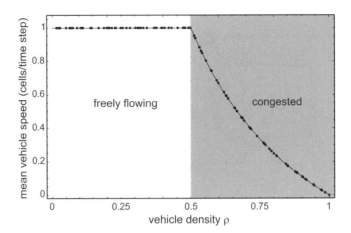

Figure 2.9 Examples of mean speed $\bar{v}$ of the vehicles as a function of the vehicle density ρ in the elementary traffic CA. The points shown in the plot are determined by running the CA for 1000 time steps with a randomly generated initial distribution of vehicles and considering only the last 500 time steps in order to let the initial transient die out. The continuous line corresponds to the relationship $\bar{v} = \frac{1-\rho}{\rho}$ obtained analytically for a cellular space of infinite length (Fuks 1997). The plot shows that at the critical 50% vehicle density there is a transition between two different kinds of traffic flow.

Of course, the fact that the model exhibits a certain global behavior does not guarantee that the behavior is an actual feature of the phenomenon from which the model is drawn. As in all scientific theorizing (see box 2.1) one must actually go back to the real phenomenon and verify that the model predictions comply with the phenomena that characterize the actual system that is modeled. In this respect, since the state set of a CA is finite, CA models have the advantage of being implementable exactly on computer, without worrying about numerical approximations and error propagation. This means that what is observed in the simulation is an actual property of the model implemented and not a numerical artifact. There is a second aspect of cellular systems that can be appreciated from the traffic CA example, that is, how cellular models help to define the model at the right level. By adopting the cellular approach one is free to chose the level of description, from the microscopic to the macroscopic. Detailed microscopic models have the advantage of reproducing the finest details of the phenomena. Moreover, they often permit a very simple and natural modeling of boundary conditions that are difficult to specify at higher levels (figure 2.10). On the other hand, very

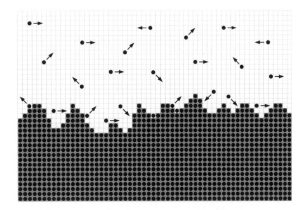

Figure 2.10 The interaction of a fluid with a fixed wall is an example of a boundary condition that is difficult to model at the macroscopic level but straightforward at the microscopic level. In the CA model represented schematically in this figure the black disks with an arrow represent moving particles and the black disks on a dark background represent fixed particles that constitute the wall (more details on cellular models representing particles will be given in section 2.5). The collisions of the moving particles with those belonging to the wall slow down the collective motion of the particles and produce a result that appears as friction once the motion of the particles is averaged over large enough portions of the space-time diagram. The amount of friction is specified directly in terms of the roughness of the wall by choosing the frequency and depth of the pits. No special boundary conditions must be specified in the CA model besides the rules defining the result of the interaction between a moving particle and a fixed particle.

detailed models require the processing of large masses of elementary entities and some space-time averaging to smooth local fluctuations and to get back to quantities that are observed at the macroscopic level. One must consider that most present-day computers are designed for efficiency in the complex manipulations of relatively few entities rather than of simple manipulations of large numbers of entities. Computers specially designed to handle cellular models do exist, but occupy a small niche of the computer market (Toffoli and Margolus 1987; Talia 2000). Moreover, when one builds a very detailed model one often discovers that many details have little or no influence on the global behavior of the system because their effect is averaged out when observation is done at a coarser scale. For example, modeling the cars at the level of their atomic constituents in the traffic CA would hardly help us to improve our understanding of the properties of traffic. By not modeling inessential details one obtains a model that displays the same global

Box 2.1: The scientific status of computational models

The scientific approach to the understanding of a system is based on the specification of a formal model for the system. Typically, the model is defined in mathematical terms so as to permit the systematic deduction of the consequences of the basic assumptions of the model. A model must include rules that specify the correspondence between some elements of the model and some observable properties of the real-world system that is modeled (Russo 2004). With respect to this correspondence, the predictions derived from the model must go beyond the observed phenomena that inspire the model. The last requirement assures that the model is not a mere formalization of what has been observed but can provide new insights about the behavior of the system.

Scientific models and theories have the additional role of allowing the development of a scientific technology, that is, of a corpus of techniques for the rational design and control of real-world systems. In some cases the laws of interaction of the elements that constitute the system are given and cannot be changed (for example, the laws of physics, or the rules of interaction of existing biological agents). However, one can still control the behavior of the system by driving it with external signals or imposing suitable constraints on it, in order to steer the spontaneous dynamics of the system in the desired direction. In this case, system modeling permits the determination of the driving signal and constraints required to achieve a predefined goal. In other cases one is free to define the rules that govern the elements of the system and their interactions (for example, by defining the control system and communication abilities of the robots in a collective robotics experiment). In this case system modeling is required also for the successful design of the rules, since it allows the analysis of the consequences of the rules, the driving signals, and the constraints that have been engineered into the system.

Before the advent of large-scale automatic computation, only analytically tractable mathematical models were considered. Analytically tractable models have many advantages. For example, they permit the prediction of the system behavior when the parameters of the model vary over continuous ranges, revealing the different regimes of operation of the system. On the other hand, these models are only a small part of all the possible mathematical models. For example, many models defined in terms of the cellular systems described in this   *(cont.)*

Box 2.1 (continued)

chapter defy analytical solution. Furthermore, for computationally irreducible models, running a simulation is provably the best way to derive
a prediction.

Now that the availability of computers permits the simulation of
analytically intractable models, we can wonder whether the results
give computational models full membership in the category of scientific models. The main problem in this respect is that the kind of insight that can be derived from a simulation is in general quite limited
in comparison with what can be obtained from an analytically tractable
model. For example, to explore the behavior of a system in a range of
parameters using a computational model one can at most run a finite
set of simulations for different values of the parameters (see figure 2.9),
and there is in general no guarantee that the explored values exhaust
the range of behaviors proper to the model. Despite these limitations,
some authors maintain that computational models deserve full membership in the scientific enterprise because, for example, they permit
proving that the details of a given model are sufficient to generate the
observed large-scale behaviors (Epstein 2006). However, other authors
challenge this opinion (Durlauf 1997; Diermeier 2007) and the scientific
community is in general divided on the value that must be attributed
to computational models in the context of scientific theories. An interesting viewpoint on the role of computational models was provided
by John von Neumann and Stanislaw Ulam at the beginning of the
computer era (Ulam 1976; Rédei 2005; Farge 2007). They observed that
in domains where mathematical models are analytically intractable in
most real-world scenarios – for example, in fluid dynamics with even
moderately complicated boundary conditions – computer simulations
can be seen as an additional experimental technique which can provide
valuable insight into the behavior of a system and thus facilitate the
formulation of a more conventional explanatory theory.

Irrespective of their scientific status, computational models of otherwise intractable complex systems are useful in an engineering context because they provide the instruments for the design and control
of these systems. However, as explained in more detail when describing the problem of design verification in chapter 1, it is important to
verify that the behavior of a system whose design is based on computational models is robust with respect to the technologically unavoidable
parameter tolerances and to the expected external perturbations.

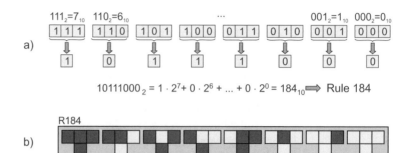

$$10111000_2 = 1 \cdot 2^7 + 0 \cdot 2^6 + \ldots + 0 \cdot 2^0 = 184_{10} \Longrightarrow \text{Rule } 184$$

Figure 2.11 *a*) The mechanism that associates Wolfram's rule code to the elementary CA. *b*) The conventional graphical representation of the transition table of elementary CAs: cells in the $s = 1$ state are represented as dark squares and cells in the $s = 0$ state are represented as white squares.

behavior and is simpler to implement and simulate and whose results are simpler to analyze. Focusing on the right model is more an art than a science, and sometimes requires experimenting with models of different level of detail. Finally, note that cellular systems are not used only to model existing phenomena. As will become clear later, cellular systems are also a tool for system design and for the definition and analysis of artificial "universes" with synthetic laws that permit the emergence of complex phenomena.

2.4   Some Classic Cellular Automata

Among all cellular systems there are two instances of CAs that enjoy a special popularity: the class of the one-dimensional *elementary CA* and the two-dimensional *game of life CA*. In this section we give a short description of the main characteristics of these CAs. In later sections we will use them to illustrate several aspects of cellular systems and of their applications.

2.4.1   Elementary CAs

As explained in section 2.2 there are 256 binary one-dimensional CAs with a neighborhood of radius $r = 1$. They are called *elementary CAs*. In a seminal study Wolfram (1983) analyzed these CAs and proposed a numbering WOLFRAM'S RULE system known as *Wolfram's rule code* which associates an integer from 0 to CODE 255 with each elementary CA, as illustrated in figure 2.11a). The association

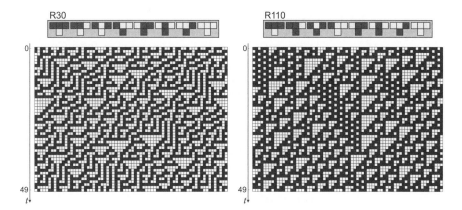

Figure 2.12 Examples of space-time diagrams for the elementary CA with rule codes 30 (*left*) and 110 (*right*). Both diagrams were obtained with a randomly generated initial state with about 50% of cells in the state $s = 1$, and periodic boundary conditions. Cells in the $s = 1$ state are represented as black squares and cells in the $s = 0$ state are represented as white squares.

is obtained by ordering the entries of the transition table according to the state of the neighborhood interpreted as a binary integer. The ordered binary digits giving the next state of the center cell are then interpreted as a binary integer, which is Wolfram's code for the CA. The transition table of an elementary CA is often represented with the stylized diagram shown in figure 2.11b).

There are two fundamental reasons to be interested in elementary CAs. First, some of them constitute the simplest cellular model of many phenomena. For example, the traffic CA described in the previous section is actually the elementary CA with rule code 184, as can be verified comparing the transition table of figure 2.11 with that of figure 2.7. A second reason of interest is related to the fact that there exist only 256 of them and that it is therefore possible to study in detail all of them, trying to derive some general conclusions about CAs and their classification. For example, looking at figure 2.8 and figure 2.12 which show samples of space-time diagrams of the elementary CAs with rule codes 184, 30, and 110, we see that quite different spatiotemporal patterns can be generated by elementary CAs. This observation will be developed further in section 2.9 when discussing the issue of the analysis and synthesis of cellular systems.

2.4.2   Conway's Life Game

In the early 1970s the mathematician John Conway defined a two-dimensional binary CA that, following its popularization in two of Martin Gardner's columns in *Scientific American* (Gardner 1970, 1971), became rapidly one of the most well-known and most studied CAs. Conway's goal was to define a very simple CA that could produce the most interesting and surprising behavior. To this end he explored extensively the space of two-dimensional CAs in search of the right rule. Eventually, after almost two years of experiments, Conway focused on a binary CA with the Moore neighborhood (figure 2.2) obeying the following outer totalistic transition rule (Berlekamp et al. 2004):

- A cell that is in the state $s = 0$ at time t switches to the state $s = 1$ at time $t + 1$ only if exactly three of its eight outer neighbors are in the state $s = 1$ at time t.

- A cell that is in the state $s = 1$ at time t remains in this state at time $t + 1$ only if two or three of its eight outer neighbors are in the state $s = 1$ at time t.

Conway gave a snappier description of the rule by calling the cell in the state $s = 0$ *dead* cells, and calling *live* cells those in the state $s = 1$. The CA rule can then be rephrased as

- **Birth rule**. A dead cell becomes a live cell only if exactly three of its eight outer neighbors are live cells.

- **Survival rule**. A live cell remains a live cell only if two or three of its eight outer neighbors are live cells. A live cell dies by *isolation* if it has fewer than two live neighbors, and dies by *overcrowding* if it has more than three live neighbors.

The CA following these birth and survival rules was renamed *The Game of Life*, or simply the *Life CA*.

Running several times the Life CA starting from a random initial configuration reveals the existence of different kinds of "objects." Some correspond to *stable* configurations that remain unchanged from one time step to the next. Conway and his coworkers called these objects *still-life* configurations. Figure 2.13 shows three common stable configurations, called by Conway the *block*, the *pond*, and the *beehive*. The second kind of common Life objects are

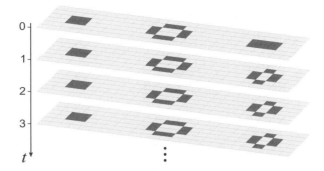

Figure 2.13   Three static objects in Conway's Life Game: (*left*) the *block* , (*center*) the *pond*, and (*right*) the *beehive*. Live cells are represented in dark gray. The figure shows also how a configuration at time step t can be generated by different configurations at time step $t - 1$. In this case, a beehive can be generated by another beehive or by two adjacent rows of live cells.

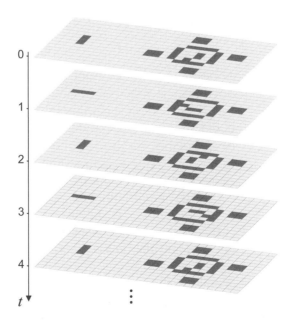

Figure 2.14   Two oscillators in Conway's Life Game: (*left*) the *blinker*, which has period two, and (*right*) the *clock II* with period four.

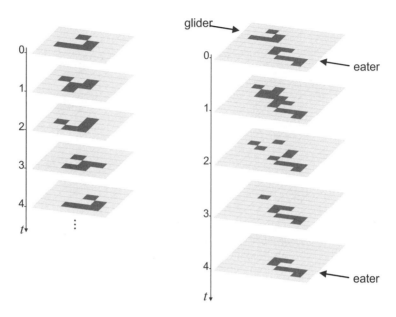

Figure 2.15 *Left*: The glider is the most common moving object in Conway's Life Game. *Right*: An eater can annihilate a glider and repair itself in four time steps.

oscillators, or *life cycles*. These are configurations that repeat themselves with a period greater than one time step. Figure 2.14 shows the *blinker*, which has period two and is the most common kind of oscillator and can be easily produced starting from a random initial condition, and the *clock II* which is a hand-designed oscillator with period four (Poundstone 1985). A third kind of common configurations in Life are *moving objects*. Figure 2.15 shows the

GLIDER   simplest and most interesting of them: the *glider*. The figure shows that gliders move diagonally by one cell each four time steps. The direction of the motion depends on the initial orientation of the glider.

The existence of moving objects like the gliders suggested to Conway that Life could be interpreted as a synthetic universe where it is possible to send signals between places. This prompted him to investigate the possibility of building more complex configurations capable of processing information. This investigation was encouraged by the discovery of the *glider gun*, a configuration designed by R.W. Gosper that is able to produce a new glider every 30 time steps (figure 2.16). A fundamental role in Conway's endeavor

EATER   was also played by the discovery of the *eater*, a static structure that is able

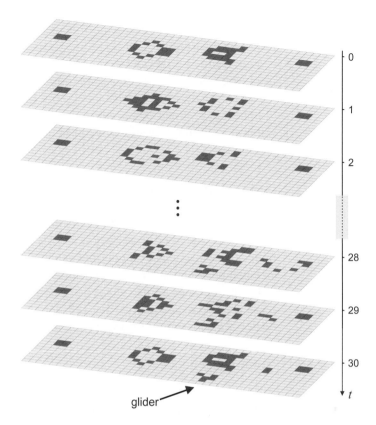

Figure 2.16   A glider gun generates a glider every 30 steps. Note that besides producing the glider the gun regenerates its initial configuration (the two spurious live cells on the right will die of isolation at time step 31).

to annihilate a suitably directed glider in four time steps while repairing the damages inflicted by the encounter with the glider (figure 2.15, right).

Later we will explain how these structures were used by Conway to define within Life structures such as computers and self-reproducing automata. For the time being let us just remark how the glider gun and the two oscillators shown in figure 2.14 illustrate two different approaches to Conway's Life Game. The first approach looks at what kind of configurations are commonly produced by the automaton's rule starting from a random state; the second approach tries to design initial states that produce a desired sequence of configurations. Another interesting observation is that Life's rule cannot REVERSIBILITY   be run backward; it is *not reversible*. This can be inferred from figure 2.13, which shows that the beehive can be obtained after one time step starting from two adjacent rows of three live cells. This proves that in the Life CA a given configuration can have more than one precursor and thus cannot be used in general to trace back the automaton history. Finally, note that being based on an outer totalistic rule, the Life CA preserves the symmetry of the configurations. Thus, any asymmetry existing in the configuration of states at a certain time step is a consequence of an asymmetry in the initial conditions.

2.5   Other Cellular Systems

CAs are just the simplest of the many possible cellular systems based on the ingredients described in section 2.1. A CA has a discrete time variable, a finite state set, a neighborhood that is homogeneous in both space and time, a transition function that is deterministic and homogeneous in space and time, and updates all its cells synchronously. By changing one or more of these characteristics many other cellular systems can be obtained. In this section we will present the most common of these other cellular systems.

2.5.1   Nonhomogeneous CA

A CA whose neighborhood or transition rule is not the same for all cells and all time steps is called a *nonhomogeneous* (or *nonuniform*) *CA*. Nonhomogeneity in space can be useful, for example, to model regions with different material properties. Nonhomogeneity in time can be used to define a sequence of different transformations of the initial CA state. For example, in image processing, the pixel values of an image can be assigned as the initial state of the CA cells, and a collection of filters can be defined in terms of transition rules. The sequential application of the different filters can be obtained by specifying that the different transition rules are valid for different intervals of time (Worsch 1999). Note that a CA with boundary cells can be interpreted as a nonhomogeneous CA because a different transition rule applies to the boundary cells.

Given a nonhomogeneous CA, an equivalent homogeneous CA can be easily obtained by extending the state set and the transition rule of the nonhomogeneous CA with an additional variable that determines what transition rule and neighborhood must be used for a given cell at a given time step. This example shows that some cellular systems that are not defined as standard CAs can be interpreted as such by complexifying the state set or the

Figure 2.17   An example of a transition rule for a binary one-dimensional *mobile CA*. In addition to specifying how the state of the center cell changes from one time step to the next, the rule specifies also the motion of the token (represented here by a disk) that identifies the active cell.

transition rule. However, even when this can be done, it is usually not very useful to adopt this point of view because it complicates the system and it mixes elements that belong to the cellular model of interest with spurious elements that are required just to implement it as a CA.

2.5.2   Asynchronous CA

In a standard CA all the cells of the cellular space are updated synchronously. This implies the presence of a global synchronization signal that is distributed to all the cells and weakens the intended local nature of the CA concept. When collective phenomena are observed in a CA it is therefore important to consider the role played by the synchronous update in the existence and nature of these phenomena. To this end one can consider the *asynchronous CA*, that is, a CA where the updating of the cells is done asynchronously. By experimenting with an asynchronous CA one can test the robustness of the observed phenomena with respect to the update policy (Fatès and Morvan 2004; Chopard and Droz 1998; Schonfisch and de Roos 1999).

The abandonment of synchronous update creates the problem of the choice of the updating scheme (Schonfisch and de Roos 1999). A simple solution is to number the cells and proceed cyclically to the ordered update determined by the cell number. For example, in a two-dimensional CA one can number the cells by rows and columns and update according to the row and column number (line-by-line sweep). The problem of this fixed ordered update scheme is that it tends to produce spurious effects due to the fixed correlation between the update times of the cells. An update policy that solves this problem consists in assigning to each cell a probability of update and, at each time step, in applying the transition rule only with the given probability, leaving the state unchanged otherwise. An alternative is to update only one cell of the automaton – the *active cell* – at each time step. Of course one must also define a rule that determines which cell is active at each time step. This can be done by defining a token which characterizes the active cell, adding to the transition rule the specification of the motion of the token (figure 2.17). The resulting asynchronous CAs are called *mobile CAs* (Wolfram 2002).

2.5.3   Probabilistic CA

In the definition of CAs given in section 2.2 the transition rule is deterministic, that is, the state of a cell at the next time step is uniquely determined by the current state of the cells in its neighborhood. In the *probabilistic* CA (also called a *stochastic CA*) a given state of the neighborhood cells at time t can lead to different states of the center cell at time $t + 1$, according to a predefined probability for each possible successor state.

A typical example of a probabilistic CA is the *forest fire CA*. This is a two-dimensional CA with the Moore or von Neumann neighborhood that is intended to provide a simplified model of the spread of fires and the subsequent regrowth of trees in a forest. In the model, a cell can be in one of three states: it can contain a green tree; it can contain a burning tree; or it can be empty. The probabilistic transition table stipulates that

1. a cell containing a burning tree becomes an empty cell;

2. a cell containing a green tree that has at least one of its neighboring cells containing a burning tree becomes a cell with a burning tree;

3. a cell containing a green tree that has no neighboring cells containing a burning tree becomes a cell containing a burning tree with probability f, and remains in its current state with probability $(1 - f)$. This provision can be assumed to simulate the ignition of trees by lightning and, correspondingly, f is called the *probability of lightning*;

4. an empty cell becomes a cell containing a green tree with *probability of growth* g, and remains empty with probability $(1 - g)$.

Figure 2.18 shows an example of a space-time diagram of the forest fire CA with values of $g = 0.05$ and $f = 0.0003$. The figure shows that with this choice of parameters regions of the "forest" where the density of green trees is sufficiently high a fire front starts to propagate, clearing large zones of forest. Since the probability of growth is much higher than the probability of lightning, trees have time to regrow in the regions cleared by the fire.

Changing the values of the parameters f and g one can simulate the phenomenon in a continuum of conditions. In other words, the presence of the

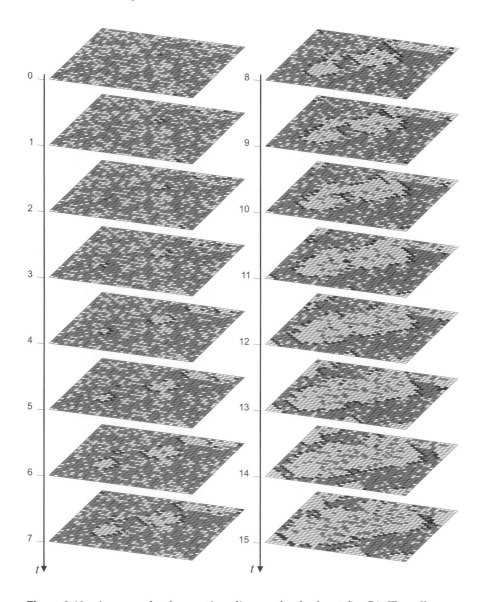

Figure 2.18   An example of space-time diagram for the forest fire CA. The cell space was initialized with a 50% density of cells containing a green tree (represented as gray cells), a few cells containing a burning tree (represented as black cells), and the remaining cells empty (white cells). The diagram was obtained with periodic boundary conditions, the von Neumann neighborhood, a probability of growth $g = 0.05$, and a probability of lightning $f = 0.0003$. The propagation of fire fronts is clearly visible in the final steps of this diagram.

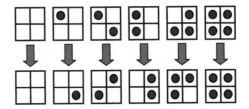

Figure 2.19 The transition table of a block rule modeling particle motion using the Margolus neighborhood. The block is constituted by a square of four adjacent cells. The transition table shown here does not contain all possible entries for the states of the block. The missing entries can be obtained by rotating the blocks of the existing entries.

probabilistic parameters injects a certain degree of continuity in the otherwise completely discrete universe of the CA. This is very useful, because it permits the fine-tuning of the properties of the simulation, which are otherwise rather constrained. This means that probabilistic CAs substantially enlarge the range of phenomena that can be modeled using CAs with simple transition rules. With probabilistic CAs one can study the different collective regimes of operation of the model for different ranges of the probabilistic parameters. For example, changing the ratio of the parameters f and g one can observe collective phenomena and space-time structures quite different from those illustrated in figure 2.18 (Drossel and Schwabl 1992, 1994; Maddox 1992; Gaylord and Nishidate 1996).

2.5.4   Particle CA

One of the most widespread applications of CAs, especially in physics, is the modeling of systems of moving particles. In its most simple form this can be obtained by interpreting the state of the cells as defining the presence or absence of a particle in the cell (more complex models admit the presence of more than one particle in a cell). The transition rule specifies the conditions under which the particles move between cells. Such a CA is known as a *particle CA* or *lattice-gas automaton*. If you consider the elementary traffic CA defined in section 2.2 you can check that it corresponds to the definition of a particle CA, with the cars playing the role of particles. Thus, strictly speaking, a particle CA is just an ordinary CA. However, there are good reasons to consider particle CAs as a separate class of CAs. The first, rather philosophical reason is that in particle CAs the state is more fruitfully thought of

LATTICE-GAS
AUTOMATON

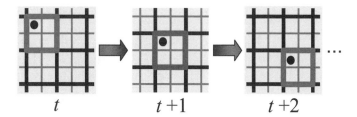

Figure 2.20   The motion of a particle determined by the transition rule of figure 2.19 induced by the alternation of the position of the blocks between time steps. The 2×2 blocks that partition the cellular space (represented here by the thick lines) are shifted by one cell in each direction at each time step. This allows the motion in time of the particles beyond the limits determined by the blocks.

as something that is exchanged between cells rather than as something that resides in the cells as in ordinary CAs. The second, more pragmatic reason is that there is the specific problem of the conservation of particles to be addressed, and this requires the adoption of specially defined transition rules.

In the one-dimensional case exemplified by the elementary traffic CA the realization of particle conservation was a simple matter because we had to consider the state of a neighborhood of only three cells and the particles could move only in one direction. However, already in a two-dimensional particle CA the number of possible combinations of particle positions and motions can become difficult to manage. A first solution is the substitution of the traditional transition rules with the so-called *block rules*. The idea is to partition the cellular space in small, disjoint, uniform blocks and consider each block as a separate entity (for this reason particle CAs defined in this way are also called *partitioning CAs*). The block rule is defined for the whole block rather than for the single cell as in ordinary transition rules. Since the block is smaller than the typical neighborhood it is easier to define a block rule. Moreover, since each block is considered as isolated from the other blocks, the rule can be defined without worrying about particles belonging to other blocks. Figure 2.19 shows an example of the block rule defined for the so-called *Margolus neighborhood*.

The problem with a block rule as defined so far is that by stipulating the isolation of distinct blocks it does not allow the exchange of particles between them. This is clearly not acceptable because particles must in general be free to move in the whole cellular space. To obviate this problem the definition of the blocks is changed from one time step to the next, letting a given cell

BLOCK RULES

PARTITIONING CA

MARGOLUS
NEIGHBORHOOD

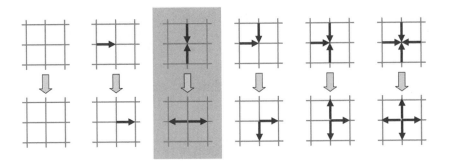

Figure 2.21 The transition table of a particle CA can be defined in terms of particles moving between cells. This transition table corresponds to the HPP gas and does not show all possible configurations of particles. The missing entries can be obtained by rotating the existing entries. Note that the only entry of the transition table that corresponds to an interaction between particles is the one drawn on a shaded background. This fact can be more easily appreciated in this representation than in the representation of figure 2.19.

exchange its particles with cells from which it was previously isolated. Figure 2.20 shows an example of the alternation of the blocks of the Margolus neighborhood in successive time steps.

An alternative approach to the definition of a particle CA is to adopt the philosophy of objects moving between cells and use as state information the position and direction of the particles. Figure 2.21 shows an example of the

HPP GAS

transition rule for the *HPP gas* (Hardy, de Pazzis, and Pomeau 1976) defined using this approach. Note that the transition rule of figure 2.19 is an alternative way to define the HPP gas.

2.5.5   Coupled Map Lattices

In a standard CA the state set is finite. If we let the state be a continuous variable we obtain a *coupled map lattice* (CML). CMLs were first studied by Kaneko in the 1980s (Kaneko 1992). Besides their interest for the study of complex spatiotemporal dynamics and chaotic phenomena, CMLs are very useful for the modeling of systems in terms of variables that represent directly macroscopic quantities at the cell sites. In this way it is not necessary to average over large regions of cellular space to obtain the macroscopic behavior of the system. Another advantage is that one can define at each cell additional continuous fixed variables to be used as parameters of the transition function. This permits a greater flexibility in modeling with respect to CAs. An example of CMLs applied to the study of particle systems is the

LATTICE BOLTZMANN MODELS

lattice Boltzmann models (Chopard and Droz 1998), where the state variables represent the concentration or the probability of presence of the particles in the cell. Lattice Boltzmann models find application in particular in fluid dynamics simulations. Note that most discretized versions of physical problems involving partial differential equations (PDEs) (see box 2.2) can also be

PARTIAL DIFFERENTIAL EQUATION (PDE)

considered CMLs. Of course, when a CML is implemented on a computer the state variables are finite rather than continuous. Thus, strictly speaking, these implementations are always CAs rather than CMLs. Still, it is more useful to consider them as approximate implementations of CMLs rather than CAs because the state space is typically too large to make the CA point of view useful, whereas the results of the CML perspective still hold, albeit in an approximate way.

2.5.6 Cellular Neural Networks

Cellular neural networks (CNNs) were introduced by Chua and Yang (1988b). They are cellular systems where both the state and the time variable are continuous. The transition function is defined in terms of ordinary differential equations and the whole cellular system corresponds to a system of coupled differential equations. In CNNs each cell resembles one of the dynamical artificial neurons that will be described later in chapter 3. The only difference between a CNN and the more general neural networks described in that chapter is that in CNNs the interactions between neurons are limited to neighboring neurons, whereas they can be arbitrary in a generic neural network. The most important application of CNNs is as *analog processors* which can be implemented in hardware to perform in real-time tasks such as image processing and pattern recognition (Chua and Yang 1988a; Fortuna et al. 2001; Arena et al. 1997). CNNs are also a useful model for the study of nonlinear dynamical systems and complex spatiotemporal phenomena.

2.5.7 Cellular Systems with Multiple Cellular Spaces

Sometimes it is useful to consider systems obtained combining several inter-

MULTILAYERED CA

acting cellular systems. A first example is the *multilayered CA* (Bandini and Mauri 1999). Multilayered CAs are composed of a hierarchy of CA layers (figure 2.22) where each cell of a CA of layer $i > 0$ corresponds to a whole CA at the next lower layer in the hierarchy. The transition function for the

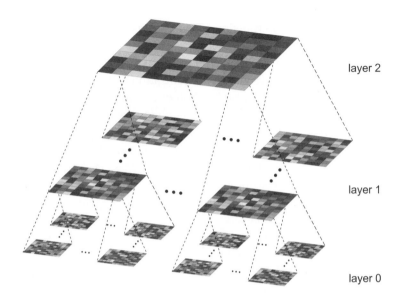

Figure 2.22 A schematic representation of a multilayered cellular system. A cell at level $i > 0$ in the hierarchy of layers corresponds to a whole CA at the next lower level.

state of cells at level i depends on a *horizontal neighborhood* composed of cells at layer i and by a *vertical neighborhood* composed of cells of the automata at the adjacent layers. In this way one can define hierarchical cellular models with rules defined at different levels of abstraction. Of course, the idea of multilayered CAs can be extended to the other kinds of cellular systems discussed above.

Another example of cellular systems with coupled cellular spaces arises in the numerical modeling of physical fields (Mattiussi 2002, 2000). To model this kind of system it is often useful to employ two cellular spaces and associate different kinds of physical quantities as states of the cells of the two cellular spaces (for example, the electric and the magnetic field in an electromagnetic numerical simulation). The dynamics of the system is assigned in terms of two coupled map lattices that are updated alternately. The transition function of each coupled map lattice takes into account a neighborhood composed of cells on the same cellular space at the current time step and cells on the other cellular space considered at the previous time step.

Box 2.2: Numerical methods and cellular systems

To assess the quality of a model and use it to design technological artifacts scientists and engineers need to check that a model corresponds to the modeled phenomenon and to evaluate the effect of given initial and boundary conditions. To this end, they use the model to predict the system behavior. For systems defined in both space and time, and characterized by local interactions the traditional approach to modeling and prediction is based on partial differential equations (PDEs). To write the set of PDEs (and the corresponding initial and boundary conditions) for a phenomenon one starts by focusing on a level of description where the quantities of interest can be considered continuous almost everywhere in space and time. The equations are then obtained by writing in differential form the laws that govern the local behavior of these quantities according to the given model (Potter 1973).

The mathematical study of PDEs has a long history. For PDEs describing linear phenomena it has produced powerful methods that give predictions in closed form using symbolic manipulation. For PDEs describing nonlinear phenomena, however, analytical solutions are available only in a handful of cases and provided the initial and boundary conditions are sufficiently well-behaved. For most nonlinear PDEs no analytical solutions are known and predictions can be obtained only by integrating the equations numerically. The numerical solution is often obtained by first discretizing the PDEs so as to obtain a cellular system (typically, a coupled map lattice), and then running the cellular model (Potter 1973; Mattiussi 2000).

The use of PDEs as the starting point for the discretization is a heritage from a time when large-scale numerical computation was not feasible and PDEs and their analytical solutions were necessarily the tool of choice for modeling and prediction of space-time phenomena. Now that powerful computers are available it is reasonable to consider the possibility of defining directly cellular models. For example, for a large class of physical problems that are traditionally modeled with PDEs it has been shown that the definition of a cellular model is indeed the most natural way to proceed in view of their simulation (Shashkov and Steinberg 1995; Mattiussi 1997, 2000, 2002; Teixeira and Chew 1999; Tonti 2001; Hiptmair 2001; Steinberg 2004; Bochev and Hyman 2006; DiCarlo et al. 2007). Going directly to a cellular model becomes a necessity for phenomena that are difficult to model with PDEs, such as those where it is not clear how to define meaningful quantities that are continuous in space and time.

2.6   Computation

CELLULAR COMPUTERS

One of the uses of cellular systems is as input-output devices that perform computation. The input of these *cellular computers* is the initial state of the cellular space and the output is the state of the cellular space (or that of a portion of the space-time diagram) once a predefined stopping condition is met. A cellular computer can be implemented in hardware as an array of analog or digital processors and has a number of advantages over a conventional computer, both in the realization of the hardware and in performance. Since the interaction between the cells is local there is no need of long-range wiring of the processors (except possibly for a global synchronization signal) or complex communication and coordination strategies between different parts of the computer. Moreover, since the transition rules are typically simple, each processor can be simple and needs a small amount of memory. Finally, since the processors work in parallel, large amounts of data can be processed very rapidly.

To program a cellular computer one must define the ingredients of the cellular system so as to have it perform the required computation. If the required processing is intrinsically local this is easy. For example, an image filter that performs a local average of pixel values can be easily implemented as a CA whose state set is the possible pixel values and whose transition rule averages the values of the cells in a suitable neighborhood. However, if the computation concerns some global property of the input data, devising a suitable cellular system can be tricky. The main problem is that there is no general systematic approach to designing a local transition rule that produces a given global behavior. Each problem must be considered separately rather than approached with a general programming methodology as in conventional computing.

MAZE CA

An example a of cellular computation strategy defined for a global problem is the CA that solves mazes. Given a maze like the one shown in figure 2.23, the problem consists in finding a path from the entrance to the exit. The conventional approach to the solution of this problem is to tread the maze following a set of rules that ensure that blind alleys are explored only once, until the exit is discovered. The cellular solution (Nayfeh 1993) starts by mapping the maze to the initial state of a binary CA with one state representing the presence of a wall and the other representing free space. Figure 2.24 at time $t = 0$ shows the result of this operation for the maze of figure 2.23. The transition function is based on the von Neumann neighborhood and specifies that a free cell surrounded by three wall cells is trans-

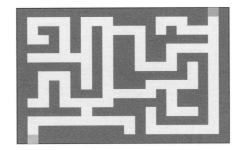

Figure 2.23   The maze used to illustrate the cellular approach to the solution of mazes. Free space is shown in white, walls are shown in dark gray, and the entrance and exit are shown in light gray.

formed into a wall cell. To fill isolated holes, free cells surrounded by four cells are also transformed into wall cells. In all other conditions the state of a cell remains unchanged. Since we know that the boundary cells are either entry or exit cells, which must remain free, or wall cells, which will not change state, there is no need to apply the rule to the boundary cells and thus we need not specify boundary conditions. The CA must be run until the state of the cellular space does not change, at which point the state of the cellular space corresponds to the maze solution (if one exists). Figure 2.24 shows this process for the maze represented in figure 2.23. Note that, similarly to the conventional approach, the transition function is defined so as to have the CA trace back blind alleys starting from their end. The difference with the conventional maze-solving approach is that the process is carried out in parallel for all the blind alleys until none are left. For a finite cellular space the process is guaranteed to converge in a finite number of time steps, since at each time step (before the stopping condition is met) at least one free cell is converted to a wall cell, and no new free cells are generated.

There is an obvious interest in the definition of cellular computers for other computation problems. For example, it would be useful to be able to implement parallel arithmetic computation on CAs. This is a particular example of the more general problem of how information that is distributed in space can be processed using local rules. This requires the ability to propagate the information and synchronize its processing. To better understand this problem, researchers have defined a set of benchmark problems that abstract its

FIRING SQUAD
SYNCHRONIZATION
PROBLEM

basic challenges (Mitchell 1998), such as the *firing squad synchronization problem* (FSSP), and the *density classification task* (DCT). In the FSSP the cells of

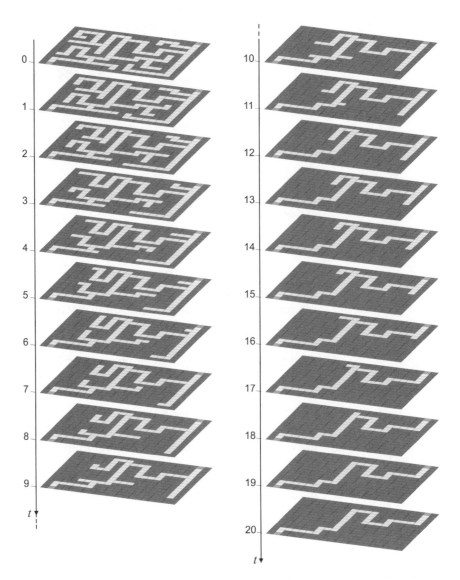

Figure 2.24   The cellular maze-solving algorithm starts with the maze encoded as the initial state of a CA. Cells in the "wall" state are represented in dark gray, cells in the "free" state are represented in white, and the entrance and exit are represented in light gray. The transition function transforms into a wall cell all free cells at the end of a blind alley. Thus, at each time step blind alleys are traced back, until only cells along the path from entrance to exit remain in the "free" state. From that point onward the state of the system no longer changes, as shown in the transition from $t = 19$ to $t = 20$.

a one-dimensional CA are initialized in a quiescent state. At a certain time step t_g one cell (the "general") is externally assigned a special state that corresponds to a "command to fire." The problem consists in defining a state set and transition function that result in the broadcasting of the command to all the other cells (the "soldiers") so that at the time step $t_f > t_g$ they all go to another special "firing" state that they must not have previously assumed. In the DCT a binary CA is assigned an initial configuration with a density ρ_0 of 1s. A "threshold density" value ρ_t is also assigned. The problem consists in defining a transition function that in t_d time steps takes and keeps the state of all cells to 0 if $\rho_0 < \rho_t$, and to 1 otherwise. Both problems are easy to solve for a system with global communication but challenging for a system with only local communication. An interesting result of the efforts devoted to the solution of these problems is the realization that it is useful to think in terms of signals and try to define a state set and transition function that allow the propagation of these signals across the cellular space without mutual interference.

DENSITY CLASSIFICATION TASK

The approach based on signals has been also used to investigate the possibility of implementing a general-purpose computer within a given cellular system, that is, a computer that allows the execution of an arbitrary finite algorithm. When this is possible the cellular system is said to be capable of *universal computation*. For example, it has been shown that this is the case for Conway's Life CA (Berlekamp et al. 2004) and the rule 110 elementary CA (Wolfram 2002). In the case of Conway's Life CA the computation is based on the use of streams of gliders as binary signals. Conway has shown (Berlekamp et al. 2004) that a suitable positioning of glider guns and eaters permits the implementation of all the basic building blocks required to assemble a digital computer, such as logic gates (figure 2.25), delay lines, and storage devices. This proves that a general-purpose computer can be actually built with Life's cellular space.

UNIVERSAL COMPUTATION

The interest of this kind of proof is not the actual implementation of a computer, which would cover an enormous region of the cellular space, would be inefficient, and would be almost impossible to program. The point is to show that even simple cellular systems can be *computationally irreducible*, that is, that given an initial state of the system there is in general no simpler way to predict the unfolding of the state than to actually run the system. This follows from some results of computational theory referring to the general impossibility of predicting the result of a program run (Hopcroft et al. 2006). This means that cellular systems can be used to produce highly nontrivial behaviors, which cannot be predicted by just looking at the state set and

COMPUTATIONAL IRREDUCIBILITY

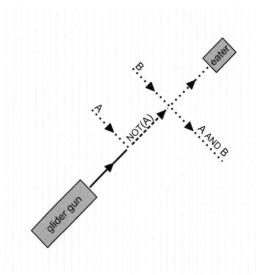

Figure 2.25   A schematic representation of the implementation of a logic gate in Conway's Life CA. The binary signals A and B correspond to streams of gliders (the presence of a glider in the stream corresponds to a logic 1, and its absence to a logic 0). The gliders produced by the glider gun are annihilated when they crash into a glider belonging to A, so that a stream corresponding to NOT(A) emerges from the interaction. The same kind of interaction between the streams B and NOT(A) produce a stream corresponding to A AND B and some residual gliders that are annihilated by the eater.

transition function. In particular, cellular systems can be used as synthetic universes in which to observe and investigate the emergence of complexity and the evolution of artificial life forms.

2.7   Artificial Life

The origins of the interest in cellular system models can be traced back to the investigations of the mathematician John von Neumann concerning the growth of complexity. In the late 1940s von Neumann was interested in proving formally that there exist machines which can produce machines more complex than themselves (von Neumann 1966; McMullin 2000). In our everyday experience a machine typically builds only machines simpler than itself. For example, the robots of a car assembly line are more complex than the parts of the car that they shape and assemble. In other words, complexity appears at first sight to be a degenerative property of machines. The evidence of the evolution of complex biological organisms from simpler forms of life suggests, however, that it must be possible for a system to give birth to a system more complex than itself.

Von Neumann realized that the growth of complexity in machines could be based on the concept of self-reproduction. Imagine a machine that is able to build a copy of itself using simpler parts that are available in its environment. This already corresponds to a machine that can build machines as complex as itself. Thus, self-reproduction dispels the persuasion of the degenerative nature of complexity. If we add the requirement that the process of building the copy be subject to random mutations, then we can expect that some of the mutated copies are actually more complex than the original machine, thus realizing von Neumann's goal (figure 2.26). All this discussion rests of course COMPLEXITY on the definition of complexity for a machine. Von Neumann was well aware of the importance of this concept but did not attempt a formal definition of it. Rather, he defined it in intuitive terms as "effectivity in complication, or the potential to do things." Von Neumann explained that he was "not thinking about how involved the object is, but how involved its purposive operations are. In this sense, an object is of the highest degree of complexity if it can do very difficult and involved things" (von Neumann 1966, p. 78). In the context of this section we will adopt this heuristic definition of complexity which we could tentatively dub *purposive complexity*.

Originally, von Neumann had in mind a self-reproducing machine built of actual, physical parts such as girders, motors, sensors, and computer elements. The parts would be floating on the surface of a pond and the self-reproducing machine would have organs to recognize, grasp, and assemble the parts required to build a copy of itself. Von Neumann soon realized that KINEMATIC MODEL the actual manipulation of physical parts in this *kinematic model* increased the difficulty of the task without benefiting its conceptual side. Following a sug-CELLULAR MODEL gestion of Stan Ulam, von Neumann thus switched to a *cellular model*, where the environment of the self-reproducing machine is a two-dimensional CA. The mathematical study of cellular systems was born.

SELF-REPRODUCING AUTOMATON The first problem that von Neumann faced in defining his *self-reproducing automaton* was the choice of the state set and transition function of the CA, that is, of the elementary "objects" of his synthetic universe and their rules of behavior. On the one hand these objects ought not be so complex as to be already endowed with the properties that von Neumann was about to investigate. On the other hand, these objects could not be so simple as to require

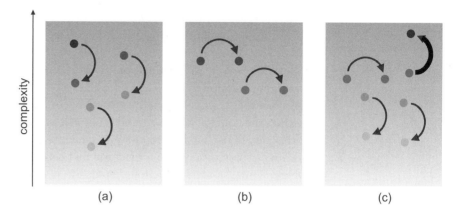

Figure 2.26   An abstract representation of the role of self-reproduction in the growth of complexity of machines. Machines are represented here as dots in a space ordered by complexity. The arrows represent the production of a machine by another machine. (*a*) The common experience is that the production of machines from the part of other machines is degenerative from the point of view of complexity. (*b*) Self-reproducing machines give birth to machines with the same level of complexity as themselves. (*c*) Self-reproduction with random mutations can give birth to machines of a different level of complexity with respect to the original machine. In particular, it can produce machines more complex than the original one (thick arrow, *top right*).

an excessive number of them to build anything interesting from the point of view of the investigation. Von Neumann settled on a state set containing 29 states (von Neumann 1966): one *vacuum state* corresponding to the absence of "matter" in the cell; several *ordinary* and *special transmission states* both existing in either a *quiescent* or *activated* condition; and some additional auxiliary states. The transition function is defined so as to permit the transformation of the vacuum state of a cell into other kinds of states, and vice versa. In other words, the cellular model permits the creation and annihilation of "matter" within the cellular space (instead of carrying it from other places, as in the kinematic model). Ordinary transmission states are used to perform logic operations, whereas special transmission states are used for growth operations. The transition function is defined so as to let the automaton operate on its own matter under the control of its built-in logic: it permits the definition of a self-modifying computer.

The next problem was to define the self-reproduction strategy. Von Neumann discarded from the outset the idea of reproduction by self-inspection for fear of logical paradoxes and of potential undesired interactions between

the activity of the automaton and that of the self-inspection organ. He relied instead on the use of a quiescent description of the automaton in the
form of a *tape*. His automaton (figure 2.27) is composed of a *universal constructor* capable of reading the tape and producing a quiescent copy of the
automaton described in it, and a *tape copier* capable of creating a copy of the
uninterpreted tape. The activity of the universal constructor and tape copier
is coordinated by a *control unit* that, when activated, can activate them in sequence and bring them back to a quiescent state. The fundamental idea for
achieving self-reproduction is to use a tape that contains the description of
all the parts of the automaton except the tape. In this way, upon activation of
the control unit the automaton will produce a quiescent copy of itself. Once
the copy is completed, the control unit of the original automaton activates
the control unit of the copy, starting again the self-replication process.

UNIVERSAL
CONSTRUCTOR

This approach to self-reproduction has several interesting properties. First,
it avoids the problem of the infinite regress of a description that contains its
own description, since the tape describes all the automaton but itself. Second, the use of a universal constructor makes self-reproduction *robust* with
respect to mutations of the tape. The universality of the constructor must be
intended in the sense that the constructor is able to interpret the description
and build any automaton that can be described on a tape. In particular, this
means that if the tape is mutated into a description of a different machine,
the reproduction process will generate the machine described by the mutated
tape with a copy of the mutated tape attached. Thus, by the line of reasoning
illustrated in figure 2.26, this structure realizes von Neumann's stated goal of
proving that the growth of complexity is possible. An interesting side result
of self-reproduction as illustrated in figure 2.27 is the possibility of defining
structures that move in the cellular space. To this end it is sufficient to stip-
ulate that upon activation, the newly formed automaton proceeds to destroy
the original automaton by extending an arm that resets all its cells to the vacuum state. At the end of the whole process the result is an automaton that
has moved from its original position to the position of the copy. In this way
the synthetic universe is populated by systems that can move, reproduce,
and evolve. Von Neumann's investigation can thus be considered as the first
formal study of the fundamental principles of life, which later became the
subject of study of artificial life (Langton 1996).

ROBUST
SELF-REPRODUCTION

ARTIFICIAL LIFE

Von Neumann's work on self-reproducing automata was left unfinished
and the edited manuscripts published after his death give only an outline of
the structure of some of the organs required by the automaton. A computer
implementation of all the crucial organs of von Neumann's self-reproducing

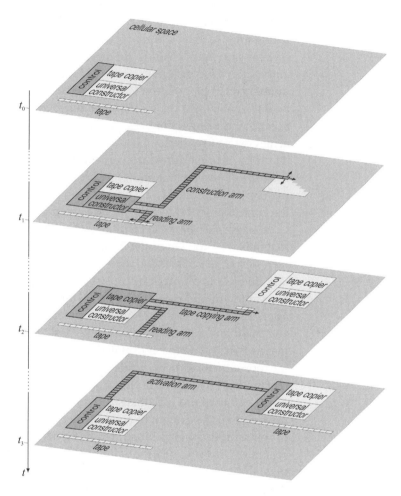

Figure 2.27   A schematic representation of the operation of von Neumann's self-reproducing automaton. The automaton is constituted of a control unit, a universal constructor, a tape copier, and a tape. The self-reproduction process starts when the control unit is activated. The control unit activates the constructor, which grows two mobile arms: one to read the tape and one to construct a copy of the automaton (except the tape) by interpreting the tape. When the construction is finished, the control unit activates the tape copier, which attaches to the newly built automaton a copy of the uninterpreted tape. Finally, the control unit of the original automaton activates the control unit of the copy, and the self-reproducing process starts anew. Note that the various part of the automata are not drawn to scale.

automaton was realized only many years later by Nobili and Pesavento (1996), with further details given in (Pesavento 1995). To date, no one has attempted to design the tape required to complete von Neumann's automaton and, therefore, there exists no running implementation of the self-reproduction process as conceived by von Neumann. Von Neumann estimated that an automaton and tape realizing the robust self-reproduction process illustrated in figure 2.27 using his 29-element state set would have a size of about 200,000 cells. Its implementation would therefore be a quite daunting task. To overcome this problem some researchers (Codd 1968; Burks 1970; Langton 1984; Sipper 1998) have tried simplify the definition of the automaton. In particular, Langton (1984) was able to define a very simple self-reproducing au-

LANGTON'S LOOP   tomaton known as *Langton's loop* using a state set with only eight elements (figure 2.28). However, to attain this result Langton was forced to drop the requirement that the automaton contain a universal constructor. This choice opens the question if Langton's loop is capable of robust self-reproduction, although some results show that this automaton is to some degree capable of evolution (Salzberg et al. 2004).

Another interesting result was obtained by Conway, who proved that a self-reproducing automaton in the spirit of von Neumann's can be implemented in a CA as simple as the Life Game (Berlekamp et al. 2004). Conway's proof is based on the demonstration that glider collisions can produce any Life structure. Thus, one can build in Life a machine containing a universal constructor that reads a tape and generates suitable configurations of colliding gliders that reproduce the machine according to the scheme illustrated in figure 2.27. Given the simplicity of the state set of the Life Game CA and the fact that the Life transition function is not explicitly designed to achieve self-reproduction, we can expect that Conway's self-reproducing automaton is even larger than von Neumann's. Poundstone (1985) estimated a scaling factor of about 100 million, so that an actual implementation of Conway's self-reproducing automaton would require about 10^{13} cells. Both von Neumann's and Conway's CA can contain a general purpose computer and thus are capable of universal computation. However, this property is not essential to achieve robust self-reproduction. The embedding of a universal computer in the self-reproducing automaton is just used to simplify the process of encoding and interpretation of the tape. In general, we can expect a tradeoff between the complexity of the tape decoder and that of the tape, more complex decoders allowing more compact descriptions, and vice versa.

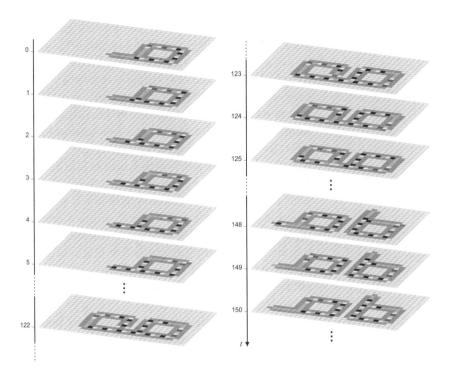

Figure 2.28 Langton's self-reproducing automaton is composed of a signal cycling in a loop, protected by a "sheath" of cells in a quiescent state. The signals can propagate outside the loop to build a copy of the original loop. At the end of the process the two loops separate and each starts a new cycle of self-reproduction in a direction that does not interfere with the other loop. When a loop is completely surrounded by other loops (not shown) it stops replicating. Since the information necessary for self reproduction circulates, there is no need of tape-reading machinery as in von Neumann's automaton. Moreover, the transition table of the CA is defined in such a way that the circulating signal creates the new structure without the need of explicit constructor machinery.

2.7.1   Correspondences with Biology

When von Neumann was formulating his formal model of robust self-reproduction the molecular details of the process of reproduction in living cells were not known. Now that molecular biology has revealed many aspects of the working of cells we can compare von Neumann's formal model with its biological counterpart. It turns out that von Neumann's model anticipates many of the solutions discovered in biological organisms. In the process of reproduction of a single cell the genome of a cell plays the role of the tape in von Neumann's automaton. The molecular machinery which interprets the genome to synthesize new proteins corresponds to the universal constructor, and the molecular machinery which copies the genome without interpreting it corresponds to the tape copier.

Note that the correspondence between what is observed in biology and von Neumann's model is not complete. In von Neumann's model the whole reproduction process is guided step by step by the instructions issued by the control unit and by those read in the tape. On the contrary, in biological cells the control of the self-reproduction activity is only in part controlled by dedicated molecular machinery and by the genome. In biology an essential role is played by processes of *self-organization*, which ensure that in the physicochemical environment of the cell several steps of the structuring of the newly created molecules will self-assemble in the right way. In other words, in biological cells some of the details of the self-reproduction process are not explicitly encoded in the machinery of the cell but follow implicitly from the laws of physics. This does not happen in von Neumann's model because the "laws" of the synthetic universe represented by the state set and transition function of von Neumann's CA are defined at a higher level than the laws of physics holding at the microscopic scale and which produce the self-organization observed in biological cells.

SELF-ORGANIZATION

A useful property of the formal model proposed by von Neumann is that it lets us better understand the role and the information represented in the various parts of biological cells and dispel some myths related to the genome and to its role. For example, it lets us appreciate the fact that a bare genome without its cellular environment does not specify a living entity, exactly as the automaton tape does not mean much without the constructor machinery capable of interpreting it. This is evident in the case of viruses, which are composed almost exclusively of genetic information without the molecular machinery required to interpret and copy it. Thus, a virus should not be considered a self-reproducing organism but merely a free-floating tape with some minimalistic machinery that allows it to invade cells and force them to use their tape copier to replicate the virus genetic information.

2.8   Complex Systems

Many systems of interest in the physical and biological sciences are composed of many simple units that interact nonlinearly. Study has revealed that at the global level these systems can display behaviors and phenomena that

look very complicated despite the simplicity of their components and interactions (figure 2.29). For this reason systems with many nonlinearly interacting units are called *complex systems* (of course, systems with few interacting
elements can also show complex behavior). The idea of complexity that we
are considering here is obviously different from the idea of purposeful complexity described in section 2.7. Following Crutchfield (2003) we could call it
structural complexity.

When the elements that form a complex system interact only locally, cellular systems are often the tool of choice for modeling. Earlier in this chapter
we have already met some examples of cellular models for complex systems. The traffic CA considered in section 2.2 is just the simplest example
of a plethora of traffic cellular models. More complex models take into account many more aspects of real-world traffic such as the stochastic nature of
the driver's behavior; the possibility of cars moving at different speeds, traveling on several lanes and changing lanes probabilistically; the presence of
road junctions; and many others (Nagel and Schreckenberg 1992; Wolf 1999;
Maerivoet and De Moor 2005; Chopard and Droz 1998). The same can be
said of the forest fire CA considered in section 2.5, which is a simple example
EXCITABLE MEDIA   of a whole class of cellular models for *excitable media*, that is, media whose
elements can store and release energy upon excitation. The elements can cy-
cle in a sequence that starts in a *resting* or *receptive* state, goes to an *excited*
state upon reception of a stimulus exceeding a given threshold, proceeds to
a *refractory* state once the stored energy is reduced below another threshold and needs to be replenished, and finally goes back to the resting state.
Besides forest fires, these kinds of models are useful for many other applications, from chemical reaction-diffusion phenomena, to pattern formation
in physical and biological systems (see, for example, the description of Turing pattern formation in chapter 4), and modeling of contagion in epidemics
(Nijhout 1997; Deutsch and Dormann 2005).

Another rich class of cellular models are inspired by phenomena where
agents distributed in space reproduce and move according to their reciprocal interactions. The study of evolution in a spatially structured population
is a natural application for this class of models. Simple models of evolution
assume that each individual may compete and mate with any other individual in the population, and that the offspring of an individual may replace
any other individual. In reality, it is often the case that these interactions are
CELLULAR   constrained by the spatial structure of the population. For this reason *cellu-*
EVOLUTIONARY   *lar evolutionary models* have been developed, where each individual interacts
MODELS   only with its neighbors for selection, mating, and replacement. Simulations

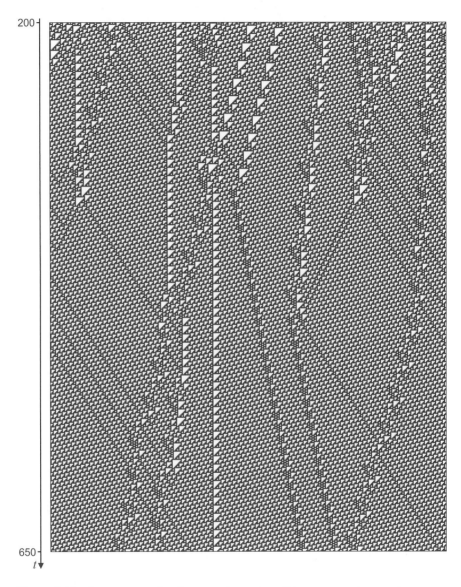

Figure 2.29 A system as simple as an elementary CA can generate complex space-time patterns. The figure shows the space-time diagram generated by the rule 110 CA (see figure 2.12 for the transition table) using a random initial condition at time $t = 0$ and periodic boundary conditions. The initial transient up to $t = 200$ is not shown.

based on these models have revealed that the constraints on individual interactions can greatly influence the evolutionary dynamics, for example, slow-
ing down the invasion of the population by fitter individuals (Giacobini et al.
2005; M.A. Nowak 2006).

AGENT-BASED MODELS

Another area of application of cellular models with interacting agents is
the study of spatial effect in social dynamics. In this context these models
are often called *agent-based models*. In a classic work Sakoda (1971) (see also
Hegselmann and Flache 1998 for a recent discussion) formulated a model
where a population of agents belonging to two distinct classes are distributed
on a two-dimensional cellular space, with some cells being left empty. The
agents represent the members of two groups of individuals and the cellular space represents, for example, a residential neighborhood or a recreation
ground. According to the model, the agents have an *attitude* (represented by
a numerical value) toward the agents that belong to their own class and an
attitude toward the agents that belong to the other class. In the model each
agent assigns a *value* v_i to its current position and to the position of the empty
cells that exist in the cellular space. The value is defined by the formula

$$v_i = \sum_{j \neq i} \left(\frac{a_{ij}}{((x_i - x_j)^2 + (y_i - y_j)^2)^{0.25}} \right)$$

where a_{ij} is the attitude of the ith agent toward the jth agent, x and y are
coordinates in the cellular space, and the sum is extended to all the agents in
the population. At each time step all the agents are considered one by one
in a random order that is renewed at each step. For each agent, if there are
in the Moore neighborhood empty cells with a value greater than the value
assigned to its current position, the agent moves to the cell having the highest
value. Otherwise, the evaluation process is extended to a 5×5 neighborhood.
If no position with a value greater than the current position is found in the
enlarged neighborhood, the agent stays in its current position.

Sakoda considered the evolution in various scenarios, starting with a random distribution of agents. In a scenario that Sakoda called *segregation*,
agents have a positive attitude $a_{own} = +1$ toward agents of their own class
and a negative attitude $a_{other} = -1$ toward agents belonging to the other
class. As expected, the simulations show that the agents tend to form several homogeneous clusters in the cellular space (figure 2.30, left column).
However, simulations where agents have a neutral attitude $a_{own} = 0$ toward agents of their own class and a negative attitude $a_{other} = -1$ toward
those of the other class (a scenario that Sakoda called *suspicion*) produce even
larger and more separated classes than the segregation scenario (figure 2.30,

segregation (own → +1, other → −1)   **suspicion** (own → 0, other → −1)

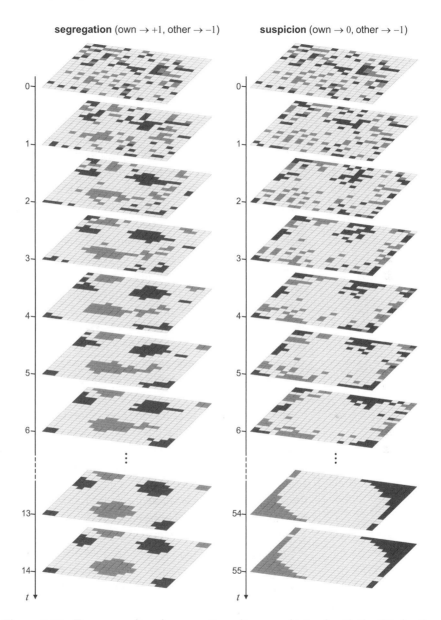

Figure 2.30 Two examples of a space-time diagram obtained with the CA for the simulation of social dynamics considered by Sakoda (1971). Cells represented as light and dark gray correspond to agents belonging to two distinct groups of 50 individuals each. Initially, the agents are distributed randomly on a cellular space of 20×20 cells with impenetrable boundaries. The empty positions are represented as white cells. See the text for the description of the CA rules and dynamics.

right column). In other words, a scenario where agents have a neutral attitude toward "strangers" results in more spatial segregation than a scenario where agents have a negative attitude toward strangers. This result can be explained considering that the positive attitude toward one's own class tends to produce the early formation of homogeneous clusters that cannot be easily broken. However, the result looks surprising at first and illustrates the difficulty of guessing the global consequences of local rules and the usefulness of this kind of model to understand social phenomena. The left column of figure 2.31 illustrates the result of a simulation with a scenario that Sakoda dubbed *boy-girl*, where agents have a negative attitude $a_{own} = -1$ toward agents belonging to their own class and a positive attitude $a_{other} = +1$ toward those of the other class. The right column of figure 2.31 illustrates a variant of this scenario where agents have a neutral attitude $a_{own} = 0$ toward agents belonging to their own class. Once again we observe outcomes that look reasonable in the light of the experiment setup, but whose details are difficult to predict from the knowledge of the parameters alone.

The CA defined by Sakoda's model has a further interesting aspect of relevance for the theory of CAs. In Sakoda's model the motion of the agents is local but the assessment of the value v_i concerns the whole cellular space. The agents move to the position that is surrounded by the configuration of other agents that is closest to the ideal configuration of agents with the highest attitude value. However, nearby agents have a larger impact on v_i than more distant ones. Toffoli (1984) has shown that a point of view of this kind is the natural one for the study of the CA dynamics in the sense that it can be used to give a definition of CAs that is an alternative to the conventional one given at the beginning of this chapter. This point of view determines as natural for the space of CA configurations a metric and topology where two configurations are made more similar by moving away from the current position the nearest cell where the configurations differ.

CA NATURAL
TOPOLOGY

Considering now the domain of physical modeling, the HPP gas example of particle CA presented in section 2.5 is also the simplest of a class of particle cellular models. For example, it has been shown that using a cellular space that allows six directions for the motion of particles, one obtains a model (the *FHP gas*) that approximates much better the actual behavior of fluid at the macroscopic level (Frisch et al. 1986; Chopard and Droz 1998). Another example is the combination of the particle and probabilistic concepts to model the behavior of *granular media* (Karolyi and Kertesz 1999; Chopard and Droz 1998). Figure 2.32 shows an example of a probabilistic particle transition rule that can be used to model the flow of sandpiles and other granular media.

GRANULAR MEDIA

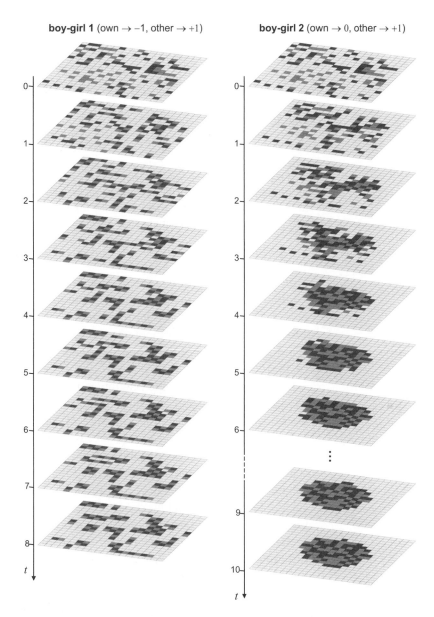

Figure 2.31 Two additional examples of a space-time diagram obtained with the CA for the simulation of social dynamics defined by Sakoda (1971). The representation, cellular space, boundary conditions, and initial conditions are the same as in figure 2.30. See the text for the description of the CA rules and dynamics

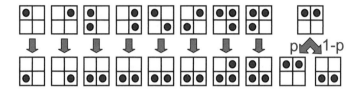

Figure 2.32   The transition table of a CA rule for probabilistic particle motion which can be used to model granular media (Chopard and Droz 1998). The basic idea is to have the particles fall downward to the nearest empty cell, possibly by toppling over other particles. The probabilistic aspect appears in the rightmost entry of the table, which represents the possibility that, with probability p, two particles are prevented by friction from falling. The transition table for the Margolus neighborhood shown here does not contain all possible entries for the states of the block but only those where the configuration changes from one time step to the next.

This is a typical example of a phenomenon that is difficult to model with traditional approaches based on PDEs but quite natural using a cellular approach. Note that this model represents friction at a level that is detailed enough to allow the representation of particles and yet abstract enough to avoid the modeling of the molecular detail of the interaction that produces friction.

CAS IN PHYSICS   Besides modeling specific phenomena such as fluid flow, cellular models have found application at a more fundamental level in physics. If we did not know that the space-time diagram of figure 2.29 is generated by rule 110, we could be tempted to interpret what we see in terms of the motion and interaction of different kinds of particles characterized by their local space-time pattern, much as physicists interpret what they observe at the subatomic scale as the result of the propagation and interaction of families of subatomic particles (Ilachinski 2001). This kind of observation has prompted physicists to investigate the possibility of producing a CA model that would underlie the laws of physics as they are currently formulated. The idea is certainly fascinating, in particular in the light of the discrete nature of physical phenomena as revealed by quantum physics. However, the production of a convincing CA model complying with all the observed physical facts – including isotropy of space and the relativistic nature of space-time – appears difficult, despite some intriguing preliminary speculations (Zuse 1982; Minsky 1982; Fredkin 1992; Wolfram 2002).

CA RANDOM NUMBER   An interesting spinoff of the complex systems side of cellular models is the
GENERATORS   possibility of using them to produce sequences of pseudorandom numbers

(Toffoli 1999; Wolfram 2002). The idea is to see cellular systems as dynamical systems which can display chaotic behavior and thus sequences of states that appear random. For example, the rule 30 elementary CA has been shown to be able to produce sequences of binary digits that can withstand the most stringent tests of randomness. Since the state set and the cellular space are finite, there is a finite number of global configurations that the cellular automaton can assume. Thus, the sequence of pseudorandom numbers will cycle after a finite number of time steps, which, however, can be very large.

What has been presented in this section gives, of course, just a short overview of some of the relationships of cellular systems with complex systems. The use of cellular models for complex systems is so widespread that it is difficult even to compile a list of existing applications. The list of suggested readings at the end of this chapter gives some pointers to works surveying the various aspects of cellular complex system modeling and can be used as a starting point for further investigations in this fascinating field.

2.9 Analysis and Synthesis of Cellular Systems

Cellular systems are defined in terms of local rules that result in global behaviors. There are thus two main questions arising in cellular modeling. The first goes from the local to the global, takes as a starting point the specification of the cellular system, and asks for a characterization of the global behavior. This is the *direct* problem and calls for the *analysis* of the cellular system properties. The second question goes from the global to the local, asking how the local rules must be defined in order to obtain a desired global behavior. This is the *inverse* problem and concerns the *synthesis* of cellular systems.

DIRECT PROBLEM

INVERSE PROBLEM

We have already considered these two problems in various contexts in the previous pages. It should be clear by now that the properties of a cellular system that one can be interested in investigating depend much on the particular system that is considered and on the application that is being envisaged for it. For example, in dealing with coupled map lattices used for the numerical modeling of physical fields one is in general interested in assessing the stability of the system dynamics and its correspondence with the dynamics of the continuous models formulated as PDEs (Potter 1973), but this preoccupation is typically absent in complex system modeling. In view of this variety of objectives it is impossible to discuss in general terms the analysis and synthesis problems for cellular systems. Instead, we will focus on some representative problems that have received much attention and have been considered from several different points of view.

2.9.1   Analysis

The reference problem that we consider to illustrate the subject of cellular system analysis is the characterization of the dynamic properties of CAs. This problem was first tackled systematically by Wolfram (1984), taking as inspiration the classification of the asymptotic behavior of continuous dynamical systems. Wolfram used computer simulations to explore the behavior of a large number of one-dimensional CAs. In particular, he explored exhaustively the class of one-dimensional elementary CAs for a large ensemble of initial conditions. Wolfram conjectured that the results of his observations on a small subset of the universe of CAs have general validity and apply to all

WOLFRAM'S CA   CAs. The observation of the resulting space-time diagrams led Wolfram to
CLASSES   propose a classification into four qualitative classes of behavior (figure 2.33).

- *Class I* CAs are those that for almost all initial conditions evolve in a finite number of time steps to a uniform state over all the cellular space.

- *Class II* CAs are those that for almost all initial conditions and after a short transient either produce a stable nonuniform structure in cellular space, or start to cycle over a small set of simple structures.

- *Class III* CAs are those that for almost all initial conditions produce random-like "chaotic" sequences of states that result in fractal-looking patterns in the space-time diagram.

These first three CA classes correspond to three kinds of asymptotic behavior observed in continuous dynamical systems, namely limit points, limit cycles, and chaotic attractors. Wolfram observed, however, a fourth class of CA behavior that has no correspondence in the theory of continuous dynamical systems.

- *Class IV* CAs are characterized by long-lived localized structures that can propagate on a crystal-like background that covers the cellular space (see also figure 2.29). Wolfram conjectured that class IV CAs are capable of universal computation and that thus, in general, for the determination of their long-term behavior there is no shortcut to the explicit simulation.

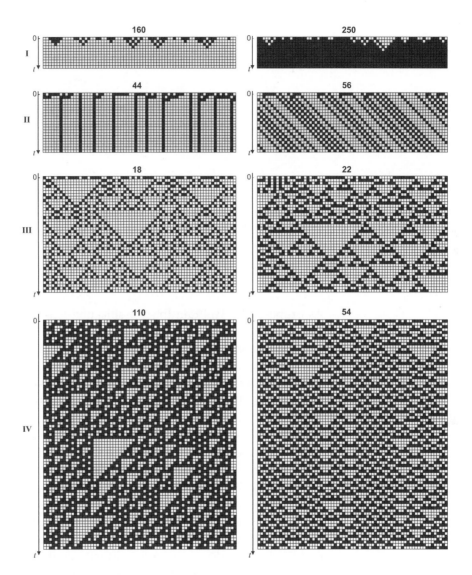

Figure 2.33 An illustration of the four qualitative classes of CA behavior identified by Wolfram. For each class the figure shows two examples of a space-time diagram generated by a one-dimensional elementary CA starting from a random initial condition with a 50% density of 1s.

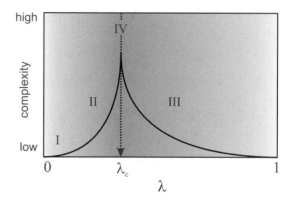

Figure 2.34 A schematic representation of the correspondence between the values of the characteristic parameter λ and Wolfram's CA classes conjectured by Langton (adapted from Langton 1990). According to this conjecture, class IV CAs correspond to the critical value λ_c at which there is a phase transition between order (classes I and II) and disorder (class III) which results in the most complex CA behaviors.

As proposed by Wolfram this classification is only phenomenological. To assign a CA to one of the four classes one must run a CA several times starting from different initial conditions and then guess the class membership by observing the resulting space-time diagrams. Attempts to formalize this process of guessing are hampered by the vagueness of the characterization of the structures that identify classes III and IV and by some theoretical results (Culik and Yu 1988) proving the impossibility of determining for an arbitrary CA the eventual attainment of quiescence from all initial configurations. Despite these reservations, Wolfram classes remain a useful heuristic concept that has inspired much of the subsequent work on the classification of CA.

A well-known example is provided by Langton (1990), who adopted an approach inspired by the physics of phase transitions. In many physical systems one can define a characteristic parameter (e.g., temperature) that identifies different phases of the system (e.g., solid and liquid) characterized by different degrees of order. Moreover, the system exhibits the most complex behaviors in the regions of the parameter where the phase transitions occur. Langton proposed to define a characteristic parameter λ for CA as follows. One element s_o of the state set is chosen as the *quiescent state*. Then one defines $\lambda = (N - n_o)/N$, where N is the total number of entries in the transition table and n_o is the number of entries leading to the quiescent state. Based on the observation of many simulations of CA with different

LANGTON'S
PARAMETER

values of λ Langton proposed a correspondence between Wolfram's classes and ranges of values of λ (figure 2.34). According to this correspondence class I and class II CAs would correspond to an ordered phase characterized by values of λ below a critical value $\lambda_c \approx 0.3$; class III CAs would correspond to a disordered phase characterized by values of λ above the critical value; class IV CAs would correspond to $\lambda = \lambda_c$, that is, to the transition between the two phases. Even if this correspondence is proposed by Langton only in terms of the average dynamical behavior of the CAs and does not refer to the dynamical behavior of each single CA having a given value of λ, some researchers have questioned the validity of this classification (Mitchell et al. 1993; Mitchell 1998). More recently, in a series of papers (Chua et al. 2002, 2003, 2004, 2005) Chua and coworkers have proposed an interesting new approach to the classification of the elementary CAs. The approach is based on the association of an ordinary differential equation with each elementary CA and on a geometric interpretation of the parameters that define the differential equations. This approach permits the formalization of some of Wolfram's original conjectures and observations. Many other approaches to the characterization of CAs can be found in the literature, based on concepts borrowed from various fields, such as dynamical system theory and statistical physics. The reader is referred to (Ilachinski 2001; Deutsch and Dormann 2005) for further discussion and an overview of the relevant literature.

2.9.2 Synthesis

The synthesis of a cellular system is similar to the activity of the scientist trying to formulate a scientific theory for an observed phenomenon. The designer of the cellular system may consider different local rules, analyze their consequences in terms of global behavior, and compare them with the desired behavior.

A first example of a synthesis problem that has received some attention concerns cellular systems for the modeling of physical phenomena. In this context Toffoli (1994) has observed that to select a set of plausible candidates one can start with the set of all automata that have a reasonable state set and neighborhood, and restrict the set by imposing the properties of invariance that are known to hold in the physical system. For example, one can impose particle conservation, momentum conservation in different directions of the cellular space, rotational invariance, reversibility in time, and so on. By so doing one reduces dramatically the huge universe of CA rules. Moreover, one discovers that some aspects of the local rules have no effect at the global level because their averaged effect is negligible, whereas other aspects survive the averaging process (in mathematical parlance, they *commute* with it) and are felt at the global level. These latter aspects are crucial, not only because we must ensure that all the important local aspects are present but also that no spurious aspects that survive averaging are present in the cellular model. For example, the HPP gas model presented in section 2.5 (figure 2.21) is plagued by the spurious conservation of the total momentum along each horizontal and vertical line of the cellular space, which results in a macroscopic behavior that differs from that of a real fluid (Ilachinski 2001).

When the desired global behavior is given without an accompanying prescription for the local behaviors that can be considered plausible, the synthesis of cellular models becomes even more problematic. This is the case, for example, of cellular systems that are required to perform some global computation (see discussion in section 2.6). In these cases an interesting alternative to the design by hand of cellular models consists in evolving them (Packard 1988; F.C. Richards et al. 1990; Mitchell et al. 1993, 1994; Mitchell 1998; Sipper 1996; Sipper et al. 1997). For example, Mitchell et al. (1994) evolved the rule table of binary CAs for the solution of the density classification task on cellular spaces of fixed, predefined size (see section 2.6, p. 135 for the definition of the density classification task). The evolution is based on a genetic algorithm whose genome consists of binary strings representing the output bits of the transition table. Given the threshold density value ρ_t, the quality of an individual is assessed by running several times the rule defined by its genome starting with randomly generated initial conditions with density randomly chosen with uniform probability in the $[0, 1]$ range. Each initial condition is tested by running the CA until the state is stable or a maximum number of time steps (varied randomly from run to run to avoid evolving solutions tailored to a specific value) is attained. The fitness of a transition rule is defined as the fraction of initial conditions that are classified correctly by the rule when the CA is stopped. The experiments lead to the discovery of rules that classify correctly about 95% of the initial configurations, which is close to the 97% correct classification rate of the best-performing hand-designed classification rule (Mitchell et al. 1994). Other interesting results with the evolutionary approach have been obtained not only for computational tasks but also for the determination of CA rules capable of reproducing the experimentally observed behavior of complex physical systems (F.C. Richards et al. 1990).

This short survey suggests that both the analysis and the synthesis of cellular systems are in general quite complicated problems and many issues are

CA EVOLUTION

far from being resolved. Again, the problem stems from the nontrivial relationship between the local rules and the resulting global behaviors. In the case of analysis one can at least always resort to running and observing the behavior of the system. For the synthesis, the most precious resource is often the designer's experience and insight into the phenomenon that must be modeled, or the sophistication of some search algorithm.

2.10 Closing Remarks

Lest the reader that has been exposed to all the examples of applications of cellular systems presented in the previous pages is left with the feeling that cellular models are good for everything and should be used on all occasions, let us now mention their principal limitation. We have seen that many prescriptions can be relaxed in the basic cellular system represented by CAs to produce a variety of systems adapted to various contexts. One thing remains, however, fixed, namely, the rigid geometrical backcloth imposed by the cellular space and by the not very flexible structure of the neighborhood. If we are interested in modeling systems composed of elements having very different amounts of connectivity with each other, then it is probably better to think different models (Bithell and Macmillan 2007), for example one of those discussed in chapter 7. The same is true if the connectivity of the systems of interest varies dynamically and rapidly.

Despite this limitation, the examples given in this chapter should have convinced the reader that cellular models are a very useful tool. They afford an appreciation of how simple local rules can produce complex global behaviors. They permit the sorting out of the essential ingredients of the local rules that are responsible for the global behavior from the nonessential details that are averaged out at the collective level. They let one investigate the robustness of the global phenomena to changes in the topology of the interaction, in the nature of the local rules, in the initial and boundary conditions. They can be used also to define parallel computational devices that can solve practical tasks in a very efficient way. In short, cellular models are essential for the scientific understanding and analysis, design, and evolution of a large class of complex systems.

2.11   Suggested Readings

Wolfram's *A New Kind of Science* (Wolfram 2002) is a massive and richly illustrated book which presents many examples of cellular systems in a very accessible way, relegating all technical details to the endnotes. It is a good starting point to get familiar with cellular systems by observing hundreds of space-time diagrams while reading some daring speculations about the role of CAs in physical and mathematical modeling. In (Toffoli and Margolus 1987) one can find a lively discussion of many methodological issues related to cellular modeling. Although the notation used in this book is unusual (referring to a specialized cellular computer designed by the authors), the book makes good reading for its many valuable insights. (Ilachinski 2001) is a treatise with encyclopedic ambitions which discusses in particular several formal approaches to the analysis of cellular systems.

A good overview of the cellular approach applied to physical modeling, with a detailed treatment of particle CAs and lattice Boltzmann models, can be found in (Chopard and Droz 1998). Gaylord and Wellin (1995), Gaylord and Nishidate (1996), and Gaylord and D'Andria (1998) illustrate cellular modeling with examples from several disciplines, from physics, to biology, to socioeconomic sciences. The examples are accompanied by computer code for the generation of fascinating space-time diagrams and animations. (Deutsch and Dormann 2005) is an interesting work that discusses the role of cellular modeling in the study of pattern formation in biological systems. Nowak (2006) discusses the modeling of evolutionary dynamics and uses cellular models to illustrate the effect of the spatial constraints on the interactions of the individuals forming the evolving population. A classic example of an agent-based model is the *sugarscape* system, described in detail in (Epstein and Axtell 1996). This book and the collection of papers in (Epstein 2006) discuss the role of computational models in the study of complex social and economic phenomena. Ball's *Critical Mass* (Ball 2004) is an excellent nontechnical overview of complex system modeling (not only cellular-based) with chapters on traffic modeling and social dynamics.

Conway's description of the Life Game and its varied fauna of objects can be found in (Berlekamp et al. 2004). This book sketches also the proof of the universal computation capabilities of Life and of the possibility of building a self-replicating automaton in Life. The same topics are treated in a more detailed and accessible way by Poundstone (1985). The original manuscript describing von Neumann's work on robust self-reproduction was edited and published as *Theory of Self-Reproducing Automata* (von Neumann 1966). The actual description of the self-reproducing automata is a bit technical, but this material is preceded by the transcript of five talks given by von Neumann which reveal his ideas on the subject. The history of the reception of these ideas by the scientific community has been reviewed critically by McMullin (2000) and Sipper (1998). McMullin's remarks on the sociology of science can be usefully complemented by the excellent work of Russo (2004) which discusses the role of models in the definition of a scientific theory and presents compelling evidence that the current viewpoint on this subject is not a modern creation – as is typically assumed – but can be found in the writings of Hellenistic scientists who date back to the third century B.C.

Finally, to go back to the starting point of this chapter, (Harold 2001) is a very readable overview of the properties of biological cells.

3 *Neural Systems*

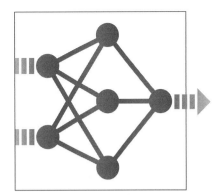

Behavior, thought, and emotions are mediated by a complex network of cells known as neurons. In the second half of the twentieth century, the availability of electronic computers allowed scientists to recreate and study *in silico* networks of lifelike neurons. Over the last 50 years, the development of formal and computational models of neural systems has been pursued mainly in two areas of research.

COMPUTATIONAL NEUROSCIENCE

Computational neuroscience, which finds its roots in the detailed mathematical model of neuronal membrane dynamics, first described by Hodgkin and Huxley (1952), attempts to understand the functioning of living brains. The main questions addressed by computational neuroscience include the type of communication used by neurons, the effects of chemicals on neuronal behavior, the dynamics of neuronal assemblies, and the theoretical capacity of neuronal computation, to mention a few.

NEURAL ENGINEERING

Neural engineering, which can be traced back to the logic-level description of a neuron given by McCulloch and Pitts (1943), instead aims at reproducing the functionalities of brains in order to engineer intelligent machines. Issues addressed by neural engineering include robust control for robotic systems, learning algorithms and high-level architectures that could reproduce cognitive abilities, and implementation of neural models in hardware.

Before describing neural systems, it is worth asking what potential advantages neuron-like systems could bring to an animal or machine. Some single-cell organisms, such as the paramecium (figure 3.1, left), do not have neurons, but can still do quite a lot of things, such as eating, moving toward light, escaping from aversive situations, and even changing behavior after repeated stimulation. They do so by means of chemical processes that affect the electrical potential across their membrane and modify the shape of

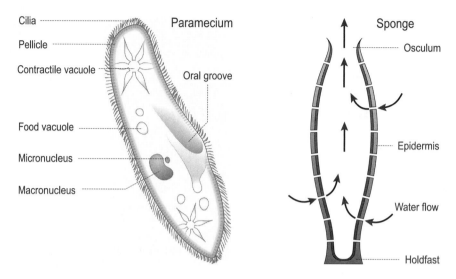

Figure 3.1 The paramecium and the sponge do not have neurons, but can react to environmental stimuli by means of chemical reactions.

the constituent proteins. Chemical processes regulate also the behavior of sponges (figure 3.1, right), which are multicellular animals without a neural system. Sponges absorb nutrients from water that is pumped through their bodies by contractile cells that respond to chemical, thermal, and mechanical stimulation.

Neural systems bring at least two advantages compared to brainless organisms: (1) selective transmission of signals between distant parts of the body and (2) adaptation by means of synaptic plasticity. These two features allow for more complex bodies and better coping with partially unpredictable environments that change too rapidly for evolution to catch up with.

Neural systems have been discovered and shaped over millions of years by evolution. At the same time, the additional adaptation provided by neural systems could improve the survival of the organism and thus affect the evolution of the species. The influence of lifetime adaptation on evolution was recognized by Darwin and by Lamarck (Lamarck 1914). Although they had different views on this issue, they both believed that characteristics acquired during life were transmitted to offspring. However, as we have seen in chapter 1, we now know that the process of gene expression does not allow a modification of the genotype by the phenotype.

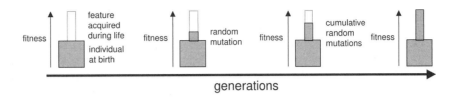

generations

Figure 3.2 Illustration of the Baldwin effect. Individuals (square blocks) undergo an adaptation process during life that makes them taller and gives higher fitness. This adaptation process has also an evolutionary cost. By effect of random mutations, some offspring are born taller and require less adaptation effort; therefore, their evolutionary cost is smaller and they have higher probability of reproducing. Gradually, evolution will select individuals that are born taller and taller by the effect of incremental mutations, thus decreasing the role of lifetime adaptation. A similar effect may apply to neural adaptation.

At the end of the nineteenth century, Baldwin (1896), Morgan (1896), and Osborn (1896) independently suggested that learning can affect evolution without assuming that characteristics acquired during life are directly transmitted to offspring. Baldwin's argument, in particular, was that learning accelerates evolution because suboptimal individuals increase their baseline reproduction rate by acquiring during life the necessary characteristics for survival. However, lifetime learning often involves a cost because the individual may be at risk at an early stage of its life or it may modify its behavior and morphology in ways that are not functional for its survival, to mention only two examples. Therefore, Baldwin suggested that evolution tends to select individuals who already have at birth, by effect of random mutations, some of the useful features that would otherwise be learned (figure 3.2). This

BALDWIN EFFECT indirect assimilation of learned characteristics, also known as the *Baldwin effect*, is still a debated issue, but is nowadays better accepted than the theory of direct assimilation of lifetime modifications by evolutionary mechanisms (Behera and Nanjundiah 1995).

The morphology and behavior of individual neurons are very similar across animal species and also across phylogenetic history, as far as one can infer from comparative biology (G.H. Parker 1919). It seems therefore that evolution of the brain occurred mainly at the level of the architecture, or connectivity, of the neuronal networks. Indeed, the first systematic classification of neurons given by the neuroanatomist Ramón y Cajal at the end of the nineteenth century was indeed based on the patterns of connectivity specific to each type of neuron (Ramón y Cajal 1911).

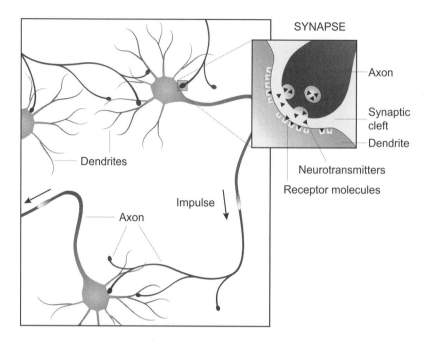

Figure 3.3 Biological neurons with detail of a synapse.

In this chapter, we start with an overview of the most salient principles and mechanisms of biological neural systems and then proceed to describe how those elements are modeled in artificial neural networks. We will also cover major learning algorithms and conclude with artificial evolution of neural networks.

Artificial neural networks are often implemented as software objects in standard microcontrollers or desktop computers. However, there are situations where the computational demands of the artificial neural network or the physical constraints (e.g., power, dimension) of the application require the development of novel hardware solutions. The final section of this chapter describes three major approaches to neural hardware, namely digital, analog, and hybrid systems interfaced to cultures of living neurons.

As in other chapters of this book, the closing remarks provide some insights into future promising directions as well as some criticism of established models.

3.1 Biological Nervous Systems

Nervous systems consist of assemblies of interconnected neural cells (figure 3.3). Neurons communicate by means of electrical signals that travel mainly in one direction along connections. Neuronal cells consist of a body with ramifications, known as *dendrites*, that receive signals from several other neurons, and of a single filament, known as an *axon*, that carries the outgoing electrical signal emitted by the neuron. The axon of a neuron branches out to establish contact with several other neurons. At the same time, a neuron or muscle cell can receive signals from several other neurons.

DENDRITES

AXON

The transmission of electrical signals between neurons is mediated by electrochemical devices, known as *synapses*, that are located at the contact point between the axon of the emitting neuron and the dendrite of the receiving neuron. Incoming electrical signals at the synaptic point trigger the release of chemical substances, known as *neurotransmitters*. Neurotransmitters open molecular gates on the dendrites of the receiving neuron that let electrically charged particles (ions) flow in. These ions generate a voltage difference across the membrane that travels from the dendrite to the body of the receiving neuron, thus affecting the voltage difference between the interior of the neuron body and the external environment. That voltage difference is known as the *activation level* or *potential* of the neuron. Sending and receiving neurons are also referred to as presynaptic and postsynaptic neurons, respectively.

SYNAPSES

NEUROTRANSMITTERS

ACTIVATION LEVEL

For the sake of simplicity, most computational models assume that the electrical voltages contributed by different synapses are summed up when they meet at branching points of the dendritic tree on their travel toward the body of the neuron. However, incoming signals can be combined also in other ways, that can be described by Boolean logic functions (figure 3.4). For example, active inhibitory synapses tend to block signals that travel from higher parts of the dendritic tree, in a way similar to the NOT logic function. Furthermore, the closer an inhibitory synapse is to the body of the neuron, the stronger its effect is on silencing the neuron altogether, no matter what other inputs are.

A neuron propagates an electrical signal along its axon when the voltage difference (activation level) across its membrane is larger than its threshold (figure 3.5). For example, the resting level of pyramidal neurons in the mammalian cortex is approximately -65 mV. The reception of incoming excitatory signals brings the activation level toward positive values (depolarization), causing an electrical discharge, also known as the *action potential*, when it

ACTION POTENTIAL

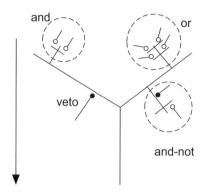

Figure 3.4 Synaptic interactions on the dendritic tree. Empty disks are excitatory synapses, black disks are inhibitory synapses. An inhibitory synapse can block all signals traveling from preceding synaptic points (veto effect). and = both synapses must be active for a signal to get through; or = one synapse is sufficient to send a signal; and-not = the excitatory synapse can transmit a signal only if the inhibitory synapse is not active.

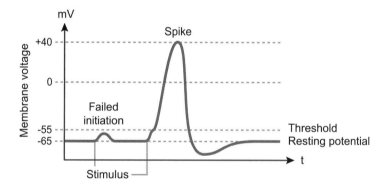

Figure 3.5 Schematic drawing of membrane dynamics. An outgoing electrical discharge (spike) is generated when the action potential reaches a positive threshold.

reaches the threshold of -55 mV. The action potential travels along the axon toward the postsynaptic neurons. The onset of an action potential is followed by a phase of hyperpolarization where the neuron goes back to negative values that are larger than the resting level. Finally, the neuron goes back to its resting potential of -65 mV. This fundamental cycle, which was first described by Hodgkin and Huxley (1952), is caused by the inflow and outflow of different types of ions across the body membrane and can last between

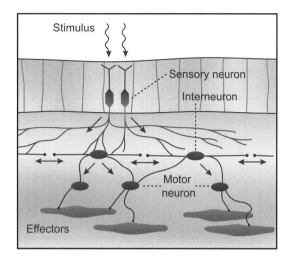

Figure 3.6 Sensory neurons, motor neurons, and interneurons.

3 and 50 ms, depending on the type of ion currents. The output discharge has a very sharp temporal envelope and therefore is often called a *pulse* or *spike*. It is also common to say that the neuron has fired. Most neurons can emit up to 250 to 300 pulses per second and even in the absence of incoming signals they display a resting activity of approximately 10 pulses per second. Brains therefore display spontaneous activity even in the absence of external stimuli, a fact that is rarely taken into account in artificial neural networks.

SPIKE

3.1.1 Neural Typology

EXCITATORY

Neurons come in two flavors: excitatory and inhibitory. Excitatory neurons establish synaptic connections that tend to increase the activation of post-synaptic neurons. Inhibitory neurons instead establish synaptic connections that tend to decrease or block the activation of postsynaptic neurons. It is estimated that only 16% of synapses in the mammalian cerebral cortex are inhibitory (Kandel et al. 2000). However, their strategic location and strong effect on postsynaptic activation provide crucial regulation of brain activity and prevent the onset of epileptic activity (Braitenberg 1984).

INHIBITORY

SENSORY

There are three major classes of neurons (figure 3.6). *Sensory neurons* are peripheral cells that have an input detector exposed to the environment and an output connection that can diverge to make contact with several other neurons or effectors (muscle cells). This arrangement allows simultaneous broadcasting of the signal across parallel neural pathways where the same sensory signal can be analyzed and combined in different ways with other signals. Where sensory neurons are directly connected to effectors, the divergent signal can synchronize the response of effectors that are distributed across distant parts of the animal body and ensure a rapid and coordinated response.

MOTOR   *Motor neurons* are peripheral cells that send signals directly to effectors or to other neighboring motor neurons. Motor neurons receive input from sensory neurons or from other neurons in the system. They allow sensory or control signals to converge from various sources before being transmitted to one or more effectors. A neural system with both sensory and motor neurons can display more complex behavior by combining sensory signals in various ways so as to produce diverse behaviors.

INTERNEURONS   *Interneurons* are neural cells that establish connections with other neurons, but are not directly connected to the environment. Interneurons increase the complexity of the neural system by allowing a larger number of connection topologies and by adding further stages of signal transformation. For example, interneurons can change the intensity of the signals transmitted from the input neurons to the output neurons or let the signals through only in the presence of signals from other input neurons. Also, inhibitory interneurons can transform an incoming excitatory signal into an inhibitory signal, thus reducing or blocking the activity of an output neuron when that signal is present.

3.1.2   Neural Communication

A recurrent issue in neuroscience is how neurons encode and transmit in-
FIRING RATE   formation. One hypothesis is that information is encoded as the *firing rate*, that is, the number of spikes within a time interval. Indeed, we know that the response of muscle cells is proportional to the firing rate of presynaptic neurons (Adrian 1928). Another hypothesis is that information is encoded as
FIRING TIME   *firing time*, that is, the time interval between spikes. For example, a neuron may fire only if it receives two input spikes with a time delay of few milliseconds between the two. This property is exploited by the acoustic nervous system to encode information about the spatial location of a sound source. Those neurons are tuned to specific time delays between spikes generated by sensory neurons on the two sides of the head (e.g., see Carr and Konishi 1988).

As we move away from peripheral neurons toward the central nervous system, it is increasingly difficult to tell whether neurons use firing rate or firing time to communicate. It has been argued that firing time may be used by inner neurons to encode not only time-dependent but also space-dependent information (Singer and Gray 1995), but the measurement of precise spatiotemporal patterns of neural activity is still technically very challenging (Villa 2000).

Neurons can affect each other in several other ways than through synapses. For example, densely packed neurons, such as those in the retina and in the hippocampus of mammalian brains, can affect the activation of neighboring neurons through direct connections between neural bodies (Makowski et al. 1977) or local electrical fields (Taylor and Dudek 1984). These types of communication seem to synchronize the activity of local neural assemblies. In addition, some neurons can also communicate by means of long-range neurotransmitters, such as nitric oxide, that diffuse over areas comprising several neurons and synapses (Garthwaite 1991). However, local interactions through synaptic joints are the most widespread type of interaction among neurons.

DIRECT CONNECTIONS

LONG-RANGE
NEUROTRANSMITTERS

3.1.3 Neural Topology

Neural systems display regular architectural patterns generated during the process of cellular growth. In addition to cell division and differentiation, during early development of the brain neural cells can move and grow axons guided by neurotrophic gradients (these are chemicals present in the brain that drive axon growth during brain development). Neural systems are symmetrically organized along the three main directions of the body: bilateral (left and right), rostrocaudal (head-tail), and dorsoventral (top-bottom). Furthermore, neurons tend to concentrate in a specific area of the body, the head of the animal, probably because of better efficiency in terms of connectivity (Ramón y Cajal 1911).

Most neurons receive connections from, and project to, neighboring neurons. This architectural feature has three consequences. The first is that the layout of neurons close to sensory areas tends to preserve the topological relations of the receptors. In other words, neighboring neurons respond to neighboring sensors. The second is that neighboring neurons tend to respond to similar patterns of stimulation, which will be explained later with a computational model. The third is that nervous systems are organized in local circuits characterized by specific patterns of connectivity, although it is

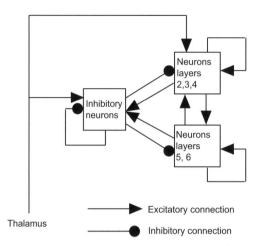

Figure 3.7   The canonical cortical circuit (adapted from Douglas and Martin 1990). Numbers within blocks indicate the layers of the cortex: 2, 3, and 4 are layers closer to the skull; 5 and 6 are closer to the center of the brain.

not yet clear to what extent this compartmentalization may correspond to specialization of the circuit function.

Topology preservation and anatomical modularity are very evident in the mammalian brain. Neurons in the visual, auditory, and tactile sensory areas of the cortex are arranged so as to reflect retinotopic and tonotopic, and somatotopic relations (Knudsen et al. 1987; Merzenich and Kaas 1980). Those topological relations are maintained throughout several layers of neurons, although they become increasingly distorted with increasing distance from sensory neurons. Sensory areas of higher interest for the animal receive attention from a comparatively larger number of neurons. For example, there are more neurons allocated to photoreceptors in the fovea (the central part of the retina) than to photoreceptors in other parts of the retina. Similarly, there are more neurons allocated to fingertips than to the dorsal part of the hand. Mammalian brains also display several modules characterized by specific architectures, such as the thalamus, hippocampus, cerebellum, and cortex, to mention a few (Shepherd 1990). Even within those modules, one can find further levels of modular organization. For example, the visual cortex seems to be organized in several areas characterized by area-specific inter- and intraconnection patterns and by area-specific neural response (DeYoe and van Essen 1988; van Essen and Maunsell 1983). On an even smaller scale, it has been suggested that the cortical areas consist of several modules with the same local design (figure 3.7), known as the *canonical circuit* (Douglas and Martin 1990). Furthermore, inhibitory interactions seem to occur only within local circuits.

CANONICAL CIRCUIT

Neural systems endow organisms with the ability to adapt to their environment during their lifetime. In this book, we define *adaptation* as the set of modifications occurring during the interaction between the organism and the environment that are functional to increasing its probability of survival. At a macroscopic (behavioral) level, adaptation has many manifestations. It includes habituation, formation of associations, memorization of items and places, and reinforcement learning, to mention a few. At a microscopic (neurophysiological) level, adaptation is provided mainly by processes that affect the strengths of synaptic connections between neurons. This process is also known as synaptic plasticity.

ADAPTATION

The important role of synaptic plasticity in adaptation was emphasized more than 50 years ago by the Canadian psychologist Donald Hebb, who suggested a possible mechanism for associative learning:

> When an axon of cell A is near enough to excite cell B or repeatedly or consistently takes part in firing it, some growth or metabolic change takes place in one or both cells such that A's efficiency, as one of the cells firing B, is increased. (Hebb 1949, p. 62)

HEBB'S RULE

This mechanism, now known as Hebb's rule, has been measured by several experimental studies (e.g., Kelso et al. 1986) and is at the core of most computational models. Hebb's rule explains only long-term strengthening (also known as *long-term potentiation* and abbreviated as LTP) of synaptic connections. More recent experimental studies highlighted variations of this mechanism that can explain also long-term weakening (also known as *long-term depression* and abbreviated as LTD) of synaptic connections when the activations of two connected neurons are not correlated (e.g., Stent 1973; Singer 1987; Stanton and Sejnowski 1989). Hebb's rule has two major implications. First, it implies strong locality of the plasticity: the modification of a synapse depends only on the presynaptic and the postsynaptic neurons. Second, it introduces the concept of activity-induced reinforcement or weakening of the synapse.

LONG-TERM POTENTIATION (LTP)

LONG-TERM DEPRESSION (LTD)

In the context of a spiking neuron, Hebb's rule implies a temporal relation because a presynaptic spike that is responsible for the emission of a spike by the postsynaptic neuron necessarily occurs earlier in time. Recent experimental measurements indeed showed that presynaptic spikes arriving a

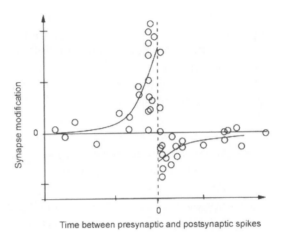

Figure 3.8 Spike time-dependent plasticity. The percentage of synaptic modification depends on the difference between presynaptic and postsynaptic spikes $(t^{pre} - t^{post})$. The temporal difference in the figure ranges between -100 ms and +100 ms. When the postsynaptic neuron fires after arrival of the presynaptic pulse (positive difference, also known as causal relation), the synaptic weight is increased; instead, when the postsynaptic neuron fires before the arrival of the presynaptic pulse (negative difference, also known as anticausal difference), the synaptic weight is depressed. From Gerstner and Kistler (2002) plotted on data from Bi and Poo (2001).

few milliseconds before a postsynaptic spike tend to increase (potentiate) the synaptic strength, while presynaptic spikes that arrive a few millisecond later tend to decrease (depress) the synaptic strength (Zhang et al. 1998; Bi and Poo 2001). The time window for synaptic potentiation and depression spans only a difference of a few milliseconds between pre- and postsynaptic firing and is asymmetric, as shown in figure 3.8. The apparently different descriptions of spike time-dependent plasticity and of correlation-based plasticity described by Hebb in fact explain the same phenomenon when neural activity is considered over extended periods of time. In that case, systematic earlier firing of the presynaptic neuron implies a causal and positive correlation, whereas late firing of the presynaptic neuron implies an independence and negative correlation.

SPIKE
TIME-DEPENDENT
PLASTICITY (STDP)

There are several other neurophysiological processes that contribute to adaptation, such as dynamic growth and death of connections as well as modifications of the molecular properties of the neuron membrane. Those processes typically occur over a longer time window, and are less frequent, than the activity-dependent synaptic process described above. They are also less frequently incorporated in computational models of neural systems, but may still play fundamental roles in the definition of brain architecture and repair. We will describe some of these processes in chapter 4.

3.2 Artificial Neural Networks

Artificial neural networks are computational models implemented in software or custom-made hardware devices that attempt to capture the behavioral and adaptive features of biological nervous systems. An artificial neural network is composed of several interconnected units, or neurons (figure 3.9). Some of these units receive information directly from the environment (input units), some have a direct effect on the environment (output units), and others communicate only with units within the network (internal, or hidden, units).

Each unit implements a simple operation that consists in becoming active if the total incoming signal is larger than its threshold. An active unit emits a signal that reaches all units to which it is connected. The connection, or synaptic point, operates like a filter that multiplies the signal by a signed weight, also known as synaptic strength.

Whereas biological neurons are either inhibitory or excitatory and have the same effect on all neurons which they send signals to, artificial neurons can emit both negative and positive signals and thus the same neuron can establish both negative and positive synaptic connections with other neurons. There are two reasons for this difference. The first is that artificial neurons are mathematical objects that are not constrained by the physiological properties of biological neurons in order to achieve the same functionality. The second is that an artificial neuron often models the average response of a population of biological neurons, which may include both excitatory and inhibitory neurons.

The response of an artificial neural network to an input from the environment depends on its architecture and pattern of connection strengths. The knowledge of the network is distributed across its connections. The behavior of the network is given by the pattern of activations of the neurons, which in some models can self-sustain and change over time even in the absence of input from the environment.

Neural networks learn by modification of synaptic strengths when presented with stimulation from the environment. Usually, learning requires

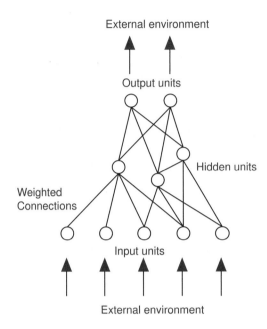

Figure 3.9 Generic neural network architecture.

several repeated presentations of the set of input patterns. There are several types of learning rules, each displaying specific functionalities and applicable to specific architectures. Typically, all synaptic connections within the artificial neural network change according to the same learning rule.

In addition to the ability of learning by exposition to examples (learning by demonstration), neural networks are often appreciated in engineering applications also for the following features.

- *Robustness*. Neural networks are robust to various types of signal degradation, such as input noise or malfunctioning of connection and unit operation in hardware implementations. As the noise level increases, neural networks display graceful degradation by increasing the error rate more or less uniformly across the entire input domain or by making errors for specific input patterns while maintaining a correct response for all other patterns. Furthermore, neural networks can be incrementally trained to compensate for signal noise or damage to their components.

- *Flexibility*. Neural networks are not domain specific, that is to say that a neural model can be applied to several types of problems (however, that does not mean that any type of neural network can be applied to any type of problem). Neural networks can be used to tackle problems for which there is not an analytical solution, but this presents the risk of giving up the effort of understanding the problem to find the comfort of a neural solution that does not increase our knowledge (Dewdney 1993).

- *Generalization.* Neural networks trained on a limited number of examples can provide the correct response to input patterns that share some similarity with training patterns, but were never seen before. This ability comes from the fact that neural networks store a larger number of input-output associations than the number of available synaptic strengths by extracting invariant features of the patterns. The ability of the network to generalize the response to a new pattern depends on the extent to which the new pattern can be described by the learned invariant features. The extraction of invariant features is also a common property of biological neural systems that allow them to operate consistently in continuously changing environments. From an engineering perspective, the ability to generalize to novel input patterns is very useful for those applications where it is impossible to obtain an exhaustive list of all situations that the system may be exposed to.

- *Content-based retrieval.* Neural networks retrieve memories by matching contents and can do so even when the input patterns are incomplete or corrupted by noise. In some neural models, such as those derived from adaptive resonance theory (Grossberg 1987), retrieval resembles the way in which humans operate: more familiar patterns are recognized faster than items that are different or seen less frequently. Instead, in conventional computer systems, data are retrieved using the address of the electronic memory cells. If that number is corrupted or lost, the entire memory is lost.

3.3 Neuron Models

An artificial neuron is characterized by a set of connection strengths, a threshold, and an activation function (figure 3.10). If we ignore transmission delays, the effect of a set of input signals $\vec{x}$ on the postsynaptic neuron is equal to the product $\vec{w} \cdot \vec{x}$, where $\vec{w}$ are the synaptic weights and can take any

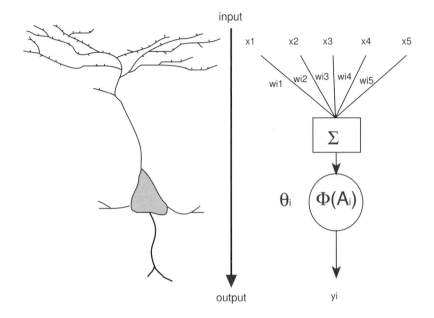

Figure 3.10   Schematic representation of a biological (pyramidal cell) and artificial neuron.

real value (both negative and positive). The net input, or activation a_i, of a neuron i is the sum of all weighted inputs from presynaptic neurons j:

$$a_i = \sum_{j=1}^{N} w_{ij} x_j$$

The output signal y_i is a function of the net input and of the neuron threshold ϑ_i, which is usually subtracted from the sum of weighted inputs:

$$(3.1) \quad y_i = \Phi(a_i) = \Phi \left(\sum_{j=1}^{N} w_{ij} x_j - \vartheta_i \right)$$

The activation function $\Phi(\cdot)$ describes the response profile of the neuron and can take several different forms (figure 3.11). In the original formulation by McCulloch and Pitts (1943) neurons have a binary output (0 or 1) and the threshold is used as a hard delimiter to tell whether a neuron emits a signal.

McCulloch and
Pitts model

$$\Phi(a_i) = \begin{cases} 1 & : \quad \sum_{j=1}^{N} w_{ij} x_j > \vartheta_i \\ 0 & : \quad \text{otherwise.} \end{cases}$$

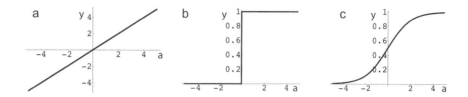

Figure 3.11 Some common activation functions (neuron output as function of total weighted input): *a*, linear function with $k = 1$; *b*, step function with $\vartheta = 0$; *c*, sigmoid function with $k = 1$.

A variation of this function is the bipolar activation where

(3.2)
$$\Phi(a_i) = \begin{cases} 1 & : \quad \sum_{j=1}^{N} w_{ij}x_j > \vartheta_i \\ -1 & : \quad \text{otherwise.} \end{cases}$$

Here the neuron can be in only one of two states and transmit only one bit of information. More information can be transmitted if the neuron could be in several states, as is the case for continuous activation functions. The output of a continuous activation function is a real number. In biological terms, this number could be interpreted either as the firing rate of the neuron over a short time window or as the sum of all excitatory and inhibitory outputs of a population of neurons at a given instant. The simplest continuous function is the linear model

$$\Phi(a_i) = ka_i$$

where k is a constant. In undesirable situations where the output of the neuron could grow indefinitely (for example, if it has a positive feedback connection), this activation function can be constrained to operate within a given interval, such as $[0, 1]$ or $[-1, 1]$.

There are also several continuous and nonlinear activation functions that are used in complex neural architectures. One of the most common nonlinear SIGMOID FUNCTION functions is the sigmoid, or logistic, function

(3.3)
$$\Phi(a_i) = \frac{1}{1 + e^{-ka_i}}$$

where k is a scaling factor that determines the inclination of the slope shown in figure 3.11, c (for $k \to 0$ the function approximates a linear function; for $k \to \infty$ the function approximates a step function). The sigmoid function

Figure 3.12   Representation of relationship between weight and input vectors of a neuron.

tends asymptotically to 0 and 1. A similar function is $\tanh(kA)$, which tends asymptotically to -1 and 1.

The activation of a neuron is proportional to the familiarity, or similarity between its weight vector and the input vector (figure 3.12). Resorting to linear algebra, we can describe the similarity between the two vectors with the cosine of the angle α between the vectors:

$$\cos \alpha = \frac{\vec{w} \cdot \vec{x}}{\| \vec{w} \| \| \vec{x} \|}, 0 \le \alpha \le \pi$$

where $\| \vec{v} \|$ is the length of vector v. Consequently, we can express the output of the neuron as

$$y = \vec{w} \cdot \vec{x} = \| \vec{w} \| \| \vec{x} \| \cos \alpha.$$

This means that, provided that the lengths of the two vectors are constant, the output of the neuron changes with the cosine of the angle between the two vectors. Its magnitude is inversely proportional to the angular distance between the input and weight vector within the same quadrant. Therefore, the output will be the smallest when the two vectors are orthogonal ($\cos 90° = 0$) and the largest when the two vectors are aligned ($\cos 0° = 1$). In a neural network with several output units, the activation levels tell which neurons have synaptic weights closer to the input pattern, but only if the weight vectors are normalized (that is, have the same length). If the activation function is binary, the neuron output can be used to discriminate between two classes of inputs, those whose angular distance is smaller than 90 degrees and those whose angular distance is larger than 90 degrees to the weight vector of the neuron.

Another way to look at the classification abilities of a neuron is to imagine that the neuron traces a separation line in the input space between the

SEPARATION LINE

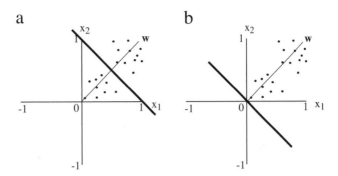

Figure 3.13 The separation line (thick line) of input space by a neuron with two input units, shown in figure 3.12, both set to 1, is always perpendicular to the neuron's weight vector (thin line). The example shows a distribution of data points that such a neuron could discriminate by responding differently depending on whether the input is above or below the separation line (see text). *a*, neuron with threshold set to 1; *b*, neuron with threshold set to 0.

classes of input patterns that produce different responses (figure 3.13). Imagine, for example, describing a sample of patients on a bidimensional graph according to their weight and cholesterol that corresponds to the activation values of two input neurons. Imagine now that these patients happen to cluster mostly into two small regions, one corresponding to high weight and cholesterol values and one corresponding to low weight and cholesterol values. If the neuron has a suitable set of weight values from the two input units, its output value (e.g., 0 and 1) will be different depending on which side of the separation line the input is. In other words, the neuron will classify patients into two classes of response according to this virtual separation line. The separation line becomes a plane if the neuron has three input units and a hyperplane for more input units. If the input patterns can be divided by such a line, plane, or hyperplane, we say that the classification problem

LINEARLY SEPARABLE   is *linearly separable*. This condition represents an important criterion for the choice of neural architecture, activation function, and learning rule, as we will see later.

For the moment, let's imagine that the neuron is presented with a linearly separable classification problem, such as the one shown in figure 3.13. The separation line is perpendicular to the weight vector of the neuron. For a

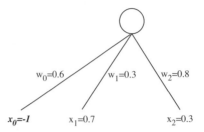

Figure 3.14   The threshold of a neuron can be represented as an additional weighted input x_0 with a constant negative value. This representation has the advantage that the threshold weight w_0, also known as bias, can be modified during learning as all other weights in the network.

neuron whose activation is given by $y = \sum_{j=1}^{N} w_j x_j - \theta$, the line passes through the points that describe the relationship

$$w_1 x_1 + w_2 x_2 - \theta = 0$$

and can be traced by reformulating the same equation as

(3.4) $$x_2 = \frac{\theta}{w_2} - \frac{w_1}{w_2} x_1.$$

This equation shows that if the threshold is not used ($\theta = 0$), the separation line must pass through the origin of the input space and therefore, no matter what values the synaptic weights take, the neuron may not discriminate classes of inputs that fall, for example, within one quadrant, as is the case in the second panel of figure 3.13, b). A nonzero threshold is also useful in cases where the output neuron is required to provide an output in the absence of input, as could be the case for a robot that must continue to move even without sensory input.

Considering the additive properties of the neuron activation (see equation (3.1)), the threshold is conveniently expressed as a synaptic weight connected to an additional input unit with constant activation set to -1 (figure 3.14). This additional synaptic weight is often called *bias* and the corresponding input unit is called *bias unit*. Within this convention, the neural activation can be expressed in the simpler form

BIAS

$$y = \Phi \left(\sum_{j=0}^{N} w_j x_j \right)$$

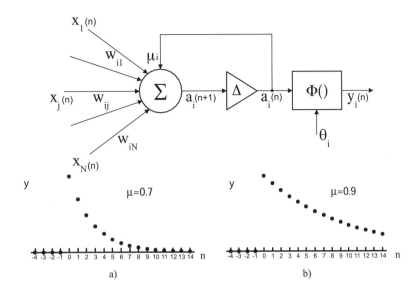

Figure 3.15   A simple model of a discrete-time, dynamic neuron consists in adding a delay element and a recurrent connection that brings the activation back to the input. If the weight of the recurrent connection is $0 \leq \mu \leq 1$, the evolution of the neural activation, when presented with an instantaneous input presented only at time $n = 0$, corresponds to a gradual decay.

where $x_0 = -1$ and w_0 can be modified during the learning process as any other synaptic connection in the network.

The neuron model described so far is static because its output is determined by its current input only. However, there are several types of problems, such as detection or generation of time sequences, where dynamical behavior is necessary. Dynamical behavior can be obtained by coupling static neurons with feedback connections. Another way of achieving time-dependent functionalities consists in using dynamic neuron models. The underlying principle of those models consists in introducing a temporal delay with a feedback loop between the input and output of the neuron.

DISCRETE-TIME DYNAMIC NEURON MODEL   Consider for example the discrete-time, dynamic neuron model shown on the top of figure 3.15, which incorporates two novelties with respect to the static model described in figure 3.10. The first element is a temporal delay Δ between the update from the current activation level $a_i(n)$ to the activation level at the next time step $a_i(n + 1)$. This implies that the neuron can hold its current activation level for some time Δ. The second element is a recur-

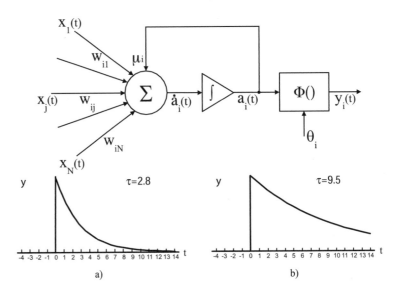

Figure 3.16   A continuous-time recurrent neuron where the delay integrates the derivative of the activation and transforms it into the activation itself. In this case, the μ_i takes the form of e^{-t/τ_i} and the evolution of the neural activation, when presented with an instantaneous input presented only at time $t = 0$, corresponds to a gradual decay when $\tau_i > 0$.

rent connection with a weight μ_i that brings the current activation back to the input used to compute the next activation level. For $0 \leq \mu \leq 1$, this corresponds to a spontaneous activation decay (figure 3.15, bottom). For each update step n, we compute first the neuron activation

$$(3.5) \quad a_i(n+1) = \mu_i a_i(n) + \sum_{j=1}^{N} w_{ij} x_j(n)$$

and then the neuron output signal

$$y_i(n) = \Phi\left(a_i(n) - \theta_i\right).$$

CONTINUOUS-TIME
RECURRENT NEURAL
NETWORK (CTRNN)

A widely used variation of this model includes continuous-time dynamics, and networks of such neurons are often called *continuous-time recurrent neural networks*, or CTRNNs (Beer and Gallagher 1992), as shown in the top of figure 3.16. In the continuous-time case, we are interested in the variation

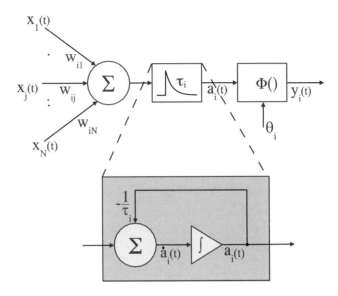

Figure 3.17 A compact representation of the neuron model of figure 3.16 where the exponential dynamics are described by a single element between the activation and the output.

of the activation over time and the activation update of equation (3.5) can be approximated by

$$\frac{da_i(t)}{dt} = -\frac{1}{\tau_i}a_i(t) + \sum_{j=1}^{N} w_{ij}x_j(t)$$

where τ_i is a time constant whose magnitude (for $\tau_i > 0$) is inversely proportional to the decay rate of the activation (figure 3.16, bottom). These types of models display rich dynamics and represent a first approximation of the time-dependent processes that occur at the membrane of biological neurons. LEAKY INTEGRATOR Sometimes these neural models are also called *leaky integrators*, with reference to electrical circuits, because the equation describing the neuron activation is equivalent to that describing the voltage difference of a capacitor, where the time constant of the exponential and synaptic weights can be approximated by a set of resistors (Haykin 2007).

Both the discrete-time and continuous-time neuron models described above generate exponential dynamics. This behavior can be described more simply by using a single exponential function between the activation and the

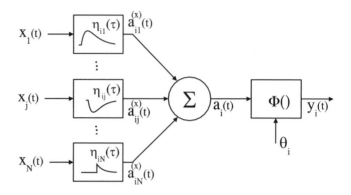

Figure 3.18 Generalization to a model where the exponential dynamics are anticipated to the input and can therefore be different for each input line.

output of the neuron, as shown in figure 3.17. The dynamics of the neuron are described uniquely by the parameter τ_i (μ_i for the discrete-time model).

To add even more flexibility to the model, we can imagine anticipating the temporal dynamics to the input level so that each input line can not only have its own different parameter value but also have different temporal dynamics that are not necessarily exponential, as shown in figure 3.18. This generalized model approximates even better the biological case where synaptic contacts on the dendritic tree of the neuron can have different temporal dynamics and effects on the neuron activation. In this case the neuron activation is the integral (or sum in the discrete-time case) of the contributions of all input lines. Each contribution is given by the temporal convolution of the input $x_i(t)$ with a function $\eta_{ij}(t) = w_{ij}\epsilon(t)$, also known as the synaptic kernel (core unit of computation). For example, a commonly used exponential function for $\epsilon(t)$ (Gerstner 1999; Floreano and Mattiussi 2001) is given by

$$\epsilon(t) = \begin{cases} e^{-\frac{t-\tau_d}{\tau_m}} & : \quad t \geq \tau_d \\ 0 & : \quad \text{otherwise} \end{cases}$$

where τ_m is the temporal constant described above and τ_d is the temporal delay necessary for the signal to travel from the presynaptic to the postsynaptic neuron. This generalized model is not only capable of producing a large class of time-dependent responses (Haykin 2007) but is also amenable to further modeling of synaptic dynamics. For example, recent work in neurophysiology and modeling emphasizes that synaptic kernels change depending on the pre- and postsynaptic activity. Those activity-dependent modifications could be easily incorporated in the generalized model described above.

In the neuron models described so far, the temporal dynamics contribute only to the definition of the neuron activation. The actual output of the neuron is a function (usually the sigmoid function) of its activation. Therefore, the information exchanged between neurons is basically the instantaneous value of the neuron activation. However, as we have seen in the previous section, most biological neurons communicate through action potentials, or spikes, which are punctual events that result from a process taking place at the output of the neuron. One of the reasons why biological neurons communicate spikes instead of activation states is that the former are large perturbations that can travel along big distances in the brain, whereas the latter are small voltage differences that degrade rapidly along the axons and consequently may alter the information transmitted. As a matter of fact, neurons that are densely packed and have short-range local connections, as is the case in the retina, transmit activation states instead of spikes.

SPIKING NEURONS   In spiking neurons the activation state, which corresponds to an analog value, can be approximated by the firing rate of the neuron. That is, a larger number of spikes within a given time window would be an indicator of higher activation of the neuron. However, if that is the case, it means that spiking neurons require a relatively longer time to communicate information to postsynaptic neurons. This hints at the fact that spiking neurons may use other ways to efficiently encode information, such as the firing time of single spikes or the temporal coincidence of spikes coming from multiple sources (Singer and Gray 1995; Rieke et al. 1997). It may therefore be advantageous for engineering purposes to use models of spiking neurons that exploit firing time in order to encode the spatial and temporal structure of the input patterns with fewer computational resources.

There are several models of spiking neurons that describe in detail the electrochemical processes that produce spiking events by means of differential equations (e.g., Hodgkin and Huxley 1952). A simple way of implementing a spiking neuron (figure 3.19, top) is to take the dynamic model described in figure 3.17 and substitute the output function with an element that compares the neuron activation with its threshold followed by a pulse generator that takes the form of a Dirac function. In other words, if the neuron activation is larger than the threshold, a spike is emitted. In order to prevent continuous spike emission, we must also add a strong negative feedback r_i so that the neuron activation goes below threshold immediately after spike emission.

INTEGRATE AND FIRE   This model is often known as an *integrate and fire* neuron.
NEURON

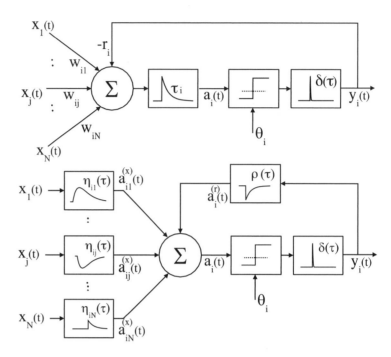

Figure 3.19 *Top*: A simple model of an integrate and fire spiking neuron is obtained by substituting the output function with a comparator followed by a pulse generator and a strong negative feedback r_i that brings the activation below threshold, thus preventing another immediate spike (see also figure 3.17). *Bottom*: Generalization to a model where the exponential dynamics are anticipated to the input (therefore can be different for each input line) and the inhibitory feedback is also described by an exponential dynamical process (see also figure 3.18). This model is also known as the spike response model (Gerstner 1999).

A more sophisticated model of a spiking neuron (figure 3.19, bottom) would anticipate, and possibly differentiate, to the input lines the temporal dynamics, as in the generalized model described in figure 3.18. Similarly, the negative feedback necessary to prevent a continuous series of spikes can be approximated by an exponential kernel. The exponential kernel is such that whenever a spike is emitted the neuron activation is brought to a very negative value from where, in the absence of incoming spikes, it exponentially SPIKE RESPONSE goes back to a resting potential. This model is known as the *spike response* MODEL *model* (Gerstner 1999).

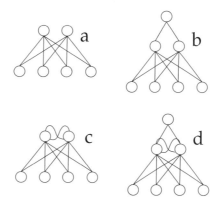

Figure 3.20   Frequently used architectures (lower layer represents input units). *a*, feedforward perceptron; *b*, feedforward, multilayer perceptron; *c*, recurrent connection among output units; *d*, recurrent connections among hidden units.

3.4   Architecture

The architecture of a neural network affects the functionality and type of learning algorithms that can be used. The simplest architectural type consists of *feedforward* networks where the output of each neuron depends uniquely on the output of neurons in lower layers. Networks with a single layer of synaptic connections (figure 3.20, a) are sometimes called *perceptrons*, from the name given by the first person who studied their computational properties as pattern classifiers (Rosenblatt 1962). These networks are used to perform classifications of linearly separable input patterns or to extract statistically significant information from the distribution of input vectors.

FEEDFORWARD

More complex classifications, where the input patterns fall into classes that are not linearly separable, can be achieved with multilayer networks (figure 3.20, b). Feedforward networks are input-output devices that cannot detect or produce temporal sequences, unless they are composed of dynamic neurons.

One way of providing temporal dynamics to neural networks of static neurons consists in adding recurrent connections from neurons in the same layer or from neurons in upper layers. In this case, the output of a neuron with recurrent connections is given by

RECURRENT
CONNECTIONS

(3.6) $$y_i^t = \Phi \left(\sum_j^N w_{ij} x_j^t + \sum_l^M r_{il} q_l^{t-1} - \vartheta_i \right)$$

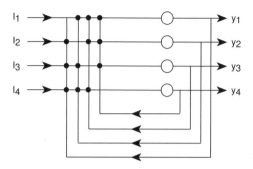

Figure 3.21   An auto-associative network. Dots indicate that lines are connected.

where q_l^{t-1} are the outputs of neurons in the same layer (possibly including the neuron i) at the previous time step and r_{il} are the synaptic connections. In two commonly used architectures, recurrent connections occur at the output layer (figure 3.20, c) or at the hidden layer (figure 3.20, d). The latter type of architectures is also called Elman networks, from the name of the person who studied their computational properties (Elman 1990) and can display more complex time-dependent behaviors. Even more complex dynamics can be achieved by using recurrent connections in networks of dynamical neurons (such as CTRNNs) or of spiking neurons described earlier.

AUTOASSOCIATIVE
NETWORKS
Autoassociative networks are used to memorize and reconstruct patterns and typically consist of a single layer of fully connected neurons (figure 3.21). An input pattern is presented to the input lines of the network, and its synaptic weights are updated so that when a corrupted or incomplete version of that pattern is presented again, the neurons can reconstruct the memorized version. In this case, the neuron outputs are computed iteratively until they stabilize according to

$$y_i^t = \Phi \left(I_i + \sum_{j \neq i}^{N} w_{ij} x_j \right)$$

where I_i are the values of the pattern presented to the network. These networks usually do not have a threshold and do not have self-connections.

ECHO STATE
NETWORKS
Echo state networks (Jaeger and Haas 2004) are neural networks with a large number (typically between 50 and 1000) of randomly interconnected hidden neurons, a layer of input connections, and a bidirectional layer of output connections (figure 3.22). Hidden neurons are wired with a low probability

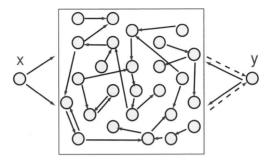

Figure 3.22 Echo state network. Solid connections are randomly prewired and fixed; dashed connections can be modified by learning.

(typically, 0.01) so as to create a reservoir of many loosely coupled subnetworks and corresponding dynamical behaviors that can be excited (as echo functions) by the input and output feedback connections. In echo state networks only the connections from the reservoir of hidden units to the output units are modified so that the output of the network approximates a desired input-output function (this can be done using a supervised learning algorithm that will be described below). All other connections are prewired and cannot be modified.

LIQUID STATE MACHINES   A simpler version of echo state networks, also known as *liquid state machines* (Maass et al. 2002), does not include feedback connections from output units to hidden units. The idea here is that connections from hidden to output units operate as readout functions of the network dynamics that have been triggered by the input, just like the perturbations of a liquid that have been triggered by an external event (a stone, wind, or other). Liquid state machines can have several output units that detect multiple perturbations of the network in parallel. However, the output units cannot affect the dynamics of the hidden units as in echo state networks.

3.5   Signal Encoding

The encoding of input and output patterns is largely determined by the problem to be solved, but there are some criteria to be considered. Let us consider, for example, the case of a neural network that must classify a set of objects. The objects can be represented using *local encoding*, where each neuron en-

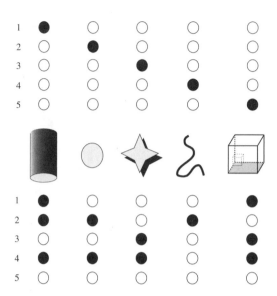

Figure 3.23 Local and distributed encoding with five input units of five objects shown in the central row of the image. Each column shows the activation of the input units for the corresponding object with local (*top*) and distributed (*bottom*) encoding. Filled disks indicate active neurons. It should be noted that distributed encoding does not need all five neurons to represent the five objects.

codes one particular object; or using *distributed encoding*, where several units participate in the representation of each object.

LOCAL ENCODING   Local encoding (top part of figure 3.23), although very simple, has some drawbacks. It demands a number of units equal to the number of objects to be represented and consequently requires the knowledge of the number of objects to be classified. Also, it is fragile in the sense that noise or malfunctioning of a neuron results in the loss of the corresponding object.

DISTRIBUTED
ENCODING   With distributed encoding (bottom part of figure 3.23) neurons encode the presence of a certain feature, such as depth, roundness, corners, edges. Whenever an object is presented to the network, the neurons tuned to some features of the object become active. In distributed encoding objects are characterized by a unique combination of active neurons. Notice that a neuron can participate in the description of several objects. Distributed encoding can describe many more objects than available units and is more resistant to noise. Furthermore, similar objects will activate a similar set of units, making the classification task easier for postsynaptic neurons.

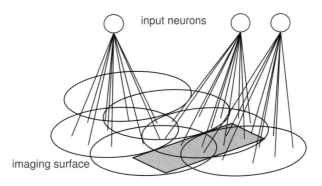

Figure 3.24 Partially overlapping receptive fields encourage a distributed representation of the information.

One way of enforcing distributed encoding throughout multiple layers of a network consists in using restricted and overlapping connections. In this case, neurons are arranged in space (for example, on a line or on a surface) and each neuron receives connections from a set of neighboring presynaptic neurons partly shared with other neighboring postsynaptic neurons. The

RECEPTIVE FIELD   set of incoming connections is also known as the *receptive field* of the neuron. Consider, for example, a layer of neurons that receive information from an imaging device (figure 3.24). If the receptive fields have a circular shape and are uniformly distributed, the level of detail that the neurons can encode is proportional to the number and radius of the receptive fields that span the imaging device (Rumelhart et al. 1986a). Jacobs and Kosslyn (1994) studied the influence of receptive field size and weighting functions on the type of operations that the network can perform on the image. They showed that networks with relatively small and nonoverlapping receptive fields learn more easily to classify images and to detect relationships (e.g., above and below) whereas networks with larger and partially overlapping receptive fields learn more easily specific features of the images and precise metrics (e.g., red triangle). These data, where corroborated by experimental results, showing that when the radius of receptive fields can be modified by the learning process, networks trained to classify images develop smaller receptive fields than networks trained to detect the presence of specific images.

Feedforward neural networks have been used to model the detection of symmetry and of other spatial relationships (Kosslyn et al. 1992; Enquist and Arak 1994), but it is questionable whether such networks do actually perceive spatial relations. It has been argued that feedforward networks simply learn to classify data sets, but cannot perceive spatial relations because neurons on the same layer are not interconnected and because the spatial arrangement of incoming inputs to a neuron does not modify the activation of that neuron, which is a sum of weighted inputs (Cook 1995; Cook et al. 1995). In other words, changing the order of input units, which would break the spatial relation, would not affect the activation of the output unit. However, this criticism does not hold if neurons on the same layer are interconnected.

A frequent problem with input data is that the magnitudes of the values may not be homogeneous. For example, sensors of a robot may return values that operate on different scales or display high peaks. These differences do not necessarily reflect significant differences in the environment, but may affect the network response by saturating the activation of a neuron or by masking the activity of other sensors. A frequently used strategy consists in NORMALIZATION normalizing the input vectors so that the length, or norm, of the vector is always equal to 1. For each input vector, this is done as follows:

$$x_i' = \frac{x_i}{\sqrt{\sum_{j=1}^{N} x_j^2}}$$

This operation also facilitates the functionality of the network because, as we have seen earlier (figure 3.12), the output of a neuron is proportional to the cosine of the angle between its weight and input vectors when both are normalized.

In a spiking neural network, a single spike is a binary event that can en- ENCODING WITH code only the *presence* or absence of a stimulus. Figure 3.25 shows three ways SPIKES to map the *intensity* of sensory information onto spiking neurons. A classic method (figure 3.25, a)) consists of mapping the stimulus intensity to the firing rate of the neuron. This method is based on the hypothesis that a neuron increases its firing rate to indicate stronger stimulation. For example, Adrian (1928) experimentally showed that the firing rate of muscle stretch receptors of frogs approximates a monotonically increasing function of the strength of the stimulation and saturates near the maximum firing rate of the neuron. More detailed studies indicate that this function is best described by a power function of the general form $R = KS^n + C$ where R is the observed firing rate, K is a constant of proportionality, S^n is the strength of the signal elevated to the power of n, and C is a constant given by the spontaneous firing rate of a neuron in the absence of stimulation (Mountcastle et al. 1963). Another method (figure 3.25, b)) consists in encoding the sensory stimulation across several neurons and mapping the intensity of the stimulation into the number of neurons that spike at the same time. This method is based on

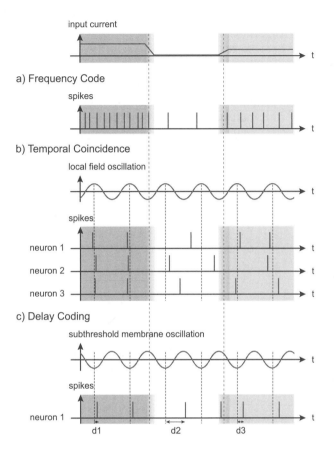

Figure 3.25   Some models for encoding sensory information in spiking neurons. Fictive spike trains recorded from five imaginary neurons. The different stimulus intensities (represented by the gray scale) are converted to different spike sequences. *a*) In the frequency code hypothesis (Sherrington 1906), neurons generate a different frequency of spike trains as a response to different stimulus intensities. *b*) In the temporal coincidence hypothesis (Singer and Gray 1995), spike occurrences are modulated by local field oscillation (gamma). Tighter coincidence of spikes recorded from different neurons represent higher stimulus intensity. *c*) In the delay coding hypothesis (Hopfield 1995), the input current is converted to spike delay. Neuron 1, which was stimulated stronger, reached the threshold earlier and initiated a spike sooner than neurons stimulated less. Different delays of the spikes (d1-d3) represent relative intensities of the different stimulus.

the hypothesis that the brain represents complex information by means of synchronized spiking activity across several neurons (Singer 1990). This hypothesis has been supported by measurements in the visual and temporal cortex of monkeys (Abeles 1991; Singer and Gray 1995). A recently suggested method (figure 3.25, c)) consists in encoding the strength of the stimulation in the firing delay of the neuron with respect to a baseline signal (for example, the so-called oscillatory theta rhythm). The underlying hypothesis is that neurons that receive stronger stimulation fire earlier than neurons receiving weaker stimulation. This hypothesis has been supported by measurements in the olfactory neurons (Hopfield 1995).

3.6   Synaptic Plasticity

As we stated in the introduction to this chapter, adaptation is a major feature of nervous systems. It allows organisms to modify and develop behaviors in order to maintain or improve their survival probability in partly unknown and dynamic environments.

UNSUPERVISED LEARNING

Artificial neural networks can adapt according to several algorithms that can be classified in two major families. *Unsupervised learning* includes algorithms that allow the network to extract statistically significant information from the distribution of input patterns or to memorize and reconstruct those

SUPERVISED LEARNING

input patterns. *Supervised learning* includes algorithms that guide the training process of the network by taking into account the desired answer that the network should provide for a given set of training patterns. Supervised learning also encompasses reinforcement learning algorithms that modify the network according to positive or negative feedback received from the environment at irregular intervals.

In the following sections we will describe the major features of unsupervised and supervised learning. We will also describe ways to combine learning and evolution. Before delving into the details of the algorithms, it is worth noting some common features of synaptic plasticity.

HEBB'S RULE

Most learning algorithms are based on, or inspired by, Hebb's rule of synaptic plasticity described earlier, which, in the case of neurons without temporal dynamics, can be formalized as

(3.7) $\Delta w_{ij} = \eta y_i x_j$

where Δw_{ij} is the amount of change of the connection strength w_{ij}, η is the learning rate, and y_i, x_j are the activation values of postsynaptic and presynaptic units, respectively. Therefore, synaptic modification is based on the correlation of the unit activations.

In the case of spiking neurons, the Hebb rule becomes a function of the temporal difference between the reception of a spike from the presynaptic neuron and the emission of a spike by the postsynaptic neuron. Therefore, Hebbian learning in spiking neural networks is also known as *spike time-dependent plasticity*, or STDP for short. An example of the STDP rule (Gerstner and Kistler 2002) is given by

SPIKE
TIME-DEPENDENT
PLASTICITY (STDP)

$$\Delta w_{ij} = \begin{cases} A_+ e^{s/\tau_1} & : \quad s \leq 0 \\ A_- e^{-s/\tau_2} & : \quad s \geq 0 \end{cases}$$

where $s = (t^{pre} - t^{post})$ is the time difference between arrival of the presynaptic spike and emission of the postsynaptic spike, and A_+, A_-, τ_1, τ_2 are constants tuned to approximate the neurophysiological data (Bi and Poo 2001; Zhang et al. 1998) shown in figure 3.8. It has been shown that the integral of the area corresponding to depression of the synapse should be larger than that corresponding to potentiation in order to obtain stable learning (Song and Abbott 2000).

In this book we won't explore in further detail STDP rules, but we refer interested readers to the excellent review article by Turrigiano and Nelson (2004) that lists a number of mechanisms that ensure dynamic stability in networks of biological and artificial spiking neurons with STDP. In the rest of this chapter, we will focus on variations of correlation-based Hebbian learning rules, which up to now have had most success in engineering applications.

All learning algorithms display some common features. Initial synaptic weights are often randomly assigned within a small range (e.g., $[-0.1, 0.1]$). Learning consists in the repeated presentation of a set of input patterns, also known as training patterns. In unsupervised learning, training patterns are the input vectors presented to the network. In supervised learning, training patterns are pairs of vectors that represent the input and the desired output of the network for that input. The modification of the synaptic weights Δw_{ij} is computed after each presentation of a training pattern and corresponding network activation (online learning) or after the presentation of all training patterns and network activations (offline learning). The new synaptic weights are computed by adding the modification Δw_{ij} to the current synaptic weight:

$$w_{ij}^{t+1} = w_{ij}^t + \Delta w_{ij}^t$$

Learning rules therefore are concerned only with the computation of Δw_{ij}. Learning in neural networks consists in the addition of new knowledge Δw_{ij} to preexisting knowledge w_{ij}. In order to reduce interference between previous and new knowledge, learning takes place incrementally by presenting the same training patterns several times and adding to the old weights only a small fraction of the modification computed by the learning rule. The magnitude of this fraction is a value $0 \geq \eta \leq 1$ that controls the learning rate of the network. The learning rate η is used to ensure the stability of the learning process.

TRAINING PHASE

TESTING PHASE

In most learning algorithms, there is a distinction between training and testing phase. The training phase is stopped when a certain learning criterion is reached. The criterion could be a minimum error between the desired response and the response of the network, or the stabilization of the synaptic weights. The testing phase consists in the presentation of new input patterns and readout of the network output. With the exception of a few models that do not require a distinction between training and testing, this distinction is problematic when the environment where the network operates changes. In those situations it is not always clear when to resume learning and how to protect the knowledge already developed by the network from potential disruptions due to learning of new patterns.

3.7   Unsupervised Learning

In unsupervised learning the neural network learns some properties of the input pattern distribution without any feedback from the environment or from the user. Learning typically consists in the extraction of information, such as the detection of common or distinctive features that allow the network to classify the input patterns. From a mathematical perspective, unsupervised learning performs statistical operations such as computation of correlation indices, estimation of parameters of the probability density function of the input patterns, and principal component analysis, to mention a few. In order to carry out those operations, the input pattern distribution must be redundant so as to allow the detection of structure (Barlow 1989). The models described in this section derive from studies of the self-organization of the visual or auditory cortex whose plasticity seems to be mainly driven by the input stimulation.

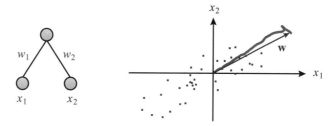

Figure 3.26   Extracting the first principal component of the input pattern distribution. The network on the left was trained with the Oja learning rule for 1000 cycles on input patterns randomly drawn from a uniform distribution of 40 patterns centered on 0. The learning rate was set to 0.1. The graph on the right shows the input patterns (dots) and the development of the synaptic weight vector during learning. The direction and length of the final weight vector is indicated by the arrow. It is located along the direction of maximum variance of the input pattern distribution.

3.7.1   Feature Detection

Let us consider, for example, a feedforward network with a single output unit whose activation is a linear function of the weighted input (figure 3.26, left) and whose synaptic weights are modified by the Hebb learning rule (equation (3.7)). A learning cycle consists in presenting an input pattern, computing the activation of the output unit, computing the modification of the synaptic weights, and adding the obtained modification to the initial weight values. This cycle is repeated for each pattern and all patterns are presented in random order several times.

Synaptic weights corresponding to the input units that have been active more frequently will become stronger than other synaptic weights. Consequently (see also figure 3.12), the output neuron will become more active when presented with input patterns that include the most frequent, or familiar, components. Since those components generate a stronger activation of the output unit, the Hebb rule will further reinforce the corresponding weights, causing an even stronger output and weight modification when those components are presented again.

A simple way to prevent this self-amplification process, which would result in synaptic weights and outputs of potentially unbounded value, consists in normalizing the synaptic weights after every learning update. In doing so, synaptic weights tend to grow, but are constrained to remain within bounds that depend on the sum of all synaptic values. This creates a sort of competition mechanism that results in the convergence and stabilization of the weight vector along the direction of maximal variance of the input distribution.

It can be mathematically shown that the synaptic weight vector moves toward the eigenvector corresponding to the largest eigenvalue of the correlation matrix of the input patterns, also known as the *principal component* of the input distribution (Diamantaras and Kung 1996). As a consequence, the output unit will tend to display stronger activation for the input patterns that lay along the direction of maximum variance of the input distribution (figure 3.26, right).

The normalization process suggested above is not biologically plausible because the update of a synaptic weight requires the global knowledge of

OJA RULE   all other weights in the network. Oja (1982) suggested a small modification to the Hebb learning rule where the synaptic weight vector tends to norm 1 without need of global normalization after every synaptic update. This is achieved by taking into account the current strength of the synaptic weight being updated:

$$\Delta w_j = \eta y (x_j - w_j y)$$

where the term $w_j y$ acts as a forgetting factor that limits the growth of the synapse as a function of its current value. A functionally similar mechanism seems to operate also in biological synapses (Stent 1973).

If the input patterns are drawn from a distribution centered around 0, the synaptic weight vector will align along the direction of maximal variance of the input distribution (figure 3.26, right) and correspond to the first principal component of the input pattern distribution. Recalling what we said earlier about the properties of an artificial neuron, such a simple network could discriminate between a familiar input pattern (one that belongs to the training distribution and thus forms a very small angle with the weight vector) and a new pattern (that belongs to a different distribution and may thus lie farther away from the weight vector).

Later, Oja (1989) suggested an extension of his learning rule that can be applied to N output units (figure 3.27, a) lying on the same layer:

$$\Delta w_{ij} = \eta y_i \left(x_j - \sum_{k=1}^{N} w_{kj} y_k \right)$$

where i, k are indices that both point to the output units. In this rule, the forgetting factor $\sum_{k=1}^{N} w_{kj} y_k$ takes into account all weights of the network,

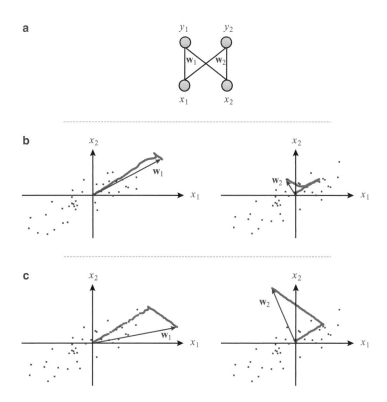

Figure 3.27 Extracting the first two principal components of the input pattern distribution with two output units. *a*, Neural network. *b*, Development of synaptic weight vectors of the network shown on top when trained with the Oja rule for N output units. *c*, Development of synaptic weight vectors of the network shown on top when trained with the Sanger rule.

thus making the rule less local. The synaptic weight vectors of this model tend to converge to the subspace spanned by the eigenvectors corresponding to the first N eigenvalues of the correlation matrix of the input distribution (figure 3.27, b).

SANGER RULE   Sanger (1989) proposed a further modification to the learning rule so that the synaptic vectors tend precisely toward the eigenvectors of the first N eigenvalues, thus extracting the first N principal components of the input distribution (figure 3.27, c):

$$\Delta w_{ij} = \eta y_i \left(x_j - \sum_{k=1}^{i} w_{kj} y_k \right)$$

This is achieved by incorporating into the forgetting factor $\sum_{k=1}^{i} w_{kj} y_k$ only the weights of the output units k up to the output unit i being currently considered. In this model, the weight vector of the first output unit develops first and aligns along the direction of maximal variance (first principal component), then the weight vector of the second unit aligns along the direction of maximal residual variance (second principal component), and so forth for any remaining output unit.

However, Sanger's rule introduces small and increasing distortions for further output units because, although all synaptic weights are always updated in parallel and continuously, the weights of further output units can develop properly only after the weights of previous output units have stabilized. Therefore, the model is efficient only when used with few output units with very large input dimensionality.

An interesting property of principal component extraction with a neural network is that input patterns can be reconstructed optimally with respect to the mean square error between the original and reconstructed input patterns. For example, once an Oja's or Sanger's network has been trained on a series of images, the pixels of each image can be reconstructed by summing the products between the final weight values and the output unit values for that particular image:

$$x_j^{\mu} = \sum_{i}^{N} w_{ij} y_i^{\mu}$$

where x_j^{μ} is the value of the pixel to be reconstructed and y_i^{μ} is the value of output unit i for image μ. Consequently, one may save quite a lot of memory by storing only the final synaptic weights and the activations of the output units for each image instead of all pixels of all images. The fidelity of the reconstructed image is proportional to the number of output units used. When only few output units are used, the reconstructed images will resemble the average of the images.

An amusing exercise is to train Sanger's network on photographs of faces of friends, making sure that each face occupies approximately the same area of the image. It is also advised to use the logarithm of the pixel intensities in order to allow the network to work on contrast rather than on light intensity and to subtract from each value the average value of all pixels of all images in order to center the distribution around zero. It is then interesting to observe how many output units are necessary for recognition of the friends in the reconstructed images.

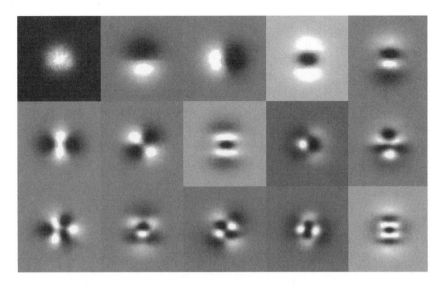

Figure 3.28   Receptive fields of the first 15 neurons of a Sanger network trained on natural images. Each box plots the strength and sign (negative = white, positive = black) of all synaptic weights to a corresponding neuron arranged as a square. The neuron will fire maximally when presented with an input pattern that is similar to its receptive field. Image courtesy of P. Hancock.

The family of unsupervised neural networks described above can be used as a preprocessing layer to reduce the dimensionality of the input patterns and provide distinct output activations to be used as input to other types of neural networks, such as the supervised models that we will describe in the next section.

Considering that these neural networks use biologically plausible learning rules and architectures, one may wonder whether mammalian brains perform something similar to principal component analysis in order to efficiently store information necessary for recognition and action. Hancock and colleagues (Baddeley and Hancock 1991; Hancock et al. 1992) trained a Sanger network on a set of natural images drawn from a collection of typical visual scenes (nature, animals, houses, etc.) and showed that the neurons developed receptive fields qualitatively similar to those found in the mammalian visual cortex (figure 3.28). Neurons corresponding to the first few principal components became sensitive to center-surround patterns and to oriented edges, which in biological visual systems are found in the lateral geniculate nucleus and in the early layers of the visual cortex (Hubel and

Wiesel 1963). Neurons corresponding to less important principal components developed receptive fields that resembled those of complex cells found in later stages of the visual cortex.

However, this explanation of cortical computation has been criticized by Field (1994), who argues that visual neurons employ a sparse and distributed activation code where each neuron has approximately the same probability of becoming active for all visual inputs, but very low probability of becoming active for a specific visual input. Within this perspective, input redundancy is not reduced (as implicitly claimed by supporters of the principal component principle), but transformed into redundancy of output unit activations.

Whatever the actual case is, it should be noted that principal component networks do not develop neurons sensitive to different spatial frequencies, while there is psychophysical evidence that mammalian visual brains do extract information about spatial frequency at several stages of sensory information processing (Watt 1991).

Another limitation is that linear methods for extraction of principal components cannot separate signal sources when the input activations are generated by a combination of independent signals. The *cocktail party* case is an example of this situation where acoustic signals arriving to the ears are a combination of independent sources, such as voices of different persons and music. Independent component analysis has been recently developed for dealing with those situations (Hyvärinen et al. 2001).

3.7.2   Multilayered Feature Detection

The models described in the previous paragraphs are characterized by a single layer of neurons that process information from sensory units. However, in biological nervous systems information is processed through a series of neural layers and cortical modules. For example, in the visual system sensory stimulation is successively relayed from the retina to the lateral geniculate nucleus and superior colliculus; and from there to cortical area V1 and further on to higher cortical areas. Neurophysiological evidence suggests that simple static properties of the visual world, such as contrast and edge orientation, are detected in the early stages of neural processing, whereas more complex properties, such as composite or 3D shapes, are detected in later stages. Other aspects of the visual world, such as color and movement, are analyzed in parallel to geometric features through independent neural pathways (DeYoe and van Essen 1988). Neurons responding to specific features, such as the orientation of inclined edges, are already found in the brain

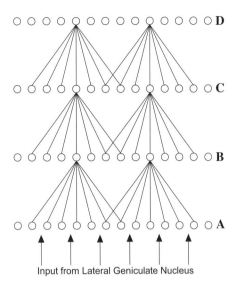

Figure 3.29 Simplified architecture of the multilayer model by Linsker. Each layer is arranged as a 2D matrix of neurons whose connection probability with neurons of the previous layer is described by a 2D Gaussian distribution.

at birth, suggesting that they may develop in the womb (another possibility is that they are genetically specified, but this is unlikely for complex mammalian brains because that implies that the genetic code specifies the exact strength and location of individual synapses of all neurons involved in that computation).

Linsker (1986, 1988) showed that restricted connectivity and Hebbian plasticity are sufficient to explain the formation of those receptive fields even in the presence of random input. The neural network is composed of four layers of neurons that receive connections only from a small set of neighboring neurons from the previous layer (figure 3.29). The output of all neurons is characterized by the linear activation function

$$y_i = k_1 \sum_j w_{ij} x_j$$

and the learning rule is given by

$$\Delta w_{ij} = k_2 y_i x_j + k_3 x_j + k_4 y_i + k_5$$

where k_{1-5} are constants and only $k_2 > 0$. Synaptic weights are clipped to $[w_{ij}^+, w_{ij}^-]$ in order to avoid infinite growth. Linsker analyzed the behavior of this network when the input is completely random, simulating the developmental conditions that may occur when the baby is still in the womb. Initially, only the synapses in the first layer are adapted until they reach stability; then the synapses of the layer above are adapted until stability, and so forth until the last layer of synapses is stabilized.

After training, neurons become sensitive to features of increasing complexity as we proceed from lower to higher layers. Neurons in layer B develop excitatory weights and thus their activation reflects the average stimulus intensity in their receptive fields. Neurons in layer C respond maximally to a strong positive signal surrounded by negative signals (similar to on-off biological cells) and to a strong negative signal surrounded by positive signals (off-on cells). Neurons in layer D respond maximally when the network is presented with bars at specific inclinations.

Although the similarity between these response profiles and those found in the mammalian visual cortex (Hubel and Wiesel 1977) is striking, the model is not intended to closely emulate the neurophysiology of visual cortical areas; rather, it is intended to show that simple architectural constraints coupled with Hebbian learning are sufficient to explain major organizational structures of the cortex. Linsker also showed that, from a mathematical perspective, the development of the synaptic weights in his model correspond to the maximization of the variance of the outputs of the neurons.

3.7.3 Self-Organizing Maps

In addition to a layered architecture that captures increasingly more complex properties of the sensory stimulation, the mammalian cortex displays two types of topological organization. The first type, which has already been mentioned in the first section of this chapter, is such that neighboring neurons reflect the activity of neighboring neurons in earlier layers and sensory receptors. This type of topological organization can be explained by the restricted connectivity between layers as in Linsker's model. The second type of topological organization is such that neighboring neurons respond to similar properties of the stimulation, such as bar orientation or acoustic tone. This type of topological organization can be explained by the presence of lateral connections between neurons on the same layer.

A cortical neuron establishes three types of connections to neighboring neurons: (a) connections within a radius of approximately 50 to 100 μm are excitatory; (b) connections within an outer ring that extends up to 200 to 500 μm are inhibitory; (c) connections that extend even farther away up to a few

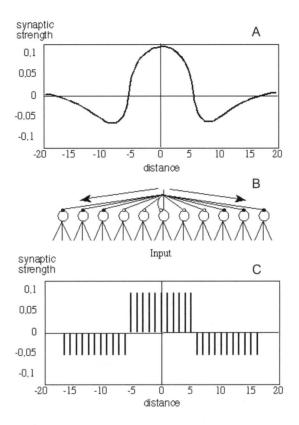

Figure 3.30 Lateral connection strength as a function of distance from projecting neuron. *A*, Mexican hat function; *B*, neural architecture: white circles = excitatory connections, black circles = inhibitory connections; *C*, stepwise approximation of connection strengths.

centimeters are weakly excitatory. When observed along one dimension, this configuration takes the form of a Mexican hat that can be approximated by a stepwise distribution, if we ignore the weak long-range connections (figure 3.30).

Kohonen (1982) showed that a layer of neurons characterized by this pattern of lateral connections will display "bubbles" of activity organized around the unit with the strongest input signal. In other words, the units in the neighborhood of the unit with the initially strongest activation will gradually become more active and those farther away less active. Provided that the output of the units is limited to a maximum value, the final stable state after some iterations of activity computation will consist of a region of units with maximum activation while all the other units will be inactive. The center of the bubble corresponds to the unit with the initial strongest activation while the size of the bubble (neighborhood) depends on the relative strength of excitatory and inhibitory connections.

Following Kohonen (1989), let's now see how this property coupled with Hebbian learning can develop topological maps of the sensory space, like those found in the cortex of mammals. This computational model is also known as *self-organizing maps*. Consider a set of neurons that occupy a specific position in a geometric arrangement (for example, in a line) and are all connected to a set of input units. Let's start by finding the neuron i^* with the highest activation for a given input pattern as the neuron with the smallest square difference between its input and weight vectors (see explanation of neuron operation in section 3.3),

$$(3.8) \quad i^* = i \left| \min_i \left(\sum_j (x_j - w_{ij})^2 \right) \right. ,$$

and let's call i^* the winning neuron. Instead of using lateral connections and iteratively computing the output of all neurons until stability, let's capitalize on Kohonen's observation that such a network will generate a bubble of activity in the neighborhood of winning neuron i^* and assume that neurons have a binary output

$$(3.9) \quad y_i = \Phi(A_i) = \begin{cases} 1 & : \quad \text{within neighborhood of} \quad i^* \\ 0 & : \quad \text{otherwise.} \end{cases}$$

In order to limit the unbounded growth of synaptic weights mentioned in section 3.7.1, let's modify the Hebb rule by adding a forgetting factor so that

$$(3.10) \quad \Delta w_{ij} = \eta y_i x_j - \Psi(y_i) w_{ij}$$

where

$$(3.11) \quad \Psi(y_i) = \begin{cases} \psi & : \quad y_i = 1 \\ 0 & : \quad y_i = 0 \end{cases}$$

where ψ is a small constant larger than 0. As for Oja's rule, the forgetting factor limits the uncontrolled growth of synaptic weights and ensures that the weight vectors have length 1. By combining formulas (3.8) through (3.11), you will notice that now the synaptic modification is 0 when the postsynaptic neuron is not active:

$$\Delta w_{ij} = \begin{cases} \eta x_j - \psi w_{ij} & : \quad y_i = 1 \\ 0 & : \quad y_i = 0 \end{cases}$$

which means that only the synaptic weights of the neurons in the neighborhood of the winning unit are modified. We can simplify even further the learning rule by assuming that $\psi = \eta$ and rewrite it as

$$(3.12) \qquad \Delta w_{ij} = \begin{cases} \eta(x_j - w_{ij}) & : \quad y_i = 1 \\ 0 & : \quad y_i = 0. \end{cases}$$

STANDARD
COMPETITIVE
LEARNING

This learning rule is also known as *standard competitive learning* to indicate the application of the Hebb rule to a set of neurons that compete for activation by means of lateral connections. If we write the final weight strengths in vectorial notation,

$$\vec{w}_i^{t+1} = \begin{cases} \vec{w}_i^t + \eta(\vec{x} - \vec{w}_i^t) & : \quad y_i = 1 \\ \vec{w}_i^t & : \quad y_i = 0, \end{cases}$$

we notice that competitive learning consists in adding to the weights of a winning unit (and of its active neighbors) the difference between its input vector and its weight vector. In other words, competitive learning moves the weight vector toward the input vector that caused the unit to have the highest activation.

The first consequence of this mechanism is that that unit will be even more active if the same, or a similar, input pattern will be presented to the network. The second consequence is that neighboring units will tend to become active for similar input vectors because the neighborhood relation will cause them to modify their weights in the same direction of the winning unit.

The choice of the neighborhood $\Lambda(i, i^*)$, that is, the radius r of the area surrounding the winning unit i^* where other units are active, is a major factor in the formation of good topological maps. It is convenient to formalize the neighborhood as

$$\Lambda(i, i^*) = \begin{cases} 1 & : \quad \| \vec{c}_i - \vec{c}_{i^*} \| \le r \\ 0 & : \quad \text{otherwise} \end{cases}$$

where $\vec{c}_i$ is the vector with the spatial coordinates of neuron i in the layer and $\| \vec{c}_i - \vec{c}_{i^*} \|$ is the Euclidean distance between that unit and the winning unit i^*. For example, if units are organized in a one-dimensional layer with $r = 1$, the neighborhood area includes the winning unit and the two adjacent units; instead, if the units are organized in a two-dimensional layer with $r = 1$, the neighborhood area includes the winning unit and the eight surrounding units (units can be organized in higher dimensional space, but for the sake of representation most models use up to three dimensions). We can now rewrite the standard competitive learning rule of equation (3.12) by taking into account the neighborhood function as

$$\Delta w_{ij} = \eta \Lambda(i, i^*)(x_j - w_{ij}).$$

There is no theoretical criterion for setting the magnitude of r in the neighborhood function. However, empirical observations show that the development of self-organizing maps proceeds through two stages. During the first "ordering stage," the weights move so as to cover approximately the area spanned by the input patterns. During the second "convergence stage," the weights undergo small modifications to refine their final position. The convergence stage requires many more training cycles than the ordering stage, but does not introduce major modifications in the topological map. Consequently, Kohonen (1989) suggested varying the learning rate η and the size r of the neighborhood function only during the ordering stage. A typical strategy consists of starting with $\eta = 1$ and decreasing it exponentially down to approximately 0.01; similarly, the size r initially is such to include all units in the network and exponentially decrease down to 1 or 0 (in this latter case only the weights of the winning unit are modified). Both the final learning rate and neighborhood size are kept constant during the convergence stage.

ORDERING STAGE

CONVERGENCE STAGE

Figure 3.31 shows the ordering stage of a self-organizing map composed of two input units and 20 output units trained on a C-shape distribution of input patterns. The high learning rate and large neighborhood size during the early iterations allow all the units in the network to rapidly align their weights toward the areas covered by input patterns. As the learning rate and neighborhood size are reduced, the units arrange themselves so as to reflect the topology and density of the patterns more precisely. If the neighborhood size is not sufficiently large at the beginning of training, there could be some units whose weights are so far from the input patterns that they never win the competition or are not in the neighborhood of other winning units. Consequently, such units would never have a chance to modify their weights and would be equivalent to dead neurons.

Kohonen's network develops weights that optimally represent the input distribution in that they minimize the distance between input patterns and final weights $1/M \sum_{\mu=1}^{M} \parallel \vec{x}^{\mu} - \vec{w}_{i^*} \parallel$ where M is the number of input patterns used for training and $\vec{w}_{i^*}$ is the weight vector of the winning unit for pattern $\vec{x}^{\mu}$. Therefore, these neural networks are very useful in data mining applications for representing complex databases with very high dimensionality into a smaller space and visualizing the relationships among individual patterns. Since it is not possible to tell in advance to what features the units

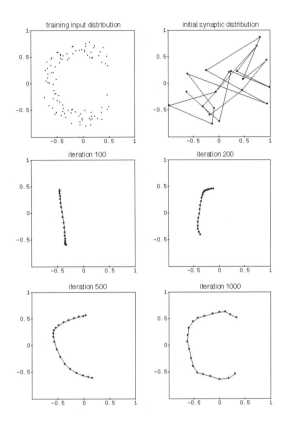

Figure 3.31   Development of a self-organizing map consisting of a one-dimensional layer of 20 neurons with two input units. The first panel on top left shows the distribution of training patterns. The second panel on top right shows the initial random distribution of the neurons on the input space. Each dot represents the synaptic vector of the corresponding neuron in the input space; the lines connect adjacent units in the one-dimensional neural layer. The other panels show the movement of the neurons in the input space during learning.

will become sensitive and which data they will represent, it is necessary to proceed to a manual labeling of the neurons after learning.

A very useful property of Kohonen's network is that it can also be trained with incomplete data where other supervised methods that we will describe in the next section typically fail (Samad and Harp 1989). For example, think of an insurance company that wishes to categorize its customers according to their personal data so as to detect potential high-risk customers. Each customer could be associated to a vector that encodes her characteristics. Since it is impossible to obtain the necessary information for all customers, some of the corresponding vectors will have empty components. In that case, the computation of the winning unit during training is obtained by comparing weight and input vectors only for the components that are known:

$$i^* = i \left| \min_i \left(\sum_{j \in P_k} (x_j - w_{ij})^2 \right) \right.$$

where P_k represents the set of input units for which the values are known. Similarly, the modification of synaptic weights occurs only for weights corresponding to known input values:

$$\Delta w_{ij} = \begin{cases} \eta \Lambda(i, i^*)(x_j - w_{ij}) & : \quad j \in P_k \\ 0 & : \quad \text{otherwise.} \end{cases}$$

In addition to that, the missing values can be inferred by presenting the input vector to the trained network, computing the winning unit, and taking the strengths of the corresponding weight components as approximate values of the input components.

3.7.4 Adaptive Resonance Theory

All models described so far, as well as most neural models, cannot develop a stable representation if the input distribution changes over time. Therefore, the response of the network to the same pattern will change over time as the input distribution changes. One way of preventing this problem consists in stopping the learning phase. The other way is to store old input patterns and continuously present them to the network along with new input patterns during learning. The former strategy does not allow the network to adapt to new situations and the latter strategy is computationally demanding and would require very large memory storage. For most applications that use a database, this problem is not very important because the input patterns are known and do not change. However, for applications in embedded systems and autonomous robotics, the problem is relevant because the input distribution may not be entirely predefined in advance. Furthermore, biological brains are capable of adapting to new situations without forgetting previously acquired knowledge and skills.

PLASTICITY-STABILITY   Grossberg (1980) named this limitation of neural models the *plasticity-sta-*
DILEMMA   *bility dilemma*: Either the neural network remains adaptive with the risk of

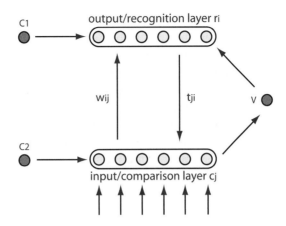

Figure 3.32   Schematic representation of ART-1, a neural network capable of developing stable representations of changing input distributions.

ADAPTIVE RESONANCE
THEORY (ART)

losing stability, or it remains stable but is no longer adaptive. To address this issue, he developed *adaptive resonance theory* (ART). ART includes a set of architectural and learning principles that a self-organizing network should have in order to remain both plastic and stable. On the basis of this theory, he suggested with Carpenter a number of neural models of increasing complexity and computational power that can develop stable representations of continuously changing input distributions. In this chapter we will briefly describe the simplest model, known as ART-1 (Carpenter and Grossberg 1987), which can operate only on binary input patterns. We refer interested readers to (Grossberg 1987) for a more detailed introduction to ART and implementation of the neural model.

ART-1

ART-1 (figure 3.32) is composed of two layers of neurons with feedforward w_{ij} and feedback t_{ji} synaptic connections. The output neurons include a pattern of lateral connections that introduce competitive dynamics. The competition results in the activation of only one winning unit for a specific input pattern. These types of competitive systems are also known as *winner-take-all* models. Capitalizing on what we explained earlier for self-organizing maps (section 3.7.3), instead of modeling the lateral connections and computing the dynamics, one can directly find the winning neuron by comparing weight and input vectors.

ART-1 also includes a set of neurons that modulate the activity of the network and, for the sake of simplicity, can be implemented as `if-then-else`

rules. The basic principle of the model is that input patterns are moved back
and forth between the two layers (resonance) until they are either recognized
as familiar or trigger a modification of the synaptic weights. When no neuron responds to an input pattern, the model creates a new output neuron
and a corresponding set of synaptic weights. ART networks do not require a
distinction between training and testing phase.

The operation of the network can be summarized in five phases. In the
initialization phase, the user allocates a maximum number M of output units
(the number of input units N is given by the size of the input patterns). The
feedback weights t_{ji} are all set to 1 and feedforward weights w_{ij} are set all
equal and proportional to the number of input units:

$$w_{ij} = \frac{L}{L - 1 + N}$$

where $L > 1$ is a constant. Finally, the values of the control units C1 and
C2 are set to 1 and the value of the so-called vigilance unit V is set between
0 and 1, with high vigilance values corresponding to high specialization of
output neurons and consequently to the use of several output neurons for
fine representation of the input patterns.

During the *recognition phase*, an input pattern $\vec{x}^{\mu}$ is presented to the network and the winning unit i^* is computed (for the first pattern, all output
units will have the same values and it is suggested to choose the neuron
with the smallest index i).

The network enters now into the *comparison phase* (the control unit C1 is
set to 0 to disable acceptance of further input patterns) during which the
feedback weights of the winning unit are "compared" to the input pattern:

$$c_j = (t_{ji^*} y_{i^*}) x_j^{\mu}$$

The ratio between the length of the resulting comparison vector $\vec{c}$ and the
length of the input vector is measured against the vigilance value and if

$$\frac{\| \vec{c} \|}{\| \vec{x}^{\mu} \|} \geq V$$

the winning neuron $i*$ is confirmed and the network proceeds to the modification of the synaptic weights; otherwise, the network enters the research
phase.

The *research phase* consists in finding another neuron more suitable for representing the current input vector. The control unit C2 is set to 0 to disable
the current winning neuron and reset all other neurons to 0. At that point

the network goes again through the recognition and comparison phase until a neuron is found that passes the vigilance check. If no neuron is found, it is possible to add a further neuron, reduce the vigilance value, or discard the input vector.

ADAPTATION PHASE   The *adaptation phase* occurs when a suitable winning neuron is found and consists in the modification of the feedforward and feedback synaptic weights of the winning neuron so as to reflect the comparison vector and the input vector respectively:

$$w_{i^*j}^{t+1} = \frac{Lc_j}{L - 1 + N + \| \vec{c} \|}, \quad t_{ji^*}^{t+1} = t_{ji^*}^t x_j^\mu$$

The feedback weights are also known as template weights because they effectively store a representation of the input vectors to which output neurons respond. From the equation above, it is easy to see that once a feedback weight becomes 0, it can never reverse to 1, which ensures the stability of the network.

ART-1 has a number of features that resemble the way in which humans learn and recognize patterns. Familiar patterns are recognized faster (i.e., with fewer resonance cycles) than new patterns. Recognition is not a simple pattern match, but context-dependent because the comparison vector is based on the length of the vectors. In other words, a pattern with a single 1 at a certain position among several 0s is recognized faster than a pattern with the same 1 among a series of 1s and 0s. The vigilance value is equivalent to an attentional mechanism that could be modulated by the output of other neural modules.

Neural competition is a very powerful principle that can explain not only differentiation and specialization of individual neurons, formation of sensory maps, and categorization of input patterns but may also explain the formation, maintenance, and update of complex concepts mediated by interconnected and competing populations of cortical neurons as advocated by Edelman (1988) in his theory of neural Darwinism.

3.7.5 Memory Formation

The models described so far are capable of developing neurons that respond to representative features of the input distribution. Hopfield (1982) showed that when the Hebb rule is applied to a network of interconnected neurons, the resulting neural network can memorize patterns and reconstruct them HOPFIELD NETWORK   from corrupted or incomplete versions. A Hopfield network (figure 3.33)

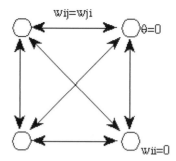

Figure 3.33   Schematic architecture of a Hopfield network.

is composed of N fully interconnected, bipolar units (equation (3.2)) where each unit functions both as input and output. The number N of units depends on the length of the patterns to be memorized. By definition, the weights are symmetric $w_{ij} = w_{ji}$, the self-connections w_{ii} and the unit thresholds θ_i are set to 0, and these conditions cannot change during learning. Before learning, all the connections are set to 0. The memorization of a bipolar pattern $\vec{p}^\mu$ consists in applying the pattern to the units of the network $x_i = p_i^\mu$ and computing the weight modification as

$$\Delta w_{ij}^\mu = x_i^\mu x_j^\mu \quad i \neq j.$$

Since the memorization of a pattern requires only one modification of the synaptic weights, the final weights of the network for M patterns are given by

$$w_{ij} = \begin{cases} \sum_{\mu=1}^M x_i^\mu x_j^\mu &: \quad i \neq j \\ 0 &: \quad i = j. \end{cases}$$

A Hopfield network recovers the memory of a pattern on the basis of its content as in biological systems, not on the basis of its memory address as in conventional computer memory. Consequently, Hopfield networks are capable of recovering an incomplete or corrupted version $\vec{p}^{\mu*}$ of a memorized pattern. The recovery consists in presenting the incomplete or corrupted pattern to the units $x_i = p_i^{\vec{\mu}*}$ and iteratively updating the output of the units using the bipolar activation function described in equation (3.2) until all units remain in the same state. This final state represents the recovered pattern. The unit update can be synchronous, in which case all units are updated in parallel at each iteration, or asynchronous, in which case a randomly selected unit is updated at each time. The latter case may require more iterations for the network to stabilize, but is more biologically plausible and is more convenient for hardware implementation because it does not require a global clock.

The memory capacity of a Hopfield network is the number of patterns that the network can memorize and recover correctly. It can be shown that this capacity depends on the number N of units. It has been formally shown that, for randomly generated patterns, the capacity of the network is 0.138 N (Hertz et al. 1991). In other words, a network of 100 units can memorize and reconstruct correctly approximately 13 randomly generated patterns. If the network is trained on more patterns, there will be interference phenomena that will disrupt the correct recovery. The capacity is reduced if the patterns are correlated. Although this capacity may seem relatively small, it is less so in realistic situations. For example a Hopfield network in theory could memorize up to 565 images with a size of 64 x 64 pixels mapped onto 64 x 64 neurons (in practice, the number of memories will be smaller if images have some degree of correlation).

SPIN GLASS THEORY Hopfield (1982) emphasized the similarity between the behavior of his neural network and *spin glass theory*, which describes the behavior of magnetic particles that, in the Ising model, can be in one of two states (spin directions). The particles tend to change spin direction depending on the spin directions of neighboring particles in order to assume a coherent overall state that is characterized by lower energy. Hopfield used this analogy in order to describe the state of the neural network by an energy value H:

$$H = -\frac{1}{2} \sum_{i} \sum_{i \neq j} w_{ij} x_i x_j$$

He formally showed that both the memorization of a pattern and the reconstruction of a corrupted pattern consists in the transition to a lower energy state (this can be easily observed in experiments by measuring H before and after the memorization or recovery process). Using the language of physics, the memorization of a pattern consists in the creation of a basin of attraction on the energy landscape and the recovery of a pattern consists in the descent toward the bottom of the closest basin of attraction.

TYPES OF ATTRACTORS Hopfield networks can have various types of attractors besides those corresponding to the original patterns. For example, for every memorized pattern $\vec{p}$ there is a corresponding symmetric basin $-\vec{p}$. In addition, there may be metastable attractors that correspond to a linear combination of an even number of patterns used to train the network (Amit et al. 1985) or to none of those patterns. Metastable attractors have a higher energy level than attractors corresponding to the original patterns and occur more frequently when the capacity of the network is exceeded.

In order to prevent stagnation in metastable attractors during the recovery phase, it is possible to use stochastic units whose state is given by a probability function

$$P(x_i = 1) = \frac{1}{1 + e^{-\beta\left(\sum_j w_{ij} x_j\right)}}$$

that takes the form of the sigmoid function described earlier (see equation (3.3) and figure 3.11). The introduction of probabilistic units increases the duration of the recovery state, but may allow the network to skip metastable attractors and converge to lower energy attractors corresponding to the memorized patterns.

Furthermore, the analogy with statistical mechanics is even more realistic because the modification of spin direction is also characterized by a certain probability when the temperature of the material is above the absolute 0. Indeed, in the Glauber model the spin transition is given by the stochastic equation described above where β, which controls the curvature of the function, is inversely proportional to the temperature. In other words, the higher the temperature, the smaller the β, and consequently the flatter the function is around 0.5, which is the point where the state change of the particles, or neurons, is completely random. Conversely, if the temperature tends to 0, β will tend to positive infinite and the stochastic function will approximate a step function between 0 and 1.

It has been suggested that the recovery process can be improved if the temperature of the units is gradually reduced along the recovery phase, a process called *simulated annealing* to signal the analogy to the temperature-lowering process used in metallurgy to ensure good quality of the final metal cast (Kirkpatrick et al. 1983).

Hopfield's network holds also for continuous units with tanh activation function that can memorize patterns with real values. Despite its simplicity, Hopfield's model seems to capture architectural and learning properties of the hippocampus, a component of the mammalian brain that is responsible for memory formation and cognitive maps of the environment (Redish 1999). The model has also been used to explain brain damage leading to deficit of face recognition (Virasoro 1989) and to reproduce the biological network of neurons that control swimming behavior in the mollusk *Tritonia diomeda* (Kleinfeld and Sompolinski 1989). The model has also been extended to include internal neurons (not directly connected to the patterns to be memorized) in order to extend the capacity of the network and the ability to memorize partially overlapping patterns (Ackley et al. 1985).

3.8 Supervised Learning

Supervised learning is characterized by the presence of a teacher that provides the response required from the network for each training pattern. Within this framework, originally proposed by Rosenblatt (1962), the synaptic weights are modified so as to reduce the error between the desired response and the response given by the network.

Consider a feedforward neural network with a single layer of synaptic weights between input and output units with linear activation function. Given a set of training patterns composed of M pairs of input vectors $\vec{x}^{\mu}$ and desired response vectors $\vec{t}^{\mu}$, we want to find a set of synaptic weights so that the response of the network corresponds to the desired response for all M patterns. The performance of the neural network can be described by the error function

$$(3.13) \quad E_W = \frac{1}{2} \sum_{\mu} \sum_{i} (t_i^{\mu} - y_i^{\mu})^2 = \frac{1}{2} \sum_{\mu} \sum_{i} \left(t_i^{\mu} - \sum_{j} w_{ij} x_j^{\mu} \right)^2$$

which represents the mean quadratic error between the desired response and the response given by the network. The error function depends uniquely on the synaptic weights and can be reduced by changing the weights in the opposite direction to the gradient of the error function with respect to the weights. Derivating the error function with respect to the weights gives us the learning rule

$$(3.14) \quad \Delta w_{ij} = \sum_{\mu} (t_i^{\mu} - y_i^{\mu}) x_j^{\mu}$$

which, when applied after the presentation of each training pattern μ, becomes

$$(3.15) \quad \Delta w_{ij} = \eta (t_i - y_i) x_j$$

where η is the learning rate. Learning can be performed either in *batch mode* or *online* by updating the weights with equation (3.14) after presentation of each pattern randomly extracted from the set of training patterns M. In both

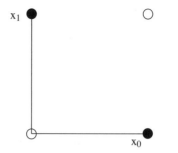

x_0	x_1	t	
0	0	0	○
1	1	0	○
1	0	1	●
0	1	1	●

Figure 3.34 The XOR function has two inputs and one output. It is not linearly separable because it is impossible to separate the two groups of patterns with a line and thus it cannot be learned by a network with a single layer of connections. *Left*: Geometric visualization of the function. Each axis represents one input. The color of the dot represents the desired output. *Right*: Tabular visualization.

cases, the initial synaptic weights are initialized to small random values centered around zero and the training patterns are presented several times until the error function is minimized. It is standard practice to assess the learning progress by monitoring the total E_W after each presentation of the complete training set M.

DELTA RULE   This algorithm is also known as the *Widrow-Hoff* rule from the names of its authors (Widrow and Hoff 1960), or most often as the *delta rule* to emphasize the role of the difference $\delta = t - y$ between the desired and the produced response. The delta rule is often written as

$$\Delta w_{ij} = \eta \delta_i x_j$$

and it can be mathematically shown that it finds a set of synaptic weights that minimize the error function if the training patterns are linearly separable.

The delta rule is applicable only to networks with one layer of connections. Those networks can learn only linearly separable mappings. Two sets of points are said to be linearly separable if it is possible to draw a line between them. We have seen earlier in this chapter that the weights of a neuron effectively correspond to the line that separates the response of the neuron in two classes (see equation (3.4)). There are cases where the patterns are not linearly separable, such as in the XOR problem shown in figure 3.34. In order to learn nonlinearly separable mappings it is necessary to add internal units (hidden units) to the network. Internal units recode the input vectors into a set of linearly separable representations so that the output units can produce the desired output. The output function of the internal units must be nonlinear because linear transformations (operated by output units) of linear transformations (operated by internal units) remain linear transformations. In other words, multilayer networks with linear output functions can always be reduced to a network without internal units and therefore can learn only linearly separable mappings. A typical choice of output function for these networks is the sigmoid function (equation (3.3)) because it is nonlinear, continuous, and differentiable. Alternatively, one may use tanh function, which has the same properties but spans a range between -1 and +1. Neural networks with hidden units and nonlinear output functions in theory can perform any mapping between input and output, provided that it is equipped with a suitable architecture and connection weights.

3.8.1 Backpropagation of Error

BACKPROPAGATION OF ERROR

GENERALIZED DELTA RULE

Training a network with hidden units and nonlinear output functions requires two modifications to the delta rule: consideration of nonlinear behavior introduced by the output function, and a method to compute the contribution of hidden units to the error measured at the output units. Although the computational potentials of multilayer networks of nonlinear units have been known since the 1960s, it remained unclear how one could compute the contribution of the hidden units to the error of the output units in order to modify the synaptic weights between input and hidden units. The method of *backpropagation of error* (Rumelhart et al. 1986a,b), also known as the *generalized delta rule*, provided a solution that could be applied to any networks with an arbitrary number of neurons and connection layers.

The core of the algorithm consists in computing the error contribution of hidden units by transmitting the error computed at the output units back to the hidden units through the same weighted connections used to forward activation signals from hidden to output units (hence the name of backpropagation of error). Backpropagation of error performs gradient descent of the error function E_W (equation (3.13)) for networks of arbitrary numbers of layers and neurons. The algorithm is so powerful and general that is has become one of the most used methods to solve computational and engineering problems with neural networks. It turned out later that this method had already been discovered previously in different contexts (e.g., see Bryson and Ho 1969; D.B. Parker 1985; Werbos 1974).

Consider the multilayer network shown in figure 3.35 and the symbols associated to its elements. The values of the input units are determined by

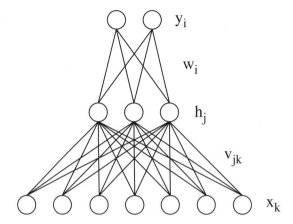

Figure 3.35 A multilayer, feedforward neural network with sigmoid activation functions.

the input pattern. The output of the hidden and output units is computed using the sigmoid function described in equation (3.3). The online version of the algorithms proceeds as follows:

1. Initialize all weights (including weights from the bias unit) to small random values centered around 0.

2. Set the values of the input units $\vec{x}$ to the current training pattern $\vec{s}$:

$$x_k^{\mu} = s_k^{\mu}$$

3. Compute the values of hidden units:

$$h_j^{\mu} = \Phi\left(\sum_k v_{jk} x_k^{\mu}\right)$$

where $\Phi()$ is the sigmoid activation function (equation (3.3)).

4. Compute the values of output units:

$$y_i^{\mu} = \Phi\left(\sum_j w_{ij} h_j^{\mu}\right)$$

5. Compute the delta error for each output unit. Notice that the error is multiplied by the first derivative (denoted by a dot) of the output function of that node because we are using a nonlinear function. (This was not necessary in the delta rule shown in equation (3.15) because the output function was linear.)

$$\delta_i^\mu = \dot{\Phi} \left(\sum_j w_{ij} h_j^\mu \right) (t_i^\mu - y_i^\mu)$$

The first derivative of the sigmoid function can be conveniently expressed in terms of the output of the unit itself. In the case of the output units, for example,

$$\dot{\Phi} \left(\sum_j w_{ij} h_j^\mu \right) = y_i^\mu (1 - y_i^\mu).$$

6. Compute the delta error of hidden units by propagating the delta errors at the output backward through the connection weights. Notice that the index in the sum of the weighted deltas is i and refers to the output units. As before, the delta error must be multiplied by the first derivative of the output function of the unit:

$$\delta_j^\mu = \dot{\Phi} \left(\sum_k v_{jk} x_k^\mu \right) \sum_i w_{ij} \delta_i^\mu$$

7. Compute the modifications to the synaptic weights of the two layers by multiplying the delta errors at the postsynaptic units by the output of the presynaptic units:

$$\Delta w_{ij}^\mu = \delta_i^\mu h_j^\mu$$
$$\Delta v_{jk}^\mu = \delta_j^\mu x_k^\mu$$

8. Finally, update the weights by adding a portion η of the modifications:

$$w_{ij}^t = w_{ij}^{t-1} + \eta \Delta w_{ij}^\mu$$
$$v_{jk}^t = v_{jk}^{t-1} + \eta \Delta v_{jk}^\mu$$

where η is the learning rate and is usually smaller than 1.

Box 3.1: Bayesian supervised learning

There is an alternative, probabilistic way of looking at supervised neural network learning (Bishop 1995; Neal 1996; MacKay 2003). The idea is to consider supervised learning as a process where the training data $T = \{\mathbf{t}^\mu\}$ represents noisy samples from an unknown function $f(\mathbf{x})$ and is used to update one's state of information about f. The mathematical tool that formalizes this concept is *Bayes' theorem:*

$$p(f \mid T, I) = \frac{p(T \mid f, I)\, p(f \mid I)}{p(T \mid I)}$$

According to Bayes' theorem the *prior probability distribution function* (pdf) $p(f \mid I)$ – which represents our knowledge of f taking into account all our background information I but not the training data T – is updated through the *likelihood* $p(T \mid f, I)$ – which takes into account the effect of the training data – with an additional normalization factor $p(T \mid I)$ called the *evidence*. The result is the *posterior pdf* $p(f \mid T, I)$, which represents our probabilistic assessment of f when we take into account both our prior information and the training data.

This formulation refers to an abstract representation of the space of functions f. The correspondence with the traditional approach is established by considering a neural network as a parameterized representation $f(\mathbf{w}; \mathbf{x})$ of the functions, with the network weights $\mathbf{w}$ acting as parameters. Then, Bayes' theorem can be rewritten as

$$p(\mathbf{w} \mid T, I) \propto p(T \mid \mathbf{w}, I)\, p(\mathbf{w} \mid I)$$

leading to an interpretation of neural network supervised learning as an update through the likelihood $p(T \mid \mathbf{w}, I)$ of the prior pdf of the weights $p(\mathbf{w} \mid I)$ to obtain a posterior pdf of the weights $p(\mathbf{w} \mid T, I)$.

There are advantages and disadvantages in the probabilistic viewpoint with respect to the conventional neural network approach. A first advantage is that the probabilistic context helps to keep in mind that, in general, many functions f could have generated the training data and that, consequently, the data assign different degrees of plausibility to the candidate functions rather than determine a single function. This is reflected in the better prediction model for further samples, which, in the probabilistic approach, permits the estimation of the uncertainty of the prediction and lets one perform "active learning" by directing the acquisition of new data samples to regions of the input space where the uncertainty is larger (MacKay 2003).　　　　　　　　　　*(cont.)*

Box 3.1: (continued)

A second advantage of the probabilistic viewpoint is that it helps to realize that neural networks are just a way to represent parametrically a set of functions and that other representations are possible. In particular, one can represent directly the pdfs over the space of functions without parameterizing the functions, as exemplified by the approach to supervised learning based on *Gaussian processes* (Neal 1996).

The probabilistic viewpoint reveals also that the minimization of the quadratic error function prescribed by the conventional neural network approach corresponds to the maximization of the likelihood in the hypothesis of uniform independent Gaussian noise in the training samples. This realization opens the way to the extension of the learning algorithm to nonuniform and non-Gaussian noise. Moreover, it permits understanding of the phenomenon of overfitting and poor generalization that plague the conventional approach, with its perplexing reference to the prediction performance on hypothetical samples that have not yet been observed and thus cannot possibly influence the current learning process. The probabilistic perspective reveals that by focusing on the sole likelihood, the conventional approach takes into account only the training data, disregarding any prior information about the different degree of plausibility of the candidate functions f, which should instead be balanced against the goodness of fit to the data.

A related question is the choice of the structure of the neural network. In the conventional approach there are no prescriptions for the comparison of the performance of different network structures. In the probabilistic perspective one can instead apply to this problem the principles of Bayesian *model selection* (MacKay 2003; Sivia 2006; von Toussaint et al. 2006). The resulting formulas provide a quantitative way to balance the goodness of fit of a model with a measure of its simplicity.

Summing up, the probabilistic viewpoint provides much insight into supervised learning and gives a sound foundation to some problematic aspects of the conventional neural network approach. These advantages must be balanced against two main disadvantages of the probabilistic viewpoint. The first is the difficulty of expressing formally the prior information in the form of a pdf. The second is the complexity of the expressions of the pdfs that can be produced even for simple learning problems and network architectures. This latter point implies that sophisticated numerical techniques must be used in general to extract the desired information from a probabilistic formulation (MacKay 2003; Jaynes 2003; Sivia 2006).

Every pair of input-output pattern in the training set is presented several times in random order until the total sum squared error computed over all output units i and all training pattern μ,

$$TSS = \frac{1}{M} \sum_{\mu}^{M} \left(\frac{1}{N} \sum_{i}^{N} (t_i^{\mu} - y_i^{\mu})^2 \right),$$

reaches a small value.

3.8.2   Using Backpropagation

Provided a suitable architecture and connection weights, it can be shown that, in principle, multilayer networks can represent any arbitrary mapping between input units and output units. Finding the right architecture is a major issue and there are no theoretical guiding principles for that. Using a large neural network in the attempt to learn complex mappings is not necessarily a good solution because larger networks require larger training sets that may not be available. The need for a larger training set is easy to understand when we consider the learning process as a parameter estimation problem. The weights are the parameters that must be estimated and the training patterns are the sample data. It is known from statistics that the problem is ill-defined if the size of the sample is smaller than or equal to the number of the parameters to be estimated. In other words, a rule of thumb is that the number of training patterns should be larger than the number of weights in the network.

Another strategy consists in starting from a suboptimal architecture and adapting the topology along with the modification of synaptic weights. Several methods have been developed that automatically increase (e.g., Frean 1990; Fahlman and Lebiere 1990) or decrease (e.g., Chauvin 1989; Scalettar and Zee 1988) the size of the network by adding and deleting neurons and connections. Furthermore, as we will see in chapter 4, the architecture of the network can be designed by a developmental process whose rules can be genetically encoded and evolved, as is the case for biological brains.

Once a supervised network has satisfactorily learned to produce the correct response for the patterns in the training set, it can be used on new patterns without any further modification of the synaptic weights. The generalization performance of the trained network consists in the ability to produce satisfactory responses to patterns that were not included in the training set. Within this perspective, a very small residual error on the training set

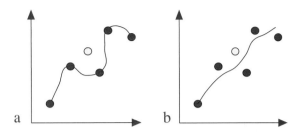

Figure 3.36   Learning to represent the training patterns (filled disks) as a fitting problem. The fitting error is the mean square error between a data point and the nearest point of the fitting function. *a*, Overfitting the training data may result in a large error for a new data point that was not included in the training set; *b*, a larger residual error on training data points may correspond to a smaller error for new data points.

after learning does not necessarily translate into a good generalization performance. This can be understood by considering supervised learning as a data-fitting problem, as shown in figure 3.36,*a*. When the network has more free parameters (synaptic weights) than training data (filled disks), each data point in the training set could be perfectly covered by the network output (fitting curve). However, when a new data point (empty disk) is presented, the network output will result in a relatively large error. This problem of

OVERFITTING   overfitting, or hyperspecialization, can be prevented either by choosing a more appropriate architecture, which is not a trivial issue, or by halting the learning phase earlier.

A widely used procedure consists of subdividing the patterns for which a desired response is known into two sets: a training set used for changing the synaptic weights and a validation set to assess the generalization performance of the network (figure 3.37). During learning, the total sum squared error is separately computed for the training set and for the validation set, but the weights are modified using only the error on the training set. Assuming that the training set and the validation set are extracted from the same data distribution, the error will initially decrease on both data sets, but at some point it will start to increase on the validation set while it will continue to decrease on the training set. That point represents the moment when the network begins to overfit the data and the learning phase should be halted to guarantee optimal generalization performance (figure 3.36, b).

Learning in supervised systems is equivalent to finding the weight values corresponding to the minimum of the error function E_W described in equa-

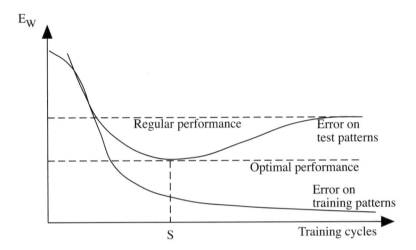

Figure 3.37 Relationship between error reduction on training and validation sets during learning. The optimal performance level is obtained if learning is halted when the error on the validation set begins to increase (S).

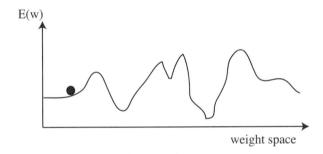

Figure 3.38 Schematic representation of the error surface for a multilayer neural network trained on nonlinearly separable patterns. The circle represents the position of the network on the error surface.

tion (3.13). This amounts to navigating over the error space by descending the error gradient in search of the absolute minimum. The error space for a single-layer network trained with the delta rule on a set of linearly separable patterns –when visualized in two dimensions– resembles a bowl. The bottom of this bowl can be easily reached by the learning procedure. However, in the case of multilayer networks with nonlinear activation functions trained with backpropagation of error on patterns that are not linearly separable, the error space is much more complex (figure 3.38) because it may present several local minima and, in some cases, flat areas where the learning process could stagnate.

A local minimum is an area of the error space where a modification of the weights in any direction corresponds to an increment of the error, but it is not the area with the lowest possible error. Small local minima can be skipped by using higher learning rates, so that single weight updates move the network over larger distances in the error space. However, if the learning rate is too high, the network may end up in areas characterized by higher error and miss the path to the lowest minimum. Another way of avoiding stagnation MOMENTUM in local minima consists in adding a *momentum*, or inertia, to the movement of the network in the error space. This is achieved by adding to the weight update a fraction of the previous weight update:

$$\Delta w_{ij}^t = \eta \delta_i x_j + \alpha \Delta w_{ij}^{t-1}$$

where $0 \leq \alpha \leq 1$ is the momentum constant. Momentum tends to reduce the oscillations due to high learning rates and improves movement across flat areas of the search space where there is no gradient information (i.e., the derivative of the network output is 0).

For some problems, such as the XOR shown in figure 3.34, the error space displays very large flat areas (Kolen and Pollack 1991) where not even the momentum can help. That is why for the XOR problem, depending on the initial weight values, sometimes it is not possible to find the optimal solution. In those circumstances, Fahlman (1989) suggested adding a small constant to the derivative of the output in the computation of the delta error:

$$\delta_i^\mu = \left(k + \dot{\Phi} \left(\sum_{j=0} w_{ij} h_j^\mu \right) \right) (t_i^\mu - y_i^\mu)$$

where the constant k ($k = 0.1$ in Fahlman's experiments) generates a movement of the network even when the derivative is 0. This is equivalent to the network skidding over the flat surface until an area with some gradient is found where the derivative is again nonzero. Several other modifications to supervised algorithms that improve their performances are described in a book by Reed and Marks (1999).

The neural networks described so far for supervised learning have a feed-forward architecture. The output of these networks depends only and entirely on the pattern currently presented as input. In some cases, such as in time series analysis, it is important to detect time-dependent features in the

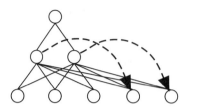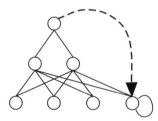

Figure 3.39 Two types of architectures with discrete time delay. *Left*: Elman architecture. The memory units hold a copy (dashed lines) of the activations of hidden
units at the previous time step. *Right*: Jordan architecture. The memory unit computes its activation by combining a copy of the output unit at the previous time step
with its own previous state weighted by a self-recurrent connection.

sequence of input patterns. One way of doing that is to expand the input
layer in order to present several patterns at the same time to the network.
One can visualize the input layer as a window looking at several successive
patterns and shifting across them one at a time. These so-called *time delay
neural networks* (TDNNs) require that the user know the appropriate size of
the window necessary to extract time-dependent features.

TIME DELAY NEURAL
NETWORKS (TDNNs)

Another strategy consists in adding recurrent connections from neurons
in the same and upper layers. These connections transmit activations with a
time delay, as described in equation (3.6). Figure 3.39 shows two types of recurrent architectures that can be trained with backpropagation of error. The
strategy consists in adding extra input units that hold a copy (memory) of
the activations of other units at the previous time step. These extra *memory
units* are connected to hidden units with feedforward connection. In the architecture proposed by Elman (1990) (figure 3.39, left) the memory units hold
a copy of the values of hidden units at the previous time step. Since hidden
units encode their own previous states, this network can detect and reproduce long sequences in time. In the architecture proposed by M.I. Jordan
(1989) (figure 3.39, right), a copy of the values of output units at the previous time step is combined with the weighted activation of the memory units
themselves at the previous time step (recurrent connection).

MEMORY UNITS

3.8.3   Sample Applications of Backpropagation

Although the delta rule and the backpropagation of error are not biologically
plausible at the neurophysiological level, they can be used as a tool for mod-

eling cognitive and neural functions where emphasis is put on the constraints and features of the transformation from input to output rather than on the precise physiological mechanisms that may lead to those transformations.

Incidentally, the delta rule is very similar to the Rescorla-Wagner rule that was independently developed in psychology to model behavioral choice of animals under classic conditioning (Gluck 1991). The Rescorla-Wagner rule describes the way in which subjects modify their preference in multichoice tasks when provided with binary feedback (yes-no) after their choice. For example, it was shown that a simple neural network trained with the generalized delta rule could develop transitive inference (after training on examples A>B, B>C, C>D, D>E, the network can generate all remaining relationships in the series, such as B>D) (De Lillo et al. 2001), which seems to contradict Piaget's hypothesis that transitive inference requires complex logical skills.

As an example of neural modeling, Zipser and Andersen (1988) used the backpropagation algorithm to train a network with two layers of connections to map the position of an object from retinal coordinates into head-centered coordinates, an operation that is performed by the mammalian brain in order to obtain the position in space of objects with respect to the head of the person. In principle, this transformation requires only a linear addition of the vector with the retinal position of the object and of the vector with the pointing direction of the eyes in head-centered coordinates. Therefore, this mapping could be learned by a network with a single layer of connections trained with the delta rule. However, since the biological system employs several layers of intermediate neurons, the authors used a multilayer neural network trained by backpropagation of error in order to study those intermediate representations.

The trained neural network displayed patterns of hidden unit activation that closely resembled the activation of biological neurons, which helped the authors to understand the computation carried out by individual intermediate neurons. Furthermore, the artificial neural network developed the same patterns of activations also when a different encoding of the input and output vectors was used, indicating that what matters in the formation of the intermediate neural representations is the type of operations involved in the coordinate transformation and not the detail of the synaptic rules and interface with other neurons.

Multilayer networks trained with backpropagation of error have been used also to model higher-level cognitive functions, such as language (Seidenberg and McClelland 1989), dyslexia (Plaut and Shallice 1993), semantic representations (Farah and McClelland 1991), and several other functions

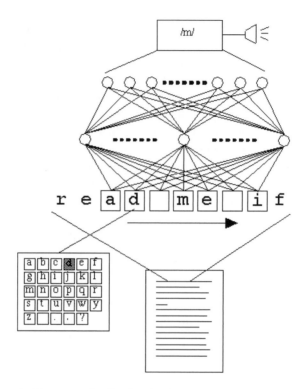

Figure 3.40   Architecture of NETtalk, a backpropagation-trained network that learns to read aloud written text (Sejnowski and Rosenberg 1987).

that interested readers can find in thematic books (e.g., Clark 1989; Churchland and Sejnowski 1992).

Supervised learning, and in particular backpropagation of error, is often used in engineering applications to learn input and output relations that are too difficult to capture by other optimization techniques, to develop control systems for plants that operate in noisy and partly unpredictable environments, and more generally in all those situations where the system must be easily customizable for new situations. Supervised neural networks are used in medicine, image processing, plant control, market analysis, data mining, signal processing, and character recognition, to mention a few fields.

NETTALK   NETtalk (figure 3.40) is a historical example of an application where a neural network is trained to read aloud written text in the English language (Sejnowski and Rosenberg 1987; Anderson and Rosenfeld 1998). The input layer

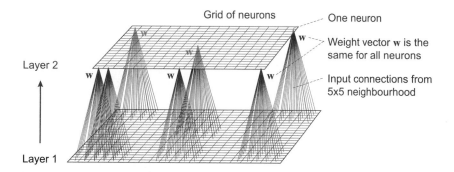

Figure 3.41 The principle of weight sharing. Several neurons in the network have the same set of synaptic weights, although they receive signals from different parts of the input layer. The synaptic modification is computed for only one neuron and the resulting weights are cloned for all the other neurons.

of the network consists of a seven-slot window moved over the text one character at a time. The network must learn to activate the output neurons that produce the phonetic correspondence of the characters currently present in the central input slot. The six surrounding slots are used only to provide contextual information on the correct pronunciation of the central character. Each input slot is represented by 29 neurons that locally code the letters of the English alphabet and other signs, such as comma, period, semicolon, question mark, etc. The 26 output units instead use a distributed representation of the phonemes in the English language and activate a sound generator. The network also contains 80 hidden units fully connected to all input units and to all output units.

After training on a text of 1024 words for only 50 cycles up to a 95% accuracy, the network could read any other written text in understandable language. The pronunciation could be improved by increasing the number of words in the training set. Interestingly, the network learned to read aloud in a way similar to how children learn to speak. Initially, the network learned to segment words in the text, then started to produce sounds similar to the babbling phase of babies, and then gradually produced words starting with shorter ones. The performance of NETtalk is inferior to that of the best systems that are commercially available today, but the results are still remarkable considering the little effort required to develop the network and its relatively small computational requirements.

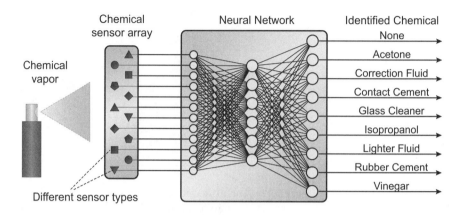

Figure 3.42 Odor discrimination with a neural network. An odorant stimulates an array of chemical detectors whose activations are fed to the input of the neural network trained to classify them (Keller et al. 1994).

OPTICAL CHARACTER
RECOGNITION (OCR)

WEIGHT SHARING

Neural networks are also employed in optical character recognition systems where the diversity of human writing and the distortion of images demand adaptation and generalization. Here we describe an academic implementation because the commercial versions that use neural networks are not disclosed. Le Cun et al. (1990) used a network with four layers of neurons to recognize handwritten postal code numbers. The network receives its input from a preprocessing system that finds the sequence of numbers on the envelope, corrects for size and inclination, and projects each number onto a 16 x 16 pixel matrix. The input layer of the network is composed of 12 sets of 8 x 8 neurons, each covering a portion of the input matrix with an overlap of two pixels per side. Each set of 8 x 8 neurons has exactly the same weight values (a technique known as *weight sharing*, shown in figure 3.41), so that it is necessary to train only one set of weights and clone it for every set of inputs. Another layer of neurons is composed of 12 sets of 4 x 4 units that map in a similar way the neurons in the preceding layer and use weight sharing too. Another layer of 30 neurons is fully connected to all units of the preceding layer and projects connections to the output layer of 10 neurons, each corresponding to a digit. The authors used a weight-pruning technique while training the network on 7000 handwritten digits that produced a 99% correct generalization to 2000 new digits.

Signal processing is another fertile area of application for neural networks because sensors are often noisy and nonlinear, which makes them hard to process. For example, smell detection is a multibillion dollar industry that includes cosmetics, food, medicine, and environmental applications. The human brain can recognize millions of odors by combining the response of only 10,000 receptors, but its performance varies with habituation, fatigue, time of day, and previous experience. For those reasons, there is considerable effort in developing artificial noses that could recognize and classify odors.

ARTIFICIAL NOSE   An artificial nose is composed of an array of sensors that, when exposed to odor molecules, selectively change their physical properties (color, size, resistance, depending on the technology employed). Keller et al. (1994) used a multilayer neural network trained with backpropagation of error to classify odors on the basis of the activation pattern of an array of 12 chemical sensors (figure 3.42). The neural network was composed of 12 input units whose activation was set equal to the corresponding sensor value, and 6 hidden units and 9 output units, each corresponding to a specific odor, such as acetone, glass cleaner, vinegar, correction fluid, etc. Each of those odors activated several sensors in various amounts. After training on a set of sample odors, the network could correctly classify various instances of the odors with very high precision.

3.9 Reinforcement Learning

Despite their computational power, networks trained with the delta rule or with backpropagation can be used only in those situations where one knows the correct response for all input patterns in the training (and validation) set. This is not always the case for agents that operate in partially unknown environments where the feedback (if any) available from the environment is usually rare and generic. Unsupervised learning does not require the specification of the correct response but pays this flexibility with its inability to discriminate the statistical regularities that are potentially useful to the agent and are thus worth learning from those that are not worth learning. Agents that must operate in a partially unpredictable environment require therefore another kind of learning that is neither purely supervised nor purely unsupervised. The solution devised by evolution consists in equipping biological agents with a kind of learning that is linked to the *consequences* of the agent's behavior.

The exact mechanism that implements this kind of learning in biological neural systems is the subject of much research and is not yet completely understood. The existing evidence points to the combined action of evolved *value systems* (Pfeifer and Scheier 1999) and *neuromodulatory* effects (Bailey et al. 2000; Fellous and Linster 1998). The value system has the task of discriminating the behaviors according to their reinforcing, punishing, or negligible consequences. This leads to the production of neuromodulatory signals that can activate or inhibit synaptic learning mechanisms.

Algorithms inspired by this kind of approach have been developed by the machine-learning community. For example, *reinforcement learning* is a class of learning algorithms that attempt to estimate, explicitly or implicitly, the value of the states experienced by the agents in order to favor the choice of those actions that maximize the amount of positive reinforcement received by the agent over time (Sutton and Barto 1998).

CREDIT ASSIGNMENT PROBLEM These algorithms must solve at the same time two formidable credit assignment problems. The *structural assignment* problem is about which action, among all those available, should be credited for a given reinforcement value. The *temporal assignment* problem is about the distribution of credit among all actions involved in a sequence that ended in a single reinforcement value. In order to solve these problems, the agent must explore several combinations of input-output patterns, also known as state-action pairs. In order to reduce the state-action space and make learning feasible in reasonable time, these algorithms are often applied to discrete simulated environments, that is, grid worlds with only a few possible actions (move forward, turn right, stay, etc.) and highly abstract sensory information (Kaelbling et al. 1996).

ACTOR AND CRITIC A simple neural architecture that implements reinforcement learning consists of two modules, the Actor and the Critic, shown in figure 3.43 (Barto 1995; Sutton 1988). Both modules receive information on the current sensory state (State). In addition, the Critic receives information on the current reinforcement value (External reinforcement) from the environment. The output of the Critic generates an estimate of the weighted sum of future rewards (Value or Internal reinforcement). The output of the Actor instead is a probability of executing a certain action. Using the output of the Actor as a probability to execute a certain action allows the system to perform an exploration of the state-action space.

Both modules are trained with a supervised learning algorithm (e.g., backpropagation of error). The Critic is trained to minimize the error between the current reinforcement value produced by the network and the current external reinforcement summed to the discounted (i.e., multiplied by a constant $0 < \gamma < 1$) value computed by the Critic for the next sensory state. In other words, the Critic module learns to estimate the sum of the future rewards

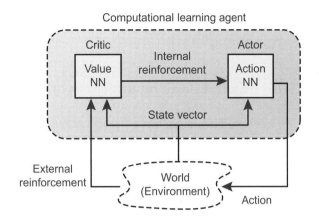

Figure 3.43   The Actor-Critic architecture for reinforcement learning.

for the current state given a certain action of the Action module. The same error is used to modify the output of the Action network, which corresponds to increasing the probability of the current action if the internal reinforcement value plus the external reinforcement is larger than the current internal reinforcement value, and decreasing that probability if the converse holds. This methodology is also known as *temporal difference reinforcement learning* because it compares reinforcement values at different points in time.

Much of current research is aimed at generalizing these algorithms to continuous input and output domains and translating them into models of neural networks (e.g., Mizutani and Dreyfus 1998; Doya et al. 2001). Many observations suggest that the mammalian brain uses learning strategies closely related to reinforcement learning algorithms (Montague et al. 1996; Schultz et al. 1997), but it is not clear whether a distinction between Actor and Critic modules exists.

In this chapter we shall not enter into more detail on this family of algorithms, but we refer interested readers to the already cited book by Sutton and Barto (1998). However, in chapter 6 we will come back to this topic in embodied systems that display reinforcement learning-like behavior by means of different architectural and learning structures. Note that reinforcement learning and evolutionary algorithms attempt to solve a similar class of problems, although in different ways. Despite this, to the best of our knowledge there are not yet objective comparisons between the two families of algorithms.

3.10   Evolution of Neural Networks

The characteristics of neural networks can be encoded in artificial genomes and evolved according to a performance criterion. The advantages of using an evolutionary algorithm are that several defining features of the neural network can be genetically encoded and coevolved at the same time, that the definition of a performance criterion is more flexible than the definition of an energy or error function, and that evolution can be coupled with any of the learning algorithms described above or even used to generate new learning algorithms. In what follows, we review some ways of evolving neural networks and we expand some of these topics in later chapters of this book.

It has been argued (Schaffer et al. 1992; Radcliffe 1991) that evolving neural networks may not be trivial because the population may include individu-
COMPETING   als with *competing conventions* (figure 3.44). This refers to the situation where
CONVENTIONS   very different genotypes (conventions) correspond to neural networks with similar behavior. For example, two networks with inverted hidden nodes may have very different genotypes, but will produce exactly the same behavior. Since the two genotypes correspond to quite different areas of the genetic space, this ambiguity generates two peaks on the fitness landscape instead of only one as would be the case for the error space in the context of backpropagation. Furthermore, crossover among competing conventions may produce offspring with duplicated structures and low fitness. Although experimental studies have shown that in practice this is not a noticeable problem (Hancock 1992), it may still be wise to use small crossover rates (much less than 100%) when evolving neural networks.

The most common way of evolving neural networks consists in encoding and evolving the synaptic weight values. Even in this simple case, there are at least two reasons for using an evolutionary algorithm instead of a learning algorithm: (a) there are no constraints on the type of architecture; (b) it is not necessary to have a detailed description of the network response for each pattern, as in supervised learning methods.

The synaptic weights (including bias weights and time constants, if applicable) are directly encoded on the genotype either as a string of real values or as a string of binary values with a given precision. In the latter case, Schrau-
DYNAMIC ENCODING   dolph and Belew (1992) suggested the use of *dynamic encoding* (see also the

Box 3.2: Learning Classifier Systems

Learning classifier systems (LCSs) are a family of problem-solving techniques that combine reinforcement learning and evolutionary algorithms. Initially proposed by Holland (1976), they have been recently simplified and modified to make them applicable to a wide range of problems where only sparse reward from the environment is available. In the simplified, or zero-level version (ZCS) proposed by S.W. Wilson (1994), an LCS is composed of a population of binary strings that encode the rule IF state THEN action, where state and action are fixed-length strings of bits. An initial population of different rules is created by randomly initializing the values of the joined state-action bit string. The rules whose state portion better matches the signals from the environment are selected as candidates to produce an action in the environment. Since the pool of candidates for a given state may include rules with very different actions, the choice of the rule that will be applied at that specific time step is made on the basis of the fitness of the rules by means of a roulette wheel selection. If at a given time step, the environment provides a reward, this is added to the fitness of the rule that was used. In addition, a fraction of that fitness if shared among all rules in the pool of candidates that had the same action string, as well as among all the rules that had the same action string at the previous time step. Furthermore, the fitness of all other rules that were in the candidate pool, but did not have the same action string, is decreased by a certain value. At certain intervals, a steady-state evolutionary algorithm is applied to the population of rules. Two rules with the best fitness selected by means of a roulette wheel are crossed over and mutated to produce two offspring that replace two rules with the worst fitness selected by means of a roulette wheel. The two offspring receive also half of the fitness of the parents. S.W. Wilson (1995) then extended the algorithm by computing the fitness of the rules on the basis of their ability to *predict* the reward from the environment (instead of the received reward). This extended version (XCS) encourages rules to better map the entire problem space and is currently one of the most widely used forms of learning classifier systems with very good performance in a variety of real-world problems (Lanzi et al. 2000). The choice of rules and fitness attribution in learning classifier systems is inspired by economic policies where winning members of a team share their reward with teammates who proposed similar actions and where team members that proposed other actions are taxed. Interested readers will find a good introduction to various forms of reinforcement learning in (Bull and Kovacs 2005).

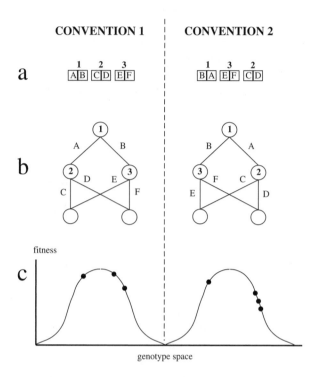

Figure 3.44 Competing conventions. Two different genotypes (*a*) may encode networks that are behaviorally equivalent, but have inverted hidden units (*b*). The two genotypes define two separate hills on the fitness landscape (*c*) and thus may make evolutionary search more difficult. (Adapted from Schaffer et al. 1992.)

section on genetic encodings in chapter 1) whereby the bits allocated for each weight are used to encode the most significant part of the binary representation until the population has converged to a satisfactory solution. At that point, those same bits are used to encode the less significant part of the binary representation in order to narrow the search and refine the performance of the evolutionary network.

Montana and Davis (1989) compared the performance of synaptic weight evolution with that of the backpropagation algorithm on a problem of sonar data classification. The results indicated that evolution finds much better networks and in significantly fewer computational cycles than backpropagation of error (the evaluation of one evolutionary individual on the data set is equivalent to one set of training cycles on the data set). These results have been confirmed for a different classification task by other authors (Whitley et al. 1990).

The architecture of a network can significantly affect its ability to solve a problem. Artificial evolution is an interesting way to find suitable architectures because the space of possible architectures for a given task to be learned is huge and noisy. When evolving architectures, it is common practice to encode in the genotype only some characteristics of the network, such as the number of nodes, the probability of connecting them, the type of activation function, etc., but not the synaptic weights. This strategy is also known as *indirect encoding* to differentiate it from *direct encoding* of all network weights and parameters (Yao 1993). When the connection weights are not specified in the genotype, the decoded neural network is trained with a learning algorithm.

For example, in pioneering work by Harp et al. (1989), the genetic string encodes a blueprint to build a network. This blueprint is composed of several segments, each corresponding to a layer of the network. Each segment has two parts. One part defines node properties, such as the number of units, their activation function, and their geometric layout, and the other part defines properties of the outgoing connections, such as the connection density, learning rate, etc. Once decoded, the network weights are trained with backpropagation. Crossover takes place only between corresponding parts of the segments. The authors showed that when the fitness function included a penalty for the number of connections, the best networks had very few connections. Instead, when the fitness function included a penalty for the number of learning cycles used to reach a predefined error threshold, the best networks learned almost 10 times faster, but used many more connections.

NEUROEVOLUTION OF AUGMENTING TOPOLOGIES (NEAT)

Neuroevolution of augmenting topologies (NEAT) is a method for genetically encoding and evolving the architectures and weights of neural networks (Stanley and Miikkulainen 2002). The approach makes use of genetic operators that can introduce new genes and disable old ones. NEAT was designed to avoid the problem of competing conventions, allowing meaningful crossover between individuals with different genetic length, produce networks of increasing complexity starting from simple ones, and protect topological innovations that may initially display lower fitness but later develop into powerful solutions.

The main insight of NEAT is that genes sharing the same origin are more likely to encode a similar function. In order to keep a genetic historical record, whenever a new gene is created, it is assigned a marker (global innovation number) that corresponds to its chronological order of appearance in

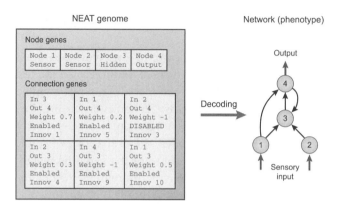

Figure 3.45 Genetic encoding of a network topology within NEAT. Genetic operators can insert new genes or disable old genes. When a new gene is inserted, it receives an innovation number that marks its inception. (From Stanley and Miikkulainen 2002.)

the evolving population. When genes are reproduced and transmitted to offspring, they retain their original markers. The marker number is used to find homologous genes that correspond to alignment points between genotypes of different length, to prevent crossover on competing conventions, and to detect the genetic similarity of individuals in order to create subpopulations of similar individuals. Selective reproduction operates on individuals within the same subpopulation and the fitness of an individual is divided by a number proportional to the number of individuals that are genetically similar.

This last feature is useful for preventing the competition for reproduction between old individuals, which have a relatively high fitness, and individuals with topological innovations (genes with high innovation numbers), which may display relatively low fitness. Since the two types of individuals will be genetically different, they will compete separately for reproduction. NEAT starts with an initial population where genotypes correspond to neural networks of minimal size. The genetic operators can modify the genotypes by inserting new genes that correspond to larger networks. If those larger networks provide a competitive advantage, they are retained and compete with networks of different size.

ANALOG GENETIC ENCODING (AGE)   *Analog genetic encoding* (AGE) is an approach for genetically encoding and evolving topologies of any type of analog network (Mattiussi and Floreano 2007). In chapter 1, we described the application of AGE to the representa-

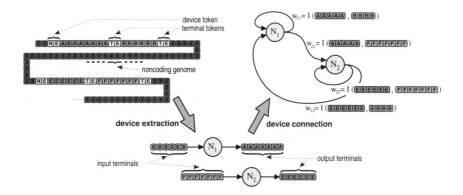

Figure 3.46   Neurons can be represented as symbolic devices with two regulatory (terminal) sequences: one for the output connection and one for the input connection. The device extraction process obtains them from the genome by assigning the sequences of characters between the device token (NE) and the terminal token (TE), and between two terminal tokens, to the respective connection. The terminal sequences of the different neurons are then used to determine the synaptic weights of the network. The interaction map $I(s_1, s_2)$ assigns a weight to a pair of sequences, such as $w_{11} = I(s_{11}, s_{12})$. The entire weight matrix can be calculated by doing this for all pairs of terminal sequences in the network. (From Dürr et al. 2006.)

tion and evolution of analog electronic circuits. Let us now see how AGE operates on neural networks, which are an instance of the general class of analog networks (Dürr et al. 2006). (For the sake of clarity, we will repeat some generalities of the method.)

The genome is constituted of a sequence of characters from a finite genetic alphabet, for example the characters of the ASCII alphabet (figure 3.46). Genes encode devices (neurons) and have both a regulatory and a coding region. The experimenter defines the type of devices that evolution is allowed to manipulate, such as one or more types of dynamic neurons. For each type of gene, a unique device token (a short sequence of characters) signals the beginning of the gene. The device token is followed by a sequence of characters that specify a regulatory region (named terminal) that can interact with the regulatory region of other genes. The end of the regulatory region is specified by a terminal token (a short sequence of genetic characters). Depending on the type of encoded device, a gene can have one or more regulatory regions. For example, a neuron is a device that can interact with other neurons via an input and an output connection; therefore, the genetic representation of a neuron will include two regulatory regions, or terminals.

During the decoding process, the genotype is scanned in search of device tokens and if one is found, the fragment of genome following the token is scanned for the necessary terminal tokens. If a device token in the genome is not followed by the required number of terminal sequences, the gene is considered invalid and the decoding continues with the next device token in the genome. Instead, if the gene is valid, it is decoded into a neuron with an input and an output terminal. The connections and strengths are allocated by allowing the terminals of all decoded neurons to interact with each other, just like regulatory regions of biological neurons interact with each other during gene expression (see chapter 1). Each possible pair of terminals is matched and the quality of the match is scored by an interaction map. The output of the interaction map defines whether a connection exists and, if it does, its strength.

AGE allows for genotypes of variable length and no special protection is needed to manipulate the genetic material. There is actually no apparent distinction between tokens, coding, and noncoding regions of the genome. Therefore AGE allows for all biologically plausible operators, such as character or fragment deletion, insertion, and substitution; genome duplication; homologous crossover; and gene insertion; to mention a few (see chapter 1). When compared to NEAT, AGE reported better performance on a nontrivial dynamic problem (Dürr et al. 2006). However, what matters most is not a comparison of a specific problem, but rather the generality of the encoding approach which allows the evolution of topologies composed of mixed types of neural devices.

Artificial evolution of architectures sometimes involves a growth process that takes place instantaneously or is extended while the neural network interacts with the environment. Since that approach takes inspiration from the way in which biological systems grow and adapt through development, we will describe them in chapter 4.

3.10.1 Evolution and Learning

The combination of evolution and supervised learning provides a powerful synergy between complementary search algorithms (Belew et al. 1992). Since backpropagation is very sensitive to the initial weight values, which may significantly affect the quality of the trained network, evolutionary algorithms can be used to find the initial weight values of networks to be trained with backpropagation. The fitness function is computed using the residual error of the network after training with backpropagation on a given task. Notice that the final weights after supervised training are not coded back into the genotype, i.e., evolution is Darwinian, not Lamarckian, as explained in chapter 1.

Experimental results consistently indicate that networks with evolved initial weights learn significantly faster and better (by two orders of magnitude) than networks with random initial weights. The genetic string can also encode the values of the learning rate and of other learning parameters, such as the momentum in the case of backpropagation. In this case, Belew et al. (1992) found that the best evolved networks employed learning rates 10 times higher than values suggested by common wisdom (i.e., much less than 1.0), but this result may depend on several factors, such as the order of presentation of the patterns, the number of learning cycles allowed before computing the fitness, and the initial weight values.

EVOLUTION OF
LEARNING RULES

Evolutionary algorithms have been employed also to evolve learning rules. In its general form, a learning rule can be described as a function of a few variables, such as presynaptic activity x_j, postsynaptic activity y_i, and the current value of the synaptic connection w_{ij}:

$$\Delta w_{ij} = \Phi\left(x_j, y_i, w_{ij}\right)$$

Chalmers (1990) suggested describing this function as a linear combination of the products between the variables weighted by constants. For example, if we take into consideration only first- and second-order products, the function above becomes

$$\Delta w_{ij} = a_1(x_j,x_j) + a_2(x_j,y_i) + a_3(x_j,w_{ij}) + a_4(y_i,y_i) + a_5(y_i,w_{ij}) + a_6(w_{ij},w_{ij})$$

where the constants a_n can assume discrete values $-1, 0, 1$ or continuous values in the range $[-1, 1]$. These constants are encoded in a genetic string and evolved. The neural network is trained on a set of tasks using the decoded learning rule and its performance is used to compute the fitness of the corresponding learning rule. The initial synaptic weights are always set to small random values centered around zero.

For example, Chalmers employed a fitness function based on the mean square error. A neural network with a single layer of connections was trained on eight linearly separable classification tasks. The genetic algorithm evolved a learning rule similar to the delta rule described earlier. Similar results were obtained by Fontanari and Meir (1991). Dasdan and Oflazer (1993) employed a similar encoding strategy to evolve unsupervised learning rules for classification tasks. The authors reported that evolved rules were more powerful than comparable, human-designed rules. Baxter (1992) encoded both the architecture and whether a synapse could be modified by a simple Hebb rule (the rule was predetermined). Floreano and Mondada (1996b) allowed evolution to choose different learning rules for each synaptic connection and evaluated the approach for mobile robot control using behavioral fitness function. This method, and further modifications by other authors, will be explained in more detail in chapter 6.

As we have seen at the beginning of this chapter, it has been known for a long time that learning may affect natural evolution (Baldwin 1896). Empirical evidence shows that this is the case also for artificial evolution when combined with some form of learning (Nolfi and Floreano 1999b).

Hinton and Nowlan (1987) proposed a simple computational model that shows how learning might help and guide evolution. The authors considered the case where a neural network confers added reproductive fitness on an organism only if it is connected in exactly the right way. In this worst case, there is no reasonable path toward the good network and a pure evolutionary search can only discover which of the potential connections should be present by trying possibilities at random. The good network is "like a needle in a haystack" (p. 495).

In their computational explorations, Hinton and Nowlan use genotypes with 20 genes corresponding to a neural network with 20 potentials connections. A gene can take three possible values: 0, 1, and ?, which represent, respectively, the absence of the connection, the presence of the connection, and a modifiable state (absence or presence of the connection) that can change its value according to a learning mechanism. In the authors' model, the learning mechanism is a simple random process that keeps changing modifiable connection weights until a good combination (if any) is found during the limited learning time of the individual.

In the absence of learning (i.e., when genes can only have 0 and 1 allelic values), the probability of finding a good combination of weights would be very small given that the fitness landscape looks like a flat area with a spike in correspondence to the good combination of genes (figure 3.47, thick line). On such a surface genetic algorithms do not perform better than a random search algorithm. However, if learning is enabled, it is more likely that some individuals will achieve the good combinations of connection values at some point during training time and start to collect fitness points.

The addition of learning, in fact, produces a smoothing of the fitness surface area around the good combination of genes (weights), which can be discovered and easily climbed by the genetic algorithm (figure 3.47, dashed line). This is due to the fact that not only the right combination of genes but

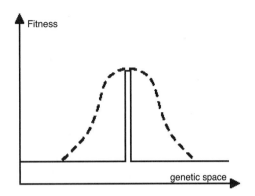

Figure 3.47 Fitness landscape with and without learning. In the absence of learning, the fitness landscape is flat, with a thin spike in correspondence to the good combinations of genes (thick line). When learning is enabled (dashed line), the fitness surface displays a smooth hill around the spike corresponding to the gene combinations which have in part correct fixed values and in part unspecified (learnable) values. The continuous line represents the fitness for each possible combination of two alleles ([0, 1]) while the dashed line represents the fitness for each possible combination of three alleles [0, 1, ?]). Redrawn from Hinton and Nowlan (1987).

also combinations, which in part have correct genes and in part have unspecified (learnable) genes, report an average fitness greater than 0. Notice that the fitness of an individual is proportional to the number of fixed correct values because the time needed to find the correct combination through learning is inversely proportional, on average, to the number of learnable values. Hinton and Nowlan claimed that evolution with learning "is like searching for a needle in a haystack when someone tells you when you are getting close" (1987, p. 496).

This simple model also accounts for the Baldwin effect that postulates that characters that are initially acquired through learning may later be fixated in the genotype. Once individuals with part of their genes fixed on the correct values and part of their genes unspecified (learnable) are selected, individuals with fewer and fewer learnable genes tend to be selected because the fitness increases monotonically by decreasing the number of learnable genes. In other words, characters that were acquired through learning in early generations tend to become genetically specified in later generations.

Hinton and Nowlan's model is very simple and elegant, but has some limitations: (1) learning is modeled as a random process; (2) there is no distinction between the learning task and the evolutionary task; (3) the environment

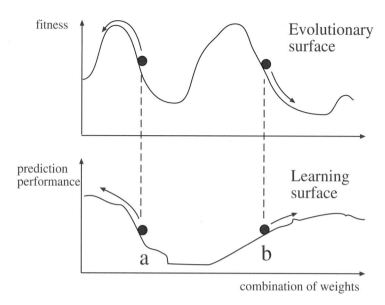

Figure 3.48 Fitness surface for the evolutionary task (food reaching) and performance surface for the learning task (sensory prediction). Movements due to *learning* are represented as arrows. Point a is in a region where the two surfaces are dynamically correlated. Even if a and b have the same fitness on the evolutionary surface at birth, a has more probability to be selected than b since it is more likely to increase its fitness during life than b.

does not change; (4) the learning space and the evolutionary space are completely correlated. The two spaces are correlated if genotypes which are close in the evolutionary space correspond to phenotypes which are also close in the phenotype space.

By systematically varying the cost of learning and the correlation between the learning space and the evolutionary space, Mayley (1996) showed that: (1) the adaptive advantage of learning is proportional to the correlation between the two search spaces; (2) the assimilation of characters first acquired through learning is proportional to the correlation between the two search spaces and to the cost of learning (i.e., to the fitness lost during the first part of the lifetime in which individuals have suboptimal performance); (3) in certain situations learning costs may exceed learning benefits.

Despite those caveats on the correlation of the evolutionary and learning landscapes, Nolfi et al. (1994a) found significant benefits by combining evolution and learning even when the two processes attempted to improve two different tasks and therefore may have had partially uncorrelated search landscapes. In their work, networks were evolved to drive an agent toward food items while parts of their synaptic weights were trained to predict the sensory consequences of their actions. The authors explained the added benefit of evolution and learning by suggesting that evolution may select individuals that happen to be in areas of the landscapes that are dynamically correlated, that is, displacement in one direction induces the same effect (positive or negative) on the two landscapes.

To understand this, imagine two different search surfaces, an evolutionary surface and a learning surface (figure 3.48). Modifications of synaptic weights due to learning produce a movement of the individual phenotype both on the learning and on the evolutionary surface. However, since learning tries to maximize performance on the learning task, individuals will move toward higher areas of the learning surface. Given that the way in which individuals move in weight space affects their fitness (the total fitness of the individual is the sum of the fitness values received during such displacements on the weight space), evolution will tend to select individuals located in areas where, by increasing their performance on the learning task, they also increase their performance on the evolutionary task.

Consider, for example, two individuals, a and b, that are positioned in two distant locations in weight space and have the same fitness at birth, i.e., the two locations correspond to the same height on the fitness surface (figure 3.48). Individual a is located in a region where the fitness surface and the learning surface are dynamically correlated, that is, a region where on average movements that result in an increase in height on the learning surface result in an increase in height on the fitness surface too. Individual b, on the other hand, is located in a region where the two surfaces are not dynamically correlated. If individual b moves in weight space, it will go up on the learning surface but not necessarily on the fitness surface. If learning is enabled, the two individuals will move during their lifetime in a direction that improves their learning performance, i.e., in a direction where their heights on the learning surface tend to increase. This implies that individual a, which is located in a dynamically correlated region, will end up with a higher fitness than individual b and, therefore, will have a better chance to be selected. The final result is that evolution will have a tendency to progressively select individuals that are located in dynamically correlated regions. In other words, learning improves exploration of the search space by allowing evolution to select individuals that improve their performances with respect to both the learning and the evolutionary tasks.

Harvey (1997) proposed another explanation for the observed benefits of learning and evolution on different tasks by using a geometric argument. He assumes that the synaptic weights of neural networks selected for reproduction are displaced from a point of high fitness by random mutations. Then, whatever learning mechanism and task are applied, the trajectory of the synaptic weights being modified has a high likelihood to transit closer to that point of high fitness, thus raising the fitness of the individual. However, some of the predictions of this elegant model were not confirmed in further experiments by Nolfi (1999).

LAMARCKIAN
EVOLUTION

One may wonder whether Lamarckian evolution (i.e., an evolutionary process where characters acquired through learning are directly coded back into the genotype and transmitted to offspring) could be more effective than Darwinian evolution (i.e., an evolutionary process in which characters acquired through learning are not coded back into the genotype). Ackley and Littman (1992), for instance, claimed that in artificial evolution, where inherited characters can be easily coded into the genotype given that the mapping between genotype and phenotype is generally quite simple, there is no reason for not using Lamarckian evolution. Indeed, the authors showed that Lamarckian evolution is far more effective than Darwinian evolution in a stationary environment (where the input-output mapping does not change). However, Sasaki and Tokoro (1997) showed that Darwinian evolution largely outperforms Lamarckian evolution when the environment is not stationary or when different individuals are exposed to different learning experiences.

3.11   Neural Hardware

As soon as the first models of learning neural networks became available at the end of the 1950s, considerable effort was invested in designing parallel machines with adaptive elements that could emulate neurons and synaptic weights. One of the first neural computers, the Mark I Perceptron, was composed of several hundred potentiometers individually controlled by electric motors that played the role of adjustable synaptic weights (Mark I Perceptron 1960). The potentiometers could be connected in arbitrary ways to implement several types of neural architectures that could be trained with an early version of the delta rule in order to recognize characters projected on a screen.

In the late 1980s and early 1990s, a decade that marked the rapid expansion of neural network research and large availability of personal digital computers, research efforts in neural hardware were mainly aimed at design of parallel computers from an assembly of several digital computers. The research in that period followed two major approaches. One approach consisted in designing specialized neural computers whose choice of components and wiring matched the specificities of one or more families of neural models (Ienne et al. 1996). Another approach consisted in using arrays of general-purpose processing units that could take the form either of a parallel coprocessor attached to the serial port of a desktop computer or of a standalone parallel computer with several processing units. In both cases, using those parallel computers required special programming languages and techniques to fully exploit the parallel features of the hardware.

TRANSPUTERS For example, transputers (transistor computers) by INMOS were composed of microcontrollers (a CPU with memory, input/output facilities, and a clock that can operate as a standalone computational device) specially designed to be interconnected by fast serial links to other transputers and operate in parallel. Often these transputers were attached to a desktop computer for programmability and input/output data storage on hard disks. After an initial success, transputers gradually disappeared because they could not match the decreasing price and increasing speed of desktop computers.

CONNECTION MACHINE Another example of general-purpose digital devices used for simulating neural systems was the Connection Machine by Thinking Machines Corporation (Hillis 1987), a massively parallel computer composed of up to 65,536 processors, each with its own memory, that could be virtually wired in a huge number of configurations. The Connection Machine was originally conceived for research in artificial intelligence and artificial life, but was also used for computer graphics and other computationally intensive tasks. Eventually, it lost ground in the competition with parallel supercomputers produced by other manufacturers and was dismissed (Taubes 1995).

OPTICAL SYSTEMS Optical systems have been considered both for neural processing and associative information storage. The advantage of optical processing is that optical beams do not interfere when they cross. Therefore, neural networks with optical connections can process in parallel and at light speeds high-dimensional data. The basic idea in optical processing (figure 3.49) is to present the input pattern as multiple light beams (one per input unit). The intensity, or amplitude, of the light beam encodes the strength of the signal. Each beam irradiates a separate row of a semitransparent matrix, which plays the role of synaptic weights. The matrix can be implemented as a liq-

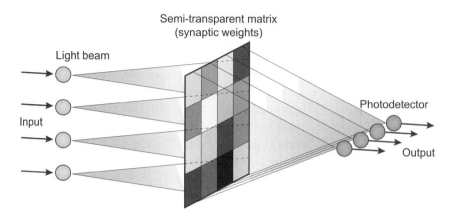

Figure 3.49 Schematic representation of an optical neural network. Input signals are encoded by light intensity; each input signal irradiates one row of the semitransparent matrix, which functions as the synaptic weight matrix. The output units are represented by a set of photodetectors, each collecting light from one column of the matrix. Learning is achieved by modifying the transparency of the matrix cells.

uid crystal device (LCD) where the position (reflectance) of the crystals at each location can be individually programmed (Casasent 1992). The readout of the weighted inputs is operated by a set of photodetectors (one per output unit) that collect light from columns of the matrix. The synaptic weight update can be implemented in a separate digital computer that modifies the LCD matrix.

A similar functionality has been demonstrated with a holographic device where the modulation strengths of the cells in the holographic grating can be individually set (Psaltis et al. 1990). In this case, learning has been implemented on the device itself by using a special mirror that feeds back the input rays on the holographic device and records at each cell the pattern of interference with the outgoing light rays in the form of an input-output correlation analogous to the Hebb rule. It has been shown that such a holographic memory can store and retrieve information similarly to the Hopfield network described earlier.

Associative memory storage and reconstruction (similar to the functionality of Hopfield networks described earlier) has been demonstrated with thick holographic devices. These devices can store multiple images that can be singly retrieved by stimulating the device with a suitable light signal, which corresponds to the memory address in electromagnetic storage devices. In addition, it has been shown that images can be retrieved by projecting a corrupted or incomplete version of the image onto the holographic device. This is achieved by a set of mirrors and threshold devices that select the brightest image (corresponding to the highest correlation with the corrupted image) returned by the holographic device and feed it back recursively until the image is fully reconstructed.

Today, entry-level desktop computers are largely sufficient to simulate and operate in reasonable time neural networks for most applications in signal processing, pattern recognition, control, and data mining. However, dedicated neural hardware is still appealing for embedded systems that demand the smallest possible size and power consumption without compromising computational speed in real-time operation.

FIELD-PROGRAMMABLE GATE ARRAY (FPGA)

Within this context, field-programmable gate arrays (FPGAs) have been used to implement artificial neural networks that take advantage of their rapid hardware reconfiguration and parallel signal processing (see chapter 1). For example, Eldredge and Hutchings (1994) suggested exploiting run-time reconfiguration to decompose the neural algorithm in several sequential stages and let the FPGA reconfigure itself after each stage. The algorithm of backpropagation of error can be decomposed into stages corresponding to the activation of each neuron layer followed by computation of the delta errors for each layer and finally by update of the connection layers. This decomposition requires less hardware because only a subset of the algorithm must be implemented in hardware at any given time and can speed up the execution of each stage by optimally matching the architecture of the FPGA to the specificities of each computational stage. Rapid hardware reconfiguration has also been exploited to implement neural systems that require modification of the neural topology during training, such as in ART networks or in networks that incorporate pruning mechanisms (Perez-Uribe and Sanchez 1996).

Dedicated neural hardware is used also for implementation of networks of spiking neurons because those models often require large assemblies of neurons and/or incorporate detailed modeling of the physiological mechanisms regulating the membrane dynamics that demand significantly higher computational power.

In the case where the membrane dynamics are greatly simplified and do not model the dynamics of ion exchange across the membrane, a spiking neuron can be reduced to a digital device whose input/output streams are composed of 1s and 0s (spike, no spike) and whose membrane dynamics can be approximated by elementary digital operators, such as a spike ac-

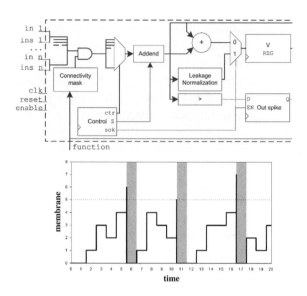

Figure 3.50 *Top*: Cell architecture implementing a spiking neuron. The main parts are a register holding the membrane potential value, a connectivity mask block, an addend block, a leakage and normalization block, and a control unit. *Bottom*: Effect of incoming spikes on the membrane potential. Each time a spike is received the membrane potential is increased (a constant decrement is also added to model leakage). When the neuron threshold is reached the neuron emits a spike and the membrane potential is reset and maintained at 0 for some time (gray column).

cumulator, a decrement operator for leakage, a threshold comparator, and a reset unit (Floreano et al. 2002). FPGAs, which are composed of thousands of simple transistors with low-band connectivity, offer an ideal platform for implementing large assemblies of such spiking neurons (Roggen et al. 2003b), as shown in figure 3.50, that can operate on a time scale similar to biological neurons.

Instead, if one needs to better approximate the neuron membrane dynamics, it is possible to exploit the nonlinear physical properties of transistor operation in analog very large-scale integrated (aVLSI) circuits. This approach, also known as *neuromorphic engineering*, was pioneered by Mead (1989, 1990) and has been constantly expanding over the last few years. There are at least three advantages with reproducing neural circuits in analog VLSI. The first is that certain operations typically used in neural computation, such as addition and multiplication, can be implemented with far fewer transistors than

NEUROMORPHIC
ENGINEERING

in digital VLSI. The second is that most variables involved in neural computation are described by real-valued numbers and continuous functions, which can be mapped more efficiently in terms of accuracy and silicon surface in analog VLSI than in digital VLSI. The third is that analog VLSI can interface directly to the real world, which is intrinsically analog, without the need of additional analog-to-digital converters used in digital VLSI.

Neuromorphic engineering makes use of standard metal oxide semiconductor field effect transistors (MOSFETs). In conventional analog and digital circuit design a MOSFET is considered active when operating in the so-called above-threshold region and is considered OFF when operating in the so-called subthreshold region. The subthreshold region is characterized by extremely low voltage differences. On the contrary, neuromorphic engineering relies on circuits that exploit the whole subthreshold region as an active operating region. This permits the realization of complex analog computation and of functions typical of neural computation at very low power consumption (Vittoz 1985; Sarpeshkar 2006), albeit at the cost of reduced noise tolerance.

Analog VLSI circuits have been used to realize artificial retinas (see Indiveri and Douglas 2000 for a review), cochleas (e.g., Lazzaro et al. 1994; van Schaik et al. 1996; Sarpeshkar 2006), and various networks of firing-rate and firing-time neurons. Some challenges of neuromorphic engineering are the implementation of global learning rules (where the weight modification does not depend only on pre- and postsynaptic activity) and storage of the values of synaptic weights when the circuit is not powered, but recent work has pointed to various promising solutions. Often neuromorphic chips include a combination of analog and digital circuitry. For example, analog VLSI may be used to reproduce dendritic integration and membrane dynamics while digital VLSI may be used to manage spike generation and broadcast.

A major obstacle to scaling up of both digital and analog neural circuits is connectivity. Biological brains extend in three dimensions and each neuron can be connected to several thousand other neurons. Electronic circuits instead are limited to two dimensions where connection wires require not only a surface exponentially growing with the number of neurons but also solutions to avoid wire crossing. Furthermore, in neuromorphic sensory circuits only a fraction of the neurons can be connected by wires to external devices.

A possible solution is to distribute silicon neurons across several chips, each implementing a small population of neurons with local connectivity, and to use high-speed data buses to interconnect those chips. *Address-event representation* (figure 3.51) is a simple and yet powerful interchip communica-

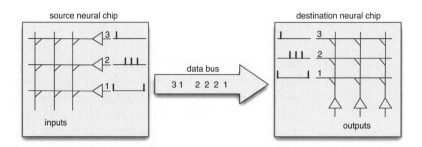

Figure 3.51 Schematic representation of address-event representation for communication between two neural chips with three spiking neurons each. The numbers represent the unique address of the neurons. Adapted from Deiss et al. (1999).

tion protocol that has been developed to encode and broadcast asynchronous spikes among neurons sitting on different chips, as well as communicate between sensors and neural chips (Mahovald 1994; Deiss et al. 1999). The idea is to share a single data bus, which functions as a universal multiplexed axon, to connect several neurons. Whenever a neuron emits a spike (event), it immediately takes control of the data bus and sends its own address. The receiving neural chip decodes the addresses and distributes the events to the neurons. In this representation, the information is encoded in the temporal sequence of spike events sent over the data bus. A problem may emerge if there are several neurons that spike at intervals too close to be queued properly on the data bus. However, considering that the time constant of silicon neurons is approximately 1 ms (similar to biological neurons), a data bus operating at 1 MHz could support asynchronous firing of up to 1000 neurons per chip. Since in practice not all neurons fire every millisecond, the number of neurons per chip could be one or two orders of magnitude larger.

3.12   Hybrid Neural Systems

MULTIELECTRODE
ARRAY (MEA)

It is now possible to study cultures of neural tissues comprising hundreds of interconnected neurons living on multielectrode arrays (figure 3.52, left). A multielectrode array (MEA) is a biocompatible material that integrates an array of microelectrodes arranged in a regular grid (matrix, hexagonal, circular, etc.). The MEA can administer electrical stimulation to, and record electrical activity from, a living culture of neurons and glia cells extracted from a ner-

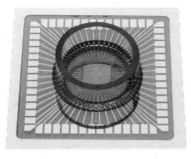 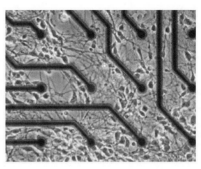

Figure 3.52 *Left*: Multielectrode array (model shown by MultiChannel Systems GmbH) with transparent culture dish. *Right*: Detail photograph of living culture of neurons and connections on the MEA surface (Potter and DeMarse 2001). The recording/stimulating electrodes appear as small dots spaced every 200 μm (image courtesy of Steve Potter).

vous system. Although MEA technology has been available for more than 25 years (Pine 1980), the relevance of the data collected was limited by the fact that neural tissues degraded in only a few hours on the MEA surface.

Potter and DeMarse (2001) surmounted that problem by using culture dish lids that form a gastight seal and incorporate a transparent hydrophobic membrane that is selectively permeable to oxygen and carbon dioxide, but relatively impermeable to water vapor. This prevents contamination by pathogens in the air and greatly reduces dehydration of the neural tissue. The authors showed that cortical cultures extracted from rat embryos and deposited on the sealed MEA (figure 3.52, right) exhibited robust spontaneous electrical activity after more than one year.

MEA technology is now also used to create hybrid systems comprised of neural wetware and hardware with the goal of combining the best of both worlds: the adaptive and regenerative properties of living neural networks and the programmability and computational power of electronic chips. Although it is not yet clear to what extent a neural culture dissociated from a brain could retain its functionalities and what is the best way to communicate with it, it has been recently shown that a computer-controlled environment

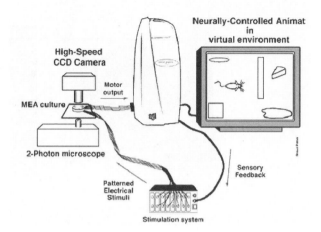

Figure 3.53 Experimental settings for the neurally controlled animat. A network of hundreds of mammalian cortical cells (neurons and glia) are cultured on a transparent multielectrode array. Neural activity is recorded and clustered into patterns used to control the movement of an artificial animal within a simulated environment. Sensory input to the animat is translated into patterns of electrical stimuli sent back to the network (DeMarse et al. 2001) (drawing courtesy of Steve Potter).

can guide a neural culture to selectively display desired patterns of activity for certain stimulation (Shahaf and Marom 2001).

Le Masson et al. (2000) interfaced a living thalamic neuron with simulated and analog VLSI neurons to study the effect of feedback inhibition in decorrelating thalamic activity from sensory activity, which happens in sleep onset.

DeMarse et al. (2001) interfaced a culture of rat cortical neurons with a simple artificial animal, or animat (S.W. Wilson 1987), in a simulated environment (figure 3.53). The spiking activity of the neural culture was sampled every 200 ms by a grid of 60 electrodes spaced every 200 μm. Each electrode channel recorded activity from a population of nearby neurons, which was transformed into a real number by integrating the spike count over time and passing the resulting number through a tanh squashing function. The vector of 60 obtained values was then categorized according to an adaptive clustering technique (similar to self-organizing maps described in section 3.7.3) and an arbitrary motor action of the animat was associated to each cluster. The animat had five types of sensory events (collision with a wall and four types of movements) that were mapped to five electrodes in the MEA that

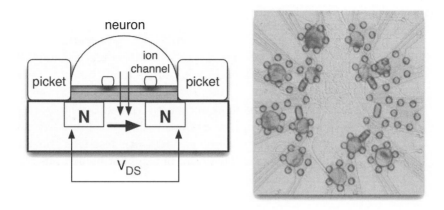

Figure 3.54 Neural interface with field-effect transistors. *Left*: Schematic representation of neuron-chip connection. *Right*: Eight leech neurons with axons and dendrites on chip. From Fromherz (2003).

were previously shown to elicit reproducible spikes in the biological network. Sensory feedback from the animat was provided to the neural tissue with 100 ms delay after reading the network activity. The authors showed that the neural tissue increased the number of different activity patterns for 50 minutes, which resulted in a good exploration of the environment by the animat. Although this behavior cannot be defined as adaptive, it describes a novel methodology to investigate neural activity in close loop with an environment.

A similar approach was used by Fleming et al. (2000) to connect the brain of a lamprey to a mobile robot placed in a circular arena. The reason for using an entire brain instead of a neural culture is that the former has a well-defined bilateral structure that, when properly connected to the sensors and motors of the robot, results in predictable light-dependent behaviors. The manipulation of the environmental properties allowed the authors to study specific adaptive modifications of the behavior produced by the living brain.

FIELD-EFFECT
TRANSISTORS (FETs)

An alternative technology makes use of field-effect transistors (FETs). The advantage of this technology is that it capitalizes on widely available semiconductor techniques whose miniaturization allows the simultaneous recording and stimulation of tens of thousands of neurons. The transistor is coated with silicon dioxide, which is a biocompatible and inert insulator, and with a matrix of proteins (figure 3.54, left). The neuron is placed on the gap between drain and source and adheres to the chip surface by means of protruding proteins that bind with those on the chip matrix. When the neuron emits a spike, it causes an extracellular potential that polarizes the silicon dioxide and modulates source-drain current. Conversely, when a voltage is applied to the chip, the capacitive current through the silicon dioxide generates a potential that opens voltage-sensitive ion channels on the neuron membrane. This arrangement allows precise recording and stimulation of single neurons, whereas the work described above with MEAs always interacted with small groups of neighboring neurons.

Fromherz and collaborators (Fromherz 2003) perfected the FET-based technology and demonstrated its potentials for hybrid computational systems. Fromherz and coworkers showed that two disconnected neurons can be put into communication by an electronic circuit that functions as an artificial synapse. Conversely, they showed an example where the circuit stimulates one of two interconnected neurons and reads the output of the other neuron at a distance. An improved version of this technology includes the formation of small fence pickets around the neuron cells to prevent displacement of the neuron from the transistor during the outgrowth of connections to other neurons (figure 3.54, right).

More recently, they progressed toward the creation of neuronal networks with desired topologies, for example in configurations that resemble the Hopfield networks described above, by encouraging neurons to grow axons along predefined paths and establish synaptic connections with other neurons. This was achieved by laying matrix protein on the silicon dioxide only along predefined paths that provide a chemical guide to the outgrowing neural connections. Neurons successfully established a functional synapse when their axons met. However, the limitation of this technique is that mechanical forces pull the connections out of their path. To counteract those forces, Fromherz and collaborators deposited a sheet of polyester on the surface of the transistor and used photolithography to carve microscopic pits and channels for the neurons and their outgrowing connections.

Hybrid neural hardware is an exciting recent technology that may significantly advance our understanding of neural systems both in circuits as they appear in nature and in novel, task-specific circuits that we may wish to build. However, its use as a machine that partly delegates computation to the living neural tissue should be taken with care. On the one hand, as soon as the functioning principles of the neural tissue are understood, they can be implemented in a purely software or hardware system and there is no reason for continuing to use the delicate biological system. On the other hand, if the principles of the neural tissue are not fully understood, then it may not be wise to delegate part of the computation to it.

3.13 Closing Remarks

In this chapter we offered a small window onto the vast landscape of neural computation and decided to take a perspective that is often adopted in major textbooks on the topic. That perspective looks at the brain as a parallel distributed system of discrete units connected by modifiable wires, differentiates the phase of neural activation from that of synaptic modification, and presents adaptation as a data-driven learning process. Although these classifications may serve as a guide to approach the rich field of neural computation, they can be misleading if taken too strictly.

Models of the brain tend to build on analogies with dominant technologies of their time: water pipes, telephone network, computer, and the Internet (Kirkland 2002). Today's models are based on the analogy with distributed systems whose units can talk only to other interconnected units. However, biological brains seem to rely also on more global processes, such as direct exchange of ions across membranes, interaction with glia cells, and gases emitted by active neurons that diffuse across cellular structures. In order to explore the potentials of these global processes, Husbands et al. (1998) developed and evolved a class of neural networks denoted as *GasNets* inspired by the modulatory effects of freely diffusing nitric oxide gases that affect the response profile of affected neurons. The neurons of GasNets, which are spatially distributed over a 2D surface, emit "gases" that diffuse through the network and modulate the profile of the activation function according to the local concentration (more detail on the application of GasNets to robotics will be given in chapter 6). The authors showed that the modulation caused by gas diffusion introduces a form of plasticity in the network without synaptic modification.

GasNets

On a similar note, the distinction between neural activation and synaptic modification is often taken to imply that a neural network has two functioning modalities: a modality during which it activates the neurons so as to operate in the environment using the knowledge embedded in the pattern of synaptic values and a modality during which it acquires new knowledge by modifying the pattern of synaptic values. Then the question arises of when should the network acquire new knowledge and when should it operate in the environment. Most models (with the notable exception of ART) address this question by imposing a time limit on the training phase. If the environment changes and the network performance is no longer suitable, the network is trained anew in the new conditions. However, living brains seem capable of continuously adapting to the environment and there is no evidence for a separation between an activation phase and a learning phase.

Furthermore, it has been shown that (1) a network with dynamic neurons and without synaptic plasticity is still capable of displaying learning-like behaviors (Blynel and Floreano 2003); and (2) a network with synaptic plasticity and static neurons is capable of using fast synaptic modification to change behavioral response without acquiring new knowledge and/or skills (Urzelai and Floreano 2001). In both cases, the parameters of the neural networks, the time constants and the learning rules, respectively, were genetically encoded and evolved instead of using available learning models.

These two examples challenge the more or less implicit assumption that neural activations are responsible for behavior and synaptic change is responsible for learning. An alternative perspective is to consider the brain as a dynamical system characterized by several time constants associated to various processes, such as the integration time of neuron membranes, the modification rate of synaptic strength, and the time delay of signal propagation through axons and through extracellular gases, to mention a few. Within that perspective, what matters is the relative rate of change of the various processes. For example, if the modification rate of synaptic strength is equal to or faster than the integration time of neuron membranes, a Hebbian process can serve as a mechanism to change the output of the neuron without necessarily storing new information. Similarly, a neural network whose synaptic modification rate is infinitely small (equivalent to no modification), but whose neurons have feedback interconnections and a suitable combination of integration times, could modify permanently its behavior after the occurrence of a specific event. Such a dynamical system perspective does not require us to make a mechanistic distinction between behavior and learning of the network.

The notion of learning itself is questionable. Learning is frequently understood as a process of experience-driven change that allows an organism to improve by incrementally acquiring new skills, knowledge, and memories (Reisberg 1999). We strongly doubt that all modifications fall within that type of learning. For example, the development of sensitivity to certain types of visual patterns, but not to others, is an example of adaptation to the environment that does not correspond to incremental accumulation

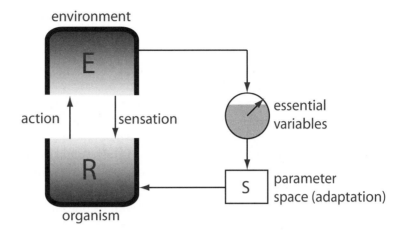

Figure 3.55 Components of an adaptive system as a *homeostat*. The organism and the environment form a unique dynamical system that can be divided for the sake of illustration into external and internal variables interconnected by motor and sensory channels. Modifications of the environment due to actions by the organism or other factors affect essential variables of the organism. If those essential variables fall outside a normal range (gray area), a modification of the parameter space S occurs, which affects the variable phase space defining the activity of the organism. That modification in S has the effect of bringing the essential variables back into the normal range (adapted from Ashby 1960, p. 83).

of new knowledge. Furthermore, a teleological interpretation of learning as change toward the better implies that the organism knows in advance what will be good for its survival in the future, a vicious circle that we already highlighted in chapter 1.

More than 40 years ago, Ashby (1960) proposed an original and still interesting interpretation of adaptation as a homeostatic process that brings an organism back to its equilibrium point following some perturbation of its environment. In Ashby's perspective, "adaptive behavior is equivalent to the behavior of a stable system, the region of stability being the region of the phase-space in which all essential variables lie within their normal limit" (Ashby 1960, p. 64). The essential variables include temperature, blood pressure, energy, and other factors necessary to the survival of the organism. Ashby developed this framework in order to encompass a large number of adaptation phenomena, from habituation all the way to trial-and-error learning, and built an electromagnetic device, named *homeostat*, that displayed some of those adaptation properties (figure 3.55). Although Ashby did not attempt to establish analogies between components of his theory and physiological processes in the brain, recent work on the evolution of learning rules for dynamical neural networks exposed to changing environments showed that evolved systems behaved according to the theory suggested by Ashby (E.A. DiPaolo 2000).

Another important aspect of Ashby's theory is that the functioning of a brain can only be understood if studied in the context of a body and of an environment. Translated into the context of this chapter, it means that neural models should be developed and studied in the context of an organism that behaves in an environment. We shall address the full implications of this issue in chapter 6, but we wish to emphasize here that if a neural network can affect the environment with its output, then the probability distribution function of the input data is not likely to be uniform as assumed by most neural models described in this chapter. Consequently, when these neural models are used in behavioral systems, either the models must be modified to account for learning skewed data distributions or the data distribution must be preprocessed and corrected to ensure convergence of the learning algorithms.

Finally, it should be noted that this chapter presented only a few *generic models*, that is, architectures and learning algorithms that can be applied to a variety of application domains. As a matter of fact, the literature on neural computation abounds in interesting *specific models*, that is, architectures and learning mechanisms that are either tailored to a specific application domain or are intended to reproduce the functionalities of specific parts of the nervous system, such as visual object recognition, spatial orientation and navigation, hand-eye coordination, selective attention, and so forth (see Arbib 1995 for a rich collection of such models). These latter types of models are often composed of modular and heterogeneous architectures and use different neuron models and learning rules in different parts of the network. Unfortunately, these models are rarely compatible because they are based on different assumptions and levels of abstraction, which makes it almost impossible to integrate them into a single artificial brain.

Leaving aside the question of the suitable level of abstraction for a neural model (Herz et al. 2006), it is likely that a great majority of the manifestations of learning and adaptation in biology are subsumed by the same physiological mechanisms, such as long-term potentiation and depression of synaptic connections based on local electrochemical events. If we subscribe to this hypothesis, the big open question is how those two mechanisms are regulated and combined in a complex neural system to provide a variety of learning abilities. In order words, the challenging question is how neurons are connected together and how do they affect, not only their activity but also the type, onset, and offset of synaptic mechanisms. The importance of connectivity recognized by Ramón y Cajal (1909, 1911) more than a century ago still remains a central issue for understanding and re-creating the brain.

3.14   Suggested Readings

Swanson (2003) presents a very clear and nicely illustrated description of the architectural and functional components of brains, starting from the simplest circuits found in invertebrates all the way to the macromodules that make up a mammalian brain. The compact format, highly readable style, and annotated bibliography make this book a strongly recommended introduction to nervous systems. More detailed information on the architecture and physiology of synaptic connectivity can be found in the book edited by Shepherd (1990), which is organized in chapters dedicated to the major components of the mammalian brain. Churchland and Sejnowski (1992) instead provide an overview of biological brains with emphasis on computational roles of circuits and modules. The book brings together biological, cognitive, and computational models into a uniform framework. In a similar vein, but on a much larger scale, is *The Handbook of Brain Theory and Neural Networks*, edited by Arbib (1995), which collects more than 250 articles especially written by leading experts in their fields. Each article is sufficiently short and comprehensive to be read at a single sitting and yet provides a good coverage of the topic.

After more than 50 years of neural computation theory and practice, there are several books on the market that offer good introductions to artificial neural networks. We recommend the two seminal volumes on *Parallel Distributed Processing* (Rumelhart et al. 1986b; McClelland et al. 1986) that in 1986 resuscitated the interest in artificial neural networks among computer scientists, cognitive scientists, and engineers. The two volumes offer not only a historical perspective on the field but also a very clear explanation of several models that are still in use today. Although the authors write for a general readership, their emphasis on cognitive models makes the books particularly appealing for computational psychologists. Hertz et al. (1991) instead offer an introduction to neural networks from the standpoint of physics. For example, readers will find there one of the best and concise explanations of neural models inspired by models of spin glass theory. The book covers the major neural models with great clarity and marks mathematical sections that can be skipped without major loss for those who are mainly interested in the implementation of the algorithms. For readers who have a background in computer science and information theory, we recommend the second, revised edition of the comprehensive book by Haykin (2007), which not only includes very recent models but also highlights the similarities between neural network techniques and statistical theory. Finally, for readers with a background in engineering, we recommend the book by Eliasmith and Anderson (2003), who do an excellent job at explaining what artificial neurons can do and how they can be used to solve various classes of problems, instead of focusing on neuron and learning models as most other books do (including this chapter).

We also suggest the rich collection of milestone papers on neural computation edited and annotated by Anderson and Rosenfeld (1988), where readers will find precursory writings from psychologist William James, seminal papers by McCulloch and Pitts, Hebb, Hopfield, Grossberg, Rumelhart, and Kohonen, to mention a few, along with several other gems that are still very inspirational after several years. Readers interested mainly in unsupervised learning may consider the edited collection on that topic by Hinton and Sejnowski (1999), which includes original descriptions of the major models. Instead, readers interested mainly in supervised learning may read the book by Reed and Marks (1999), which presents also a large number of practical solutions to improve performance.

Herz et al. (2006) offer an excellent synthesis of the levels of abstraction used in neural modeling and indicate the extent of the predictions on biological phenomena that each level can generate. They also describe the intrinsic computational properties displayed by various levels and their behavioral effects observed in animals.

Belew and Mitchell (1996) brought together an unsurpassed collection of chapters on the combination of evolution and learning. The chapters, written by experts in their fields after a brainstorming meeting in Santa Fe, New Mexico, offer an inspiring description of the principles at work in systems where adaptation takes place at several temporal and spatial scales. The chapters are preceded by introductions written by other contributors to the same book in order to make them accessible to a wide readership. The book also includes early, hard-to-find papers on the topic, such as extracts from Baldwin, Lamarck, and Morgan.

The foundations of neuromorphic engineering are described in the seminal book by Mead (1989) whereas a survey of more recent models and methods in the context of spiking neurons can be found in the book edited by Maass and Bishop (1999). For readers interested in implementations, Liu et al. (2002) provide a comprehensive tutorial on analog VLSI circuits used in the context of neuromorphic engineering.

4 *Developmental Systems*

Multicellular organisms are assemblies of cells organized in structures of amazing complexity that realize impressive feats of coordination and functionality. Some multicellular organisms are composed of a few hundred cells, as in the case of the microscopic worm *Caenorhabditis elegans,* but some are composed of an astronomical number of them; for example, many trillions of cells form a human body. The idea of an organized system composed of such huge numbers of elements and the problems posed by its construction are difficult to appreciate from the perspective of our everyday experience. You can realize the complexity of the task if you consider that there are more cells in a human finger than people in the world (Wolpert 1992).

DEVELOPMENT The assembly of the structures that constitute a multicellular organism is due to a process of *development*, which starts from a single cell – the fertilized egg or *zygote* – and builds progressively the organized structures that form the complete organism. We don't know yet all the details of the developmental process that builds biological organisms, but many aspects have been already elucidated (Wolpert et al. 2007). We know that to build the final structure, cells form patterns of organized activity in both space and time, using a variety of mechanisms. These mechanism include the exchange of signals between cells, the reactivity of cells to environmental conditions, and the ability of cells to grow, divide, die, migrate, and differentiate. All these activities are influenced by the genome and by the initial state of the zygote, that is, by the concentration and distribution of chemicals that determine the initial steps of the developmental process. The zygote, however, does not contain a blueprint of the developed organism but, rather, the instructions that – in a suitable environmental context – steer the process of self-organized construction of the organism. We can say that the zygote and

DEVELOPMENTAL
REPRESENTATION its genome constitute a *developmental representation* of the organism.

4.1   Potential Advantages of a Developmental Representation

From the properties observed in biological organisms we can infer that developmental representations have a number of favorable properties. A developmental representation provides the possibility of defining a *compact* description of potentially very complex structures. An often cited example of this property is represented by the orders of magnitude of difference between the number of human genes and the number of neurons and connections in the brain of a newborn. The conciseness of the representation is partly due to the possibility of *reuse* of the developmental programs describing the construction of substructures, as if they were program subroutines that can be invoked multiple times. The biological evidence shows that a developmental representation permits and possibly favors the definition of structures possessing a certain degree of *modularity* and *symmetry*.

The large variability in the total number of cells that compose different kinds of biological organisms proves that a developmental representation can have good properties of *scalability*. This property is related to the fact that the developmental process is a *self-organized* and *distributed* process whose activity is *decentralized* and characterized by *parallel operation* of its constituents. One of the consequences of parallel operation is that a small number of cycles of parallel division of the cells can produce a structure composed of an astronomical number of elements. The property of parallel operation and the sensitivity of cells to environmental signals permits the endowment of the developmental process with remarkable properties of *robustness* to perturbations occurring during the process and with a certain degree of *adaptability* to the environmental conditions that exist during development. Note that there is no contradiction between the robustness to environmental perturbations and the adaptability to environmental conditions, because the developmental representation can be structured so as to let the developmental process discriminate between environmental signals that must be treated as noise, and signals that can be allowed to influence the process.

The biological evidence suggests that major changes in the size and structure of the outcome of the developmental process can be achieved with minimal changes in the developmental description. For example, the fact that the developmental process is composed of several processes that unfold in parallel in time permits the control and change of the outcome of the process by controlling the rates and relative timing of the different developmental processes (a phenomenon called *heterochrony* in biology). The potential flexibility and robustness of a genetically instructed developmental process can endow systems thus obtained with greater *evolvability* with respect to more rigid genetic descriptions.

The processes that structure a multicellular organism are most apparent in the first stages of the organism's life, the so-called *embryonic phase*. However, these processes are not completely and suddenly disabled once the overall structure of the organism is in place but, rather, continue to some extent during the whole life of the organism. For example, considerable processes of *growth* and adaptation to environmental conditions can be observed in mammals after birth, and in some organisms developmental processes can be activated at any time for the *regeneration* of lost or damaged parts, for example, the tail, limbs, and retina of certain amphibians (Wolpert 2003). More generally, the developmental processes observed in biological organisms are *self-limiting* in the sense that they subside as soon as the required structure has been built, with the possibility of reactivation if the need arises.

Summing up, a developmental representation for a system has potentially many favorable properties, such as compactness, scalability, self-organization, robustness, adaptability, evolvability, fault tolerance, and self-repair. Consequently, there have been many attempts to define developmental representations for artificial systems endowed with similar characteristics. In this chapter we will describe and analyze critically some of the ideas that have been proposed to realize this objective. We will call the *artificial developmental system* the combination of a developmental representation and the rules that specify how the representation must be interpreted to build the artificial system. In this chapter we will focus mainly on the developmental systems inspired by the early phases of the developmental processes of multicellular organisms, that is, on the development of what in biology is called the *embryo*. Artificial developmental systems inspired by the processes of lifetime adaptation, learning, and regeneration of biological organisms will be described in the chapter 6.

ARTIFICIAL
DEVELOPMENTAL
SYSTEMS

In taking inspiration from biology to model developmental systems we can operate at different levels of abstraction. At one extreme we can disregard all the low-level cellular details of the developmental process, leaving only the essential elements such as cell duplication. At the other extreme we can implement the finest details of the chemistry and physics of gene regulation and cell interaction. In between there is a continuum of models of intermediate complexity and abstraction. In the next section we start by examining some of the most abstract models, called rewriting systems. We will progressively add detail and flexibility to the rewriting models considered.

After examining a representative selection of models, we will pause to consider the difficulties of the definition and usage of developmental systems. This leads naturally to a discussion of the relationship between developmental models and evolutionary approaches, which is useful also to understand the less abstract models considered in subsequent sections.

4.2   Rewriting Systems

If we strip a biological developmental process of all its low-level details, what remains is the activity of division, differentiation, and death of the cells that constitute the organism. In 1968 the biologist Aristid Lindenmayer (1968a,b) proposed to model these aspects of the developmental process using a class of formal models called *rewriting systems*. A rewriting system is a formal system that works on strings of symbols called *words*. The system defines a set of rules that specify how a word transforms into another word. Starting with an initial word and applying recursively the rules, one obtains thus a sequence of words. The original idea was to consider the sequence of words produced by the rewriting system as representing the sequence of configurations of cells produced by the developmental process

4.2.1   L-Systems

The class of rewriting system considered by Lindenmayer is now known as *Lindenmayer systems*, or *L-systems* for short. An L-system is a rewriting system that operates on strings of symbols. The system is defined by assigning an *alphabet* A of symbols, an initial string of symbols ω called the *axiom*, and a PRODUCTION RULES set $\pi = \{p_i\}$ of *rewriting* or *production rules* that specify how each alphabet symbol is replaced by a string of symbols at each rewriting step. An example of a production rule is

$$a \rightarrow abc$$

where the symbol a on the left is called the *predecessor*, and the sequence abc of symbols on the right is called the *successor*. The rule states that at each rewriting step the predecessor is replaced by the successor. If no replacement rule is explicitly specified for a symbol of the alphabet, a default rule that replaces the symbol with itself is implicitly assumed to hold.

   In the original formulation of Lindenmayer, each symbol in the L-system alphabet represents a cell in a given state and the production rules represent the change of state of a cell from one time instant to the next. Since the

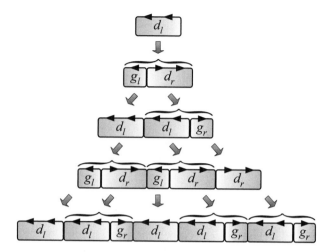

Figure 4.1   The graphical representation of the first four rewriting steps that model the development of a filament of bacteria using the L-system described in the text.

state of cells is updated in parallel in biological systems, Lindenmayer stipulated that at each iteration the production rules are applied in parallel to all the symbols of the current word. For this reason an L-system is called a *parallel rewriting system*. In general, a *stopping condition* for the rewriting process is also specified in defining an L-system. The simplest stopping condition is just the attainment of a predefined number of rewriting steps. For L-systems endowed with the property of self-limited growth, the stopping condition can be the observation that the structure has not changed between two rewriting steps.

One of the examples originally considered by Lindenmayer concerned the modeling of the development of multicellular filaments of cells observed in some types of bacteria. Simplifying the results of the observations on the actual bacteria considered by Lindenmayer, one can split their life cycle into two stages. In the first stage, which we denote with the symbol g, the cells are growing and not yet ready to divide. In the second stage, which we denote with d, the cells are ready to divide. When a cell divides it produces two daughter cells, one in the g stage and one in the d stage. The cells have one of two possible kinds of polarity that we denote with the subscripts r and l. The polarity specifies the position and polarity of the daughter cells that will be produced at the next division. Summing up, the cells can be in four possible states, g_r, g_l, d_r, and d_l. The rules that govern the development

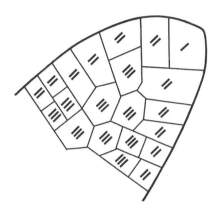

Figure 4.2   A schematic representation of the types of cells composing a moss leaf according to Nägeli (1845). Primary cells (I) are found at the tip of the leaf, secondary cells (II) are found on the margin of the leaf, and tertiary cells (III) compose the interior of the leaf.

of a filament starting from a cell in the d_r state can be represented by the following L-system:

$$A \;=\; \{\, g_r,\, g_l,\, d_r,\, d_l \,\}$$
$$\omega \;=\; d_r$$
$$\pi \;=\; \{\; d_r \to d_l\, g_r, \quad d_l \to g_l\, d_r, \quad g_r \to d_r, \quad g_l \to d_l \;\}$$

Figure 4.1 shows a graphical representation of the first steps of the simulated developmental process determined by this L-system.

Lindenmayer (1975) considered also the application of L-systems to describe the development of two- and three-dimensional assemblies of cells. For example, he considered the rules of development of moss leaves, which are composed of a single sheet of cells. According to Nägeli (1845), a moss leaf is composed of three types of cells: primary cells are found at the tip of the leaf, secondary cells are found on the margin of the leaf, and tertiary cells compose the interior of the leaf (figure 4.2). In the early stages of development a primary cell divides into a primary cell and a secondary cell, whereas a secondary cell divides into two secondary cells or into a secondary cell and a tertiary cell. Lindenmayer modeled this developmental process, representing cells as rectangles in the plane and defining the following L-system:

$$A \;=\; \{\, a, b, c, d, e, f, g, h, i, j, k, l, m, D, R \,\}$$

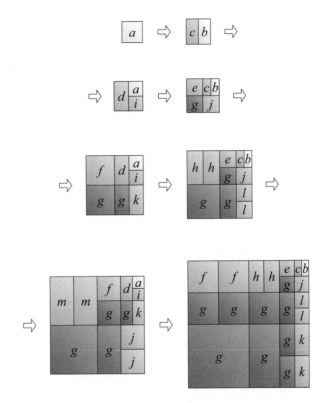

Figure 4.3   The graphical representation of the first seven rewriting steps modeling the development of a moss leaf using the L-system proposed by Lindenmayer (1975). The final result has the same topology of the leaf cells shown in figure 4.2, with rectangles marked with the letters a and b (light shading) representing cells of type I, rectangles marked with the letter g (dark shading) representing cells of type III, and rectangles marked with other letters (intermediate shading) representing cells of type II.

$$\omega = a$$

$$\pi = \left\{ \begin{array}{llll} a \to cRb, & b \to aDi, & c \to d, & d \to eDg \\ e \to f, & f \to hRh, & g \to g, & h \to m \\ i \to j, & j \to gRk, & k \to lDl, & l \to j \\ m \to fDg \end{array} \right\}$$

The lowercase symbols in the alphabet represent cell states. The D and R symbols represent the direction in the plane (down or right) along which the rectangles that represent the cells divide, as illustrated in figure 4.3. Fig-

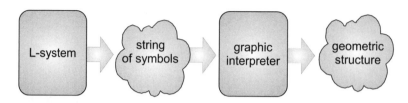

Figure 4.4   A schematic representation of the graphics interpretation of L-systems. The developmental process defined by the L-system produces a string of symbols as a result of the rewriting rules acting recursively on the L-system axiom. The graphic interpreter transforms the string of symbols into a geometric structure.

ure 4.3 shows the first steps of the simulated leaf developmental process described by this L-system. Comparing figure 4.3 with figure 4.2 reveals that the symbols a and b represent two states of primary cells, the symbol g stands for tertiary cells, and the other lowercase symbols of the alphabet represent different states of secondary cells.

4.2.2   Showing the Result: Turtle Graphics Interpretation

In the examples considered above, most of the symbols appearing in the alphabet represent cells of the developing organisms in various states. One can thus easily see the correspondence between the steps of the developmental process generated by the L-system and the phases of the development of the organs of a multicellular organism. It was soon realized that the potential of L-systems for the description of structures can be further increased by abandoning the direct correspondence between symbols and cells. The idea is to consider a more abstract interpretation of strings of symbols as sequences of drawing instructions, typically in a two- or three-dimensional Euclidean space. In other words, the strings of symbols generated by the operation of the L-system are no longer a direct representation of the structures but must be first *interpreted* by a suitably defined graphics system (figure 4.4).

TURTLE   The most popular implementation of the idea of a graphics interpretation for the strings generated by L-systems is the so-called *turtle interpretation* (Prusinkiewicz 1986; Prusinkiewicz and Lindenmayer 1990). In its two-dimensional realization, the *turtle* represents a drawing tool in the Euclidean plane. The turtle has a *position* represented by the Cartesian coordinates (x, y) and an *orientation* represented by an angle α that specifies the *heading* relative to a reference direction. The turtle is assigned an *initial position* (x_0, y_0), an

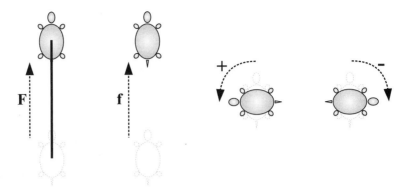

Figure 4.5 The two-dimensional turtle graphics interpretation of the symbols F, f, $+$, and $-$ of an L-system using an angle increment $\delta = 90°$.

initial heading α_o, and two parameters: the *step length* d and the *angle increment* δ. Given a string generated by an L-system, the turtle interprets the following four symbols as commands (figure 4.5):

F Move forward by a step while drawing a line.

f Move forward by a step without drawing a line.

+ Turn left by an angle δ.

- Turn right by an angle δ.

All other symbols appearing in the string are ignored by the turtle; this permits the use of auxiliary symbols that help define the growth process without interfering with the graphical interpretation. When a word generated by an L-system is processed by the turtle interpreter, it is transformed into a drawing, as illustrated in figure 4.6. Note that the turtle interpreter produces a many-to-one mapping from words to geometric structures. For example, the sequence of symbols $+-$ turns the turtle to the left and then to the right, leaving the position and heading of the turtle unchanged and without producing any drawing. Consequently, two words that differ only on a substring of this kind produce the same drawing when processed by the turtle graphics interpreter.

The turtle interpretation can be easily extended to the three-dimensional Euclidean space (Abelson and diSessa 1981; C. Jacob 2001). The position is now represented by three Cartesian coordinates (x, y, z). The orientation

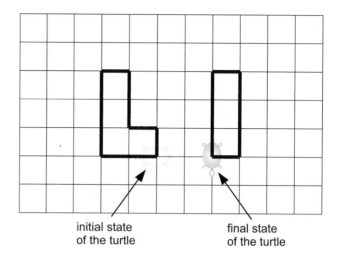

initial state
of the turtle

final state
of the turtle

Figure 4.6   An example of drawing produced by interpreting the strings of symbols
$FF - FFF - F - FF + F - F + f f F + FFF + F + FFF$ with the turtle graphics
interpreter. The step length d corresponds to the step size of the background grid,
and the angle increment is $\delta = 90°$.

of the turtle is defined by three orthogonal vectors **h**, **l**, and **u** specifying
the *heading*, *left direction*, and *up direction* of the turtle. The set of symbols
interpreted by the turtle as commands is extended as follows:

F Move forward by a step while drawing a line.

f Move forward by a step without drawing a line.

U Turn left by an angle δ.

u Turn right by an angle δ.

L Pitch down by an angle δ.

l Pitch up by an angle δ.

H Roll left by an angle δ.

h Roll right by an angle δ.

Note that the 2D and 3D turtle interpreters described above are just two
examples of the turtle system for the graphics interpretation of the strings

produced by L-systems. More sophisticated drawings can be obtained by expanding the list of drawing commands, for example, to include lines of different thickness, color, dashes, and so on. In other words, apart from the existence of a common subset of symbols, the turtle system that interprets graphically the strings of symbols is independent of the L-system that produced the strings. Different graphical interpretations with different meaning and different graphical sophistication can be applied to the same L-system, producing different geometric structures.

One of the most popular applications of the turtle graphics interpretation of L-systems is the generation of fractal structures. Fractals are geometric structures that exhibit some degree of self-similarity across a large range of scales (see box 4.1) (Mandelbrot 1982; Peitgen and Saupe 1988; Avnir et al. 1998). Fractals can be typically defined in terms of the recursive application of a mathematical rule such as, for example, a geometric rewriting rule. For this reason the combination of the rewriting properties of L-systems and the geometric rendering of the turtle interpretation is ideally suited for the description of fractal structures. Figure 4.7 shows the first steps of the construction of the fractal curve known as the *Koch curve*. The sequence of strings defining the curve are generated by the following L-system:

$$A = \{F, +, -\}$$
$$\omega = F$$
$$\pi = \{F \to F + F - - F + F\}$$

which results in the following sequence of strings:

$$\omega \quad : \quad F$$
$$\downarrow$$
$$s_1 \quad : \quad F + F - - F + F$$
$$\downarrow$$
$$s_2 \quad : \quad F + F - - F + F + F + F - - F + F - - F + F - - F + F + F + F - - F + F$$
$$\downarrow$$
$$s_3 \quad : \quad \ldots$$

To transform the resulting sequence of strings into the Koch curve, the two-dimensional turtle is given a horizontal initial heading and an angle increment $\delta = 60°$. The step length d is not specified because it is assumed that the drawing is rescaled in order to keep fixed its horizontal width. In the case of the Koch curve this can be obtained by dividing by three the step length at each iteration. Note that, contrary to this particular example, given

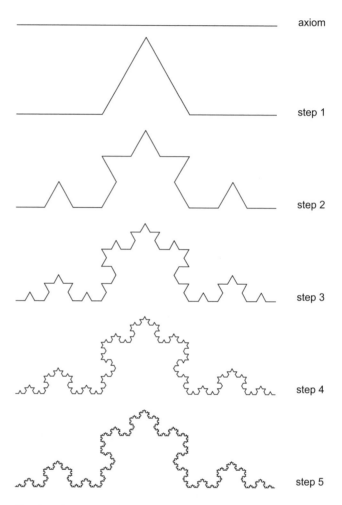

Figure 4.7   The first five steps of the development of the Koch fractal curve produced by the L-system described in the text, followed by the turtle graphics interpretation with angle increment $\delta = 60°$ and rescaling the graph at each iteration.

an arbitrary L-system there is in general no way to derive analytically from the L-system the scaling factor that is required to keep fixed the scale of the whole drawing (Peitgen et al. 1992). The solution is in general to build the drawing using a fixed step length and then to rescale it.

4.2.3 Circuits and Networks: Graph Interpretation of L-Systems

The turtle graphics interpretation of L-systems is quite powerful, as illustrated by the example above and by those that will be presented later. However, it is not very well suited to the definition of graphs containing many closed paths, such as networks and circuits. The reason is that the turtle interpreter puts on the user the burden of ensuring that the sequence of drawing commands corresponds to a closed path. To transform the sequences of symbols produced by an L-system into a network structure it is therefore better to rely on interpreters that are especially tailored to that kind of structure. An example is provided by the *graph interpretation* of L-systems proposed by Boers and Sprinkhuizen-Kuyper (2001). In the graph interpretation system the symbols appearing in the L-system alphabet A are divided into two subsets N and L. The set N is used to represent the *nodes* of the network and is in general a set of alphabetic characters. The set L is used to represent the *links* between the nodes and is a subset of the set of integer numbers. The words generated by an L-system for graph interpretation must be structured as a sequence of blocks composed by an element of N followed by one or more elements of L.

An example will clarify the actual working of this system (figure 4.8). The word

$$A\,2\,3\,B\,-1\,C\,-1\,2\,D\,0\,1\,E\,-4$$

is interpreted as a network whose nodes are denoted by the uppercase alphabetic characters A, B, C, D, E. In order to connect the nodes, they are put in a linear list according to the order in which they appear in the word, as illustrated in figure 4.8a. A number l appearing in the list of elements of L immediately after the symbol for a node denotes a directed link from that node to a node having distance l from it on the right or on the left, according to the sign of l. A value $l = 0$ corresponds to a node self-connection. Note that with this simple interpretation, the numbers that can appear after a node are constrained to correspond to the position of actual nodes. To simplify the definition of the L-system production rules, additional interpretation rules can be specified. For example, numbers pointing to node positions beyond the actual list of nodes can be truncated to the boundaries of the list, or they can be interpreted as links to fixed input and output nodes, or they can be folded within the list by evaluating the remainder of the absolute position modulo the list length. One can also specify constraints on the connectivity by specifying suitably the set L. For example, one can limit the connection to

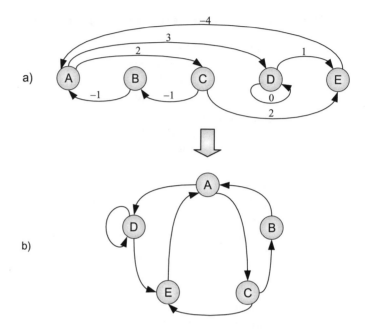

Figure 4.8   The network produced by the graph interpretation of the string of symbols $A\,2\,3\,B\,-\,1\,C\,-\,1\,2\,D\,0\,1\,E\,-\,4$. *a*) The nodes are ordered linearly to permit the interpretation of the numbers as specifying the relative position of the nodes connected by a link. *b*) Once built, the network can be redrawn for clarity removing the linear ordering of the nodes.

forward nodes by admitting only positive integers in L. Note that the graph interpretation of the word produced by the L-system is very similar to that used by Cartesian genetic programming (J.F. Miller et al. 2000) to describe electronic circuits (see chapter 1).

4.2.4   Plants, Branches, and Subnetworks: Bracketed L-Systems

Besides fractals, a very popular application of the combination of L-systems and turtle graphics interpretation is the modeling of plants. The reason is that many plants have a remarkable degree of self-similarity between their parts (see box 4.1). For example, branches growing from the main trunk of a tree (called first-order branches) may have the same structure as the whole tree, and the same property may hold recursively for higher-order branches. Thus, we could use L-systems to define a developmental process that realizes

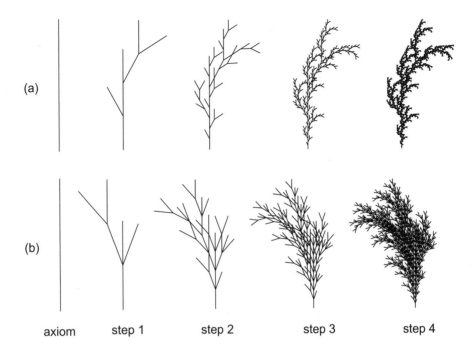

axiom   step 1   step 2   step 3   step 4

Figure 4.9   The first four steps of the development of two plant-like structures using the bracketed L-systems described in the text.

this kind of self-similarity, as we did for the fractal structures considered previously.

   At first sight, to realize this goal we could use the turtle-based graphic system introduced above. However, that simple set of drawing commands is not suited to the production of branching structures. The problem lies in the necessity to reposition the turtle at the base of a branch after the drawing of the branch itself, in order to proceed with the drawing of the structure. An elegant solution consists in the use of *bracketed L-systems*. A bracketed L-system is an L-system whose alphabet includes the following two additional symbols used to save and restore the state of the turtle.

[Save the current state of the turtle, that is, the turtle position and orientation.

] Restore the state of the turtle using the last saved state.

In other words, the $[$ symbol saves the current state of the turtle in a last-in, first-out memory, and the $]$ symbol retrieves the state from the memory. In the parlance of computer systems, the bracket symbols push and pop the turtle state to and from a stack.

Figure 4.9 (a) shows the first steps of the development of a plant-like structure generated by the following bracketed L-system

$$A = \{F, +, -, [,]\}$$
$$\omega = F$$
$$\pi = \{F \rightarrow F[+F]F[-F[+F][-F]]F\}$$

using a turtle interpretation with an angle increment $\delta = 29°$. Figure 4.9 (b) shows the first steps of the development of a plant-like structure generated by the bracketed L-system

$$A = \{F, +, -, [,]\}$$
$$\omega = F$$
$$\pi = \{F \rightarrow F[+F[+F] - F][-F]F\}$$

using a turtle interpretation with an angle increment $\delta = 21°$. These figures show that it is possible to use bracketed L-systems to generate structures that look like different kinds of plants. The realism of the result can be still improved, including production rules for the generation of leaves and using sophisticated graphics techniques that add texture and color to the generated structures (Prusinkiewicz and Lindenmayer 1990).

Bracketed L-systems can be also used in conjunction with the graph interpretation to define *hierarchically structured* and *modular networks* (Boers and Sprinkhuizen-Kuyper 2001). A sequence of symbols within brackets is interpreted as a subnetwork whose depth in the hierarchy is determined by the number of bracket pairs that enclose the symbols corresponding to the subnetwork. A subnetwork is seen as a monolithic node from the point of view of other nodes and subnetworks at the same level in the hierarchy. This requires a rule to transform inputs and outputs to a subnetwork into inputs and outputs to the actual subnetwork nodes. There are different ways to define this rule (Boers and Sprinkhuizen-Kuyper 2001). For example, the inputs to a subnetwork can be routed to all the subnetwork nodes that have no inputs within the subnetwork, with a corresponding rule for the outputs. Figure 4.10 shows the development of a network obtained applying this rule to the following bracketed L-system:

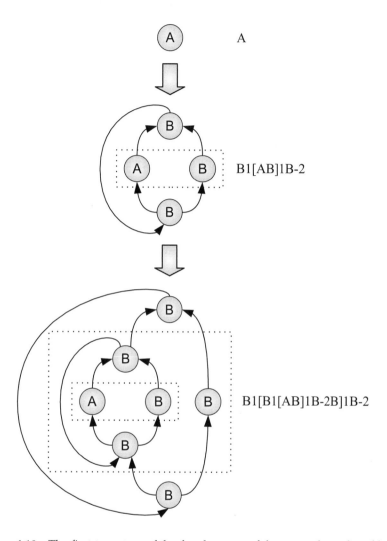

Figure 4.10   The first two steps of the development of the network produced by the graph interpretation of the strings generated by the bracketed L-system described in the text. The dotted lines enclose the subnetworks defined by the brackets appearing in the strings.

$$A \;=\; \{\,A,\,B,\,1,\,-2,\,[,\,]\,\}$$

$$\omega \;=\; A$$

$$\pi \;=\; \{\,A \rightarrow B\,1\,[\,A\,B\,]\,1\,B\,-\,2\,\}$$

4.2.5   Variations on a Theme: Stochastic L-Systems

One of the characteristics of the systems considered so far is that they are
deterministic. A *deterministic L-system* uses production rules that always replace a given symbol with the same string. Thus, for a given axiom the
rewriting process of a deterministic L-system always produces the same sequence of strings. If the graphic interpretation system is also deterministic it
always transforms a given symbol into the same graphical command. This
means that a deterministic L-system coupled with a deterministic graphic
interpreter always produces the same sequence of geometric structures. The
use of a deterministic system to build, say, a forest of trees for a virtual reality
scene, would produce a very unnatural result composed of many identical
copies of the same structure.

To add some degree of randomness to the geometric structures that are
generated we can operate at the level of the graphic interpreter or at the
level of the L-system (figure 4.4). The effect of adding a limited amount
of randomness to the graphic interpreter alone is to produce structures that
are merely variants of the same topology (Prusinkiewicz and Lindenmayer
1990). The observation of different specimens of real plants belonging to the
same species reveals, however, that in general their topology too varies from
specimen to specimen. This kind of variation is easily obtained by injecting
some randomness at the level of the L-system production rules. A *stochastic
L-system* is an L-system which can associate multiple production rules to a
single symbol of the alphabet. Typically, values of probability summing to
1 are specified for the production rules associated with a symbol, and each
rule is chosen with the given probability.

Figure 4.11 shows a few examples of plant-like structures generated after
four rewriting steps by the following stochastic L-system

$$A = \{F, +, -, [,]\}$$
$$\omega = F$$
$$\pi = \left\{ F \rightarrow \left\langle \begin{array}{ll} F[+F]F[-F]F & \text{with probability } \frac{1}{3} \\ F[-F]F[+F]F & \text{with probability } \frac{1}{3} \\ F[-FF-F]F & \text{with probability } \frac{1}{3} \end{array} \right\rangle \right\}$$

using a turtle interpretation with an angle increment $\delta = 29°$. Although
the topology of the plant-like structures shown in figure 4.11 varies, there
is a common underlying theme that makes them appear as belonging to the
same plant species.

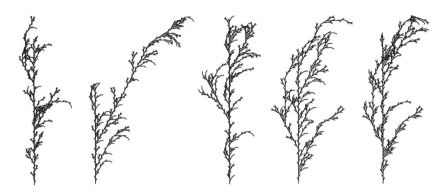

Figure 4.11 Different instances of plant-like structures generated using the same stochastic L-system.

4.2.6 Values and Conditions: Parametric L-Systems

The turtle interpretation system used so far for the transformation of strings into geometric structures is based on two fixed geometric parameters: the step length l and the angle increment δ. These two parameters define the geometric granularity of the structures that can be built by the turtle. In principle, this is not a limitation because one is free to choose very small values for the two parameters to describe small details in the system structure. With this approach, however, systems with structures defined at different scales require long strings of commands for their description. A more flexible solution is to add the possibility of having numerical parameters within the strings processed by the L-system, and the possibility of manipulating these parameters during the rewriting process. The use of parameters is useful not only to add flexibility to the definition of geometric structures but also to model biological parameters such as the concentration of chemicals in a cell, the age of a cell, or the number of divisions that it has undergone (Lindenmayer 1974).

PARAMETRIC L-SYSTEM   An L-system of this kind is called a *parametric L-system* (Prusinkiewicz and Lindenmayer 1990). In a parametric L-system, the symbol appearing on the left side of a production rule is followed by a list of parameters enclosed within a parenthesis. The parameters can be formal parameters belonging to a set X, or actual numeric or symbolic values. Examples of parametric production rules are

$$a(x) \rightarrow a(x+1)\, b(x-1)$$

and

$$a(x, y) \rightarrow a(x)\, b(y)$$

where x and y belong to the set X of formal parameters. In applying the production rule, the actual parameters appearing in the word that is rewritten are substituted for the formal parameters appearing in the definition of the rule. An additional degree of flexibility is obtained by admitting *conditions* within the production rules, as in

$$a(x) \,:\, x > 1 \rightarrow a(x + 1)\, b(x - 1)$$

where it is stipulated that the production rule must be applied only if the condition stated between the symbols ":" and "→" is true for the actual value of the parameter x. Note that the set of production rules of a parametric L-system can mix nonparametric rules and parametric rules.

When we use a parametric L-system in conjunction with a turtle interpreter, the definition of the latter must be extended to accommodate the possibility of using the parameters. For example, the commands of a parametric turtle interpreter for the Euclidean plane can be defined as follows:

F(x) Move forward by a step of length x while drawing a line.

f(x) Move forward by a step of length x without drawing a line.

+(x) Turn left by an angle x.

-(x) Turn right by an angle x.

Parametric and nonparametric symbols corresponding to drawing commands can be allowed to mix by specifying that the default values l of step length and δ of angle increment must be used in the absence of an explicit parameter. Using this parametric turtle interpreter, the developmental steps for the construction of the self-rescaling Koch curve shown in figure 4.7 can be defined by the following parametric L-system:

$$
\begin{aligned}
A &= \{\, F, +, -, (,) \,\} \\
X &= \{ x \} \\
\omega &= F(1) \\
\pi &= \{ F(x) \rightarrow F(\tfrac{x}{3}) + F(\tfrac{x}{3}) - - F(\tfrac{x}{3}) + F(\tfrac{x}{3}) \}
\end{aligned}
$$

given a horizontal initial heading for the turtle, and an angle increment $\delta = 60°$.

4.2.7   Signals and Regulation: Context-Sensitive L-Systems

In the L-systems that we have considered so far the production rules specify how a symbol is replaced by a string of symbols independently of the symbols that precede and follow it in the string that is being rewritten. In other words, the replacement does *not* take into account the *context* in which a symbol appears. If we go back to the biological developmental processes that originally inspired Lindenmayer's definition of L-systems, the independence of the symbol replacement from its context corresponds to the independence of the fate of a cell from the context in which the cell is placed. This means that the developmental future of a cell is determined only by its current state and does not contemplate the possibility of interactions with other cells.

MOSAIC AND REGULATIVE EMBRYOS
In biology, embryos where the development of cells is determined only by the current state are called *mosaic* embryos, whereas embryos whose development relies on the interaction between cells are called *regulative* embryos (Wolpert 2003). The problem with a mosaic embryo is that all the information for the structuring of the system must be transmitted through inheritance from a cell to its descendants, and there is no way to sense and correct developmental errors and anomalies. It is therefore not surprising that, although there exist some biological embryos that develop for certain periods of time as mosaic embryos, all known biological developmental processes are based at least in part on the interaction between the cells. In other words, a purely mosaic embryo is merely an abstraction without correspondence in actual biological development (Lawrence and Levine 2006).

It is therefore reasonable to consider the possibility of extending L-systems to take into account during the rewriting process the context in which the symbols appear. L-systems extended in this way are called *context-sensitive L-systems*. Correspondingly, the L-systems that we have considered so far, which do not take context into account, are called *context-free L-systems*.

CONTEXT-FREE L-SYSTEMS
To specify a context-sensitive L-system we just need to extend the definition of the production rules from isolated symbols to symbols in a context within a string (Prusinkiewicz and Lindenmayer 1990). The context is constituted by one or more symbols that must be adjacent to the left or to the right. To define context-sensitive production rules we use the additional symbol ◀ to delimit the *left context* and the symbol ▶ to delimit the *right context*. For example, the production rule

$$b \blacktriangleleft a \to g$$

specifies that the predecessor a is replaced by the successor g only if it is preceded by the symbol b, that is, if the left context of the symbol a is the symbol b. The production rule

$$b \blacktriangleleft a \blacktriangleright c \to g$$

specifies that the predecessor a is replaced by the successor g only if its left context is the symbol b and its right context is the symbol c. In general, we can define production rules that take into account several symbols on the right and several symbols on the left. In defining a context-sensitive L-system we can mix rules that do not depend on the context and rules that depend on it. We can also define several production rules of different context specificity for a given symbol, with the provision that more specific rules are tried first, until a matching rule is found.

The definition of a context-sensitive production rule can be extended to bracketed L-systems. In verifying the matching of a context-sensitive rule that contains brackets we must take into account the fact that two symbols must be considered adjacent if they are separated by a string enclosed by brackets. For example, the production rule

$$b \blacktriangleleft a \blacktriangleright c\,[\,d\,]\,e \to g$$

applies to the symbol a contained in the string

$$u\,\mathbf{b}\,[\,v\,[\,w\,x\,]\,]\,\mathbf{a}\,\mathbf{c}\,[\,\mathbf{d}\,y\,]\,\mathbf{e}\,z$$

(where the symbols appearing in the production rules are represented in boldface for clarity) because in this string the symbol a must be considered adjacent to the symbol b on the left and to the strings $c\,d$ and $c\,e$ on the right.

As said above, the motivation for the introduction of context sensitivity is the desire to model the interaction between cells and elements that take place in biological development. We can verify that, indeed, context-sensitive L-systems permit the exchange of signals between adjacent elements of a structure. Concatenating a series of local exchanges one can obtain the long-range propagation of signals. Figure 4.12 shows an example of long-range signal propagation across a branching structure defined by the following context-sensitive L-system:

$$
\begin{aligned}
A &= \{\, F\,,\, S\,,\, Q\,,\, +,\, -,\, [,\,]\,\} \\
\omega &= S\,[\,-F\,[\,-F\,]\,F\,]\,F\,[\,+F\,[\,+F\,]\,F\,]\,F\,[\,-F\,]\,F \\
\pi &= \left\{
\begin{array}{ll}
S \blacktriangleleft F \to S, & \text{ignore } +,\, - \\
S \to Q &
\end{array}
\right\}
\end{aligned}
$$

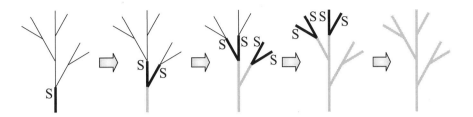

Figure 4.12 An example of signal propagation across a structure defined through a context-sensitive L-system. The reception of the signal by an element leads to the change of state of the element, represented here by thick black lines. After reception of the signal, the element is set to a quiescent state, represented here by thick gray lines.

In this simple example the geometric structure is defined from the beginning by the axiom ω, and the reception of the signal by an element is represented by the change of state of the element from F to S. After reception of the signal, the element is switched to a quiescent state Q. The "ignore $+$, $-$" clause associated with the $S \blacktriangleleft F \rightarrow S$ production rule means that the $+$ and $-$ symbols must be disregarded while determining the context of the symbol F. This clause permits the propagation of the signal to the whole structure. Without this clause the signal in figure 4.12 would propagate only vertically, leaving out the lateral branches.

Context-sensitive production rules can be also extended to parametric L-systems. This gives additional flexibility to the definition of the exchange of information between the parts of a developing system. For example, a production rule such as

$$b(y) \blacktriangleleft a(x) \blacktriangleright c(z) \rightarrow g(x + y + z)$$

realizes the transmission to the symbol a of the information of the parameters y and z associated with the symbols that are adjacent to a.

An example of the biological modeling using context-sensitive parametric L-systems is described by Lindenmayer (1974). Like the one considered in section 4.2.1, this example concerns the modeling of the development filaments of bacteria. However, in addition to the state of the cells, we now take into account the concentration of a chemical acting as inhibitor and the age of the cell. The production rules of the L-system are

$$
\pi = \left\{
\begin{array}{l}
c(s,y,v,V) \blacktriangleleft c(a,x,u,U) \blacktriangleright c(t,z,w,W) : x > 3, u < U \rightarrow \\
\qquad\qquad\qquad\qquad\qquad\qquad c(a, \frac{y+z-2x}{4}, u+1, U) \\[2ex]
c(s,y,v,V) \blacktriangleleft c(a,x,u,U) \blacktriangleright c(t,z,w,W) : x > 3, u = U \rightarrow \\
\qquad c(a, \frac{y+z-2x}{4}, 0, \mathrm{rnd}(6000,600)) c(a, \frac{y+z-2x}{4}, 0, \mathrm{rnd}(6000,600)) \\[2ex]
c(a,x,u,U) : x \leq 3 \rightarrow c(b, 1000, 0, 0) \\[2ex]
c(b, 1000, 0, 0) \rightarrow c(b, 1000, 0, 0)
\end{array}
\right\} .
$$

The sequence $c(a, x, u, U)$ stands for a cell in state a, with a concentration x of inhibitor, having age u, and maximum age U. The first production rule states that a cell in state a with an internal concentration of inhibitor above a threshold $t = 3$ and below maximum age will increase its age and update its inhibitor concentration according to the formula $\frac{y+z-2x}{4}$ This formula represents the diffusion of the inhibitor across the three adjacent cells that constitute the target and the context of the production rule (at each time step one-fourth of the inhibitor contained in a cell diffuses in an adjacent cell, so that the center cell $c(a, x, u, U)$ loses a quantity proportional to $\frac{2x}{4}$ to the adjacent cells, and receives from them a quantity proportional to $\frac{y+z}{4}$). The second production rule states that if a cell attains its maximum age with a concentration of inhibitor above the threshold, it divides into two new cells of state a and age zero. The maximum age of the newly formed cells is set to a random number with average 6000 and standard deviation 600. The third production rule states that a cell whose concentration of inhibitor goes below the threshold goes to a steady state b in which it does not age and maintains a constant concentration of inhibitor, as revealed by the fourth production rule.

According to Lindenmayer (1974) these production rules capture well the development of filaments of a class of blue-green algae, as described in (Baker and Herman 1972a,b). From a more general perspective, the additional interest of this example is that it illustrates how parametric context-sensitive L-systems facilitate the modeling of important aspects of the developmental processes such as self-limitation. This example suggests that there are at least two ways to implement developmental self-limitation in the production rules, one based on a parameter used as a counter that is decremented by the rewriting process, and the other by a threshold mechanism linked to a diffusion process based on parametric context-sensitivity.

4.2.8 Remarks on Rewriting Systems

The overview of rewriting systems presented in the previous sections has shown how it is possible to use them to define and study some abstract models of development. The high level of abstraction of the models means that one must accept some limitations in exchange for the simplicity of the model. In rewriting systems the developmental process proceeds at discrete time steps and thus one cannot observe a continuous growth process. However, by defining production rules that represent small developmental steps, this intrinsic discreteness is of limited impact. A more serious limitation is that there is no simple way to represent the motion of cells across the developing structure. The observation of biological development reveals instead that in the first phases of the developmental process a crucial role is played by the migration of the cells that compose the embryo. It is true that we could define in an L-system a multistage process mimicking the motion of a symbol across the developing structure, by deleting a symbol in one place, propagating a signal across the structure, and introducing a new symbol at the signal destination. However, this definition would appear quite artificial and difficult to implement in a general way. In the end, cell migration during development is better modeled using less abstract developmental models discussed below. Finally, rewriting systems and L-systems in particular were originally conceived as a means to describe existing developmental processes producing self-similar structures. Thus, they were not intended to be easy to synthesize in the absence of an explicit model, and we cannot expect this kind of description to be particularly evolvable. Nonetheless, as discussed later, several examples of applications of rewriting systems in conjunction with artificial evolution exist in the literature.

REWRITING SYSTEMS
VS. CELLULAR SYSTEMS

It is interesting to compare L-systems, as originally conceived by Lindenmayer, to other cellular systems, such as the cellular automata described in chapter 2. The developmental process illustrated in figure 4.1 can resemble the structure of one-dimensional cellular automata. More precisely, the cellular structure, the state of the cells, and the production rules of L-systems can recall the notions of cellular space, cell state, and the transition rule of cellular automata. There is, however, a fundamental difference. In cellular automata the structure of the cellular space is fixed and the cells cannot be created or removed. On the contrary, the process described by the L-system

Box 4.1: Self-similarity

The reason L-systems and fractals are well suited to the representation of many biological and physical structures is the fact that these structures exhibit a certain degree of self-similarity across a non-negligible range of scales. From the perspective of scientists this observation is just the beginning of the investigation, the next step being the attempt to explain why the observed self-similarity arises at all. For nonbiological phenomena the explanation must be searched in the self-organization across different spatial and temporal scales produced by the operation of physical laws (Gouyet 1996; Ball 1999). For biological systems, besides the possibilities and the constraints provided by the laws of physics, the notion of adaptation must be brought into the picture. This means that we must investigate the possibility that the use of self-similar structures provides some kind of evolutionary advantage to the organism.

Elsewhere in this chapter we discuss the possible advantages of achieving a compact genetic representation of the phenotype using a recursive developmental representation. This is a first possible evolutionary reason for the self-similarity observed in biological organisms. It is very likely, however, that there are additional reasons related to the mechanical and thermodynamic performance of self-similar structures. This aspect assumes a great practical importance because it could lead to the synthesis of better artificial systems.

When these artificial systems are generated by automatic synthesis tools such as artificial evolution, one can assume the functional usefulness of self-similar structures as a working hypothesis. In this case, one must just ensure that the genotype-to-phenotype map provides the possibility of generating self-similar structures (without imposing it), and that the evolutionary process has the possibility of accessing them. The evidence of biological organisms and the examples of artificial developmental systems described in this chapter show that the use of a suitable developmental genetic description can endow the evolutionary process with both these possibilities.

When the artificial systems are not automatically synthesized, self-similar structures must be explicitly designed into the systems and, apart from aesthetic considerations, this is justified only if it can be ascertained that this choice benefits the system performance. An example of self-similar structures derived in this way is those designed using the *constructal theory* developed by Bejan (1997, 2000). The constructal theory deduces the presence of self-similar structures in   *(cont.)*

Box 4.1 (continued)

biological and artificial systems as a consequence of the optimization of
the flows required by the operation of the systems, given the physical
constraints on the materials and shapes that can be used in their construction. For example, the structure of the bronchial tree in the lungs
is derived from the constructal theory imposing the requirement of optimal oxygen distribution and carbon dioxide removal, given the constraints imposed by the physical mechanisms available for the transport
and exchange of fluids (Reis et al. 2004; Bejan 2000).

Note, incidentally, that some critics raised objections to the actual performance of the constructal theory (Ghodoossi 2004; Kuddusi
and Eğrican 2008). What interests us here, however, is that the self-
similarity of the structures generated by the constructal theory is in part
the result of the hierarchical design strategy on which it is based, and
this reduces the explanatory power of the theory with respect to the
question of the origin of self-similarity in biological organisms.

This limitation can be understood considering that the classic design
of a structure via an optimization problem requires the establishment
of constraints that make the problem mathematically well posed and
tractable. For example, one cannot set up a mathematical optimization problem for the optimal oxygen distribution and carbon dioxide
removal in the space of all the possible geometrical shapes, because
this space is too vaguely defined. Thus, the hierarchical assumptions
of constructal theory appear as a reasonable way to define the range
of the search. In this light, the best we can hope for is probably a synthesis where the search space encompasses the largest possible variety
of system structures. Artificial evolutionary methods coupled with a
suitable genetic representation like the developmental representations
described in this chapter can be seen as a way to realize this objective.

is crucially based on the possibility of inserting or removing cells from the existing structure. Although a cellular automaton that mimics this property on
a fixed cellular space could be conceived, its specification is in general more
complicated than the specification of an L-system realizing an equivalent development. This fact is illustrated by the complexity of the self-reproducing
cellular automaton conceived by John von Neumann and described in chapter 2. Note that, despite the greater difficulty of defining a construction process in a cellular automaton, von Neumann's choice of using a cellular au-

tomaton rather than a rewriting system is justified by his desire to specify a system that could potentially evolve and complexify in a given artificial universe. The two-dimensional cellular space and relatively simple transition rules of von Neumann's construction provide such an artificial universe. It would be quite difficult to realize the same in the more rarefied context of an L-system, where a natural notion of space-time backcloth is absent.

4.3   Synthesis of Developmental Systems

A fundamental problem with the application of the developmental approach to the description and design of artificial systems is the definition of the developmental system. In some cases we have some insight into the nature of the developmental rules, for example, because these rules are actually observed in an existing biological organism. This was the case for the two examples given above of L-systems defined by Lindenmayer for the modeling of filaments of bacteria. Given the biologist's description of the cell states and developmental rules, Lindenmayer was able to abstract them in terms of L-system alphabet and production rules. In this case, assuming correctness of the biological model, the verification of the adequacy of the L-system developmental model requires just the verification of the correspondence between the elements and rules of the two models. A nonbiological instance of this scenario is given by the L-system defining the development of the Koch curve shown in figure 4.7. In that case the mathematical definition of the Koch curve is already given in terms of recursive rules and thus we need only to translate those rules into the language of L-systems and of their turtle interpretation.

INFERENCE PROBLEM   The problem becomes much more difficult if what is specified is the desired developmental outcome rather than the details of the developmental process. The synthesis of the developmental system realizing a given developmental outcome constitutes what is called the *inference problem* (or *inverse problem*) of developmental systems. The difficulty with the inference problem stems from the complex relationship that exists in general between the developmental description and the outcome of the developmental process. We met a similar problem with the definition of the state set and transition rule of cellular automata in chapter 2. For both developmental systems and cellular systems the rules are local and their consequences are global. Given the rules it is easy to compute their consequences but it is in general difficult to discover the rules required to obtain a given global outcome. As for cellular systems, there is in general no systematic way to transform global requirements into a set of developmental system specifications.

To solve the inference problem for developmental systems we can envisage a trial-and-error procedure. Given an initial guess of the developmental system properties, we can evaluate its consequences and compare them with the desired outcome. If the actual outcome is judged unsatisfactory, we go back to the first step and modify the definition of the developmental system. In some cases this procedure can be executed by hand, typically, because the determination of a good initial guess is easy or because the specification of the outcome allows some freedom in the outcome. An example of the latter type is given by the examples of plant development shown in figure 4.9 and figure 4.11. In these cases what we wanted to obtain was just a reasonably realistic plant-like appearance of the developmental outcome. The exploitation of the potential of definition of self-similar structures was thus sufficient to achieve a satisfying result. We did not even check if the result is botanically plausible in terms, say, of the relative position of the branches. This kind of loose specification of the developmental outcome is a situation often encountered in graphics modeling applications of developmental systems.

The search by hand using the iterative procedure described above is not feasible in most other cases, for example, when the required matching between the actual developmental outcome is strict, or when the target is not specified directly in terms of structure of the developmental outcome but, rather, in terms of some functionality produced by it. An example of this latter case is encountered when a developmental system is used to build a control structure such as an artificial neural network for a robot. In these cases we need a tool for the computer-based automatic synthesis of the developmental system. Not surprisingly, the most popular approach to the automatic synthesis of developmental systems is based on the use of evolutionary algorithms. This is due in part to the great representational flexibility of evolutionary algorithms, which permits the representation and manipulation of a large variety of developmental specifications. However, the relationship between artificial developmental systems and artificial evolutionary systems is complex and many-sided, and cannot be reduced to the use of evolution as a tool to synthesize developmental systems. This parallels the growing appreciation of the role of development in biological evolution. In the next section we will briefly describe some of the current views on the relations between development and evolution, and how they affect artificial evolution.

4.4   Evolution and Development

As mentioned in chapter 1, the generation of phenotypic variation (or phenotypic diversity) in the individuals of a population is one of the pillars on which the theory of evolution is founded. However, not all phenotypic variation that is generated in the offspring is equally useful. For evolution to proceed the phenotypic variation must contain some novelty with respect to the parent population, and must have the potential of being selected and to spread through the population. If these conditions are not verified, the evolutionary process stalls.

In considering natural evolution it is often assumed that the genetic mutations and the resulting genotypic population diversity produce the right kind and amount of phenotypic variation in the population. The validity of this assumption is far from obvious. This fact is well-known to the users of artificial evolutionary systems, who often observe the absence of phenotypic evolution even though the genetic operators produce abundant genotypic and phenotypic diversity in the offspring. To ensure the success of artificial evolution, that is, to ensure evolvability in the system, it is known that one must choose the right kind of genotype, the right kind of genetic operators, and the right kind of genotype-to-phenotype mapping.

Biologists are also aware of the unwarrantedness of the assumption that the generation of genetic variation is sufficient per se to guarantee the progression of evolution. They have thus begun to investigate the mechanism that generates the variation that propels the evolution of biological organisms. For multicellular organisms, one of the most promising candidates for this role is the process of development itself. This has led to the emergence of a new discipline called *evolutionary developmental biology* (or *evo-devo*), which studies the role of development in the generation of the selectable variation of organisms which leads, through the action of natural selection, to the generation of evolutionary changes in phenotypes (G.P. Wagner 2000, 2001; G.P. Wagner and Larsson 2003).

EVOLUTIONARY
DEVELOPMENTAL
BIOLOGY

From the point of view of artificial evolutionary and developmental systems there are two main messages to be gained from the discoveries of evolutionary developmental biology. The first message is that the relationship of artificial evolution and development is not limited to the possible use of evolution for the synthesis of a developmental process. On the contrary, the use of a developmental process can play a crucial role in the realization of artificial evolutionary systems that display an actual potential for the generation of complex systems. In other words, development is not just a way to describe complex systems compactly, but can be instrumental to their effective automatic synthesis via artificial evolution. The second message stems from the recent and puzzling discovery of the high degree of conservation of the basic developmental mechanisms in organisms that appear very different phenotypically, for example, insects and vertebrates (Wolpert et al. 2007; Gerhart and Kirschner 1997). Apparently, natural evolution has discovered just once a set of developmental mechanisms suitable for the evolution of complex organisms, and thereafter it has just tinkered with these mechanisms. If natural evolution – which was able to rediscover several times how to build complex structures like the eye (Land and Fernald 1992) – could not repeat this feat with developmental mechanisms, we can expect the search for an evolvable artificial developmental system to be a difficult task.

From this point of view, one of the most intriguing questions of evolutionary developmental biology is how the "right" developmental mechanisms were evolved in the first place. The simplest hypothesis (Kirschner and Gerhart 1998, 2005), working at the level of single organisms, is that the basic mechanisms of development were selected because they contributed to the robustness of the developmental process to environmental disturbances and noise (A. Wagner 2005). According to this hypothesis, many of the characteristics that ensure developmental robustness, such as the independence of modules and the potential of varying in a coordinate way the elements of a module, would – as a byproduct – also ensure the possibility of generating selectable novelty. Thus, they would endow the organism with evolvability. Of course, evolvability itself can give evolutionary advantages and be selected, but this can happen only beyond the level of the single organism, for example, at the level of a line of descent of parents and offspring.

4.5 Defining Artificial Evolutionary Developmental Systems

To simplify the analysis and discussion of artificial evolutionary developmental systems it is useful to classify them in four categories .

CLASSIFICATION OF DEVELOPMENTAL SYSTEMS

1. Systems where a parameterized developmental process is fully prescribed by the designer and evolution can thus work only on the choice of the parameters, which are encoded in the genome. This approach puts on the designer the whole burden of devising a good developmental process. In the end, the result in just the evolution of a set of parameters mediated by a nontrivial hand-designed genotype-to-phenotype map. Since this approach sidesteps the issue of the evolutionary synthesis of the develop-

mental process, it can be subsumed in those presented in chapter 1 and will not be consider further below.

2. Systems where the basic mechanisms of the developmental process are encoded in the genotype and evolved, but their order and number of applications is fixed and cannot evolve. In general, when this approach is used the goal is not to achieve evolvability but to search for a developmental system that can produce a given outcome within a given class of developmental systems. For example, as illustrated by the examples presented in the next section, this approach can be used to evolve parallel rewriting systems such as L-systems.

3. Systems where the basic mechanisms of the developmental process are hand-designed, but their order and number of applications is encoded in the genome and are subject to evolution. The designer task of choosing the right basic mechanisms is obviously nontrivial. However, the freedom given to evolution in combining these mechanisms gives potential access to a large variety of developmental processes. The typical example of this kind of approach is given by developmental systems based on genetic programming, some examples of which will be described below.

4. Systems where some aspects of both the basic mechanisms of the developmental system and of their order and number of applications is encoded in the genome and can be evolved. These systems can explore a larger space of developmental systems and have the greatest freedom in the search for systems displaying a good evolutionary potential. However, the designer of this kind of system typically is faced with the choice of many low-level details of the evolutionary environment. Moreover, the second message gained from the observation of biological systems mentioned in the previous section warns us that the problem of evolving from scratch a set of suitable basic developmental mechanisms is presumably very difficult.

This classification can be helpful in the choice of the best kind of artificial evolutionary developmental systems for a given application. This choice, however, is still more an art than a science and the suitability of a choice must be often judged, at least in part, by the results that it is able to generate. Once a kind of evolutionary developmental system is selected, one has to chose a genetic representation for it. In general, for a given kind of developmental system there are many possible options for its genetic representation. The choice of the genetic representation entails also the choice of the genetic operators that can be applied to it, and influences the exploratory properties of the evolutionary process. In general, the more abstract and formally structured the developmental model, the greater the care required to define the genetic operators so as to ensure that the mutated genome can be still interpreted as describing a well-defined developmental system.

In the following sections we will present some examples of artificial evolutionary developmental systems. Here we will deal mainly with the problem of defining the system structure. Later, in chapter 6, we will reconsider some of these examples and assess them from the point of view of the performance of the evolved systems. Since the possible combinations of elements composing an evolutionary developmental system are countless, as are the examples that can be found in the literature, we consider only some examples that illustrate the main issues. Pointers to the literature describing other interesting examples can be found in the list of suggested readings at the end of the chapter.

4.6 Evolutionary Rewriting Systems

In this section we describe some examples of the second of the four kinds of artificial developmental systems listed in the previous section. These are systems where the basic mechanisms of the developmental process are encoded in the genotype and evolved and the modality of their applications is fixed and cannot evolve. The typical example, which we will consider throughout this section, is L-systems. In L-systems the basic developmental mechanism is the rewriting of symbols and, thus, in evolutionary L-systems it is the rewriting rules that are genetically encoded and evolved. The application of the developmental mechanisms is instead fixed and corresponds to the parallel application of the rewriting rules. For this reason, we need focus only on the genetic representation of the axiom and of the rewriting rules.

4.6.1 Binary Representation

The simplest approach to the genetic representation of an L-system is the use a binary strings to encode the elements that define the L-system (Boers and Sprinkhuizen-Kuyper 2001; C. Jacob 2001). The binary genome is decoded by dividing it into chunks of fixed length (for example, eight bits) and interpreting these chunks as symbols. A special symbol such as | is chosen as a marker that delimits the parts of the genome that correspond to the axiom and to the

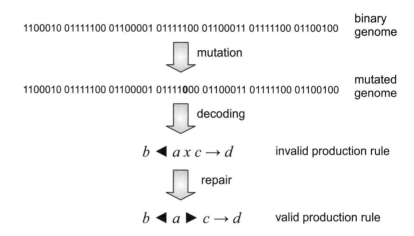

Figure 4.13   The simple binary genetic representation of L-systems can produce sequences of symbols that do not correspond to a valid L-system description. A possible solution is the implementation of a repair process after the decoding of the binary string.

production rules. For example, using the ASCII binary representation for the symbols, the context-sensitive production rule

$b \blacktriangleleft a \blacktriangleright c \to d$

is represented as follows using a binary string:

$$\underbrace{01100010}\ \underbrace{01111100}\ \underbrace{01100001}\ \underbrace{01111100}\ \underbrace{01100011}\ \underbrace{01111100}\ \underbrace{01100100}$$

↓	↓	↓	↓	↓	↓	↓
b	|	a	|	c	|	d
↓	↓	↓	↓	↓	↓	↓
b	◀	a	▶	c	→	d

The genetic operators of the evolutionary algorithm, such as mutation and recombination, will operate directly on the bits of this binary string without concern for its interpretation as high-level symbols. The difficulty in adopting this simple genetic representation for L-systems is that an arbitrary binary string is not in general a valid representation of an axiom followed by a list of production rules. This means that even if we start with a binary string that is a valid representation of an L-system, the application of a genetic operator could produce a string that is no longer decodable as an L-system (figure 4.13).

A first remedy for this problem consists in following the decoding of the bit string with a phase of *repair*. The repair transforms into a valid L-system description the string of symbols that has been originally decoded (figure 4.13). Boers and Sprinkhuizen-Kuyper (2001) have applied this approach to the evolution of neural network architectures. Their approach is based on the graph interpretation of L-systems described in section 4.2.3. In their approach, the L-system describes just the architecture of the neural network, whereas the values of the network weights are determined by a separate search algorithm that is performed on the decoded network.

4.6.2   Syntactic Representation

To avoid the complexity of the repair technique described above one can enforce some constraints on the genetic operators. The constraints must ensure that a genome that is a valid representation of an L-system remain so after the action of the genetic operators. The implementation of this approach is facilitated by the use of a genome composed of high-level symbols rather than binary symbols. For example, using the symbol $|$ as a marker, the sequence of symbols $a|b|a|c|d$ is first interpreted as follows:

$$
\begin{array}{cccccccccc}
a & | & b & | & a & | & c & | & d & | & \ldots \\
\downarrow & \downarrow & \downarrow & \downarrow & \downarrow & \downarrow & \downarrow & \downarrow & \downarrow & \downarrow \\
a & , & b & \blacktriangleleft & a & \blacktriangleright & c & \rightarrow & d & , & \ldots
\end{array}
$$

and then transformed into the following axiom and production rules:

$$
\begin{aligned}
\omega &= a \\
\pi &= \{b \blacktriangleleft a \blacktriangleright c \rightarrow d, \ldots\}
\end{aligned}
$$

The list of symbols $a|b|a|c|d$ can be used directly as a genome instead of being first transformed into a binary genome as in the example in the previous subsection. To ensure the preservation of the genome structure, the separator symbols that appear in the genome are protected from the action of the mutation operators (figure 4.14).

This approach corresponds in practice to use as the genome a high-level syntactically constrained representation of the axiom and production rules of the L-system. The various elements of the genome need not even be considered as assembled into a unique string of symbols, since in any case the separators are not touched by the genetic operators and the syntactic structure of

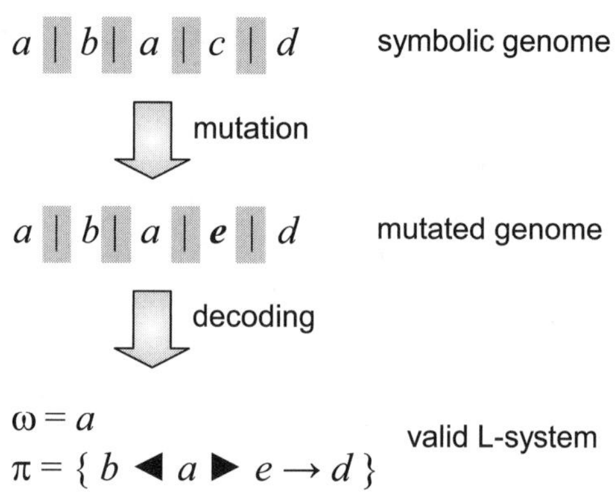

$$a \mid b \mid a \mid c \mid d \qquad \text{symbolic genome}$$

⬇ mutation

$$a \mid b \mid a \mid e \mid d \qquad \text{mutated genome}$$

⬇ decoding

$$\omega = a$$
$$\pi = \{\, b \blacktriangleleft a \blacktriangleright e \rightarrow d \,\} \qquad \text{valid L-system}$$

Figure 4.14 To circumvent the problem of nonvalid genomes produced by genetic operators illustrated in figure 4.13, the genetic operators can be constrained so as to transform a valid L-system representation into another valid representation. In this case, the separator symbols | appearing in the genome are protected from mutation, thus preserving the genome decodability. The figure illustrates this approach using strings of high-level symbols rather than binary strings as the genome. (The protected symbols are identified here by the shaded background.)

the L-system is always preserved. This is the most straightforward genetic representation of an L-system and, given the simplicity of its implementation, is thus also most often used in practice. An example of an application of this kind of approach is described in (C. Jacob 2001), where it is used to successfully evolve fractal shapes. The set of genetic operators considered by Jacob includes mutation, recombination, deletion, and duplication of rules. The decoded L-system is run for a fixed number of rewriting steps, and the outcome is interpreted with a turtle graphics system. The resulting geometric structure is then compared with the desired fractal shape to determine the fitness of the genome. Escuela et al. (2005) used a similar syntactic representation of L-systems to infer evolutionarily the structure of proteins, and Ochoa (1998) to evolve artificial plant morphologies. A syntactic representation of parametric L-systems to evolve the morphology of physical objects such as tables is illustrated in (Hornby and Pollack 2001a) and its extension to the synthesis of the morphology of robots, called L-robots, is described in (Hornby and Pollack 2001b; Hornby et al. 2003).

$$\mathcal{E} \longrightarrow \begin{bmatrix} A & B \\ C & D \end{bmatrix}$$

$$A \longrightarrow \begin{bmatrix} a & d \\ a & a \end{bmatrix} \quad B \longrightarrow \begin{bmatrix} c & b \\ b & a \end{bmatrix} \quad C \longrightarrow \begin{bmatrix} b & a \\ a & c \end{bmatrix} \quad D \longrightarrow \begin{bmatrix} a & b \\ a & d \end{bmatrix}$$

$$a \longrightarrow \begin{bmatrix} 0 & 0 \\ 0 & 1 \end{bmatrix} \quad b \longrightarrow \begin{bmatrix} 1 & 0 \\ 0 & 0 \end{bmatrix} \quad c \longrightarrow \begin{bmatrix} 0 & 0 \\ 1 & 0 \end{bmatrix} \quad d \longrightarrow \begin{bmatrix} 0 & 1 \\ 0 & 0 \end{bmatrix}$$

Figure 4.15   An example of rewriting rules for the evolution of a developmental process for neural network connectivity proposed by Kitano (1990). The symbol $\mathcal{E}$ is the axiom of the rewriting system.

4.6.3   Example: Matrix Rewriting

Matrix rewriting (Kitano 1990) is one of the first examples of evolution of neural networks using a syntactic genetic representation for a developmental system. The aim of matrix rewriting is the evolution of the connectivity of the network. Once the network connectivity is established, the actual synaptic weights are determined using a suitable neural network learning algorithm. Kitano's system corresponds to an L-system combined with a graphic interpreter that transforms the final result into a neural network. Constraints on the genetic operators ensure that the genome remains decodable as a parallel self-limiting rewriting system.

In the matrix rewriting system, the set of developmental rules consists of a collection of matrix rewriting rules as exemplified in figure 4.15. The figure shows that the rewriting system is based on a hierarchy of rewriting rules. On top, there is a rule that substitutes the primeval symbol $\mathcal{E}$ (the axiom), which stands for the embryo, with a matrix of uppercase alphabetic characters. Next in the hierarchy, there is a set of rules that substitute matrices of lowercase alphabetic characters for uppercase alphabetic characters. Finally, a set of rules substitute matrices of 1s and 0s for lowercase alphabetic characters. The 1s and 0s are used to represent the presence and absence, respectively, of a link between the nodes of the evolved network, and thus there is no further set of rewriting rules below in the hierarchy. Of course, the user is free to define additional intermediate rewriting steps.

$$\varepsilon \Rightarrow \begin{bmatrix} A & B \\ C & D \end{bmatrix} \Rightarrow \begin{bmatrix} a & d & c & b \\ a & a & b & a \\ b & a & a & b \\ a & c & a & d \end{bmatrix} \Rightarrow \begin{bmatrix} 0 & 0 & 0 & 1 & 0 & 0 & 1 & 0 \\ 0 & 1 & 0 & 0 & 1 & 0 & 0 & 0 \\ 0 & 0 & 0 & 0 & 1 & 0 & 0 & 0 \\ 0 & 1 & 0 & 1 & 0 & 0 & 0 & 1 \\ 1 & 0 & 0 & 0 & 0 & 0 & 1 & 0 \\ 0 & 0 & 0 & 1 & 0 & 1 & 0 & 0 \\ 0 & 0 & 0 & 0 & 0 & 0 & 0 & 1 \\ 0 & 1 & 1 & 0 & 0 & 1 & 0 & 0 \end{bmatrix}$$

Figure 4.16   The result of the application of the matrix rewriting rules of figure 4.15, leading to the production of a binary matrix defining the connectivity of a neural network.

Figure 4.16 shows the result of the application of the three levels of rewriting rules defined in figure 4.15. The rewriting rules are applied in parallel and, at the end of the three rewriting steps, the result is a binary matrix. Each row of the matrix corresponds to a node of the network. A 1 in row i, column j of this matrix corresponds to a connection from the ith node to the jth node of the network. Figure 4.17 shows the neural network corresponding to the connection matrix generated by the rewriting steps of figure 4.16.

In Kitano's evolutionary system, the rewriting rules are generated by an evolutionary process. Once the alphabets used at the various hierarchical levels are defined, we can define a genome that encodes the successor of each rewriting rule. For example, the rules shown in figure 4.15 can be represented with the following genome:

$ABCD$ $adaa$ $cbba$ $baac$ $abad$ 0001 1000 0010 0100

where, to facilitate its reading, we have inserted a space between groups of symbols referring to the same production rule. The mutation operator is defined so as to guarantee the substitution of characters suitable to the level to which the rule belongs.

Despite the compactness of the representation, it is still unclear if the matrix rewriting approach leads to actual advantages with respect to the direct encoding. Although Kitano (1990) reported advantages in terms of speed of convergence of the learning algorithm for the weights, which grew larger as

$$\begin{array}{c} \begin{array}{cccccccc} \mathbf{1} & \mathbf{2} & \mathbf{3} & \mathbf{4} & \mathbf{5} & \mathbf{6} & \mathbf{7} & \mathbf{8} \end{array} \\ \begin{array}{c} \mathbf{1} \\ \mathbf{2} \\ \mathbf{3} \\ \mathbf{4} \\ \mathbf{5} \\ \mathbf{6} \\ \mathbf{7} \\ \mathbf{8} \end{array} \left[\begin{array}{cccccccc} 0 & 0 & 0 & 1 & 0 & 0 & 1 & 0 \\ 0 & 1 & 0 & 0 & 1 & 0 & 0 & 0 \\ 0 & 0 & 0 & 0 & 1 & 0 & 0 & 0 \\ 0 & 1 & 0 & 1 & 0 & 0 & 0 & 1 \\ 1 & 0 & 0 & 0 & 0 & 0 & 1 & 0 \\ 0 & 0 & 0 & 1 & 0 & 1 & 0 & 0 \\ 0 & 0 & 0 & 0 & 0 & 0 & 0 & 1 \\ 0 & 1 & 1 & 0 & 0 & 1 & 0 & 0 \end{array} \right] \end{array}$$

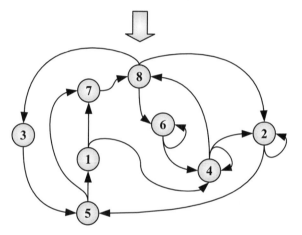

Figure 4.17 The transformation of the connectivity matrix of figure 4.16 into a neural network. The presence of a 1 in row i, column j of the connectivity matrix produces a connection from the ith node to the jth node of the network.

the network size increased, some researchers have questioned the validity of these claims (Siddiqi and Lucas 1998).

4.6.4 Tree-Based Representation

L-systems can also be genetically encoded using the tree-based representation described in chapter 1. Koza (1993) used the genetic programming tree-based representation to evolve L-systems for the turtle graphics interpreter. The goal of the evolutionary search was a bracketed L-system capable of drawing a curve matching a given curve. For example, Koza considered as

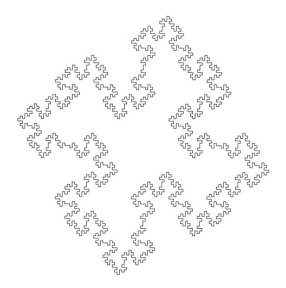

Figure 4.18 The quadratic Koch island is the fractal curve used as a target for the evolution of an L-system using a tree-based genetic representation.

target the fractal curve obtained after three rewriting steps using the production rule

$$F \rightarrow F - F + F + FF - F - F + F$$

starting from the axiom $\omega = F + F + F + F$ and using an angle increment $\delta = 90°$ (figure 4.18).

Koza used the following set of terminals (see chapter 1 for a description of the genetic programming terminology):

$$\mathcal{T} = \{F, +, -\}$$

where the symbols have the usual turtle graphics meaning described above, and the following set of functions:

$$\mathcal{F} = \{C, [\,]\}$$

where C is a function that combines its two arguments, and $[\,]$ is a function that puts its single argument within brackets. The symbols $+$ and $-$ are left unchanged during the rewriting process. Thus, F is the only symbol for which a production rule must be provided. The use of a tree and of the functions and terminals listed above ensures that the string decoded from the

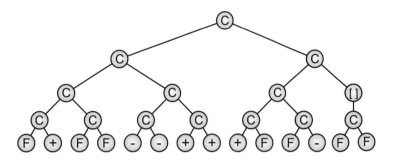

Figure 4.19 An example of tree encoding of a sequence of symbols. This sequence can be used as the right side of a context-free production rule. Thus, this structure can be interpreted as the tree-based genetic representation of a context-free L-system with the single production rule $F \rightarrow F + FF - - + + + FF - [FF]$.

tree is always a valid successor for the production rule. Figure 4.19 shows an example of a tree representing the production rule

$$F \rightarrow F + FF - - + + + FF - [FF].$$

Note that in this simple case, since the L-system is context-free and there is only one production rule, we need only encode the successor of the production rule.

To compare the structures produced by the evolved L-system with the target fractal curve, Koza used the following approach. Both the target curve and the evolved curve were enclosed within their minimal containing square. The containing square was divided into a regular grid of 10,000 cells. The similarity of the two curves was defined as the number of cells in which the behavior of the two curves did not differ. This discrepancy was used as the fitness function, maximized by the evolutionary search. The evolutionary algorithm was able to generate a curve that perfectly matches the target curve. Figure 4.20 shows an example of an individual generated at an intermediate step of the evolutionary process. Eventually, the evolutionary process found the production rule

$$F \rightarrow F - F + F - - + + + FF - F[-F] - F + F$$

which gives a curve that achieves a perfect match with the target curve. The resulting rule does not correspond formally to the production rule $F \rightarrow F - F + F + FF - F - F + F$ used to generated the target curve. However, the sequence $- - + +$ can be canceled because it does not draw any line or change

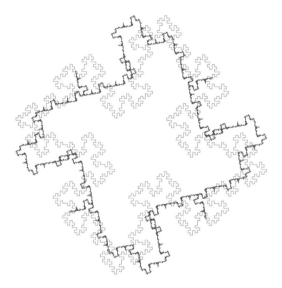

Figure 4.20   An intermediate result produced by the evolutionary process aimed at the generation of an L-system for the quadratic Koch island described in (Koza 1993). This curve in black corresponds to the production rule $F \rightarrow [F]F - F + F - - + + +$ $FF - [F][[-F]F + -F + - - F]FFFF$. The target curve is shown in the background for comparison.

the turtle status, and the sequence [-F]-F draws twice the same line and thus corresponds in practice to the sole sequence $-F$. With these simplifications the evolved production rule corresponds exactly to the production rule used to generated the target curve.

4.7   Evolutionary Developmental Programs

In this section we describe some examples of the third of the four kinds of artificial developmental systems listed above. These are systems where the basic mechanisms of the developmental process are fixed and cannot evolve, but the modality of their application is encoded in the genome and can evolve. We can interpret the fixed developmental mechanisms used by these systems as the basic instructions of a programming language, so that what is evolved can be seen as a developmental program based on these instructions. The tree-based representations discussed in chapter 1 are the most popular genetic representations for programs. Thus, typically, evolutionary developmental systems of the type discussed in this section are based on genetic programming or one of its variants. We have already described in chapter 1 an example of the application of genetic programming to the developmental description of electronic circuits. We consider here a variant of this approach called cellular encoding.

4.7.1   Example: Cellular Encoding

Cellular encoding was proposed by Gruau (1994a,b) as a developmental approach to the definition and evolution of artificial neural networks. Developmental systems are in general well suited to the description of *structured networks*, that is, networks where a certain network motif is treated as a module and reused several times or encapsulated within a higher-order motif in a hierarchy of levels of organization. As we will see, in defining cellular encoding, Gruau aimed specifically at the possibility of defining modular networks.

NETWORK
TRANSFORMATION
OPERATIONS

Similarly to what happens in genetic programming, the basic element of the cellular encoding description is a set of *network transformation operations* that alter the topology or the parameters of the neural network. Figure 4.21 illustrates some possible network transformation operations. Figure 4.21(a) represents the initial untransformed graph. The five nodes at the top of the graph represent the inputs and the five nodes at the bottom represent the outputs. The central node is the one that will be the subject of the network transformation operation. For example, the operations illustrated by figure 4.21(b,c) substitute the central node with two daughter nodes connected in series between inputs and outputs; the operations illustrated by figure 4.21(d,e) substitute the central node with two nodes connected in parallel between inputs and outputs; the operation illustrated by figure 4.21(f) removes a connection; the operation illustrated by figure 4.21(g) adds a recurrent connection to the central node; and the operations illustrated by figure 4.21(h,i) change an input and an output weight, respectively. There are several other network transformation operators that are similar to those shown here. Although these operators appear as reasonable candidates for the development of a network, the task of choosing them and of verifying that they are sufficient and not too redundant is not simple.

In cellular encoding the sequence of instructions that constitutes a cellular programming developmental program is represented in the form of one or more trees (figure 4.22). The network transformation operations shown in figure 4.21 are represented by symbols. When one of these symbols is

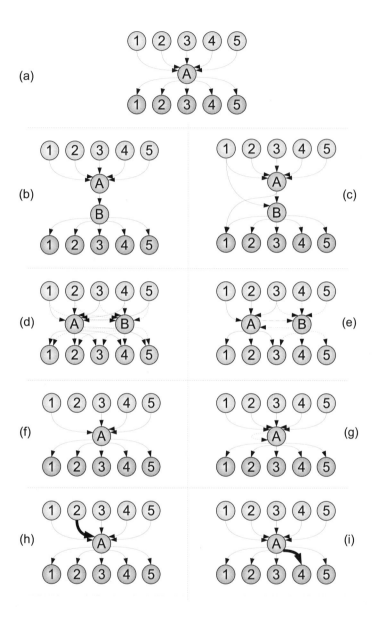

Figure 4.21   Some examples of network transformation operations that can be defined for the development of a neural network using the cellular encoding approach (adapted from Gruau 1994a). (*a*) Original network. (*b,c*) Series division operations. (*d,e*) Parallel division operations. (*f*) Connection removal. (*g*) Recurrent connection insertion. (*h*) Input weight change. (*i*) Output weight change.

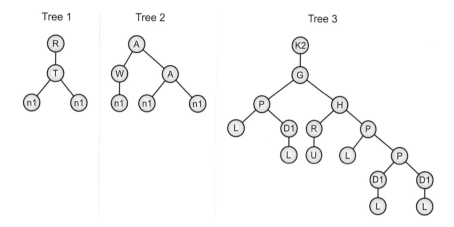

Figure 4.22   An example of cellular encoding of a neural network using the mechanism of automatic definition of neural subnetworks. The terminal symbols L and U define the nature of the neural network nodes, the terminal symbols $n1$ instruct the system to execute the developmental program represented by the next tree in the sequence, and all the other symbols correspond to network transformation operations like those represented in figure 4.21 (adapted from Gruau 1994a).

encountered while parsing the tree, the corresponding network operation is applied to a node of the developing network, starting from the unique node of the initial elementary network, as shown in the first developmental step of figure 4.23. Special terminal symbols L and U, which can appear as leaves of the tree, define the nature and parameters of the neural network nodes created by the network transformation operations. The cellular encoding developmental program is in general constituted by a list of several trees. Besides the terminal symbols L and U, the leaves of a tree can contain also a special symbol n followed by an integer d. The symbol n is interpreted as an instruction to substitute a node with the network generated by the execution of the sequence of instructions represented by the tree whose relative distance from the currently executed tree is d. For example, each symbol $n1$ present on the leaves of the first tree seen in figure 4.22 corresponds to the substitution of a node with the network generated by the execution of the instruction of the second tree, and each symbol $n1$ present on the leaves of the second tree corresponds to the substitution of a node with the network generated by the execution of the third tree. This mechanisms permits the reuse as a subnetwork of the network generated by the set of instructions

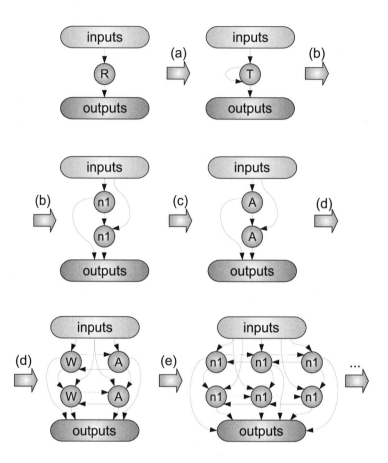

Figure 4.23   The first steps of the development of the neural network represented by the cellular encoding description of figure 4.22 (adapted from Gruau 1994a). (*a*) The symbol R in the first node of tree 1 inserts a recurrent connection. (*b*) The symbol T in the second node of the tree splits the original neuron into two neurons connected in series. (*c*) The symbols $n1$ in the terminal nodes of tree 1 start the parallel execution of tree 2 on both neurons. (*d*) The symbol A splits each neuron into two neurons connected in parallel. (*e*) The symbol W inserts a waiting step in the development of the neurons on the left, whereas the symbols A perform another parallel splitting of the neurons on the right. At this point all the neurons are assigned the symbol $n1$ appearing in the terminal nodes of tree 2, which cause the parallel execution of tree 3 on all neurons (not shown).

represented by a tree. Gruau (1994a) called this mechanism *automatic defini-tion of neural subnetworks*. Figure 4.23 shows the result of the execution of the first steps of the developmental program defined by the trees of figure 4.22.

Gruau's experiments confirm that the possibility of reusing developmental modules permits the evolution of very compact genomes that develop into modular neural networks. Moreover, as explained in more detail in chap-ter 6, the structure of the evolved networks can comply with the symmetries of the controlling task, for example, the bilateral symmetry of a six-legged insect-like robot. On the contrary, if the possibility of reusing developmental trees is disabled, the evolved networks have many more nodes and connec-tions, and do not display any structure obviously related to the structure of the controlled system (Gruau 1994a).

4.8 Evolutionary Developmental Processes

Biological developmental processes are obviously more flexible and complex than any of the models considered above, and underwent evolution both at the level of their basic mechanisms and of their combination and unfolding in space and time. In this section we describe some examples of the last of the four kinds of artificial developmental systems listed above, namely, those that – like biological systems – encode and evolve features of both the basic mechanisms of development and of their application and interaction with the environment. Since the goal is the definition of artificial developmental systems with characteristics comparable to the biological ones, it is advisable to look in more detail to the fundamentals of the latter.

A first observation is that a biological developmental process is not just a gene-driven process but is composed of the interplay of at least three ele-ments (Mahner and Bunge 1997): the genome; the cellular environment (cy-toplasm) with which the genome interacts; and the external environment in which development takes place. A second observation is that the genome does not unidirectionally control the development. For example, the cellu-lar machinery that interacts with the genome does not merely read it but can modify the genome and select some regions to actively interact with the cyto-plasm while silencing other regions. Moreover, this process depends on the cell state and on the external signals received by the cell which, in turn, are influenced by the external environment. An example of the interaction of the external environment with the developmental process is represented by the development of the vertebrate nervous system, which is influenced by the sensory signals received by the organism during development. A third observation is that the elements of the developing organism comply with a set of physical laws that determine the nature of the interactions that can exist and be exploited by the developmental process. For example, molecules can diffuse, attract, and bind to other molecules according to well-defined laws that can be exploited to build cells that divide, differentiate, migrate, and adhere to each other, to produce signals that propagate in the developing organism, and so on. Summing up, development can be seen as a dynamic process that takes as the initial condition the initial state of the cell – which includes but is not limited to the genome – and unfolds according to the laws of physics and the boundary conditions imposed by the environment.

The inclusion of development thus leads to a much more sophisticated scenario for evolution than is usually taken as inspiration for artificial evolutionary systems. In particular, it highlights the fact that what is passed from parents to offspring is not just a genome but, rather, a genome plus a cellular environment plus an external environment in which development takes place, given a fixed backcloth of physical laws determining how the first three elements interact. In setting up a fully fledged artificial evolutionary developmental system inspired by biological evolutionary development it is of course unreasonable to attempt a detailed modeling of all these aspects. We must thus decide what simplifying assumptions to adopt, that is, we must decide which of the above-mentioned features we want to implement and how. Despite many attempts and proposals, this field of investigation is still in its infancy and no consensus has yet emerged on what are the crucial aspects that must be modeled in order to achieve satisfying results. The number of possible choices and combination of elements is so large that it is impossible to mention all the approaches that have been proposed in the literature. Below we present some approaches that have been proposed for integration into an artificial evolutionary framework, some of the elements observed in the biological developmental process, such as phenotypic plasticity, morphogenesis, gene regulation and signal transduction, and cell differentiation and adhesion.

4.8.1   Example: Phenotypic Plasticity

The first example that we consider was proposed by Nolfi et al. (1994b) in order to explore the role of the interaction of the environment with the evolutionary developmental process. The basic idea is to define evolutionary agents that are characterized by a *plastic phenotype*, that is, a phenotype that

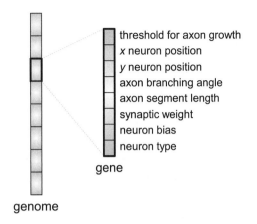

threshold for axon growth
x neuron position
y neuron position
axon branching angle
axon segment length
synaptic weight
neuron bias
neuron type

gene

genome

Figure 4.24   The structure of the genome used in the phenotype plasticity experiments. The genome is composed of a fixed number of genes, each gene encoding a fixed number of parameters, some of which influence the developmental process that builds a neural network (adapted from Nolfi et al. 1994b).

is not completely specified by the genotype but is instead in part determined by the interaction of the agents with their environment. In the model proposed by Nolfi and coworkers the agents are simulated wheeled robots and the part of their phenotype that is plastic is the artificial neural network that controls their behavior. The evolved genome contains information that can be used to control the development of a neural network according to the activity of the neurons. This means that the development of the network is not completed before the start of the simulated life of the agents but takes place in part during life.

The genome is composed of a fixed number of genes, each gene corresponding to one neuron (figure 4.24). The connections between the neurons are established using an L-system with the standard turtle graphic interpreter. The L-system is used to build in a 2D space a branching structure which represents the axon of the neuron (figure 4.25). Two numerical parameters x and y contained in each gene specify the position of the neuron in the plane. Two other gene parameters specify the branching angle and the segment length of the branching structure. To build the branching structure, a hand-designed L-system and turtle interpreter that produces binary branching is applied a fixed number of times (four in the example shown in figure 4.25). The direction of growth of the branching axon and the range of the y parameter for each neuron type are chosen so as to guarantee the generation of feedforward networks. When the growth of the branching structure is terminated, the branches that do not contribute to establishing a connection between the neuron at the root of the branching structure and other neurons are pruned. The surviving branches are assigned a value of synaptic weight specified by an additional gene parameter. The root neuron is assigned a bias and a type (sensory, hidden, output) specified by specific gene parameters. The resulting neural network is further simplified to remove the neurons that are not functional, either because they are isolated or because they do not contribute to the network output.

GENE EXPRESSION
THRESHOLD

The crucial aspect of the system, which determines the interaction of developmental process with the environment via the agent's behavior, is the existence of an additional gene parameter interpreted as a *gene expression threshold*. This parameter determines the conditions under which the branching axon is actually built. In the experiments reported in (Nolfi et al. 1994b) the axon is built only if the variability of the neuron activation during the last 10 time steps exceeds the gene expression threshold. This means that if the threshold is set to its minimal value by the evolutionary process, the axon is always grown at birth, independently of the neuron activation. If the threshold is set to its maximal value, the axon is never grown. For all intermediate values, the axon growth is conditional on the relation between the neuron activity and its threshold.

This developmental mechanism implies that the structure of the network is jointly determined by both the genetically encoded parameters of the neuron and by its environment. Moreover, the relative roles of the genome and of the environment are not fixed but can be influenced by the evolutionary process, which can chose which neuron structure must be grown at birth and which can be influenced by interaction with the environment. Note that the environment of a neuron can be either the external environment of the agent (which sends the sensory stimulation) or other neurons in the network (which stimulate the neuron with their activation). This corresponds to what happens in biological developmental systems, where both the external environment and the other cells of the organism constitute the environment of a cell. As discussed in more detail in chapter 6, Nolfi et al. (1994b) have shown that this type of developmental process endows the evolving agents with a form of phenotypic plasticity that results in a greater adaptivity of the evolved agents to changing environments.

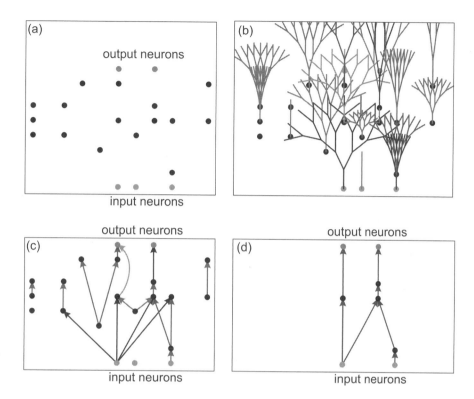

Figure 4.25 An example of development of the neural network in the phenotype plasticity experiments of Nolfi et al. (1994b). (*a*) The neurons are placed in the 2D space according to their (x, y) coordinates encoded in the genome. (*b*) Branching axons are grown from the neurons according to their activity and the value of the gene expression threshold; the parameters determining the geometry of the axon are also encoded in the genome. (*c*) The branching axons are pruned to leave only the connections between the neurons. (*d*) The nonfunctional neurons and the corresponding connections are removed from the network, producing the final neural network. This developmental process continues during the whole life of the agents whose behavior is controlled by the neural network (adapted from Nolfi et al. 1994b).

4.8.2 Morphogens

MORPHOGENESIS

One of the characterizing properties of development is the phenomenon of *morphogenesis*, that is, the robust generation of complex forms and patterns starting from embryos that appear at first look to be quite homogeneous. In 1953 Alan Turing, using a simple mathematical model, showed how patterns can emerge spontaneously by amplification of small fluctuations in an

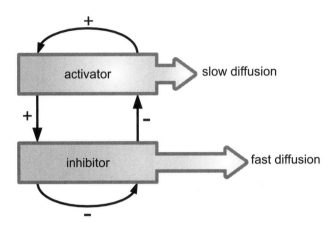

Figure 4.26   In Turing's model of morphogenesis two chemical substances act as activator and inhibitor. The activator diffuses slowly and enhances its own production and that of the inhibitor. The inhibitor diffuses rapidly and decreases its own production and that of the enhancer.

TURING MODEL

otherwise homogeneous structure. Turing's model considers a few chemical substances that react with each other and diffuse in a cellular or homogeneous tissue (figure 4.26). Turing proved that – given the right equations for the chemical reactions and the right values for the diffusion parameters – after some time the concentration of the chemical substances forms wave-like or spot-like *Turing patterns* in the structure (figure 4.27). To simplify the model and make it amenable to mathematical analysis Turing did not attempt to model the internal dynamics of the cells. One must consider that at the time not much was known about the structure of the genes and of their interaction with the cytoplasm and with the substances in contact with the cells. Moreover, Turing's goal was not to formulate a realistic model of spontaneous pattern formation in biological organisms, but just to prove the theoretical possibility of this phenomenon.

TURING PATTERNS

In 1969, Wolpert proposed a more realistic model of pattern formation in biological organisms, based on what he called *positional information*. In the simplest formulation of this model, a fixed concentration of one or more chemical substances called *morphogens* is assigned at the boundary of the embryo. Morphogens diffuse in the embryo and determine a profile of concentration within it. These profiles provide the cells with information about their position with respect to the boundaries. Morphogens can be originally

POSITIONAL
INFORMATION THEORY

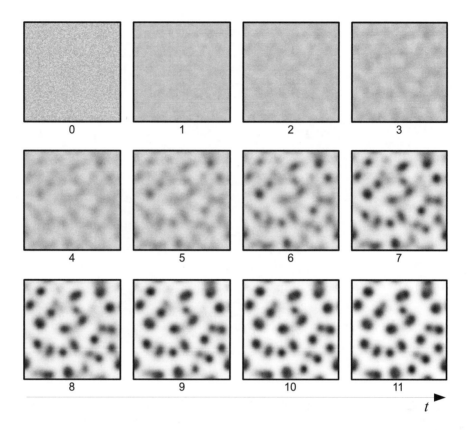

Figure 4.27 The reaction-diffusion dynamics of Turing's model amplifies random fluctuations in the initial distribution of the two substances, leading to the production of large-scale patterns of activator concentration.

present in the zygote or can be a result of the activity of some of the cells that form the developing organism. The information provided by morphogens can result in a change of state (*cell differentiation*) that influences the subsequent developmental history of a cell. For example, a morphogen can change the fate of a cell when its concentration around the cell exceeds a given threshold (figure 4.28). By using several thresholds and several morphogens, the initially undifferentiated cells of a tissue can differentiate and form a pattern. This model of pattern formation is quite robust to perturbations and changes of scale, being based only on the profile of concentration of the morphogens. For example, the final pattern of figure 4.28 is invariant

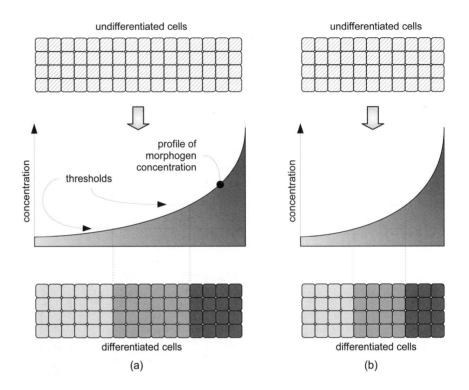

Figure 4.28 A schematic illustration of the positional information model of pattern formation. (*a*) The fate of the initially undifferentiated cells of the embryo is determined by the value of morphogen concentration. The internal machinery of the cell, instructed by the genome, compares the external concentration of the morphogen to the internally set thresholds and steers the further history of the cell accordingly. (*b*) The positional information mechanism is robust to perturbations; for example, if the profile of concentration of the morphogen remains qualitatively the same, a change of scale of the embryo does not change the pattern of differentiation of the cells.

to changes in the length of the structure and in the number of cells that it contains, provided the profile of concentration along the structure is maintained.

4.8.3   Example: Morphogenetic System

Although the mechanism based on the diffusion of morphogens is just one of several mechanism of pattern formation in biological organisms (Wolpert

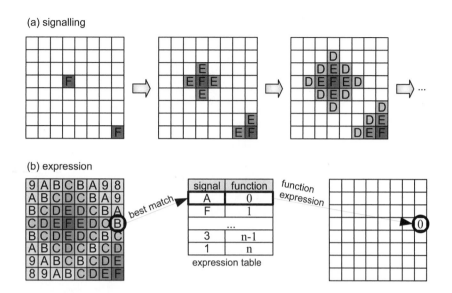

Figure 4.29 The developmental steps of the morphogenetic system (Roggen et al. 2007). (*a*) In the signaling phase the diffusers are placed in cells as specified by the genome, and the morphogens diffuse in the 2D lattice of initially undifferentiated cells. (*b*) When a cell has been reached by a morphogen, the expression table encoded in the genome is used to assign a function to it, thus modeling cell differentiation.

et al. 2007), its simplicity and robustness led to its adoption in many artificial evolutionary developmental systems (e.g., Astor and Adami 2000; Bentley 2004; Bongard and Pfeifer 2003; Dellaert and Beer 1996; Eggenberger 1997b; Fleischer and Barr 1994; Hampton and Adami 2004; Jakobi 2003; P. Kennedy and Osborn 2001; Kitano 1995; J. Miller and Banzhaf 2003; Roggen et al. 2007; Stanley and Miikkulainen 2003). By way of example, we will describe the *morphogenetic system* devised by Roggen et al. (2007) and the POEtic architecture (Tempesti et al. 2002; Tyrrell et al. 2003) that is ideally suited to its hardware implementation.

MORPHOGENETIC SYSTEM

The morphogenetic system is based on a 2D cellular lattice of elements, where each element can be thought of as a cell of the developing organism. Each cell contains an artificial genome that determines the functionality of the cell according to the signals present on the 2D lattice, which are interpreted as morphogens. The system models morphogens as generated within cells by *diffusers*. Morphogens diffuse in the cellular space so as to decrease their concentration according to a simple law. The position of the diffusers –

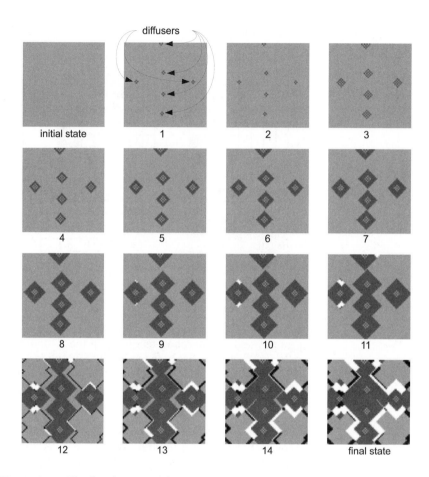

Figure 4.30   The development of a morphogenetic system evolved to match the Norwegian flag pattern (figure 4.31). At the beginning, the diffusers are placed in the lattice of initially undifferentiated cells. The activity of the diffusers starts to be visible at step 1. At step 15 all the cells have been assigned a functionality and development stops (adapted from Roggen et al. 2007).

one for each type of signal defined in the system – is encoded in the genome. The functionality of the cells is determined by an an *expression table* that is also encoded in the genome. The expression table associates the identity and concentration of the morphogens with the cell function, using a mechanism of pattern matching (figure 4.29). The development of the system starts with all the cells in an undifferentiated state (figure 4.30). The diffusers are placed

Figure 4.31   The Norwegian flag and CA-generated reference patterns used in the pattern-matching experiments with the morphogenetic system (from Roggen et al. 2007).

in the lattice according to the genome prescriptions and diffuse at discrete time steps in the so-called *signaling phase*. When the signaling phase is finished each cell has been reached by at least one morphogen. This activates the *expression phase*, where the functionality of the cells is assigned according to the prescriptions of the genetically encoded expression table. When a functionality has been assigned to all cells, the development is complete.

The system has been tested first for its ability to evolve the development of given patterns, for various sizes of the cellular lattice. In this batch of experiments the cell functions correspond to the displaying of a color on the cell surface, and the fitness is the proportion of cells that match the color of the prescribed pattern. Figure 4.30 shows an example of an evolved developmental process aimed at the production of the Norwegian flag pattern (figure 4.31). The authors compared the performance of the morphogenetic system with that of an evolutionary system using a direct genetic encoding of the cell colors. The results (figure 4.32) show that for small lattice sizes the direct encoding performs better, but that the developmental encoding of the morphogenetic system displays a smaller degradation of performance as the lattice size increases.

Another property of the developmental model implemented in the morphogenetic system is that it can exhibit a certain degree of recovery from perturbations. In a series of experiments, a certain percentage of the cells at the end of the development have been reset to the undifferentiated state, and the developmental process has been restarted. Figure 4.33 shows that the morphogenetic system can recover almost to the initial degree of matching

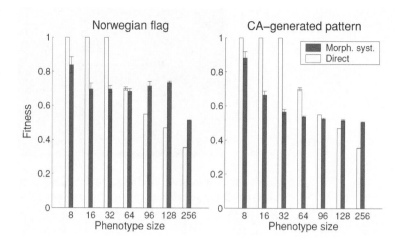

Figure 4.32 Comparison of the results of the pattern-matching experiments using a direct encoding and the developmental encoding of the morphogenetic system for various sizes of the cellular lattice (from Roggen et al. 2007). The fitness corresponds to the fraction of cells whose color matches that of the reference pattern (figure 4.31).

of the prescribed pattern for a large range of perturbation rates. By contrast, the direct encoding, which has no mechanism of recovery, suffers a large degree of degradation of its pattern-matching performance. To test the potential of the morphogenetic system in tasks more complex than static pattern matching, two series of experiments of neural network evolution were also performed. In the first series, neural network was used for character recognition, whereas in the second it was used to produce obstacle avoidance behavior in a miniature robot. In both tasks the morphogenetic system outperformed an evolutionary system based on direct encoding of the networks (Roggen et al. 2007).

The results of the morphogenetic system evolutionary experiments show that indeed the direct encoding can work well for simple systems that can be encoded in a small genome. When the size and complexity of the system increases, the genome size increases and the performance of the direct encoding falters. By contrast, the developmental encoding scales well with the problem size. However, its low performance for small problem sizes illustrated by figure 4.32 suggests that the developmental process can produce constraints on the kind of structure that can be produced. This means that if the developmental process is not suited to the kind of structures that need

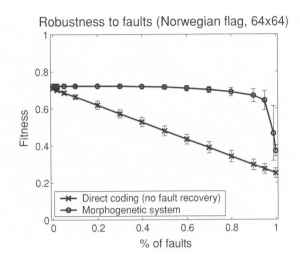

Figure 4.33 The developmental morphogenetic system can recover part of the functionality when development is re-enabled after a certain percentage of cells has been reset to the undifferentiated state. Recovery is almost complete up to very high levels of disturbance. The result for the direct encoding – which does not have a mechanism of fault tolerance – are also plotted for comparison (from Roggen et al. 2007).

ROBUSTNESS

to be produced (as, for example, L-systems are suited to the production of plant-like structures), a perfect solution to even simple problems can be difficult to evolve. One of the most interesting aspects of the results reported in (Roggen et al. 2007) is that the evolved systems displayed a robustness to perturbations even if there was no selective pressure to achieve that functionality. Developmental robustness seems to be generated as a byproduct of evolution coupled with a suitable developmental process. This suggests that when coupled with a suitable developmental process, evolution can produce developmental robustness as a byproduct. Note that in describing the relationship between evolution and development we referred to the hypothesis that developmental robustness can produce evolvability as a byproduct. In other words, the relationship between evolvability and developmental robustness seems to be a two-way affair of mutual enhancement.

4.8.4 Intrinsic Artificial Development

The property of artificial developmental systems such as the morphogenetic system of recovering from perturbation can realize its full potential only if

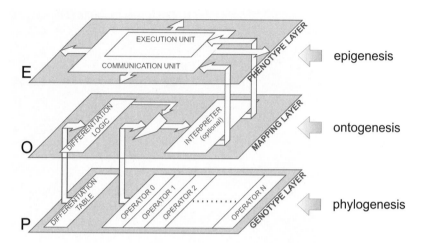

Figure 4.34   The layered architecture of the POEtic tissue. The genotype layer permits the implementation of an intrinsic evolutionary process and corresponds to the phylogenetic axis of the system. The mapping layer permits the implementation of an intrinsic developmental process and corresponds to the ontogenetic axis. The phenotype layer permits the implementation of an online learning process and corresponds to the epigenetic axis (adapted from Tyrrell et al. 2003).

the system can be implemented in hardware, on a platform permitting development and dynamic reconfiguration during the operation of the system. Using the terminology introduced for evolutionary electronics in chapter 1, the developmental process must be *intrinsic* rather than extrinsic. When this is the case, the theoretical property of recovery from disturbances can be transformed into one of run-time *fault tolerance* of the system (Roggen et al. 2007). In this scenario, the return of a cell to an undifferentiated state can correspond to the reset of a subsystem that is found to be faulty (for example, using artificial immune system techniques like immunotronics discussed in chapter 5). The subsequent recovery of the cell state corresponds then to the automatic reconfiguration of the subsystem state brought about by the developmental process.

An example of a hardware platform that permits the realization of this online developmental scenario is the POEtic tissue (Tempesti et al. 2002, 2003; Roggen et al. 2003a; Tyrrell et al. 2003). This is a custom hardware platform specially conceived for bioinspired experiments involving evolution, development, and learning. The first three letters of the name refer to the availabil-

(margin notes)

INTRINSIC
DEVELOPMENTAL
PROCESS

FAULT TOLERANCE

ity of the three axes of the adaptation process in nature: the phylogenetic axis (P), which refers to evolution; the ontogenetic axis (O), which refers to development; and the epigenetic axis (E), which refers to learning. The POEtic tissue (figure 4.34) is based on a layered reconfigurable FPGA-like architecture (see chapter 1 for a description of FPGAs). This architecture has a genotype layer that permits the implementation of an intrinsic evolutionary process, a genotype-to phenotype mapping layer that permits the implementation of an online intrinsic developmental process, and a phenotype layer that permits the implementation of an online learning process. Thus, it permits in particular the implementation of the morphogenetic system and the actual realization of its online fault-tolerance mechanism (Roggen et al. 2007).

4.8.5 Cell Physics

As pointed out by Wolpert (2003), the actual cornerstone of the complexity of multicellular organisms is the complexity of the regulatory dynamics, signaling strategies, and physical properties of the eukaryotic cell. The examples described so far in this section remain at a relatively high level of abstraction with respect to these aspects. It is thus worth exploring the possibility of modeling them in more detail, at a lower level of abstraction. The drawback of this approach is that the computational complexity of the models will grow with the amount of detail modeled. In exchange, we can expect to obtain a greater freedom for the evolutionary process in the discovery of the right developmental mechanism. Another advantage is the possibility of exploiting the potential of a sufficiently rich dynamics to produce complex self-organized structures (Camazine et al. 2001), without being forced to explicitly encode all their detail in the genome.

An interesting series of explorations along these lines was performed by Eggenberger (1997a,b, 2003, 2004a,b). Eggenberger defined an evolutionary ARTIFICIAL developmental system that he named AES (artificial evolutionary system). EVOLUTIONARY AES models some aspects of gene regulatory dynamics, and a simple model SYSTEM (AES) of cell signaling, differentiation, division, adhesion, motion, and death (figure 4.35). The gene regulatory and cell-cell signaling dynamics is defined by differential equations that take into account the interaction between gene products and their diffusion in the developing structure. The physics of cell-cell interaction is modeled using a physical simulator that implements Newton's laws and provides the possibility of defining elastic and viscous forces acting on cells. Eggenberger showed that AES is capable of evolving a developmental process that creates aggregates of cells displaying morphologies

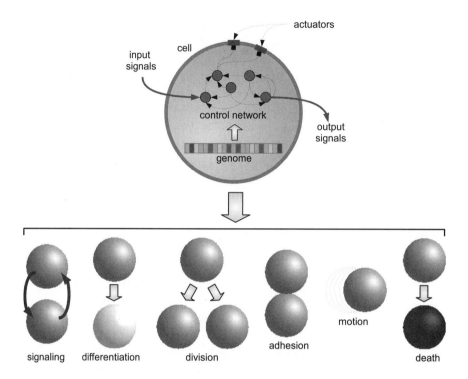

Figure 4.35 The basic properties of the artificial cell model defined by Eggenberger (1997a) in the artificial evolutionary system. The genome determines the structure of the control network which governs the internal dynamics of the cell and its interactions with the environment and with other cells. The cell can be also endowed with mechanical properties and actuators controlled by the internal network. This cell model permits the evolution of cell functions such as cell signaling, differentiation, division, adhesion, migration, and death.

similar to those observed in the early stages of vertebrate embryo development. AES was also used to successfully evolve functional structures such as optical lenses and neural networks for robot control.

Taking inspiration from AES, Bongard and Pfeifer (2001, 2003) defined an artificial evolutionary developmental system called artificial ontogeny (AO), with the aim of exploring further the possibility of using such a system to co-evolve the behavior and the morphology of artificial agents. To this end, they integrated in the system a developmental model for neural networks based on Gruau's cellular encoding described above in this section. The spherical

ARTIFICIAL ONTOGENY
(AO)

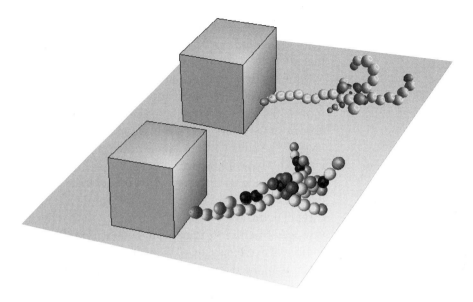

Figure 4.36 Two examples of structures evolved for block pushing in the experiments with the artificial ontogeny system. The body of the agents is composed of spheres and develops in a virtual environment under the control of the evolved genome contained in each sphere. The developmental process determines how the spheres are connected, controlled, and actuated, and what kind of sensors they possess. In the figure, light gray spheres have both sensors and actuators, dark gray spheres have only sensors, and black spheres are passive building blocks lacking both sensors and actuators (adapted from Bongard and Pfeifer 2003).

basic building blocks of the agent's body are endowed with sensors and actuators that can send and receive signals from sensor and motor neurons of the neural network. The actuators operate on the links connecting the building blocks and permit their relative motion. This means that, compared to Eggenberger's system, the building blocks are no longer intended as models of biological cells but, rather, correspond to macroscopic elements that can be used to build the agent's body and limbs. Moreover, the agents are no longer evaluated with respect to their morphology but with respect to their ability to perform behavioral tasks in a virtual physics environment. The evaluation of the evolved structure is performed after the development of the agent is completed. Bongard and Pfeifer were able to obtain encouraging results in the evolution of agents capable of directed motion and block pushing (figure 4.36). They also observed phenomena similar to those observed in biological organisms, such as the repeated convergence of the agents, to similar body plans for a given task, and the greater impact of genome mutations that affect the early phases of the developmental process.

Running both the AES and the AO evolutionary models is computationally expensive. However, the experiments performed with these systems proved that it is already computationally feasible to model many low-level aspects of the biological developmental process, provided one remains at the level of the cells without going down to the molecular details of the chemical and physical interactions. Moreover, these experiments confirmed that the developmental representation scales well with the size of the problem, in the sense that once a critical level of complexity in the genotype dynamics is exceeded, the genome is almost independent of the number of cells in the developed structure. It remains to be elucidated, on the one hand, the level of complexity of the systems that can be obtained with this approach while remaining in the realm of the computationally feasible and, on the other hand, which of the many low-level aspects of biological developmental processes that are modeled by AES and AO are essential to the result obtained.

4.9   Closing Remarks

Developmental processes are a powerful tool for the description and synthesis of artificial complex systems. Thanks to the scalability provided by the possibility of reuse of substructures and modules, more complex structures do not require a correspondingly more complex developmental description. However, the indirect and possibly opaque relation between the description and the resulting system can entail some difficulty in the definition of a developmental system producing a given result. This is especially true in the absence of some clues concerning the basic developmental mechanism required to achieve the result, as can be obtained, for example, from a recursive definition of a geometric object.

In this respect, the observation of biological developmental processes is an important source of inspiration for the definition of artificial developmental systems. However, the multiplicity of mechanisms at work in biological development poses a serious difficulty in the choice and combination of the basic elements. Some guiding principles and a better understanding of the role and integration of the elements of biological developmental processes are required before we can think of building by hand a developmental system using these elements as building blocks. Artificial evolution can provide a powerful tool for the exploration of the space of the combinations of these building blocks. Here we are, however, faced with a dilemma. On the one hand, modeling the biological phenomena at a very low level and with plenty of detail in the hope of letting evolution discover and exploit their essential property is computationally very expensive. On the other hand, choosing a set of high-level mechanisms entails the risk of exploring the wrong subset of developmental processes, with the risk of imposing unwanted constraints on the possible result and of hampering evolvability. For example, few of the examples that can be found in the literature give much relevance to the "negative" phenomenon of controlled cell death and catabolism, which appears instead to be a fundamental component of biological processes (Marijuán 1996).

The issue of evolvability in artificial evolutionary experiments is another crucial contribution that development could bring to the synthesis of artificial systems. We do not mean by this that development must be assumed as the only way to obtain evolvability. For example, the genetic regulatory networks of unicellular organisms are certainly also evolvable, without being the result of a developmental process. However, development appears as the most promising way to evolve complex structures, systems, and behaviors, thanks to the combination of its potential impact on evolvability and its power of providing a compact and scalable representation for complex systems.

In biological systems the crucial step toward the combination of these aspects seems to have been the evolution of the complexity of the eukaryotic cell, with its complex regulatory dynamics, signaling, and mechanical properties (Wolpert 2003). One of the challenges for artificial evolutionary developmental systems is thus the identification and abstraction at the right level of the fundamental properties that characterize the eukaryotic cell and its dynamics in multicellular development. Given the possible role of phenotypic plasticity in fostering the evolution of evolvability, another challenge is the realization of evolutionary experiments where environmental variability and noise require the evolution of developmental plasticity. The examples described in the second part of this chapter attest to the efforts that are currently being devoted to answering these challenges and bringing to artificial systems the combined power of evolution and development.

4.10   Suggested Readings

A good textbook on biological development, which lists as first author the originator of the theory of positional information, is (Wolpert et al. 2007). For a more colloquial description of the mechanisms of development and their genetic control, we recommend (Nüsslein-Volhard 2006). For a crash course in evolutionary developmental biology, see (Wolpert 2003). An interesting methodological overview of developmental biology, with a formal definition of development, can be found in (Mahner and Bunge 1997). An extensive collection of examples of developmental mechanisms with a detailed discussion of their relationship to evolvability is (Gerhart and Kirschner 1997). The same authors reformulated and extended these themes in terms of a *theory of facilitated variation*, which is explained in simple terms in (Kirschner and Gerhart 2005). The importance of phenotypic plasticity and its relation to evolvability is extensively discussed in (West-Eberhard 2003).

(D.W. Thompson 1941) is a classic on the relation between biological form and the constraints and possibilities created by physical laws and by the requirements of geometric consistency. A recent edited and abridged reprint of this work is (Thompson 1992). Although not focused on development, (Camazine et al. 2001) and (Ball 1999) present a modern perspective on the role of self-organization in biological and natural pattern formation. Deutsch and Dormann (2005) give an overview of mathematical models of biological pattern formation, with a special emphasis on cellular automata modeling and a chapter devoted to Turing's model of morphogenesis.

The classical introduction to L-systems and the turtle graphics interpretation, with some history of the field and many examples and illustrations is (Prusinkiewicz and Lindenmayer 1990). A more technical work, focused on the modeling of plants is (Deussen and Lintermann 2005). Examples of programs that implement L-systems, their 2D and 3D turtle graphics interpretation, and an evolutionary encoding can be found in (C. Jacob 2001). The collection of contributions edited by Kumar and Bentley (2003) gives a good overview of artificial evolutionary developmental systems of the type described in the second part of this chapter, with some background on biological developmental processes.

5 *Immune Systems*

To survive and reproduce, living beings need suitable materials and energy and must find these resources in their environment. Since all known naturally evolved living beings are composed of the same basic building blocks, they are potentially a rich source of high-quality matter and energy for each other. For this reason, living organisms must protect themselves from the attempt of other organisms to exploit their resources. In some cases the size of PATHOGEN the would-be exploiter – the *pathogen*– is many orders of magnitude smaller HOST than that of its target – the *host*. For example, viruses, bacteria, fungi, protozoans, and some kinds of parasitic worms in the initial stage of their life cycle are much smaller than the typical vertebrate. Due to the greatly different spatial scales of pathogen and host, the organs that the latter uses to interact with the environment are typically poorly suited to the detection and elimination of potential pathogens. The countermeasures that the host can take against the pathogens at the spatial scale of its body are mostly aimed at reducing the probability of getting in contact with the pathogens, for example by staying away from environments that can potentially house them. The host can also have physical barriers that reduce the probability of entry into its body of pathogens with which it has come into contact, or change the environment represented by its own body to make it less suitable for the survival and replication of the pathogen. However, since an organism cannot be completely isolated from its environment, and cannot perturb too much its own bodily variables, these kinds of countermeasures can be only partially effective. Another consequence of the difference in size between the pathogen and its host is that the pathogen can reproduce much faster and can generate easily populations that are orders of magnitude larger than the typical host population. This means that pathogens can rapidly evolve ways to neutralize the countermeasures adopted by the host to keep pathogens outside of its body or to render its body a hostile environment for the pathogens. In order to balance the struggle the host needs a set of countermeasures which operate on the same scale and which can keep up with the evolutionary pace of IMMUNE SYSTEM the pathogen. This collection of countermeasures constitutes the *immune system* of the host. Its function is to detect the pathogens once they have entered the host body and to eliminate them with minimal cost in terms of resources employed and damage done to the host. The immune system represents also a protection against the possibility of malfunctioning and failure of individual host cells. For example, cancer cells can behave as independent entities that pursue their own agenda and entail for an organism problems that are similar to those represented by pathogens.

Many human-built systems face the same kind of problems of biological organisms targeted by pathogens. For example, computer systems represent computational resources and contain data that attract nonauthorized users in the form of computer viruses and network intrusion attempts (Nachenberg 1997; Mukherjee et al. 1994). Typically the nonauthorized operations take place at a low level in the hierarchy of software levels of the computer system so that their effect is not immediately apparent at the scale of the computer user or network administrator interface. The countermeasure consisting in the isolation of the computing system is seldom an option in times of widespread networking. The addition of built-in protections to the operating system does not always solve the problem, because the frequency of update that is reasonable for the operating system is quite low when compared with the speed with which the attack modality can change. Currently, the most common solution is the use of frequently updated antivirus and intrusion detection programs. However, the implementation and update of these protection programs requires a substantial effort, and the effectiveness of the protection can be compromised if a communication failure or an oversight results in the omission of an update. A better solution would be a protection system capable of autonomously detecting and opposing the attempts at intrusion and exploitation. Human-built systems must also be protected against malfunctioning and failures of their subsystems. As mentioned above, the strategies used to automatically fight exploitation attempts can be also used for the detection and cure of faults, that is, to obtain systems ARTIFICIAL IMMUNE with built-in fault tolerance. *Artificial immune systems* (AISs) are the result of SYSTEMS (AIS) an effort to implement protection against external attacks and internal faults explicitly inspired by the workings of biological immune systems. More generally, an AIS is any artificial system that implements some of the processes that are found in biological immune systems. The applications of an AIS can thus go beyond system protection and fault tolerance to encompass other functions such as pattern recognition, noise reduction, function optimization, and biological modeling. To pave the way to the understanding of AISs, in the next two sections we describe the structure and operation of biological immune systems. Then, we proceed to show how the concepts inspired by biological immune systems can be applied to the definition of artificial immune systems, and describe some examples of their application to computer and network protection and to fault detection in electronic systems.

5.1   How Biological Immune Systems Work

Biological immune systems are exceedingly complex and are typically formed by several components that work in coordination. Given this complexity, in this section we provide an abstract description of biological immune systems, in order to permit the appreciation of their logic of operation, uncluttered by all the implementation details. The next section will describe to some extent the low-level implementation details, that is, the molecules and cells that form the vertebrate immune system. It is useful to distinguish between *innate immunity* – which refers to the immune countermeasures that do not change during the lifetime of the host – and *adaptive immunity* – which refers instead to the countermeasures that can change during its lifetime – although the two systems work in strict coordination.

5.1.1   The Innate Immune System

Once a pathogen has breached the physical barriers of the host and entered its body, the immune system of the host must recognize and destroy the pathogen, or at least interfere with its activity so as to render it harmless. Since a pathogen has an evolutionary goal that is different from that of the host, there exist in general some molecular structures which are found in the pathogen but not in the healthy host. If these structures are accessible to the immune system, it can use them to identify the pathogen as such. For this reason the working of an immune system can be based in part on a collection of *immune detectors* distributed in the whole body of the host (figure 5.1), which carry *pattern recognition receptors* (PRRs). Any structure that carries a pattern that can be potentially recognized by a PRR is called an *antigen*. The structure of an antigen that is recognized by a PRR is called an *antigenic determinant* or *epitope*. A single pathogen can have multiple epitopes and can thus

PATTERN RECOGNITION
RECEPTORS (PRRs)
ANTIGEN

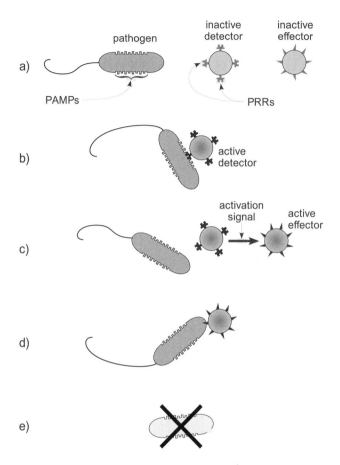

Figure 5.1   A schematic representation of the operation of the innate immune system. *a*) When a pathogen enters the host body it finds itself in an environment patrolled by inactive immune detectors and effectors *b*) If the pattern recognition receptors (PRRs) of an immune detector match the pathogen-associated molecular patterns (PAMPs), the detector becomes activated. *c*) The activated detector activates an effector. *d*) The activated effector attacks the pathogen and *e*) kills it. Note that, although logically distinct, the detector and effector functions can be implemented by the same immune elements and share parts of the same physical structure.

AUTOANTIGEN   be recognized by several distinct PRRs. Note that both the structures of the host and those of the pathogens can be antigens. To distinguish the two, the former are referred to as *autoantigens.* To ensure that the activity of the innate immune system is focused on the pathogens, its PRRs are structured so as to

PATHOGEN-
ASSOCIATED
MOLECULAR PATTERNS
(PAMPs)

recognize *pathogen-associated molecular patterns* (PAMPs), that is, patterns that are peculiar to the pathogens. When an antigen is recognized by an immune detector as belonging to a pathogen, the detector is activated and proceeds to activate the elements of a collection of *immune effectors*, which attack and destroy the pathogen.

The PRRs of the innate immune system are genetically encoded and cannot change during the lifetime of the host. Therefore, they must target PAMPs that the pathogen cannot easily modify or hide in order to escape detection. For example, good targets of the innate PRRs are patterns that belong to structures which are required by some crucial function of the pathogen and that are necessarily exposed, like the flagella of bacteria, or some molecules that are essential constituents of their cell wall. Different pathogens can require different strategies of attack, which means that the effectors must be also tailored to the characteristics of the pathogen. The innate immune system includes many kinds of effectors, and links the recognition of a pathogen to the activation of the effectors which are most effective against that pathogen. Besides the direct elimination of pathogens, the effectors also play a role in mobilizing the immune system. If the host is subject to a massive attack of pathogens that cannot be rapidly eliminated by the effectors present in place, the effectors produce an inflammatory response which enhances and focuses the activity of the immune system at the site of the attack.

5.1.2   The Limits of Innate Immunity

Innate immune systems can be very effective, as witnessed by the fact that invertebrates seem to rely almost exclusively on them for their protection from pathogens. However, there are limitations on what can be achieved with a system that cannot change during the lifetime of the host. The problem is that pathogens can evolve and change the patterns that are accessible for inspection by immune detectors. One could imagine building an immune system capable of recognizing and attacking all the structures carrying patterns that are not found in the healthy host. This strategy, however, entails a number of problems. The first is that the number of required PRRs could be very large, even if one assumes that – as is typically the case – each PRR is able to recognize many PAMPs. Since the PRRs of the innate immune system are encoded in the genome of the host, the size of the genome would grow unacceptably. Moreover, an excessively large genetically encoded repertoire of PRRs would have a high probability of producing PRRs that recognize autoantigens as a consequence of random genetic mutations. Another problem with this strategy is that it limits the possibility that the host can change, or that different individuals of the host population can exchange and recombine their patterns in forming new individuals. For example, new autoantigens typically appear during development, with aging, and during pregnancy, facts that are incompatible with a hypothetical fixed universal repository of PRRs. Summing up, an all-encompassing implementation of innate immunity appears problematic because the memory requirements of the immune system become excessive and the evolutionary and developmental flexibility of the host is unduly constrained. A solution to this problem consists in endowing the protection system with the possibility to change during the lifetime of the host.

A first strategy for adding adaptivity to the host defenses is to outsource part of the protection activity to other organisms. For example, vertebrates and invertebrates can associate mutualistically with harmless bacteria that interfere with pathogen invasion. Typically, the protecting bacteria line the surfaces where exchange between the host and its environment takes place (Loker et al. 2004). Since the mutualistic bacteria operate at the same spatiotemporal scale of the pathogens, they can fight them effectively. Mutualistic bacteria, however, do not entirely share the evolutionary fate of the host and therefore cannot be trusted too much in their protection role. The ultimate strategy for overcoming the limitations of innate immune systems is therefore based on the deployment of an adaptive arm of the immune system which complements and integrates the activity of the innate system.

5.1.3   Monitoring of Subsystems

An additional difficulty for the operation of the immune system is that the pathogens could hide within subsystems of the host which are not accessible to the immune detectors and effectors. For example, the detectors and effectors of the vertebrate immune system are cells and molecules that do not enter the cells of the host. To permit the elimination of infected subsystems whose internals are not accessible, the vertebrate immune system adopts two strategies that require the active collaboration of the subsystems. The first strategy is implemented by innate immunity and consists in distributing in the body of the host effectors which are ready to destroy everything that comes into contact with them, unless inhibited by special signals. The inhibitory signals are provided by all healthy host subsystems but are not easily mimicked by pathogens. The second protection strategy (figure 5.2) is more sophisticated and requires the cooperation of the adap-

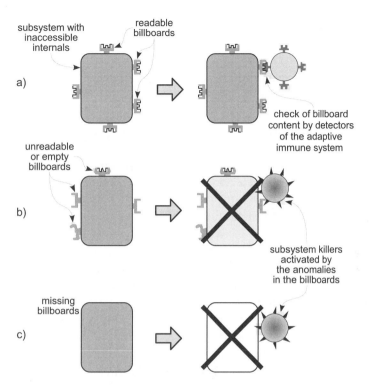

Figure 5.2 The "billboard" strategy used by the immune system to deal with inaccessible subsystems. *a*) The subsystem reports on its internal activity on the billboards. The detectors of the adaptive immune system inspect the content of the billboards. *b*), *c*) The billboards are either empty, unreadable, or missing. These conditions activate specialized elements of the innate immune system which destroy the subsystem.

tive immune system. This strategy consists in requiring that the subsystems report on their internal activity using specialized interfaces or "billboards" (Sompayrac 2003). The adaptive immune detectors that will be described below scan the billboards and target for destruction the subsystems that display patterns that appear to be the product of pathogen activity. To neutralize the obvious pathogen countermeasure consisting in the disabling of the billboards, subsystems which fail to report because the billboards are either missing or unreadable are assumed as potentially invaded by a pathogen and destroyed by the innate immune system. To make the task more difficult for the pathogens, there is a large variability in the way the immune systems of different individuals of a population process and affix on the billboards the results of the internal activity of their subsystems. This means that even if a pathogen evolves an escape from the "billboard detection system" of the immune system of one individual, this escape is almost certainly ineffective for the immune system of other individuals of the population.

5.1.4   The Adaptive Immune System

Just like innate immune systems, adaptive immune systems use a collection of detectors and effectors. The novelty of adaptive immune systems is that both can change during the lifetime of the host. The problem thus becomes the definition of the strategy for the generation of detectors and effectors that are effective against pathogens but do not interfere with the normal activity of the host tissues.

A simple way to generate a large variety of PRRs for the adaptive detectors consists in producing random structures. Adaptive immune systems use these approaches to generate an initial collection of candidate detectors and effectors. However, there are several reasons for not putting directly into use detectors with randomly generated PRRs. First, there is the problem that many of the PRRs thus produced will be targeted at autoantigens and would thus be *autoreactive,* resulting in the immune attack of the host tissues. A second problem is that many of the randomly generated detectors and effectors would not be effective against any pathogen and would thus represent a waste of resources for the host. These limitations can be obviated by subjecting detectors and effectors carrying the randomly produced PRRs to a process of *selection* before being accepted as part of the immune system and readied for action. The process of selection of the vertebrate adaptive immune system is composed of several stages. Each stage contributes to lowering the probability that the selected elements are active against the host and increase the probability that they are effective against potential pathogens. The end result is a very low probability that a pathogen goes undetected or that an autoreactive element is left free to damage the host (see box 5.1). Some stages perform a *negative selection* that eliminates the autoreactive elements. Other stages perform a *positive selection* that preserves and reproduces preferentially the elements that have the potential of being useful in the fight against pathogens. In some cases the preferential reproduction is accompanied by a random mutation of the reproduced elements, which results in the implementation of a Darwinian process of variation and selection that improves the performances of the immune elements.

NEGATIVE SELECTION
POSITIVE SELECTION

The purported ability of the adaptive immune system to perform the selection described above on the pool of randomly generated elements requires some explanation, since the antigens of the pathogens do not differ per se from autoantigens. Therefore, it is not clear at first sight how the adaptive immune system can discriminate between the two. The solution of this problem rests on the use of a notion of *context*. The innate immune system collaborates with the host tissues in marking some regions of the host as *danger zones*. Patterns belonging to antigens found in the regions that are not marked as danger zones are assumed as being associated with harmless materials and structures. Patterns belonging to antigens that are encountered in danger zones are assumed instead as potentially belonging to a pathogen. The definition of the danger zones is not static but varies dynamically according to the presence of *danger signals* (Matzinger 2002, 2007; Seong and Matzinger 2004). A first category of danger signals corresponds to PAMPs that are recognized by the innate immune system and reveal the persistent presence of pathogens which have not been rapidly removed by the innate immune system, typically due to the lack of suitable effectors. A second category of danger signals corresponds to distress signals released either intentionally or unintentionally by the host tissues. An example of a danger signal produced unintentionally is observed when a cell dies due to the attack of a pathogen. This is typically a catastrophic process that exposes structures that are normally hidden from the immune system. These structures can thus be interpreted as danger signals by the immune system. Note that the normal (apoptotic) death of a cell in the context of the physiological processes of the host is managed so as to avoid the exposure of such structures.

The production of the elements of the adaptive immune system which are activated to recognize and destroy the pathogens proceeds as follows. First, the immune system produces a collection of detectors and effectors carrying randomly generated PRRs. The newly generated elements are not yet active and are carried to specialized safe regions where they can interact with a large collection of autoantigens. The elements are subjected to a process of negative selection which eliminates all the elements that are responsive to the autoantigens present in the region. The unresponsiveness of an element of the immune system to an antigen is called *tolerance* and the process just described realizes the so-called *central tolerance*. Besides being centrally tolerized, the detectors that work by inspecting the billboards of the host subsystems are also tested for their capacity to access the billboards. A process of positive selection preserves only the detectors which are able to inspect the billboards of the subsystem present in the regions where central tolerance

DANGER SIGNAL

CENTRAL TOLERANCE

Box 5.1: The power of the product

The result of the multiplication of several real numbers smaller than unity can be an exceedingly small value. This fact has some interesting consequences for the design and evolution of systems. For example, to keep low the risk of autoimmunity, adaptive immunity is implemented so as to lead to an autoimmune attack only if several improbable and independent conditions are realized. The probability of an autoimmune attack is thus made to correspond to the product of the probabilities of each separate condition, and in this way it is reduced to sufficiently small values.

In the design of portable devices such as portable phones it is important that accidental pressing of the keypad does not result in unwanted activation of functions. For this reason these devices permit the locking of the keypad, and require a unique sequence of key pressings to unlock it. Some devices adopt as unlocking strategy the continuous pressing of a specific key for a certain time. Since the operating software of the device polls the keypad at discrete intervals, the probability of accidental unlocking corresponds to the joint probability that the unlocking key is accidentally pressed at all the polling instants. The problem is that in normal operating conditions, especially if the device is put in a pocket or in a bag with other objects, the probability that a key is accidentally pressed is not independent of its having being accidentally pressed in the near past. Thus, the joint probability does not correspond to the product of the probability of the separate events and is in fact not much smaller than the probability of a single accidental key pressing. Consequently, users of devices adopting this unlocking strategy can be expected to experience several accidental activations during the operating lives of the device (although this tends to be masked by the self-locking of the keyboard after a period of inactivity). Other devices adopt as unlocking strategy the sequence of events constituted by the pressing of different keys in a well-defined temporal sequence. With a suitable choice of the sequence, the accidental pressings of keys can be rendered practically independent of each other. In this way the probability of accidental activation becomes small enough that a user of this device has a good probability of never experiencing an accidental activation during the whole operating life of the device.

In other circumstances the smallness of the product of several terms has instead undesirable consequences. For example, in evolutionary processes the rate of mutation of the genomes is kept low to avoid producing too many nonviable offspring. If the improvement of   *(cont.)*

Box 5.1 (continued)

an evolving system depends on the simultaneous occurrence of two or more independent genome mutations, the corresponding event has a very low probability and evolution stalls.

The properties of the product can be observed also when a collection of numbers greater than 1 are multiplied. In this case the result can become rapidly very large and the phenomenon can have favorable or unfavorable consequences. A well-known unfavorable consequence is the combinatorial explosion of the complexity of problems that depend on the number of possible combinations of the elements that enter the problem. For example, this effect is present when the solution of a problem requires the sampling of a multidimensional space, since the number of samples required to cover the space reasonably well grows like the product of the samples required by each dimension. In this case the unfavorable effect is referred to as the "curse of dimensionality" (Bellman 1961) and prevents the practical solution of problems that exceed a certain dimensionality.

The complementary favorable effect of the combinatorial explosion is extensively exploited by evolution and human engineering. A typical example is constituted by the possibility of using the combination of a few elements, each taken from a small set, to represent an enormous variety of elements. This is the case of the languages based on finite alphabets and of the positional representation of numbers. In biology, we see an example of exploitation of this phenomenon in genomes, where an astronomical number of different configurations can be potentially obtained by combining the four letters of the genetic alphabet. Another example can be seen in the generation of the PRRs of the vertebrate immune systems, where the large required variety is obtained by combining the elements of several gene libraries, each providing an element selected among a few alternatives.

is established. The detectors and effectors that survive these first rounds of positive and negative selection are dispatched to an ensemble of specialized "meeting places" of the immune system. These meeting places are linked to form a network where the newly generated elements can circulate from node to node. These elements are still kept inactive when dispatched because there is still the possibility that they are reactive for autoantigens not present in the regions where central tolerance was induced. There is therefore the need for

PERIPHERAL
TOLERANCE

a further selection aimed at the elimination of the remaining autoreactive elements. This further selection realizes the so-called *peripheral tolerance* of the elements of the immune system. Finally, there is a process that activates the elements that can recognize and destroy the pathogens that have invaded the host.

ANTIGEN-PRESENTING
CELLS (APCs)

The establishment of peripheral tolerance and the activation of the immune elements are based on the activity of special agents of the immune system called *antigen-presenting cells* (APCs). APCs patrol the host body and systematically engulf the antigens that are found in the patrolled region. The APCs then migrate to the meeting places where they present the captured antigens to the detectors of the adaptive immune system, using special "professional" billboards with which the APCs are endowed. We can think of APCs as taking and carrying to the meeting places a "snapshot" of the zone where they engulf the antigens (Sompayrac 2003). In the absence of danger signals the APCs perform their duty at a low pace and remain in an inactive state. When an antigen is presented by an inactive APC it can thus be assumed as belonging to something that is not dangerous. This means that there is no need to eliminate the structures that carry that antigen but is rather the detector that must be assumed as being autoreactive. Consequently, an inactive detector or effector that recognizes an antigen presented by an inactive APC is permanently disabled or eliminated (figure 5.3). This is a crucial step of peripheral tolerization, because it can happen that an antigen that belongs to the healthy host is accidentally captured in a danger zone by an APC. However, the same antigen is more likely to be found and captured in a nondanger zone. Thus, an immune element that recognizes nondangerous antigens is more likely to be eliminated than activated. This mechanism of peripheral tolerization is known as *frequency-based* tolerization. Finally, an inactive element is eliminated if it fails to find in reasonable time an APC that presents an antigen recognized by it. In this way, the immune system is freed from elements that are not useful in the prevailing conditions.

In the presence of a danger signal the APCs switch to an active state and become much more willing to engulf antigens and migrate to the meeting places. When an antigen is presented by an active APC it can thus be assumed as belonging to something that is dangerous. If an inactive detector or effector recognizes a pattern on an antigen presented by an activated APC, it also switches to an active state. This mechanism of activation is thus based on the presence of two signals converging on a detector: the first signal is constituted by the presence of the recognized antigen; the second signal is

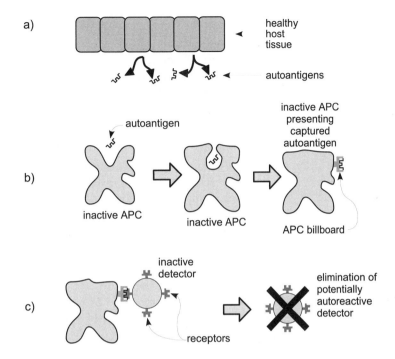

Figure 5.3 A schematic representation of the process of peripheral tolerization of the elements of the adaptive immune system. *a*) The normal activity of healthy host tissues results in the presence of autoantigens. *b*) Inactive APCs capture the autoantigens and present them to the APCs' billboards. *c*) Inactive detectors that recognize the antigens presented by an inactive APC are eliminated, freeing the adaptive immune system from potentially autoreactive elements.

COSTIMULATION known as *costimulation* and conveys the information that the antigen was found in a danger zone and belongs to something that must be presumably identified as a pathogen and destroyed. Due to the importance of the exchange of information between APCs and immune detectors, the surface of

IMMUNOLOGICAL contact between APCs and immune detectors has been called an *immunologi-*
SYNAPSE *cal* (or *immune*) *synapse* (D.M. Davis 2006; D.M. Davis and Dustin 2004; Friedl et al. 2005), with reference to the surfaces of contact through which neurons communicate. Activated immune detectors proceed to make copies of themselves, and each copy proceeds to activate the effectors that are targeted at the same antigen recognized by the detector. The choice of the effectors that are activated is guided by the kind of danger signal sensed by the APC that

has performed the initial activation, and by additional signals produced by the tissue under attack (Matzinger 2007). Figure 5.4 summarizes the logic of activation of the elements of adaptive immunity.

The activated effectors proliferate and are then distributed in the body, being especially attracted to the danger zones, where they proceed to eliminate the host structures carrying the recognized antigen. The elimination is carried out by "graceful" cell death (apoptosis) so as not to produce new danger signals which would further fuel the activity of the immune system and lead to its uncontrolled runaway. To further reduce the risk of an uncontrolled escalation of autoreactivity, the detectors and effectors of the adaptive system are only temporarily activated and must be reactivated after they have reverted to their inactive state. The automatic inactivation of active elements happens when a sufficient time has elapsed since their activation or when a sufficient number of eliminations has been performed. The active APCs themselves revert to an inactive state or die some time after entering a meeting place. In this way the information about the danger is kept updated. If a detector or an effector that has reverted to the inactive state recognizes an antigen but is not reactivated by an APC within a reasonable time, it is eliminated according to the rule mentioned above, or it becomes a so-called regulator element which proceeds to the elimination of other immune elements that recognize the same pattern.

SOMATIC HYPERMUTATION

AFFINITY MATURATION

In the process of proliferation of activated detectors and effectors, some of them can change slightly their PRRs in a process called *somatic hypermutation* because it involves nongerminal cells and rates of mutation many orders of magnitude higher than those normally observed in the genome. The elements which are the most effective against the pathogens are then preferentially selected for further activation, leading to a process of *affinity maturation* of detectors and effectors. Hopefully, this process leads to a progressively increasing effectiveness of the adaptive immune system in fighting the pathogen. Eventually, the rate of pathogen elimination surpasses the rate of pathogen replication and the host is progressively cleared from the presence of the pathogen. When the pathogen invasion has been eradicated and all the pathogens destroyed, the danger signal abates, the immune elements cease to be reactivated, and most of them are eliminated for lack of matching antigen patterns, except for a small population of long-lasting elements which is formed by the detectors and effectors that most effectively fought the pathogen. These element constitute the *immune memory* of the pathogen.

IMMUNE MEMORY

They have a increased sensitivity of recognition of the antigens and permit quick mounting of an immune response against the same pathogen if it is

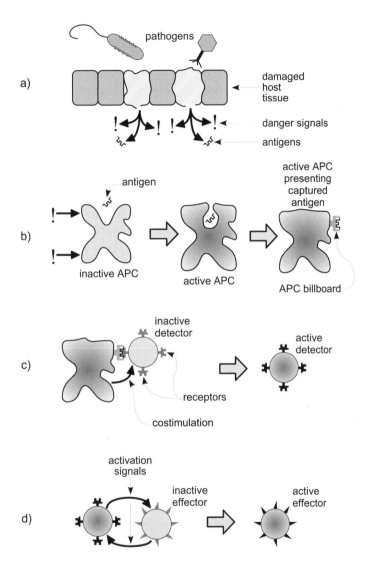

Figure 5.4   A schematic representation of the process of activation of the elements of the adaptive immune system. *a*) The presence of a pathogen causes stress or damage to the host tissues, which produce danger signals. *b*) The danger signals cause the activation of the APCs and enhance their activity of capture and presentation of antigens on the APCs billboards. *c*) Activated APCs activate the detectors that recognize the captured antigens. *d*) Activated detectors activate the effectors most suited to fight the pathogens that caused the production of the danger signals.

encountered again. The immune memory also speeds the generation of an effective immune response against new pathogens that resemble pathogens that have been successfully fought in the past.

Summing up, an inactive detector or effector is activated only when it is ascertained that it recognizes a pattern belonging to an antigen that was found in a danger zone. It is instead permanently deactivated or eliminated if it is ascertained that either it does not recognize any pattern or if it recognizes a pattern belonging to an antigen that is found in a region not marked as a danger zone.

This context-based strategy of activation of the immune elements permits the extension of the activity of the immune system to the control of errors and dysfunctions of the subsystems of the host itself, independently of the presence of pathogens. If a subsystem starts to malfunction due to some internal damage or wear, and if a danger signal is produced as a consequence of the stress induced by the malfunction, the adaptive immune system can identify the patterns that characterize the dysfunctional subsystem, destroy it, and replace it with a newly produced healthy version. For example, this is the case of cells that become cancerous, provided they lead to the early generation of danger signals. It has been hypothesized that the enforcement of control of the identity and functionality of the subsystems constituting the host was the original reason which led to the evolution of adaptive immune systems (Klimovich 2002).

5.1.5   The Limits of Adaptive Immunity

Adaptive immunity is a substantial improvement over innate immunity, since it permits keeping up with the pace of rapidly changing pathogens and damaged subsystems. However, it entails additional risks for the host, and it comes with its own limitations. A first limitation of adaptive immunity is the delay with which it generates an effective response when challenged by a new pathogen. Since the variety of possible antigens and patterns is immense, only a few PRRs that are specific to the new pathogen can be expected to be present in the host body at the time of the initial pathogen attack. The activation of the adaptive immune response requires the meeting of those rare detectors and effectors with the APCs carrying the pathogen antigens, and this may require some time. The existing PRRs that recognize the pathogen can be also expected to have initially a low affinity for the pathogen. There is thus typically a further delay which is imposed by the process of affinity maturation. This initial interval is known as the *la-*

PRIMARY RESPONSE   *tent* or *lag phase* of the *primary response* which follows the first exposure to
a pathogen. During this initial phase the innate immune system must be
able at least to contain the invasion and prevent the host from being overwhelmed by the pathogen. Since the innate immune system is unable to fight
the pathogen effectively (otherwise the adaptive immune system would not
have been called into action), this phase can be characterized by a reduced
functionality of the host. On the other hand, the mechanism of immune
memory ensures that further encounters with the same (or a very similar)
pathogen lead to a quick and strong reaction of adaptive immunity known
SECONDARY RESPONSE as the *secondary response* (figure 5.5). The speed and effectiveness of the secondary response make it typically asymptomatic, that is, no reduction of the
host functionality ensues from further encounters with the pathogen. The
limited speed with which the adaptive immune system can mount an effective response against a pathogen that it has never encountered entails also
the possibility that a pathogen escapes destruction through rapid *antigenic
variation*. For example, the human immunodeficiency virus (HIV) responsible for AIDS has a high rate of mutation of the genetic material that encodes
its surface proteins. In this way, an instance of HIV for which the adaptive
immune system has developed effective countermeasures can rapidly produce variants that are no longer recognized by the existing immune detectors
(Nowak and McMichael 1995).

Another potential problem linked to the activity of adaptive immunity
is the destruction of healthy structures of the host. The attack on the host
AUTOIMMUNE   from its immune system is called *autoimmunity* and leads to the so-called *au-
DISEASES   toimmune diseases*. A first scenario for this behavior is the activation of an
effector from the part of an APC that has engulfed and processed an autoantigen present in a danger zone (Matzinger 1998). The possibility that
such an effector exists is reduced but not eliminated by the mechanisms of
central tolerance. The mechanism of peripheral tolerance typically limits the
ensuing destruction to the few acts of elimination that an effector is enabled
to perform before requiring reactivation on the part of an APC. Therefore,
this scenario can entail extensive destruction of host tissues only if the immune system is unable to clear the host and the pathogen presence becomes
chronic. The problem becomes much more serious in the context of a second
scenario, where there is a strong similarity between a pattern found on an
antigen belonging to a pathogen and one belonging to an autoantigen that
is not present in the region where central tolerance is established. In this
case the activity of eradication of the pathogen will be accompanied by the
destruction of the host structures carrying the autoantigen that is similar to

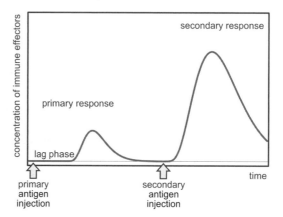

Figure 5.5   When an antigen is first presented to the adaptive immune system, a primary response is observed, which is characterized by a delay between the antigen injection and the mounting of a significant immune response. The immune memory of the first encounter permits the faster reaction to subsequent encounters with the same antigen that is observed in the secondary response.

the pathogen antigen. This phenomenon will recur each time the pathogen invades the host, leading to a progressive self-inflicted damage. For example, it has been hypothesized that multiple sclerosis and rheumatoid arthritis are autoimmune diseases produced by the similarity between some viral and bacterial antigens and some autoantigens found in healthy myelin and cartilage cells (Sompayrac 2003).

Finally, the mechanism of activation of the adaptive elements upon detection of a danger signal entails the possibility of an asymptomatic invasion of the host. This would happen if a pathogen that is not recognized by the innate immune system enters the host without producing damage and thus without resulting in the generation of danger signals. This invasion would not be fought until some damage to the host ensues. In the meantime the antigens of the invader would tolerize the elements of the immune system. Only when damage ensues would danger signals start to be produced and newly produced detectors and effectors could slowly start to be activated to fight the invader. The risk is that by that time the pathogen infection is so widespread that no effective immune response can be mounted before the host is overwhelmed. The answer seems to be a combination of the refinement of the detection abilities of the innate immune system and of the abilities of host cells to generate danger signals following even minor perturbations of their activity. This reduces the probability that a widespread pathogen invasion goes undetected, even if it does not initially produce any damage. This approach, however, is less effective against anomalous behaviors of the host subsystems, because the host antigens cannot be the target of innate immune detection. It follows, for example, that the adaptive immune system will not target cancerous cells as long as their proliferation does not harm the host, at which point the spread and speed of replication of the cancerous cells could be excessive for the resources of the immune system (Matzinger 1998). Moreover, part of the burden in the production of danger signals rests on the cells themselves and this mechanism could also be perturbed in a cancer cell (D.M. Pardoll 2003). On the other hand, this scenario opens the way to the development of techniques to mobilize the immune system against tumors by generating artificially suitable danger signals in the early phases of their growth (D.M. Pardoll 1998; Wiemann and Starnes 1994; Gilboa 2004).

5.2   The Constituents of Biological Immune Systems

In this section we will describe the constituents that implement the immune processes described in the previous section. Given the complexity of the immune function the description omits many details of our current understanding of the vertebrate immune system. Note that the operation of the vertebrate immune systems is far from being completely understood and that there are still controversies concerning not only its details but even its general "philosophy" of operation (Matzinger 2002). For each function, we describe here only the most important components, focusing on the aspects that are most relevant to the implementation of an artificial immune system. The details discussed in this section are not necessary for the understanding of the rest of the chapter. Readers not interested in these biological details can skip this section without compromising their understanding of the subsequent sections. However, even this simplified overview gives an idea of how many elements and interactions are required to implement effectively the protection strategies described in the previous section. Comparing the complexity of biological immune systems with that of the existing AISs that will be described in the following sections one can appreciate the gap that currently exists between the two and understand the limitations of current AISs.

ANALOG AND DIGITAL
RECOGNITION

The patterns recognized by biological immune systems are the molecular structures of the various substances that can reveal the presence of a pathogen. Antigens are always recognized from their three-dimensional molecular shape. We can distinguish a "digital" modality of recognition from an "analog" one. In the digital case, the recognition concerns a specific sequence of molecular substructures. A typical example is the recognition of a subsequence in the sequence of amino acids that constitutes the primary structure of a protein. In the analog case, the recognition concerns the three-dimensional shape of a part of the antigen, without a specific recognition of its separate constituents. In this case there is a continuous range of values of affinity of the receptor relative to the antigen, rather than the highly specific digital recognition of a subsequence. This implies that the analog recognition is more adaptable through a selection process, but it is also less reliable than its digital counterpart. For this reason, the activation of the immune effectors is almost exclusively entrusted to detectors operating in the digital modality.

CELLULAR AND
HUMORAL IMMUNITY

In some cases the elimination of the pathogens is done directly by the immune effector cells. In this case we speak of *cell-mediated* or *cellular immunity*. In other cases the elimination of the pathogens is mediated by molecules that are secreted and circulate in the intercellular spaces. In this case we speak of *humoral immunity*. There is a strong interaction between these two kinds of immunity. In particular, many elements of humoral immunity do not directly disable or kill the pathogens but mark them for destruction from the part of the elements of cellular immunity. The most important effectors of cellular innate immunity are a class of cells called *phagocytes*. Phagocytes work by engulfing the structures that must be destroyed and flooding them with aggressive chemicals that fragment the engulfed material into their basic constituents. Several phagocytes can cooperate in the extracellular killing of a pathogen by surrounding it and secreting special antipathogen chemicals.

PHAGOCYTES

COMPLEMENT SYSTEM

The most important element of innate humoral immunity is the *complement system*. It is constituted of chemicals that are distributed in the blood and in the tissues of the host. These chemicals tend to stick to the surface of cells and start a cascade of self-amplifying reactions that end up drilling holes in the cell membrane, thus killing the cell. The host cells produce specialized antidotes that inactivate the first step of the process and prevent the cascade of reactions from happening. The antidote itself is complicated to produce, thus making its synthesis quite demanding on the part of pathogens. Even partial inactivation of the cascade is ineffective, since phagocytes recognize and kill cells whose surface is coated with components of the complement system.

DENDRITIC CELLS   The APCs that realize the mechanism of context-based activation of adaptive immunity described in the previous section are the *dendritic cells* (DCs). To perform their function, DCs are equipped with many kinds of PRRs which permit them to discriminate between the various kinds of danger signals that can exist in the host body. DCs remain normally in an inactive state and patrol the body in search of danger signals generated by distressed host tissues or represented by PAMPs. DCs are equipped with the machinery required to capture and present antigens existing in the extracellular fluid that surrounds them. When they sense a danger signal, DCs become activated and increase their rate of capture and presentation of external antigens. Moreover, they become much more inclined to migrate to nearby meeting places to present the captured antigens to the components of adaptive immunity. When DCs are activated, they communicate this fact to the components of adaptive immunity by accompanying the presentation of the captured antigens with the presentation of an additional signal which represents the costimulation required for the activation of the elements of adaptive immunity.

The elements of the adaptive immune system are activated when they meet activated APCs (typically, DCs) carrying antigens that the adaptive elements recognize. To realize this function, there must be in the body "factories" where adaptive immune elements are produced, and a system for the distribution of the newly produced elements in the host body. Newly generated adaptive immune elements are not directly distributed in the whole body since they would be too dispersed and the probability of their meeting the right activated DCs would be too low. Newly generated adaptive immune elements are instead dispatched to specialized meeting places where the density of DCs and immune elements permits a more efficient transfer

LYMPHATIC SYSTEM   of information from the DCs to the immune elements. The *lymphatic system* is the ensemble of organs where all these processes take place. It includes *primary lymphoid organs* like the thymus and the bone marrow, where the immune elements are produced and screened; *lymphatic vessels* for (among other things) the dispatching and circulation of immune elements; and *secondary lymphoid organs* such as the *lymph nodes*, which are the meeting places mentioned above, where DCs and immune elements can meet and exchange information. The two kind of cells of the adaptive immune systems discussed below, namely, T cells and B cells, are consequently called *lymphocytes*.

As anticipated in the previous section, the scenario where the pathogens become invisible to the immune system after entering the host cells is avoided by requiring that the cells report about their internal activity using

MHC-I   specialized "billboards" present on their external surface. The most impor-

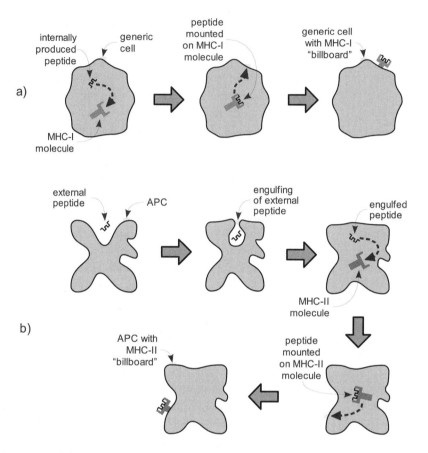

Figure 5.6 Schematic representation of the process of MHC antigen presentation. *a)* Generic cells (including APCs) use MHC-I molecules to present internally generated peptides. *b)* APCs use MHC-II molecules to present peptides they have engulfed from their surroundings.

tant and best known kind of billboards are those devoted to the reporting about the proteins that are synthesized within the cells. These billboards are implemented by the so-called *class I major histocompatibility complex* (MHC-I) molecules. The presentation of proteins based on MHC-I molecules works as follows (figure 5.6a)). Proteins are continuously synthesized within cells. To obtain a functional protein, a linear chain of amino acids is first assembled and then folded to produce a three-dimensional structure. Many of the assembled chains of amino acids do not fold correctly and are quickly disassembled into their amino acid constituents by specialized cellular machines. The same fate is reserved for proteins that have lost their functionality or are no longer needed by the cell. In order to let the immune system know what kind of proteins exist and are synthesized within the cell, not all proteins targeted for destruction are directly reduced to their constituent amino acids. A sample of protein fragments of some 8 to 10 amino acids in length is intercepted before their complete disassembly. These sequences of amino acids are mounted on a groove existing on newly synthesized MHC-I molecules. Then, the complex formed by the MHC-I molecule and the sequences of amino acids migrates to the membrane of the cell, where the MHC-mounted protein fragment is exposed outside the cell for the inspection of the immune system.

The presentation of proteins via MHC molecules has several interesting properties. First, the protein fragment is presented as a sequence of amino acids. This permits its digital recognition on the part of the immune system, thus giving the immune system the possibility of recognizing, with a low probability of error, the presence of pathogens within cells. Moreover, the cutting into pieces of the proteins before presentation exposes parts of the protein that are normally hidden from view due to protein folding. The only limitation of this modality of antigen presentation follows from the fact that the mounting of the protein fragment on the groove of the MHC molecules imposes certain constraints on the nature of some of the amino acids of the sequence. In other words, a given MHC molecule is compatible only with certain sequences of amino acids. To compensate for this limitation, each individual has several genetically encoded MHC molecules. This means that many different sequences of amino acids can be presented by the whole set of MHC molecules of one individual, especially if there is a good diversity in its set of MHC genes (Mitchison 1993). Vertebrate populations are highly polymorphic on MHC-encoding genes. This means that two randomly chosen individuals have a high probability of having different sets of MHC genes. Their MHC molecules will therefore display different parts of a given pathogen-derived protein. Thus, an intracellular pathogen which escapes recognition by the immune system of one individual is probably detected by the immune system of the other. It has even been hypothesized that this phenomenon is a major reason for the maintenance of sexual reproduction (Hamilton et al. 1990).

The mechanism based on MHC-I molecules gives to all host cells the possibility to inform the immune system about their *internal* activity of protein synthesis. Antigen-presenting cells (APCs) must perform the additional task

MHC-II

of picking up antigens from their surrounding and presenting them to the detectors of the immune system (figure 5.6b)). To this end APCs use the so-called *class II major histocompatibility complex* (MHC-II) molecules. They differ from MHC-I molecules in that they have a longer groove which can house protein fragments up to 20 amino acids long. Note that only APCs possess MHC-II molecules and the machinery to load and expose them. Other kinds of specialized APCs billboards have been recently discovered in vertebrate immune systems, which report on antigens other than proteins, for example lipids (Moody et al. 2005). Their logic of operation is the same as MHC molecules.

As discussed in the previous section, a mechanism of reporting on the internal activity of inaccessible subsystems can be effective only if the reporting cannot be safely disabled by pathogens that have entered the subsystems. In fact, many viruses have evolved a mechanism to reduce the display of MHC-I molecules loaded with antigens on the surface of the cells that they have

NATURAL KILLER CELL

infected. The countermove of the immune system is represented by *natural killer* (NK) *cells*. The main function of NK cells is to monitor the density and the quality of the MHC-I molecules that are displayed on the surface of the host cells. If the density of MHC molecules drops below a certain level, or if their structure has been modified beyond recognition, the NK cells activate a process that kills the host cell. This is the logic of operation illustrated schematically in figure 5.2.

T CELL

T_H CELLS

T_C CELLS

The adaptive immune elements that are in charge of inspecting the MHC billboards are called *T cells*. Two different kinds of T cells are devoted to the two existing kinds of MHC molecules. *Helper T cells* (T_H cells) are specialized for MHC-II molecules, whereas *cytotoxic T cells* (T_C cells) inspect the MHC-I molecules. The role of T_H cells is to orchestrate the working of the other parts of the adaptive immune system using the information gathered by APCs, while the role of T_C cells is to kill the cells that show signs of having been invaded by a pathogen. T cells constitute the cellular arm of

T CELL ANTIGEN
RECEPTORS (TCRs)

adaptive immunity. To inspect the MHC molecules T cells use specialized *T cell antigen receptors* (TCRs). TCRs are proteinic molecules that are produced within T cells and subsequently migrate to their surface to act as the PRRs of T cells (figure 5.7a)). Their capability to inspect the MHC molecules and the antigens mounted on them is due to the shape of the portion of the TCR surface which is exposed to the outside of T cells (figure 5.7b)). Part of this surface complements and recognizes the MHC molecule. Another part of the TCR is capable of sensing the sequence of amino acids mounted in the MHC groove. The parts of a TCR which complement the MHC molecules need a

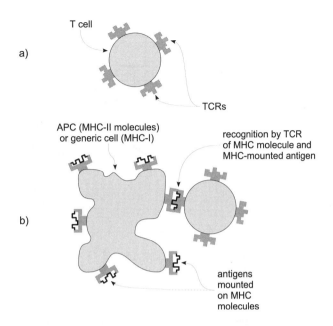

Figure 5.7   *a*) T cells are equipped with T cell antigen receptors (TCRs) that are exposed on their surface.   *b*) The shape of the TCR surface complements in part the shape of the surface of MHC molecules and in part the shape of the surface of the antigens mounted on the MHC molecules. This permits the recognition of the antigens on the part of T cells and limits the recognition to the MHC-mounted antigens.

limited variability for the small set of different MHC molecules existing in each individual. The parts of a TCR which recognize the antigen mounted in the MHC groove need instead an enormous variability. In principle, the diversity should permit the recognition of all the possible sequences of amino acids that can be presented by the existing MHC molecules. TCRs are built using the information encoded in the host genome. For each region of the TCR the host genome encodes a collection of some tens of alternative genes, depending on the variability required by the region. Each collection constitutes a a *gene library* for the definition of the corresponding TCR region. The strategy for the generation of the required TCR diversity is the random selection of elements from these gene libraries (figure 5.8). This process is carried out when T cells develop in the thymus (whence their name). At the end of this process the genome of a given T cell contains the information to build just one kind of TCR. Thus, all the TCRs of a given T cell and of all its clones

GENE LIBRARIES

MONOSPECIFICITY OF
T CELLS

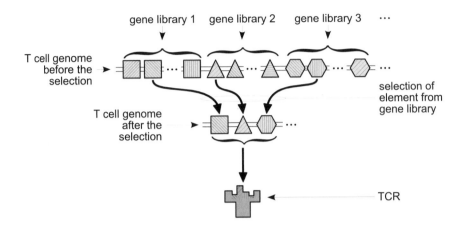

Figure 5.8 The generation of TCR diversity using random selection of elements belonging to the gene libraries of the host genome devoted to the encoding of TCRs.

that will be possibly produced are identical. We refer to this property saying that T cells are *monospecific*. Thanks to the process of random selection within gene libraries, there can be as many different TCRs as products of the number of alternative genes for each region (see box 5.1). Moreover, the process of joining the selected regions of the genome is purposefully kept loose, so as to add further variability to the result. It is estimated that, depending on the kind of T cell, the number of different TCRs that can be potentially generated varies from 10^{15} to 10^{18}, enough to recognize all possible antigens mounted on MHC molecules. Due to the random element in the process of genome reorganization, it is possible that the TCRs of a newly produced T cell are unable to access MHC molecules because the required complementarity is missing. For this reason, before T cells are dispatched to lymph nodes, a process of positive selection is executed in the thymus. This positive selection is called *MHC restriction* and preserves only the T cells that can access the MHC molecules of the cells present in the thymus. Newly produced T cells can also have TCRs that recognize autoantigens. A negative selection process eliminates all T cells that recognize the antigens that are presented in the thymus. This establishes the central tolerance of T cells. T cells that survive both selections specialize as T_H or T_C cells, (the mechanism of commitment to one of the two types has not been fully clarified yet).

    Newly produced T cells are dispatched to the lymph nodes as *inactive* T_H or T_C cells. Eventually, they will be either activated against some pathogen,

MHC RESTRICTION

or eliminated by lack of matching antigen or by the mechanism of peripheral tolerance. More precisely, inactive T_H cells that in their circulation in the lymphatic system recognize an antigen presented by an inactive DC are permanently inactivated or eliminated. The same fate awaits inactive T_H cells that after some time have not yet found a DC that presents an antigen that they recognize. Inactive T_H cells that in their circulation in the lymphatic system recognize an antigen presented by an active DC are activated, thanks to the presence of the costimulation provided by the active DC. Inactive T_C cells are peripherally tolerized in much the same way as inactive T_H cells, except that T_C cells interact with the MHC-I billboards of DCs rather than with their MHC-II billboards.[1]

Activated T_H cells proceed to clone themselves, to activate B cells as described below, to migrate to the danger zones, and to produce chemicals that stimulate the other components of adaptive and innate immunity (figure 5.9). Activated T_C cells proceed to clone themselves and migrate to the tissues, where they kill the cells that present the antigen that is recognized by their TCRs. The monospecificity of the T cells (and that of the B cells described below) is crucial for the safe working of the mechanism of activation just described, since it ensures that a T cell activated by one antigen is not active against unrelated antigens. As explained in the previous section, all activated elements of the adaptive immune system, including T_H and T_C cells, must be periodically reactivated. In the normally prevailing conditions of localized pathogen invasion, autoantigens are found much more frequently outside of danger zones then inside them. Thus, according to the tenets of frequency-based tolerization, autoreactive T cells that have been accidentally activated have a good probability of being peripherally inactivated or eliminated rather than reactivated. After the pathogen invasion is cleared, most T cells produced and activated during the attack die for lack of the antigen that they can recognize, except a few memory cells. The mechanism of choice and maintenance of memory T cells is currently poorly understood.

T cells can eliminate the cells of the host that have been infected by a pathogen. However, they do not fight the pathogens in their extracellular form, whereas the life cycle of many pathogens takes place at least in part

---

1. The fact that T_C cells interact with MHC-I rather than MHC-II billboards creates a difficulty, since the information for the activation of adaptive immune elements is possessed by activated DCs, which normally present externally captured antigens on MHC-II rather than on MHC-I molecules. A possible scenario that reconciles this difficulty works as follows. An active DC activates a T_H cell which, in turn, "conditions" the DC to present the externally captured antigens on MHC-I rather than on MHC-II molecules. Conditioned DCs can thus activate T_C cells.

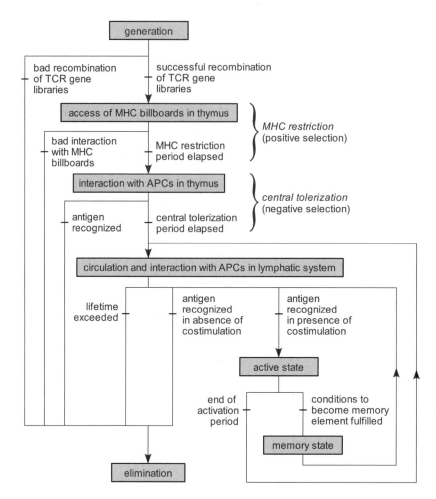

Figure 5.9   The life cycle of T_H cells. Unless explicitly marked with an upward arrow, all the flow lines proceed from top to bottom. The short horizontal lines that cut the flow lines denote conditions that must be verified before the T_H cell development is allowed to proceed along the corresponding flow line.

outside of the host cells. There is thus the need for an additional component of adaptive immunity capable of dealing with extracellular pathogens and with the products of their activity. This additional component is made up by B CELL   *B cells* and constitutes the foundation of the humoral arm of adaptive immunity. B cells do not operate by directly attacking the pathogens but by pro-

ANTIBODY ducing molecules called *antibodies.* Antibodies are proteinic molecules that circulate in the extracellular fluids of the host and are shaped so as to bind to pathogen antigens. In some cases the presence of antibodies bound to their surface targets the pathogens for destruction by the effectors of the innate immune system. In other cases this binding itself inactivates the pathogen, for example by altering a molecular structure that is essential to the activity of the pathogen. This is a very efficient strategy since molecules like antibodies are much more economical to produce than whole cells, and can be generated rapidly in large quantities. We can distinguish two regions in the anti-

ANTIGEN-BINDING REGION body molecule: an *antigen-binding region* and a "tail." The antigen-binding region is the part of the antibody that is used to recognize the pathogens and bind to them. The recognition is based on the complementarity between the three-dimensional shape of the surface of the antigen-binding region and a portion of the surface of the pathogen. Contrary to what happens for T cells, this kind of recognition has an analog nature. By slightly changing the shape of the antigen-binding region the *affinity* of the antibody for the pathogen can be modified incrementally. Moreover, the recognition applies not only to proteins but to many other kinds of molecules such as lipids and polysaccharides. The tail of the antibody molecule determines the *class* of the antibody response, that is, how the antibody operates against the pathogen. This permits the choice of different modalities of attack against different pathogens. There is a small number of classes of antibodies and B cells are instructed to switch to the production of a given class by the signals they receive during their activation.

Before a given kind of antibody is mass-produced and released in the extracellular fluids it is necessary to ascertain that it has a high affinity for the antigens of the pathogens that are present, and that it has a low affinity for autoantigens. To ensure this, antibodies are first produced in the form of

B CELL ANTIGEN RECEPTORS (BCRs) *B cell antigen receptors* (BCRs). In this role, they are constrained to remain attached to the cell, with their tail embedded in the B cell membrane and their antigen-binding region exposed at the surface of the B cell. B cells can sense the degree of affinity of the antigen-binding region for the antigens with which they come into contact. In this way a B cell can "know" when it has bound to an antigen for which its BCRs have a high affinity. Only after the B cell has been activated according to the mechanism described below the is B cell permitted to mass-produce and release its BCRs as antibodies.

BCRs – like TCRs – must be able to match the antigens of an enormous variety of pathogens. To this end, B cells must be able to produce a large variety of BCR antigen-binding regions. The mechanism of generation of this variability is similar to that used for TCRs (figure 5.8) and is based on the random selection of elements of gene libraries that are devoted to the encoding of the BCR molecules. As in the case of T cells, at the end of the process of selection in the gene libraries, the genome of a given B cell contains the information to build a unique BCR. Thus, all the BCRs of a given B cell are identical. The establishment of central tolerance for B cells is also similar to that of T cells and is based on a process of negative selection which eliminates the B cells that bind to the antigens that are present in the bone marrow. B cells that survive the negative selection which establishes central tolerance are dispatched in the host body as inactive B cells since there is still the risk that centrally tolerized B cells recognize autoantigens. Therefore, there must be a further selection leading to the elimination of autoreactive B cells and activation of B cells that recognize pathogen antigens.

The activation and peripheral tolerization of B cells is mostly based on their interaction with T_H cells (figure 5.10). When an inactive B cell senses the binding with high affinity of its BCRs to an antigen, it proceeds to engulf, process, and display the antigen on its MHC-II molecules. This process can be seen as an analog-to-digital conversion which links the analog recognition performed by the antigen-binding region of BCRs to the digital display of MHC molecules. An inactive B cell that encounters an inactive T_H cell which recognizes the antigens presented by the B cell on its MHC-II molecules is permanently inactivated or eliminated. The same fate awaits an inactive B cell that after some time has not met a T_H cell that recognizes the antigens. An inactive B cell that encounters an active T_H cell which recognizes the antigens presented by the B cell is instead activated. As said above, active T_H cells have been informed by APCs that the antigen they recognize can be found in a danger zone. With the mechanism of B cell activation just described, active T_H cells relay this information to B cells. Activated B cells

SOMATIC
HYPERMUTATION

Figure 5.10   (*facing page*) *a*) B cells carry BCRs on their surface. *b*) If the antigen-binding surface of the BCRs complements well a portion of the surface of an antigen, the antigens bind to the BCRs. Antigens bound with high affinity to BCRs are engulfed by B cells, processed internally, and presented on MHC molecules. *c*) An active T_H cell that recognizes an antigen presented by a B cell activates it. *d*) Activated B cells proliferate and undergo somatic hypermutation which modifies their BCRs. Clonal selection preserves the B cells with higher affinity for the antigens. These cells will become either memory B cells or plasma cells. The latter are specialized cellular factories that produce and secrete antibodies.

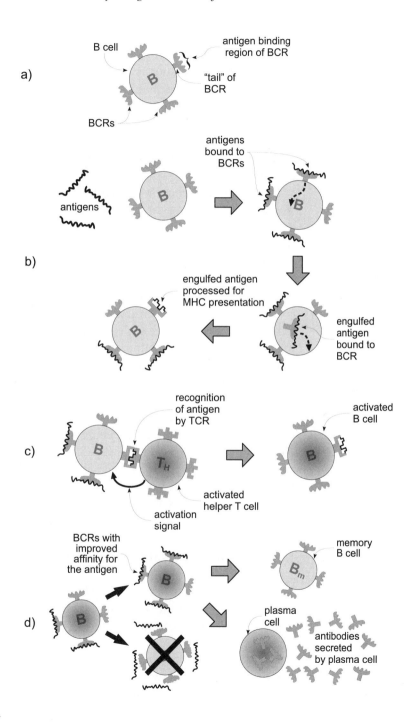

start to proliferate and when they reproduce are subject to a high rate of mutation of certain regions of their genome which code for the antigen-binding region. This phenomenon is known as *somatic hypermutation* (see box 5.2). It has been hypothesized that the rate of mutation is higher for B cells whose BCRs have a lower affinity for the antigens that are present. This focuses the adaptation process on B cells whose BCRs are relatively less efficient against the invading pathogens (Berek and Ziegner 1993). Finally, a phenomenon of *clonal selection* multiplies preferentially B cells with better affinity for the antigens.

The adaptation process brought about by somatic hypermutation and clonal selection increases the effectiveness of the antibodies that are produced. Note that the mechanism of somatic hypermutation could generate B cells whose BCRs are autoreactive. The mechanism of activation through activated T_H cells ensures that these autoreactive B cells are not activated. Contrary to B cells, T cells do not undergo somatic mutation and clonal selection after their activation. It is not clear if this absence is due to the excessive complication that would result from this additional adaptation step (with the necessity of reactivating also the mutated T cells from the part of the activated APCs), or to the minor efficiency of the evolutionary process on the much more rugged landscape determined by the digital modality of recognition of TCRs compared to the analog one of BCRs. Some of the activated B cells become *memory cells*, whereas some of them become *plasma cells* and instead of membrane-bound BCRs they start to produce and secrete antibodies. Like T cells, after the pathogen invasion is cleared, most B cells activated during the attack die for lack of their complementary antigen, except for the memory B cells which remain in circulation to speed the response to possible future invasions of the same or similar kind of pathogen.

Figure 5.11 presents a simplified schematic overview of the activity of the main actors of the adaptive immune system.

5.3   Lessons for Artificial Immune Systems

We proceed now to review what we have learned about the natural immune system, focusing on the aspects that can be expected to be relevant to the synthesis of artificial immune systems.

Performance   The immune protection is an ongoing process that is kept active by the struggle between the host and the pathogens. We cannot expect the immune system to attain a definitive success against pathogens and fail-

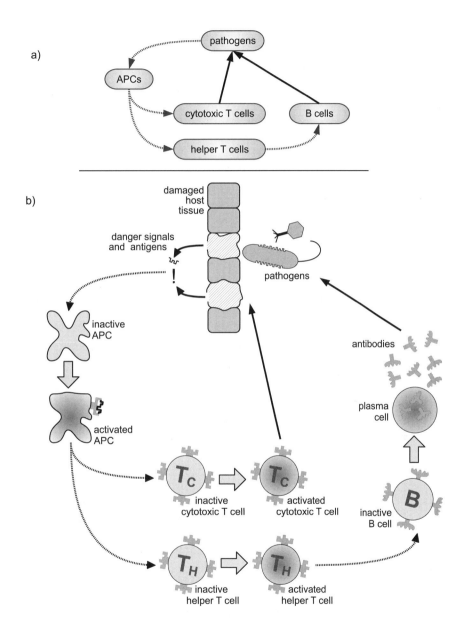

Figure 5.11   *a*) A schematic overview of the activity of the adaptive immune system and *b*) some additional detail on the activation of the most important elements involved in the immune process.

Box 5.2: Localized hypermutation

The high mutation rate that is observed in somatic hypermutation of B cells is not uniformly distributed on the whole genome but localized in the regions that encode the antigen-binding region. A similar phenomenon is observed in pathogens for the regions of their genome that encode the antigens that are most exposed to the detection of the immune system of their host. Even higher organisms, such as venomous insects, snakes, and marine snails, present a high degree of mutation from parent to offspring of the genes that encode the toxins that they use to immobilize or kill their prey. All these examples show that organisms can have different local rates of mutation in different parts of their genome. This phenomenon can be observed also with respect to other genetic operators. For example, recombination hotspots have been identified in the genome of many organisms (McVean et al. 2004). The nonuniform distribution of the genome reorganization rates focuses the generation of variability where it is required, keeping undisturbed the regions of the genome where important functions must be conserved. It is tempting to interpret this nonuniform distribution as an adaptation brought about by natural selection. However, the hypothesis that natural selection can adapt the local mutation rates is controversial. The contentious cases are those where there is no immediate advantage in either survival or reproduction for an individual. It has been argued (Dickinson and Seger 1999) that in these cases selection cannot work on the corresponding feature since this would require foresight on its part. To counter these objections it is sufficient to consider the reproductive success of the offspring in addition to that of the parent. The term *second-order selection* has been proposed for this extension (Weber 1996; Metzgar and Wills 2000; Tenaillon et al. 2001). The focusing of genetic changes on the regions where they have a greater probability of being useful operates in the light of the past trend of the environmental conditions. This increases the chances of survival and reproduction of the offspring and, other things being equal in terms of the immediate reproductive success of the parent, it increases its long-term reproductive success.

The mechanisms that permit the encoding of the local reorganization rates are only partially understood. A first hypothesis is based on the exploitation of the degeneracy of the genetic code that associates nucleotide triplets (codons) to amino acids. Degeneracy indicates that several codons correspond to the same amino acid; the same protein can be encoded using different sequences of codons. The choice *(cont.)*

Box 5.2 (continued)

of a particular codon in the set of those that specify a given amino acid can thus correspond to the choice of the mutation rate that must be applied to that particular region of the genome. This corresponds to the definition of an additional code on top of the genetic code (Caporale 1984). The hypothesis of existence of a mechanism of this kind is corroborated by the observation of highly evolutionarily conserved codon choice in the encoding of certain proteins. The idea of encoding and evolving either the global or the local mutation rates has also been considered in evolutionary computation (Bäck 1992). The results obtained so far, however, have not led to a widespread acceptance of this practice, especially in the field of genetic algorithms. An exception is represented by evolutionary strategies, where it is common to encode and mutate the variances that determine the Gaussian mutation of the real phenotypic parameters represented in the genome (Bäck 1996). Note that, for the reasons mentioned above, the self-adaptation of local rates of genome reorganization can be expected to be much more powerful than that of global rates.

---

ures of host subsystems. Rather, a balance is achieved between the effectiveness of the protection and the costs and stability of the system.

Costs An immune system can be expected to be very expensive in terms of resources. An innate immune system requires the generation and operation of many components. In addition, an adaptive immune system requires a strict selection of the adaptive elements and this increases significantly the magnitude of the effort. In the human immune system it is estimated that more than 95% of the lymphocytes are destroyed soon after their generation by the mechanism of central tolerance and MHC restriction (Palmer 2003). The mounting of a full-scale adaptive immune response is so expensive in terms of resources that it can result in a long-term reduction in the fitness of an individual. Thus, there is a tradeoff between the mounting of an immune response and the operation of other physiological systems of an organism (Hanssen et al. 2005; Svensson et al. 1998). It has even been hypothesized that organisms can modulate the strength, timing, and specificity of their immune response according to an estimate of the resources that will be available to the organism in the near future (Schmid-Hempel 2003) and that this could provide an explanation for the existence of the placebo effect (Humphrey 2002).

Damage and Regeneration   The operation of an immune system can inflict damage on the host, either intentionally, in order to destroy infected subsystems, or accidentally, due to autoreactivity. This is especially true for adaptive immunity, since the generation of danger signals required for its activation is often associated with already existing damage. This damage constitutes an additional cost for the host and it implies that the host must be able to generate new, healthy subsystems to replace the ones destroyed. This capability of regeneration is an important and demanding requirement for systems protected by immune systems, be they innate, adaptive, or artificial. For example, cells invaded by a virus are destroyed and replaced by newly created ones, just like programs attacked by computer viruses must in some cases be removed and reinstalled from trusted backups. The immune system can thus be expected to be endowed with a dual role: the first is to destroy the pathogens and the infected subsystems; the second is to initiate and manage the recovery operations required by the damages that it has inflicted while playing its first role.

Design for Immunity   The relationship between the immune system and the system it protects takes the form of a dialogue rather than of a monologue of the immune system. The host is not merely a passive receiver of the protection provided by the immune system but is instead explicitly designed to cooperate with it. This means that to realize an effective protection the protected system should be designed from the beginning with this cooperation in mind, rather than retrofitted a posteriori with an immune system. This joint design permits in particular the use of the concept of a danger signal to solve the problem posed by the inability to discriminate a priori between legitimate and unauthorized activity when both are produced by entities that are composed of the same building blocks.

Distributedness, Decentralization, Self-Protection, and Robustness   To simplify the wording of the exposition we have so far talked of the host and of the immune system as if they were two distinct entities. In fact, the immune system is itself part of the host. This means that the immune system must be built so as to be *self-protecting*. The vertebrate immune system is built as a self-organizing *distributed* system composed of autonomous agents. The control of the immune activity is *decentralized*. Small groups of immune elements are capable of initiating an immune action based on the integrated information they have collected. In this way, besides controlling the host, the elements of the immune system can control each other. An immune system realized as a centralized protection system would generate instead a *self-protection paradox*, since an additional second-order immune system would be required to protect the first immune system, leading to an infinite hierarchy of such systems. Thanks to its distributed nature and decentralized control, the immune system is *robust* not only relative to pathogen attacks but also to the malfunctioning of individual agents. To overwhelm such a system, a large number of agents must be separately attacked and disabled.

SELF-PROTECTION
PARADOX

Parallel Operation and Scalability   The distributed and decentralized nature of the vertebrate immune system is accompanied by the parallel operation of its elements. This gives the immune system a good scalability. The adaptation of the immune activity to different sizes and complexities of the host requires merely the adaptation of the number of immune elements, instead of a complicated reprogramming of their behavior and interactions.

Adaptivity, Tolerance, Autoimmunity   The example of the vertebrate immune system shows that an immune system must be adaptive if it must deal effectively with changes in the host or changes in the pathogens. It shows also that the adaptivity of the action implies the risk of autoimmunity and requires a sophisticated mechanism of establishment of tolerance to reduce this risk to acceptable levels. Two mechanisms for the establishment of tolerance are suggested by the observation of the vertebrate immune system. The first is a mechanism of central tolerance based on a process of negative selection and the second is a mechanism of peripheral tolerance based on danger signals, activation by costimulation, and frequency-based tolerization.

Dynamic Allocation of Resources and Self-Limitation   To contain the costs and adapt the response to the prevailing contingencies, the resources available to the immune system are dynamically allocated in terms of type of elements and their distribution in the host body. Different types of immune elements can be produced rapidly and concentrated at the site of the infection when needed. Moreover, a mechanism of automatic shutdown of the response leads to the rapid reduction of activity and resource usage when the emergency subsides. The lifetime of most immune elements is kept short and forces the dynamic adaptation of the tolerance and the response to the prevailing contingencies. The turnover of detectors permits the coverage of the space of antigens with a small population of detectors that is regenerated at a high rate.

Circulation of Detectors and Effectors   The circulation of the detectors and effectors of adaptive immunity in search of their antigen results in the exposure of the immune elements to a random sample of antigens. This helps the establishment of adaptive tolerance and reduces the population of immune elements that must be generated and maintained in the host system in order to ensure a reasonable probability of pathogen detection.

Adaptation of Local Sensitivity   The modality of operation of the vertebrate immune system permits knowing not only that there is a pathogen invasion but also where the threat is located in the body. The mechanism of inflammation permits changing locally the conditions required for the activation of the immune elements. In this way the local activity and sensitivity of the immune system can be adapted to the prevailing level of pathogen threat.

Generation of Diversity   The detection of the pathogens on the part of the adaptive immune system requires the generation of an enormous variety of pattern recognition receptors. The strategy used by the vertebrate immune system to generate this diversity is based on the random recombination of the elements of genetically encoded libraries of building blocks rather than on the random generation of receptors from scratch.

Strategies of Detection   The detection of pathogens is based on the recognition of the patterns constituted by antigens that are characteristic of the pathogens. The vertebrate immune system contains detectors with different modalities of recognition and different specificity. The processing and presentation on the billboards of antigens from the part of APCs is a strategy that permits the presentation of multiple "views" of the pathogen.

Choice of Effector   The information conveyed by the APCs that mediate the activation of the immune effectors concerns not only the pathogen antigens but also the kind of pathogen that is presumably associated with the antigens. The vertebrate immune system uses this information (for example, in deciding the class of the antibody response) to link the detection to the activation of the kind of effector that has the greatest probability of being effective against the estimated threat.

Learning and Memory   The vertebrate immune system can use the information conveyed by the pathogens it fights to increase the effectiveness of its adaptive detectors and effectors. This feature is based on the use of an adaptive process founded on localized hypermutation and clonal selection.

Moreover, the rate of mutation can be linked to the affinity of the matching, in order to improve the performance of the process. The information thus gathered can be retained for a long time in memory elements that permit the use of the information of the past threats to fight more effectively and rapidly the new threats.

Population Diversity The diversity of the vertebrate population relative to the genes encoding the billboards for antigen presentation points to the importance of maintaining a diversity in the immune systems of different individuals of the population. The lesson for artificial systems is that it is not advisable to protect multiple instances of a machine or computer with identical copies of a given artificial immune system. So far, this policy has been implemented by information technology departments by using different protection software or different operating systems. The implementation of an effective artificial immune system should lead to the automatic implementation of this kind of diversity.

5.4 Algorithms and Applications

The previous sections have shown that the operation of an immune system implies the presence of a variety of agents and processes. To date, no AIS tries to implement all that variety, although the ARTIS system, which will be described below, represents a first step in this direction. The majority of examples of AISs described in the literature focus instead on the implementation of one or at most a few of the concepts listed in the previous section. It has been remarked (e.g., by Garrett 2005) that when considered separately, many of those concepts and processes have some elements in common with concepts and processes derived from other bioinspired approaches, in particular in the context of evolutionary methods. Still, as explained below, in their immunological embodiment these ideas possess typically some peculiarities that make them a distinct and useful bioinspired tool, possibly for applications that are different from intrusion attack and fault protection.

Apart from the complexity of its implementation, one of the principal reasons that hinders attempts at the realization of a full-blown AIS is the paucity of systems that are designed from the beginning to operate in collaboration with an AIS. Typically, the current approach is instead to try retrofitting existing systems with immune protection. This is particularly true in the field of computer and network protection, since the existing infrastructure is too extensive to be completely scrapped in favor of newly designed systems

(Mukherjee et al. 1994). For example, very few systems implement a mechanism of reporting about the operations of their subsystems which parallels the processing and presentation of proteins on the billboards of generic cells and APCs. The only widespread exception that comes to mind is electronic systems designed to operate in conjunction with watchdog devices, which in their simplest form require that the protected subsystem send periodically a signal to the watchdog in order to confirm that the system operation complies at least with a basic temporal pattern of activity (Mahmood and McCluskey 1988).

It is also very difficult to find systems implementing a mechanism of generation of danger signals comparable to those found in biological systems protected by immune systems. To justify this absence, it must be said that the generation of useful danger signals is one of the most difficult aspects in the implementation of an AIS inspired by the immune model described in the previous pages (Garrett 2005). In the case of biological systems the danger signals are the result of the joint evolution of the host and of its immune system, whereas most of the existing systems that are candidates for the implementation of an AIS are hand-designed. This lays on the human designer the burden of devising suitable danger signals. Another obstacle to the implementation of danger models is that the activation of the protection system following the generation of a danger signal on the part of the protected system implies that some damage has possibly already occurred to the system. This prospect is somewhat alien to the mainstream engineering practices, which prefer a scenario where the protective action precedes the damage. This position is more understandable if we consider that apart from a few isolated efforts (e.g., Mange et al. 2000), current technology does not permit the regeneration of damaged subsystems. Summing up, the exploitation of the concepts revealed by the analysis of biological immune systems requires that the protected system be endowed with many of the properties of biological systems, an objective that is not yet within the reach of hardware implementations. However, this perspective is already being considered for systems implemented in software (Bentley et al. 2005). Moreover, as we shall show below, many of the concepts inspired by immune systems lead also to techniques that are useful for more conventional systems (Garrett 2005).

DANGER MODEL   The *danger model* that we have described so far is a relatively recent (and still not universally accepted) viewpoint on the operation of the vertebrate

TRADITIONAL MODEL   immune system (Matzinger 1994). The traditional model that is still prevalent in textbooks does not contemplate the generation of danger signals on

SELF/NONSELF   the part of the host tissues and is instead focused on the concept of *self/nonself*
DISCRIMINATION

discrimination. According to this view, the crucial process in the operation of the immune system is constituted by the mechanism of negative selection that establishes central tolerance. Correspondingly, the first implementations of AISs were also focused on the implementation of a mechanism of discrimination between self and nonself. This viewpoint is certainly more palatable to a conventional engineering mind, since the protective activity follows the recognition of the nonself, which can precede the occurrence of any damage to the protected system. Unfortunately, this conventional model cannot explain a number of properties observed in the operation of the adaptive immune system of vertebrates, in particular the tolerance for a changing self and for what constitutes a harmless nonself (Matzinger 2002). For this reason, the AIS community is also moving toward the adoption of the danger model viewpoint (e.g., Aickelin and Cayzer 2002; Kim et al. 2005).

IMMUNE NETWORK MODEL

There exists a third model of the immune system, the so-called *immune network model* (Jerne 1974). This model puts the focus on the activity of adaptive immune detectors, which are assumed to form a network of elements whose receptors can not only sense the pathogens but are also engaged in an ongoing activity of mutual recognition. In the absence of pathogen invasion, the network attains a condition of dynamic equilibrium which corresponds to tolerance for the healthy host. The presence of pathogens produces a perturbation of the equilibrium of the immune networks which, above a certain threshold, entails the activation of the immune response. The immune network has enjoyed a certain popularity in biological circles in the past but its validity has been questioned on the basis of more recent observations. Nonetheless, it is still largely used as an inspiration for the implementation of an AIS. We will not consider here artificial immune network models, since the additional principles of operation with respect to models already discussed are very similar to those found in artificial neural networks, even if some peculiarities exist (Dasgupta 1997; de Castro and Timmis 2002).

5.5 Shape Space

The concept of shape space (Perelson and Oster 1979) is an abstraction that gives a geometric interpretation to the process of recognition of an antigen on the part of an immune detector. The goal is to obtain a simplified model of the action of the immune system to be used in the analysis and design of artificial immune systems.

GENERALIZED SHAPE

In biological systems the actual recognition of antigens on the part of the immune detectors (both in the analog and in the digital modalities of recognition) is based on the complementarity of the geometric shape and electric charge distribution of parts of the surface of the antigen and parts of the surface of the receptors that equip the detector. We can simplify the description of this interaction by giving an abstract mathematical representation to the receptor and to the antigen. We assume that the properties of the antigen and of the receptor that are significant in the interaction can be represented with a list of l parameters, that is, an l-tuple. This l-tuple is called the *generalized shape* of the antibody or receptor (de Castro and Timmis 2002). The value of l depends on the complexity of the receptor and its interaction with the antigen. The nature of the parameters can vary according to the kind of model adopted for the molecules and the interaction. In general, they are either real numbers or symbols belonging to a finite alphabet. Summing up, the first step is the representation of the antigens and receptors as points in an l-dimensional *shape space*.

AFFINITY

The next step is the representation of the recognition of the antigen by the detector. In our representation we can ignore the complementarity that exists in the actual interaction between antigen and receptor, because from an abstract point of view what matters is the matching between the two and not how it is physically realized. We arrive in this way to a representation where a receptor $\mathbf{r}$ that matches perfectly an antigen $\mathbf{a}$ is characterized by the same l-tuple of parameters, so that $\mathbf{a} = \mathbf{r}$. More generally, the receptor will not match the antigen exactly, and we can specify a measure of the *affinity* between the receptor and the antigen which represents the strength of the binding between them. The representation of the affinity is obtained by associating with the pair $(\mathbf{a}, \mathbf{r})$ a real number $d(\mathbf{a}, \mathbf{r})$ that defines the distance in the shape space between the points that represent the receptor and the antigen. Larger distances mean more dissimilar pairs, with a null value only in the case of a perfect matching. A complementary approach is to associate with each pair $(\mathbf{a}, \mathbf{r})$ a value of similarity $s(\mathbf{a}, \mathbf{r})$ which increases with the affinity of the matching and is maximal for perfectly matching pairs.

RECOGNITION REGION

We can now say that a detector D equipped with receptors of type $\mathbf{r}$ recognizes an antigen $\mathbf{a}$ if $d(\mathbf{a}, \mathbf{r})$ is below a certain threshold θ_D. Thus, this detector permits the recognition of all the antigens $\mathbf{a}$ that satisfy the condition $d(\mathbf{a}, \mathbf{r}) < \theta_D$. The region of space thus defined is the *recognition region* of the detector. The recognition region is specified by the triple constituted by the receptor $\mathbf{r}$, the distance function $d(\cdot, \cdot)$, and the threshold θ_D. The

SPECIFICITY

value of the threshold determines the *specificity* of the detector, with larger

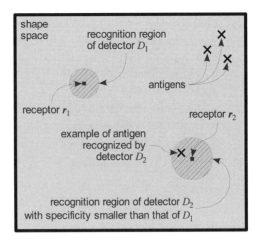

Figure 5.12 Representation in shape space of antigens, receptors, and detectors. Antigens and receptors are represented as points whose position is determined by the *l*-tuple of their parameters. The detectors are represented by their recognition region, whose position is determined by that of the receptor of the detector, and whose shape and size is determined by the distance function and by the threshold used by the detector. The figure shows two immune detectors with different specificity, one of which recognizes an antigen thanks to the fact that the position of the antigen in shape space falls within the recognition region of the detector.

thresholds corresponding to bigger recognition regions and smaller specificity (figure 5.12).

IMMUNE REPERTOIRE

The union of the recognition regions of all the detectors of an immune system corresponds to the set of all the antigens that can be recognized by the immune system. This region defines the coverage of the shape space by the *immune repertoire*. Even if the variety of antigens can be potentially infinite, physical constraints limit the range of the *l* parameters that characterize the antigens, and thus the size of the shape space. Consequently, a finite number of detectors carrying distinct receptors can cover the space of all the possible antigens. To avoid autoimmunity, however, no active detector should include in its recognition region an autoantigen. Ideally, the immune repertoire should cover all the regions of space that do not correspond to autoantigens. If this is not the case, we say that the immune repertoire has *holes* that can be

HOLES IN THE IMMUNE REPERTOIRE

potentially exploited by a pathogen to escape detection.

In this abstract scenario the problem that a natural or artificial immune system faces is thus the following: how to chose (1) the number N_D of distinct detectors, (2) the distribution of their receptors in the shape space, (3) the distance function $d(\cdot, \cdot)$ that they implement, and (4) the threshold θ_D that determines their specificity, so as to ensure that the probability P that a pathogen is recognized by at least one detector is reasonably high, while avoiding autoimmunity.

A possible choice (Perelson and Oster 1979) is to represent antigens **a** and receptors (antibodies) **r** as l-tuples of real numbers, and the affinity between them by the Euclidean distance

$$d(\mathbf{a}, \mathbf{r}) = \sqrt{\sum_{i=1}^{l} (a_i - r_i)^2} \, .$$

The recognition regions are thus l-dimensional balls centered on the points representing the antibodies. Assuming a random distribution of antibodies with uniform density in the shape space, the probability that at least an antibody recognizes a randomly chosen antigen is given by

$$P = 1 - e^{-N_D \hat{\theta}^l}$$

where $\hat{\theta}$ is a normalized threshold value that satisfies $0 \leq \hat{\theta} \leq 1$, with the upper limit attained when an antibody has no specificity and recognizes all the possible antigens. Although derived from a simplified model, this formula is useful to understand the relationship between the size of the immune repertoire, the specificity of the detectors, and the complexity of their receptors. By plotting the formula for P as a function of N_D (figure 5.13) one observes the presence of a relatively rapid transition from the near certainty of antigen escape to the near certainty of antigen recognition across a range of values of N_D. The position of the transition region shifts toward larger values when the specificity of the receptors increases. This means that using a number of distinct detectors in the range below the transition region results in a poor performance of antigen recognition, whereas using a number of distinct detectors in the range above the transition region gives a progressively diminishing return in terms of improvement of the recognition performance. The curves of figure 5.13 show also that if $\hat{\theta}$ is reduced and the specificity is consequently increased, the number of distinct detectors that is required for a given performance increases very rapidly. On the other hand, we know that the specificity cannot be reduced too much because excessively large recognition regions would tend to fill all the shape space, leaving no room for the host antigens.

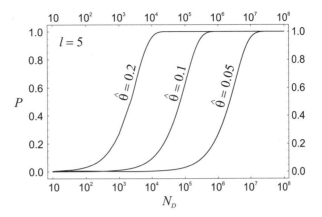

Figure 5.13 The curves of the probability that given a set of N_D detectors randomly distributed in the shape space, at least one recognizes a randomly chosen antigen (adapted from Perelson and Oster 1979). The curves correspond to the case $l = 5$ and are plotted for three different values of the normalized recognition threshold $\hat{\theta}$. Smaller values of $\hat{\theta}$ correspond to greater specificity of the detectors.

The scenario described above assumes a unique distance function and a fixed threshold for all detectors. This constraint reduces the flexibility of the immune system because it forces all the recognition region to have the same shape. This complicates the task of covering the holes of the immune repertoire without recognizing autoantigens. A technique to weaken this constraint on the shape of the recognition regions consists in implementing several distinct distance functions. In this way, given a pathogen antigen, the immune system has a greater possibility of producing a detector with a recognition region that includes the pathogen antigen without overlapping with the regions of the shape space that contain autoantigens (figure 5.14). The diversity of the billboards for antigen presentation in vertebrate immune systems can be interpreted as a way to implement this strategy of diversification of the shape of the recognition regions. This reminds us also of the fact that when the detection of a pathogen involves some processing of the antigen like the one provided by APCs and their billboards, the process itself contributes to the definition of the distance function.

The Euclidean distance is well suited to the representation of an analog type of antigen detection. This is the case, for example, for the recognition of antigens on the part of B cells and antibodies. In this case the affinity and the parameters that constitute the l-tuples which represent the receptors and

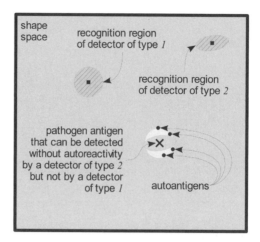

Figure 5.14   The problem of holes in the immune repertoire, and how it can be alleviated using detectors with different recognition regions. The cross and the dots represent in the shape space the position of a pathogen antigen and that of a few host antigens, respectively. The shape and size of the recognition region of the detector of type 1 implies that it cannot be equipped with a receptor permitting it to recognize the pathogen without incurring in the problems due to the recognition of the autoantigens and the associated autoreactivity. A detector of type 2 has instead a recognition region that, given a suitable receptor, permits the detection of the pathogen antigen without risk of autoreactivity.

the antigens can take a continuous range of values. In AIS applications it is often the case that receptors and antigens are strings of symbols that belong to a finite alphabet, for example binary strings. In this case other measures of distance or similarity are typically used. These measures correspond to a digital type of antigen detection and are also more suited to the modeling of the recognition of antigens on the part of T cells.

A first example of distance suited to pairs of strings of symbols of the same length is the *Hamming distance*, which is defined as the number of corresponding positions where two strings differ. Another example is a measure of similarity defined as the length of the longest sequence of contiguous symbols in corresponding positions which are the same in the two strings (the length of the longest corresponding substring). This measure of similarity leads to the *r-contiguous symbols* rule for the definition of the recognition region (Percus et al. 1993; Forrest et al. 1994). This rule establishes that an antigen is recognized by a receptor if the length of the longest corresponding

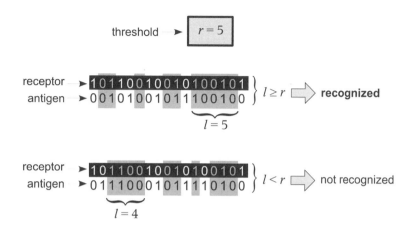

Figure 5.15   The working of the r-contiguous symbols rule for the recognition of an antigen by a receptor.

substring that the antigen has in common with the receptor is greater than or equal to r (figure 5.15). The value of r corresponds thus to the threshold that determines the specificity.

PERMUTATION MASK

Typically, given a distance function for strings of symbols, by applying a *permutation mask* to one of the two strings – that is, by reordering the string – before the application of the distance function, one obtains a different measure of similarity and a different shape of the recognition region. As explained above the availability of several different kinds of recognition regions simplifies the problem of reconciling the coverage of the regions of the shape space where pathogens are present with the absence of autoreactivity.

5.5.1   Example: Vaccine Design

The concept of shape space is used in almost all implementations of the AIS and, in particular, in the examples considered at the end of this chapter. This concept is also useful in studying biological immune systems. To illustrate this fact we consider here an example of application of the concepts just introduced to the analysis and design of vaccines for biological organisms. We will refer here to some details of the workings of biological immune systems described in section 5.2.

The modality of operation of the adaptive vertebrate immune system permits the generation of protective immune memory for a pathogen in an in-

dividual that has never been infected by the pathogen. This procedure is
called *active immunization* or *vaccination*. We can use the concept of shape
space to model how vaccination works and to predict the effects of multiple
vaccinations targeted at different pathogens. The technique of vaccination
consists in introducing into an organism a small amount of a preparation
called a vaccine which contains a certain quantity of antigens that are characteristic of the pathogen, or are very similar to the actual pathogen antigens.
These antigens are attenuated to render them harmless to the organism. The
vaccine contains also substances that represent or lead to the production of
danger signals in the region where the antigens are introduced. The adaptive
immune system of the organism reacts to the introduction of the vaccine by
mounting an immune response which is targeted at the antigens. The final
result is the establishment of an immune memory for the antigens present in
the vaccine. In particular, the vaccination will result in the generation of a
collection of memory B cells whose BCRs possess a high affinity for these
antigens. If the organism is now attacked by the pathogen, the immune
memory thus established permits the mounting of a secondary immune response that can eradicate the pathogen before it can cause any harm to the
host.

We can model this process in terms of shape space, as follows (figure 5.16,
top). An antigen present in the vaccine corresponds to a point a_v. The result
of the vaccination is the generation in the shape space of a cloud of points
corresponding to the receptors. These points are distributed around the antigen and correspond to the BCRs of memory B cells whose recognition region
includes the antigen a_v. The process of clonal selection ensures that the density of points is greater in the vicinity of a_v. Any pathogen antigen a_p that is
similar to a_v will correspond to a point of the shape space that is close to a_v.
Thus, it will also fall in the recognition region of the memory B cells induced
by a_v. Therefore, the pathogen will be attacked and cleared by those B cells.

Some pathogens use a mechanism of evasion from the control of the immune system that is based on the variation of their antigens. For example,
the influenza virus changes rapidly the antigens that are recognized by the
immune system of its hosts. Consequently, a new vaccine against influenza
must be synthesized each year, based on the antigens of the new strains that
appear more likely to produce the next epidemic. Note that the production
of the vaccine must be done well before the actual epidemic materializes, so
that the antigens included in the vaccine derive from a guess about the structure of the pathogen. Consequently, the antigens included in the vaccine can
differ somewhat from those of the actual epidemic strain.

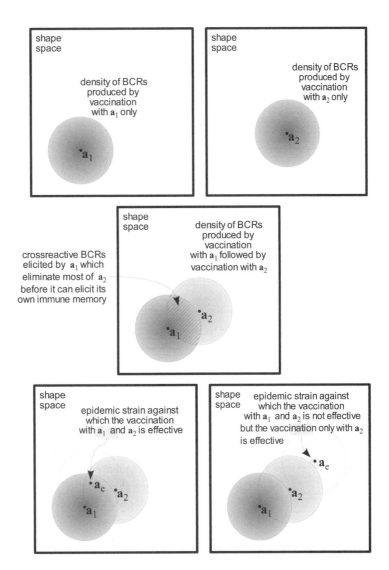

Figure 5.16 The shape space representation of the interaction between two vaccines. Each black dot represents an antigen. The shading of the disks centered on the antigens represents the density of cell receptors of the memory B cells produced by the immune response to the antigens. See the text for the other details of the explanation.

It has been observed that in some cases the efficacy of the vaccine for a given influenza strain is reduced if an individual has already been vaccinated against another influenza strain. However, in other cases no such effect is observable. The representation in terms of shape space permits the interpretation of this puzzling phenomenon. According to the model presented in (D.J. Smith et al. 1999), the influence of a first vaccination against the virus strain v_1 on the efficacy of the vaccination against v_2 when the epidemic strain v_e materializes depends on the relative position in the shape space of the antigens a_1, a_2 of the vaccines and of the antigen a_e of the epidemic strain. If a_1 and a_2 are far apart in the shape space, the two vaccines will not interact. If, instead, a_2 belongs to the recognition region of the memory B cells elicited by the vaccination with a_1, those memory cells will be activated and produce antibodies that will rapidly remove the antigen a_2 from the body. Thus, the vaccination with a_2 will not result in the generation of a cloud of receptor points centered on it (figure 5.16, center).

This does not necessarily mean that an individual vaccinated against v_2 is unprotected against v_e. Since the antigen a_e of the actual epidemic strain can be expected to be close to a_2, if a_1 and a_2 are close to each other there is a good probability that a_e also belongs to the recognition region of the memory B cells elicited by the vaccination with a_1. If this is the case, the old vaccination will provide protection also against the new epidemic strain. If, however, the antigen a_e of the new epidemic strain v_e does not belong to the recognition region of the memory B cells elicited by the vaccination with a_1, the result of the old vaccination will not protect against the new strain (figure 5.16, bottom). On the contrary, it will be the cause of the absence of the protection that could have been provided by the vaccine against v_2 alone (figure 5.16, top).

The realization of this mechanism of interaction can be used as a guide to the design of the vaccines (D.J. Smith et al. 1999). For example, if the data on the estimated coming epidemic strain v_e leave a choice between different candidate antigens a_2, the choice should fall on the antigen that is at the greatest distance from a_1.

5.6   Negative Selection Algorithm

The concept of shape space permits visualizing the concept of detection of pathogens by the detectors of the immune system. In particular, it helps to appreciate that one of the problems that an immune system must solve in the generation of an immune repertoire is the distribution of the receptors in the shape space so as to include in the recognition region of the corresponding detectors the pathogens but not the autoantigens. We have seen that one of the strategies used by the vertebrate immune system for the solution of this problem is the random generation of receptors and a process of negative selection that removes the receptors that match autoantigens. Inspired by this strategy, a *negative selection algorithm* was proposed (Forrest et al. 1994; Ji and Dasgupta 2007) for the generation of the detectors of an AIS.

The negative selection algorithm assumes that there is a collection P of fixed-length strings of symbols which must be protected from unauthorized change. For example, this collection could be an ensemble of data and program files in the memory of a computer, or the control program of an electronic device, or the patterns of operation of a machine, or the patterns of connectivity and traffic of a networked computer. In the absence of unauthorized changes P corresponds to a collection S which is called the *self*. The goal of the algorithm is to generate a set of detectors that can signal the appearance in P of any string that does not belong to S, that is, the appearance in P of any *nonself* string. Nonself strings could be generated, for example, by the presence in the system of a virus or a network intrusion. To attain this goal the algorithm prescribes the following steps:

1. Assign a similarity or *matching* function $m(\cdot, \cdot)$ for pairs of strings, a detection threshold θ_D, a mechanism of generation of candidate receptor strings, and the maximum acceptable probability P_f of detection failure.

2. Estimate the number N_D of strings required to obtain the performance specified by P_f using the recognition regions specified by $m(\cdot, \cdot)$ and θ_D, and the mechanism of generation of candidate receptor strings specified at step 1.

3. *Censoring phase* (figure 5.17): Generate a candidate receptor string r_c. If r_c matches any self-string, that is, if $m(s, r_c) \geq \theta_D$ for any string s belonging to S, discard the string; otherwise include r_c in the initially empty set of receptors R. Repeat this step until the size $|R|$ of R corresponds to N_D.

4. *Monitoring phase* (figure 5.17): Choose (either deterministically or randomly) a string s in P and a string r in R and evaluate $m(s, r)$. If $m(s, r) \geq \theta_D$, signal the detection in P of a string not belonging to the legitimate self S; otherwise repeat this step with another pair of strings (s, r).

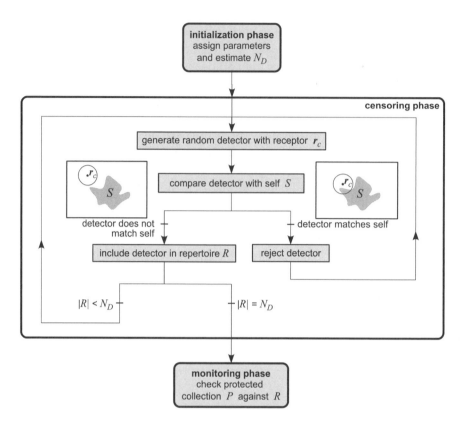

Figure 5.17   The steps of the negative selection algorithm with details of the censoring phase. A schematic shape space representation of the criterion of acceptance of newly generated detectors is also given. Unless explicitly marked with an upward arrow, all the flow lines proceed from top to bottom. The short horizontal lines that cut the flow lines denote conditions that must be verified before the algorithm is allowed to proceed along the corresponding path.

In its original formulation the negative selection algorithm used the r-continuous symbols rule for the matching function and generated randomly the candidate receptor strings r_c. The experiments reported in (Forrest et al. 1994) show that the algorithm is effective in detecting changes in protected collections of strings. However, the original version of the algorithm suffers from some limitation. The main problem lies in the complexity of the censoring phase whose time grows exponentially with the size of the protected set. For this reason, improved techniques of generation of the candidate receptors have been devised which result in a linear time complexity (D'haeseleer et al. 1996). Another problem is the difficulty of the determination of the required size of the detector set. The theoretical estimates of the required size for the original algorithm have been shown to be close to the experimental values in simple cases, but generalization to more difficult cases is not granted or may lead to excessively large estimates and detector sets.

Another limitation of the original algorithm is the use of the same recognition region for all detectors. This exacerbates the above-mentioned problem (section 5.5) of the presence of holes in the coverage of the set of nonself strings. To alleviate this problem Hofmeyr and Forrest (2000) suggested a version of the algorithm which applies a permutation to the characters of one of the strings before the evaluation of the matching function. This corresponds to the use of a different matching function and of a different shape for the recognition regions. A further increase in the flexibility could be obtained by letting different detectors use different thresholds. Note, however, that in general the holes cannot be avoided if the detection must remain probabilistic, that is, if we want to avoid the degenerate case where the detectors have maximum specificity and recognize a single antigen. This compromise is another face of the problem of generalization and overfitting that was discussed in the context of neural networks.

In the version described above, the negative selection algorithm works for a fixed protected system but cannot deal with a changing collection of protected strings S. For example, the activity of authorized users typically results in changes in the files that are stored in the hard disk of a computer. In the vertebrate immune system the process of negative selection is used just to establish central tolerance, and is complemented by the mechanism of peripheral tolerance in order to deal with a changing host. Similarly, the negative selection algorithm must be complemented by other protection mechanisms in the AIS that are designed to protect changing collections of strings (Hofmeyr and Forrest 2000). Nonetheless, the negative selection algorithm has several interesting characteristics as an anomaly detection system (Kim and Bentley 1999; Forrest and Hofmeyr 2001). Thanks to its strategy of implicit definition of the nonself using the information represented by the self, it can protect a system against a threat without the necessity of specifying the nature of the threat. The algorithm is able not only to signal the presence of an anomaly but to inform about its nature by pointing to the string that represents the anomaly. The probabilistic nature of the recognition permits a tradeoff of the probability of detection failure with the complexity of the detection. The algorithm can be easily distributed if several similar systems must be protected. In this case, by using different local sets of detectors, even small sets of detectors and high probabilities of local failure can result in very low probabilities of global failure. Moreover, the algorithm does not require a mechanism of central coordination for the generation of the detectors, thus avoiding the corresponding communication costs.

5.7   Clonal Selection Algorithm

The detectors of the vertebrate immune system that survive the process of negative selection are distributed in the body and interact with the antigens. For some of these detectors there exists a mechanism that improves their recognition performance . This mechanism of *affinity maturation* is based on an iterative process of production of clones, variation, and selection which resembles an evolutionary process. Eventually, the best performing detectors resulting from this *clonal selection* process are preserved as memory cells.

A *clonal selection algorithm* based on the characteristics of this biological process has been proposed (de Castro and Von Zuben 2002) for pattern recognition and function optimization. In the case of function optimization problems the goal is to produce a population R of receptors that constitute a collection of candidate solutions to the problem. To achieve this end the clonal selection algorithm prescribes the following steps:

1. Assign the size of the population of receptors, the selection strategy, and the number of random receptors generated at each iteration. Assign a function that transforms the value of the optimizing function into a value of affinity. Assign a function that links the affinity to the rate of mutation to be applied. Assign the function that links the affinity to the number of clones to be produced.

2. Initialize a random population R of receptors.

3. Evaluate the affinity of the receptors in the population R.

4. Select the receptors with highest affinity obtaining a collection R_H.

5. *Clone* the elements of R_H, that is, for each element in R_H create a number of copies prescribed by the function assigned at step 1.

6. *Mutate* the clones with a mutation rate given by the function assigned at step 1, obtaining a collection R_M.

7. Generate randomly a collection R_R of new receptors.

8. Select the best receptors among R_H, R_M, R_R to form the new population R of receptors and return to step 3, unless some termination criterion is satisfied.

For function maximization problems the affinity corresponds in general directly to the value of the function to be maximized. The function that gives the rate of mutation is defined so as to result in larger mutation rates for smaller values of affinity. In this way, the best receptors are modified only slightly to fine-tune their affinity, whereas low-affinity receptors are mutated substantially in order to create diversity and explore the shape space. An example of function linking the affinity to the mutation rate p is (de Castro and Von Zuben 2002)

$$p = \exp(-\rho\, \hat{a})$$

where ρ is a parameter that determines the scale of the mutation, $\hat{a}$ is a normalized affinity a/a_{max}, and a_{max} is the estimated maximum affinity.

In the case of pattern recognition, the goal of the algorithm is to produce a set of elements that can recognize the element of a set P of patterns. To this end a collection M of *memory receptors* is used in addition to the population R. The objective is to specialize the memory receptors on the different patterns (antigens) that must be recognized. At each iteration, one pattern is selected from P and the joint collection formed by M and by the set of receptors R is processed. At the end of the iteration, the receptor with the highest affinity for the current pattern is stored in M if its affinity is greater than that of the previously stored memory receptor with highest affinity for the pattern.

The results produced by the clonal selection algorithm are encouraging, especially in the case of multimodal function optimization (de Castro and Von Zuben 2002). However, it is not clear if there are substantial advantages with respect to a conventional genetic algorithm (Garrett 2005), since the only original aspect of the clonal selection algorithm is the mechanism of affinity-related mutation. Moreover, the fact of assigning a priori a function that prescribes the number of clones to be produced as a function of their affinity can result in large collections of clones, the evaluation of whose affinity can slow the execution of the algorithm.

5.8   Examples

5.8.1   ARTIS and LISYS

ARTIS (*art*ificial *i*mmune *s*ystem) (Hofmeyr and Forrest 2000; Glickman et al. 2005) is an AIS framework that models many of the processes and properties of the vertebrate immune system, including some of the concepts and algorithms described in the previous sections. The goal of ARTIS is to specify the elements of a general adaptive distributed system without reference to any specific application. These generic elements must be then particularized according to the characteristics of the application. As an example of this particularization we will describe LISYS (*l*ightweight *i*ntrusion detection *s*ystem) (Hofmeyr and Forrest 2000), a network intrusion detection system based on ARTIS.

ARTIS assumes that the system to be monitored is a distributed environment represented by nodes that can exchange information. At each node a collection of fixed-length strings is defined, which are the target of the security monitoring. For example, the nodes can be computers in a network and the collection of strings can represent the network traffic, including the traffic between the protected computers and other computers not belonging to the network protected by ARTIS. The goal of the monitoring is the detection of the appearance of anomalous strings in the collection. The performance of the system is measured in terms of the proportion of anomalous strings FALSE NEGATIVES AND that are not signaled as such – the rate of *false negatives* – and the proportion POSITIVES of legal strings that are classified as anomalous – the rate of *false positives*. Typically, the lowering of one rate conflicts with the lowering of the other.

To realize the protection, ARTIS uses a distributed version of the negative selection algorithm which also allows the protected strings to vary in time. A local collection of detectors is defined at each node. The detectors are constituted by a string (the receptor) of the same length as those of the monitored collection, by a locally defined function that specifies the similarity between pairs of strings, and by a local activation threshold. In the original formulation the similarity function is based on the r-contiguous symbols rule and the local similarity function is obtained using a permutation mask (different for each node) that is applied to the characters of one of the two strings. The local activation threshold permits choosing the sensitivity of the node and the SENSITIVITY LEVEL tradeoff between the rate of false positives and false negatives. In ARTIS the local activation threshold is varied according to a local *sensitivity level* which links the value of the threshold to the number of anomalies detected in the near past. This increases the alertness of the node when there are indications that the node is under attack and restores it to its base level when the attack subsides.

In ARTIS the strings that constitute the receptors are randomly generated. Newly created detectors are subjected to a process of *tolerization* of fixed duration T during which they are compared with a sample of the monitored strings that are present in the node. If during the tolerization period a detector recognizes a monitored string, the detector is eliminated. Detectors that survive the tolerization period continue to be compared to the monitored strings but now become *activated* if during a certain time interval they recognize a sufficient number of strings. This condition creates a bias toward the detection of temporally clustered anomalies, which is a characteristic of many kinds of system anomalies (Glickman et al. 2005). When a detector becomes activated it sends a signal to a human operator. If the human operator does not confirm within a time T_s (the *costimulation delay*) that the detected string is a harmful one, the detector is eliminated. If instead the operator confirms the dangerous nature of the detected string by sending a *costimulation signal* to the detector, the detector enters a competition with all the detectors activated by the same string. The best-matching detector becomes a *memory detector* and copies of it are sent to the neighboring nodes. Memory detectors have a lowered threshold of activation thus leading to a faster *secondary response* to returning attacks of the same kind. ARTIS implements a dynamic population of detectors. A detector that survives the tolerization period has a certain fixed probability p_d of dying at each time step. This means that each detector has a finite lifetime and is eventually replaced by a new randomly generated one. The only exception is memory detectors, which have a potentially infinite lifetime. To avoid the unbounded increase in the number of memory detectors, once a given size of the population of memory detectors is attained, the least recently used are replaced by newly formed ones. Figure 5.18 summarizes the life cycle of a detector in ARTIS.

LISYS is a partial particularization of ARTIS to the problem of network intrusion detection (Hofmeyr and Forrest 2000; Glickman et al. 2005). In LISYS the monitored nodes are computers that must be protected against unauthorized access (figure 5.19). The activation of detectors corresponds to the detection of suspect traffic that could correspond to an attack. In this case false negatives correspond to undetected attacks and the false positives correspond to unnecessary signals sent to the human operator. The strings that are monitored summarize the information about the connections that concern the nodes. Each string contains the identity (IP addresses) of the

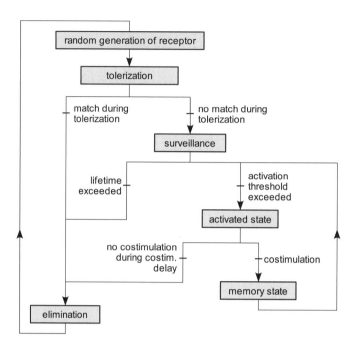

Figure 5.18   The life cycle of a detector in ARTIS (adapted from Hofmeyr and Forrest 2000). See the caption of figure 5.9 for explanation of the diagram symbolism.

connected nodes and the specification of the kind of service requested. For simplicity, it is assumed that all the information about the traffic of the whole network is available at each node. The system was tested with data collected from real computer networks which contained known intrusions and was able to detect all the intrusion attempts, apart from very short ones, with a small rate of false positives.

ARTIS implements many of the processes of the vertebrate immune system as described in the first half of this chapter, such as negative and frequency selection for tolerization, the local modulation of sensitivity observed in inflammation, the tolerization and activation of detectors, the use of a costimulatory danger signal for detector activation, and the turnover of detectors. Some processes observed in biological immune systems such as clonal selection for affinity maturation and the use of gene libraries for receptor generation, are not used in ARTIS but could be easily included in the framework. Still other concepts, such as the association of different kinds of effectors to the kind of anomaly detected, the presence of an innate immune system that

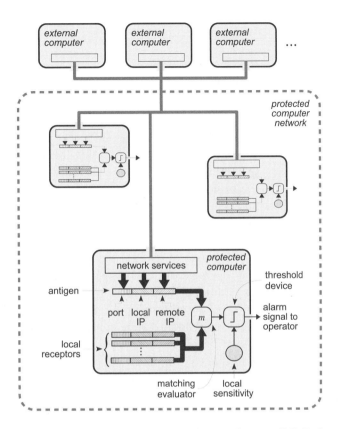

Figure 5.19 The structure of the system and of the nodes in LISYS. Each node in the protected system is a networked computer which has a local collection of receptors and a local sensitivity level. The antigens that must be monitored by the detectors are strings that contain information about the network traffic that affects the protected nodes. The detection of an anomalous string results in the generation of an alarm for a human operator.

cooperates with the adaptive system, and the fully autonomous operation of the system without the intervention of a human operator, are not implemented. Nonetheless the results obtained with LISYS are encouraging and permit, in particular, confirmation that all the implemented processes and concepts contribute to the system performances (Glickman et al. 2005). As in many bioinspired systems there are many parameters that the user can choose. If this makes the setup of the system more difficult, it permits on the other hand a balancing of the various tradeoffs of the system performance.

5.8.2 Immunotronics

FAULT DETECTION

Immunotronics (Bradley and Tyrrell 2002) is an application of AIS concepts to the detection and recovery of faults in digital electronic systems. The classic approaches to fault detection and recovery in artificial systems are redundancy and the addition of protection systems that check and possibly correct the validity of the system state. The immunotronics approach defines a system of this latter kind but applies the immune system concept of self/nonself discrimination to automate the generation of the verification criteria used by the protection system. The immunotronics approach applies to finite state machines (FSMs), a class of systems where the operation is modeled in terms of states and transitions between them and which encompasses the majority of existing electronic systems.

FINITE STATE
MACHINES (FSMs)

In the case of an FSM the self can be defined as the collection of strings that represent the legal transitions between the states of the machine. For example, the strings can be formed by concatenation of a string containing the values of the current inputs, a string containing the current state, and a string containing the next state generated by the FSM (figure 5.20). A collection of self-strings can be obtained by observing the operation of the system in its fault-free condition. By applying the negative selection algorithm, a collection of receptor strings can be created and used to monitor the subsequent operation of the system. An additional hardware device is connected to the system and monitors the incipient state transitions. The corresponding antigen strings are compared with the collection of receptors using a matching function $m(\cdot, \cdot)$ that measures the similarity between antigens and receptors. If the matching exceeds a predefined threshold θ_D, a potential anomaly is signaled in terms of an alarm signal. By using a content-addressable memory, the matching can be evaluated in parallel for all the strings of the detector collection, thus permitting the generation of the alarm signal in realtime, before the state transition is actually permitted to happen.

There is a relationship between the performance of the system in terms of probability of false-positive and negative fault detection, and the number of receptors and their sensitivity and diversity. The flexibility ensuing from the possibility of tuning these parameters is the main advantage of the immune approach with respect to traditional methods. However, the system described above applies only partially the tenets of immune systems. In particular, the protection system is not decentralized and does not include a mechanism of recovery from the detected faults. A possible solution to this latter problem would be the use of an additional collection of strings that

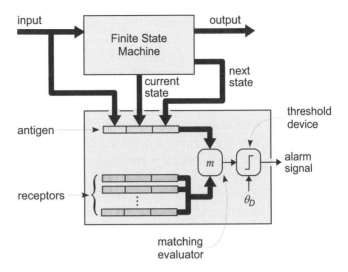

Figure 5.20   The immunotronics approach to fault detection uses negative selection to build a collection of receptor strings. These are compared to the antigen strings produced by the finite state machine under observation. A degree of match above a threshold θ_D corresponds to the detection of potential nonself and results in the generation of an alarm signal.

represents a sample of the "healthy host," that is, a sample of the legal transitions, so that an estimated correct transition can be obtained after anomaly detection.

5.9   Closing Remarks

There are in the literature many examples of AISs besides those analyzed in the previous pages, with applications that go beyond computer security and fault detection, to include pattern recognition, machine learning, optimization, control, and robotics (de Castro and Timmis 2002; Ishida 2004). These examples can differ in some detail of their implementation, but they are all based on the fundamental concepts and processes observed in the operation of the vertebrate immune system which have been presented in the first half of this chapter. The analysis of the strategies used by the vertebrate immune system shows that there are several interesting concepts that solve in novel ways many problems relevant to engineering (Glickman et al. 2005). The algorithms that these strategies inspire can be profitably exploited as separate tools. However, the full potential of the immune system concept emerges when all the elements work together.

Not many years ago the operation of biological immune systems was understood in terms of discrimination between self and nonself. Correspondingly, the first AIS models and applications were also based on that perspective. Today, in both the biological and AIS modeling, we are witnessing a shift away from a static self/nonself dichotomy toward models that emphasize the cooperation of the protected system, of the innate immune system, and of the adaptive immune system in the dynamic definition of what constitutes the self. This transition promises to take the field of AISs to new levels of performance and flexibility. However, to achieve its full potential it requires a profound rethinking of the current engineering approach to the design of systems. The availability of systems that are protected like the biological organisms are protected by their immune systems can be expected to become more and more important as the designed systems' complexity, their potential of connectivity, and the resulting number of potential malfunctions increases.

5.10   Suggested Readings

(Coico et al. 2003) is a clear and well-illustrated description of the human immune system, which remains within limits of size and detail that are more than adequate for the scope of the present chapter. Lydyard et al. (2000) renounce fancy colored illustrations in favor of a highly structured presentation organized in many short and highly focused chapters. (Alberts et al. 2002) is a classic molecular biology textbook that contains two chapters that give an overview of the immune system and of its relations with the operation of cells. (Sompayrac 2003) is a very short and readable introduction to the immune system that omits most of the low-level details and focuses on the logic of the operation of the immune system. (Hofmeyr 2001) is a compact introduction to biological immune systems which is written with AISs in mind and includes a very useful glossary. Typically, the immunology textbooks have not yet adopted the danger model perspective presented in this chapter. However, this perspective is well described in the review papers of one of the originators of the model (Matzinger 1994, 1998, 2002).

The special September 1993 issue of *Scientific American* (Piel 1993) contains several interesting and well-illustrated papers devoted to particular aspects of the human immune system. Friedl et al. (2005) discuss and illustrate the concept of the immune synapse, comparing it to the neural synapse. Engelhard (1994) gives an excellent overview of antigen processing and presentation via MHC molecules. The evidence supporting the existence of a genetic encoding of local mutation rates in natural genomes is reviewed in (Caporale 2003, 2004). Humphrey (2002) proposes a fascinating hypothesis that explains the placebo effect in terms of the cost required by the operation of the immune system.

(Segel and Cohen 2001) contains many interesting contributions that look at the biological immune system from an engineering perspective. Ishida (2004) and de Castro and Timmis (2002) provide extensive descriptions of AIS concepts and applications. In particular, the reader can find in these books a detailed treatment of the network immune models, which have been omitted from this chapter. (Garrett 2005) and (Timmis 2007) are critical overviews of the AIS concepts which compare this approach to other bioinspired methodologies with the aim of assessing the novelty and usefulness of the AIS contribution. A review of the applications of AISs to intrusion detection can be found in (Kim et al. 2007a). Finally, Glickman et al. (2005) present a critical analysis of AIS concepts from the perspective of machine learning.

6 *Behavioral Systems*

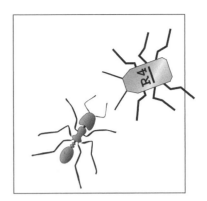

Behavior is a sequence of interactions between an organism and its environment where the actions of the organism affect its own perceptions, and consequently its future actions and perceptions. A behavioral system consists of a body with a sensory and a motor apparatus that allows the interaction with the environment, a brain to map sensory stimulation onto motor actions, and a metabolism to support its operation.

The significance of behavior and embodiment for understanding and reproducing biological intelligence has been realized only recently, as we will see in the next two historical sections. Autonomous robots are behavioral machines capable of operating in partially unknown and changing environments without, or with limited, human intervention. Assembly robots used in factory plants are not behavioral systems because their input and output sequences are strictly predefined by human designers.

AUTONOMOUS ROBOTS

The increasing availability and usability of robot technologies are contributing to the adoption of robots as models and tools to study biological organisms. The reader should not be surprised therefore that this chapter is almost entirely dedicated to biological inspiration for autonomous robots and to the use of robots for understanding biology.

After the next two historical sections, we will describe behavior-based robotics, biologically inspired robots, robots as biological models, epigenetic and developmental robotics, evolution of control systems, evolution and learning, as well as evolution and development in behavioral systems, co-evolution of body and brain, self-reproducing robots, and issues on the use of simulations and hardware in robotics. In order to better understand and compare the various methods, we will present several research examples that address comparable questions from different perspectives. The description of behavioral systems will then continue in chapter 7.

6.1   Behavior in Cognitive Science

The systematic study of human perception, also known as psychophysics, paved the way for the first theories of human mind summarized by William James (1890) in the book *Principles of Psychology*. By the end of the nineteenth century, psychology was mainly interested in conscious processes.

BEHAVIORISM

A radical shift occurred in the early twentieth century with the birth of *behaviorism*, a movement that emphasized the study of actions over introspection of mental processes. Behaviorism criticized introspection for not adhering to the scientific criteria of objective measurement, reproducibility, and testability. Behaviorism focused mainly on the laws that relate perception to actions, but largely ignored the effects of actions on perception. However, behaviorism made at least three long-lasting contributions that remain at the foundations of modern approaches to intelligent systems research.

The first contribution was the recognition of a continuity between the behaviors of animals and humans (Jennings 1906). This was in sharp contrast with the established assumption at the end of the nineteenth century that human intelligence was unique, but was perfectly in line with Darwin's observations on the evolution of the species. As a consequence, behaviorists often carried out experiments with simple animals, such as rats and pigeons, assuming that the results could be extrapolated to better understand human intelligence.

The second contribution was the emphasis on learning by reinforcement (Skinner 1938). Behaviorists developed accurate and sophisticated methods for training animals and measuring their performances, some of them to be found in today's methods for training and assessing robots.

The third contribution was the effort to translate behavioral and psychological phenomena into *operational models* (which today we would call "computer programs") that could be subjected to scientific study. These phenomena included reactive behaviors, learning, and even representation-mediated behaviors, such as navigation with the help of spatial maps (Tolman 1948).

COGNITIVE SCIENCE

By the end of the 1950s, behaviorism had slipped into the background of a renewed interest in internal states and mental processes, which resulted in the rise of *cognitive science*. Cognitive science brought together psychology, neuroscience, and computer science in the attempt to create a computational theory of mind. For example, Donald Hebb, whose contributions to learning theory were discussed in chapter 3, linked cognitive processes to specific neural mechanisms, thus giving birth to a field that later became known as *computational neuroscience* (Churchland and Sejnowski 1992).

COMPUTATIONAL
NEUROSCIENCE

At the same time, the nascent computer technology served as a model for mapping cognitive processes onto operational models (von Neumann 1958) and a tool for testing such models at different levels of abstraction, ranging from symbolic rules to neural networks.

In the early years, cognitive science was focused on identifying the mechanisms and processes that intervene between the recording of sensory data and the production of decisions and actions (Neisser 1967). Although later reformulations of the research agenda emphasized the need to better incorporate behavior as an action-perception loop (Neisser 1976), mainstream cognitive science remained focused for most of the twentieth century on internal representation, symbolic reasoning, and problem solving. The dominant assumption was that human intelligence was equivalent to a sophisticated information-processing system (Lindsay and Norman 1972).

As cognitive science became established, a number of philosophers, psychologists, and neuroscientists felt increasingly uncomfortable with an approach that essentially endorsed the primacy of mental processes over behavior. Merleau-Ponty in France was among the first to point out that sensations were affected by the physical body of the perceiver and that perception was not a passive recording of environmental stimulation, but a process of active exploration akin to body movement (Merleau-Ponty 1962). Merleau-Ponty's philosophical writings, whose value was fully appreciated only in very recent years, pointed to the necessity of considering embodiment and behavior as constituent elements of sensation and cognition.

Approximately at the same time in the United States, Gibson stressed the role of environmental situatedness and behavior for understanding visual perception. He identified a number of perceptual laws that derive from the relationship between an embodied organism and its environment (Gibson 1979). Gibson distanced himself both from the reductionist stimulus-response approach of behaviorism and from the mentalism of cognitive science where free interaction of the organism with the environment was largely

ECOLOGICAL
APPROACH

ignored. He argued for an *ecological approach* to the study of organisms as they interact with the environment and affect the stimulations that arrive to their sensing organs. Gibson claimed that the environment of an organism could not be described with the same tools used to describe the physical world. Each organism has physical characteristics that allow it to interact with the environment at certain spatial and temporal scales. This unique level of interaction creates specific opportunities that are readily detected by the perceptual system of the organism, but not by an external observer.

Gibson uncovered the limitations of the dominant view of perception, which he called "snapshot vision," where the eye is considered as a camera that records photographic images for further processing. He pointed out that arthropods display sophisticated vision-based behaviors, but their eyes do not have a lens and a projecting surface similar to a retina. Instead, their visual system consists of a set of tube-like photoreceptors pointing in different directions that signal modifications in light intensities to the brain without passing through a photographic image. Gibson argued that what matters in visual perception are indeed modifications in light intensity, mostly caused by the movement of the physical organism in the environment. In particular, he dedicated much effort to the study of *optic flow*, that is, the pattern of sensory modification that occurs when the organism or the environment moves.

Gibson's work remains unique for its scientific and philosophical force in showing the role of behavior in perception. His message caused cognitive science to revise its position (Neisser 1976), but in practice it did not have a major impact at that time, probably because computers, which were at the heart of cognitive science models, did not have eyes and legs. With the recent development of robot technologies, Gibson's approach became immediately relevant and inspired a range of algorithms and technologies for vision-based robots, as we will see later in the chapter. Gibson's approach is appealing for engineering because it suggests that vision is not necessarily a complicated and computationally expensive business if one departs from the photographic metaphor of snapshot vision.

While several parts of Gibson's theory are applicable to behavioral systems of any type and complexity, his examples were mostly concerned with relatively simple perceptual experiences and organisms. An increasing number of experimental and theoretical work is now indicating that behavior plays an important role also in human perception with important implications for representations, memories, and consciousness. This was recently summa-

ENACTIVE PERCEPTION rized by Noë (2004), who coined the name *enactive perception* to emphasize that all perceptions are actions.

Building on Merleau-Ponty's philosophy, Noë argued that "we do not represent whole scenes in consciousness all at once. Visual experiences do not present the scene in the way that a photograph does. In fact, seeing is much more like touching than it is like depicting" (Noë 2004, p.72). The enactive approach to perception has two major implications. Since vision is an exploratory act based on the sensorimotor abilities of the organism, developing a model of vision requires developing a model of the sensorimotor characteristics of the organism (O'Regan and Noë 2001) and consequently of its body and environment. Similarly, understanding the neural basis of vision requires a novel neuroscience of embodied activity in place of mainstream neuroscience of brain activity (E. Thompson and Varela 2001).

EMBODIED COGNITIVE
SCIENCE

The ensuing movement that emphasizes the role of embodiment and behavior in shaping intelligence is now known as *embodied cognitive science* (Clark 1997). In order to account for the rich set of spatiotemporal interactions between the organism and its environment, this approach often relies on dynamical systems theory (McFarland and Boesser 1993; Thelen and Smith 1994; van Gelder 1998) where a set of variables describing the brain, body, and environment continuously evolves over time. Instead of symbols and representations, dynamical system approaches resort to concepts as attractors, transients, and trajectories in a dynamical space. Often dynamical models are implemented as neural networks, such as continuous-time recurrent neural networks described in chapter 3. Within embodied cognitive science, robots have now replaced computers as the metaphor to develop and test models of biological intelligence.

6.2 Behavior in Artificial Intelligence

ARTIFICIAL
INTELLIGENCE

Artificial intelligence was formally established as a research area at a meeting organized by John McCarthy at Dartmouth College in 1956. The mission of artificial intelligence was to understand human intelligence and build intelligent machines, which at that time were intended mostly as digital computers. The legitimacy of this enterprise was based on the nascent computation theory and technology after the World War II, the realization that neurons were equivalent to computational devices (McCulloch and Pitts 1943), and the hypothesis that cognition was a form of computation (Turing 1950). Artificial intelligence developed in parallel to cognitive science and the two fields affected each other in many ways, often sharing theories, models, and tools. Until the 1980s, mainstream artificial intelligence was mainly concerned with symbolic reasoning (Newell and Simon 1972), representations (Marr 1982), and language (Chomsky 1957). The typical testbed of artificial intelligence was the game of chess, a well-defined task that requires sophisticated perceptual and cognitive abilities, but not necessarily embodiment and behavior.

Figure 6.1   The robots devised by Grey Walter and built with the assistance of Bunny Warren in Bristol, England. The motor system of the robot was like a tricycle where the front wheel was actuated by motors for regulating speed and steering. The robot shell hung from a pole and came into contact with internal sensors whenever an obstacle was encountered, thus functioning like a touch sensor. A photoreceptor mounted on a rotating head was used to measure light intensity while scanning the environment. The photograph shows a replica built by Owen Holland. Image courtesy of the Intelligent Autonomous Systems lab, University of West England at Bristol.

In contrast to that approach, in the early 1950s Grey Walter built a series of mobile robots to demonstrate that complex and purpose-driven behavior could emerge from a set of simple and interconnected neuron-like devices embodied in a situated organism (Walter 1950, 1951). Grey Walter, who is now considered the creator of the first autonomous robots (Freeman 2003), was trained as a neurophysiologist and electrical engineer. His robots were also called "turtles" because of their shelled body and relatively low speed. The electronics were composed only of analog devices (two valves, two capacitors, and two mechanical relays) and Walter insisted on the importance of using analog elements instead of the digital computation embraced by mainstream artificial intelligence. Owen Holland recently reconstructed the robots from original pieces and circuit diagrams (figure 6.1). He showed that these robots displayed complex and partially unpredictable behaviors such as exploration, negative and positive tropism, discrimination, adaptation to changing environments, and behavioral stabilization (Holland 2003).

Walter's work attracted the attention of the media at that time, but did not have an immediate impact on cognitive science and artificial intelligence.

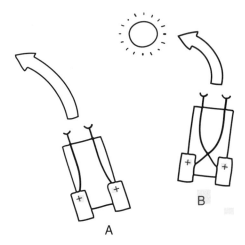

Figure 6.2 Two Braitenberg vehicles: vehicle A is repulsed by light and vehicle B is attracted. Reproduced from Braitenberg (1984).

Thirty years later, the neuroscientist Braitenberg (1984) published a booklet describing a series of imaginary vehicles with simple wiring of sensors and motors inspired by the anatomical and physiological principles of nervous systems, such as symmetry, cross-lateral connection, excitation and inhibition, time-delayed activity, and nonlinear dynamics (figure 6.2). When placed in an environment, Braitenberg's vehicles displayed a range of complex behaviors that an external observer might have labeled as aggression, love, fear, logic, foresight, egotism, and free will. He even described an imaginary experiment with artificial evolution of these vehicles.

Braitenberg's vehicles were intended to convey two major messages. The first message was that much of the complexity observed in behavioral systems stemmed from the interaction with the environment rather than from the complexity of the brain, similar to what Grey Walter intended to show with his mechanical turtles. The second message was the *law of uphill analysis and downhill invention*: "When we analyze a mechanism, we tend to overestimate its complexity. In the uphill process of analysis, a given degree of complexity offers more resistance to the workings of our mind than it would if we encountered it downhill, in the process of invention" (Braitenberg 1984, p.20). In other words, he argued that neuroscience and cognitive science could gain considerably by combining the analysis of living brains with the construction (synthesis) of embodied and behavioral circuits. Al-

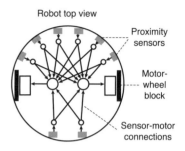

Figure 6.3 The miniature mobile robot Khepera was instrumental in the development of bioinspired approaches in robotics and artificial intelligence. The robot's software included an example of a Braitenberg architecture for collision-free navigation (Mondada et al. 1993).

though Braitenberg was mainly concerned with understanding living nervous systems, his book had a strong impact on artificial intelligence and robotics, partly because many of his imaginary vehicles could be readily implemented in real robotic systems (figure 6.3).

In the late 1980s, Rodney Brooks published a series of influential papers describing a radically new approach to artificial intelligence and robotics, BEHAVIOR-BASED which he named *behavior-based robotics* (see Brooks 1999 for an annotated col-ROBOTICS lection). Brooks claimed that intelligent behavior is generated by the direct coupling between perceptions and actions without invoking the mediation of high-level cognitive processes, such as representation and reasoning. He argued that intelligence existed only in the eye of the observer and was an emergent phenomenon resulting from the interactions among multiple perception-action modules in a situated and embodied system.

Behavior-based robotics served as a catalyst for a large-scale and widespread paradigm shift in artificial intelligence, cognitive science, and robotics. On the wave of this shift, connectionist models of the brain (Rumelhart et al. 1986b) were criticized for being oversimplified structures that learned in vacuo (Parisi et al. 1990; D.T. Cliff 1991) and were therefore inadequate to capture the principles of intelligence.

Ballard (1991) extended the behavioral approach to computer vision. He claimed that many of the problems in reproducing human visual abilities stemmed from the attempt to map "two dimensional data into a description [of] 3-D surfaces, volumes, boundaries, shadows, occlusion, depth, color, motion" (Ballard 1991, p. 57). He argued instead that biological vision is ANIMATE VISION *animate vision* because it continuously moves by means of saccadic displacements, it is governed by behavioral modules, and it does not necessarily need categorical representations of the 3D world. Along with other researchers (Aloimonos et al. 1987; Bajcsy 1988), Ballard suggested that a behavioral approach to vision may even simplify the complexity of vision-based tasks by focusing computation on fewer and simpler features at a time.

Behavior affects the type and distribution of sensory information that an embodied nervous system receives. Consequently, this may affect the formation of the synaptic connections in developing and adapting brains. However, this is almost never taken into account by computational neuroscience, which assumes uniform or well-behaved probability distribution functions of input data. It has been recently shown that the actual probability distribution function of natural images recorded from a camera strapped to the head of a cat is significantly different from uniform (Betsch et al. 2004). This holds also for a robot that moves around an environment according to an arbitrary, but not random, vision-based behavior. A comparison of the snapshots taken by the robot's camera with those taken by the same camera pointing in all possible directions and orientations shows that a behavioral system selects only a subset of the theoretically available information and that subset is not a uniform sample (Floreano et al. 2005). A control system designed, learned, or evolved by taking into account the behavior of the embodied system therefore may be simpler, faster, or more efficient (Verschure et al. 2003) because it does not need to build and rely on a complete model of the external environment.

NEW AI
GOOD OLD-FASHIONED AI

The modern approach to artificial intelligence that emphasizes behavior, embodiment, and situatedness is now known as *new AI* (Pfeifer and Scheier 1999) and is contrasted to *good old-fashioned AI* (Haugeland 1985), which focused on symbolic reasoning and planning and went under major criticism over recent years (Dreyfus 1992).

6.3 Behavior-Based Robotics

Although behavior-based robotics started off as a practical methodology to design and build efficient robots that could operate in the real world, it caused a major rethinking of priorities and approaches in artificial intelligence.

Behavior-based robotics is perhaps best understood by comparing it with the established approach to intelligent robotics that dominated until the mid-1980s, and still is mainstream in major robotics conferences. In the main-

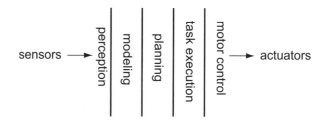

Figure 6.4   Functional decomposition of robot control. From Brooks (1986).

stream approach, the control system of a robot is typically divided into a sequence of functional modules that decompose the solution into separate units, such as perception, modeling, planning, and execution (figure 6.4).

According to Brooks (1991c), this functional division can be traced back to the history and organization of artificial intelligence. For example, perception is typically dealt with in computer vision, which is expected to provide a description of the environment in terms of objects and landmarks from large vectors of pixel intensities (e.g., Marr 1982). Computer vision is not necessarily concerned with the uses that other people make of those descriptions. A large area of artificial intelligence attempts to put together those descriptions into models of the environment that can be manipulated and exploited for planning. Symbolic reasoning, expert systems, and connectionism are a small sample of different techniques for developing those world models; they rely on the existence of perceptual descriptions and are not necessarily concerned with the implementation of the process in a real mobile robot. Hardcore robotics consists in designing the physical machine and executing the planned trajectories. Besides the mechatronic aspects of robot design, the questions of concern at the control system level consist in developing suitable automatic methods for finding sequences of joint or wheel torques that will result in the trajectory demanded by the planning system.

This functional decomposition produces precise, controllable, and predictable robotic systems that may be required in some applications, such as manufacturing and surgical robotics. It also tends to produce general architectures that can be applied to several instances of robots. However, this approach does not cope well with the noise and uncertainty that is met with by autonomous robots. Furthermore, since each functional level depends on the completion of previous levels to operate, failure at an early stage can jeopardize the functioning of the entire control system. Finally, the need for

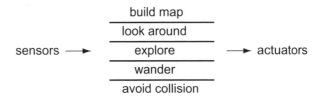

Figure 6.5 Behavioral decomposition of robot control. Adapted from Brooks (1986).

building a model and producing plans requires considerable computational power, especially when both the model and the plan are both built at the same time. As a consequence, the mainstream approach is often applied to relatively well-defined environments with little noise and autonomy, results in relatively slow and bulky robotic systems, and does not have a general strategy for coping with malfunction of some software or hardware component.

Brooks (1986) instead suggested decomposing the problem in terms of the competences, or task-achieving behaviors, that the robot requires in order to meet its goals (figure 6.5). Each behavior can have direct access to the sensors and actuators of the robot and operate in parallel to other behaviors by means of distributed coordination mechanisms.

SITUATEDNESS

EMBODIMENT

Behavior-based robotics rests on the core ideas that the control system must be (a) *situated*, that is, it must deal with the sensory and motor contingencies of the environment where the robot operates and not with abstract descriptions; and (b) *embodied*, that is, it must experience the world directly through its sensors and physically act on the environment, rather than operate in simulated worlds (Brooks 1991c). As a consequence, behavioral decomposition can lead to very different control architectures that depend on the type of robots and goals.

One can visualize a behavior-based architecture as a layered stack of parallel behaviors where the behaviors at the bottom deal with the survival of the robot and the behaviors at the top achieve desirable goals if opportunities exist. The design of a behavioral control system proceeds incrementally. The first layer is designed, tested on the robot, and refined until it is satisfactory. At this point, the robot is already operational. Then, the second layer is designed, tested, and refined. At this point the robot can make use of both types of competences. And so forth.

Behaviors in higher layers can rely on, and sometimes affect, behaviors in preceding layers. However, behaviors in higher levels do not use the be-

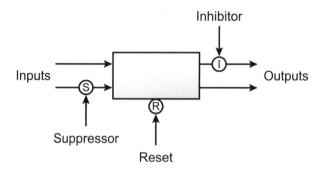

Figure 6.6   A behavior module in subsumption architecture. Adapted from Brooks (1986).

haviors in preceding layers as subroutines, but simply as a set of preexisting competences. This architecture is known as *subsumption architecture* because higher layers subsume, or encompass, early layers. Early layers can continue to operate also without the presence of further layers and need not be reprogrammed. However, if opportunities exist for higher layers to operate (for example, the detection of gas molecules activates a plume-tracing behavior), they may suppress, modify, or substitute the output of lower behaviors with their own behavior.

The incremental design of behavior-based architectures recapitulates in a broad sense the natural evolution of intelligence (Brooks 1991b) where basic survival competences, such as forage and escape, emerged before higher competences, such as communication and reasoning. Similarly, behavior-based design proceeds from the development of behaviors that ensure the viability of the robot all the way to higher goal-seeking behaviors that build upon basic behaviors.

Each competence layer can be structured differently and further decomposed in several ways, including using the mainstream approach of perception, modeling, and execution. Indeed, some implementations of behavior-based architectures include a combination of low-level behaviors with high-level reasoning (e.g., Arkin 1990).

BEHAVIOR   A behavior is a self-contained and operationally independent module with input and output lines (figure 6.6). In its simplest version, a behavior is a motor response to a sensory stimulation from the environment. The internal operation of a behavioral module cannot be modified by other modules, although some implementations of behavior-based architectures share memory registers among several behaviors. The input lines convey signals from the sensors of the robot or from other behavioral modules. The output lines send signals to the actuators of the robot or to other behavioral modules. In the subsumption architecture, input and output lines can be suppressed or inhibited by other modules. In the case of suppression, the signal from the sending module substitutes for the signal traveling on the suppressed line. In the case of inhibition, the signal traveling on the inhibited line is zeroed for some time. Furthermore, a module can receive a reset signal from another module. In doing so, higher behavioral modules subsume, or incorporate, the competences of lower behavioral modules.

Each behavioral module can be, in principle, implemented as a separate augmented finite state machine (Brooks 1989) with its own internal clock, but without global synchronization. Consequently, if a behavior does not provide an output after a certain amount of time, another behavioral module may take over the control of the actuators.

There are several ways of coordinating behaviors, each corresponding to a different behavioral architecture. In the subsumption architecture (figure 6.7, a)), the priority among behaviors is fixed by the pattern of inhibition and suppression among modules. In the action-selection architecture (figure 6.7, b)), each behavior includes an activation value and the behavior with the highest activation level is selected to take over the control of the robot (Maes 1989). In the voting architecture (figure 6.7, c)), each behavior casts a vote for a set of predefined robot actions and the action with the highest number of votes is executed (D.W. Payton 1986). For a review and comparison of these and other behavioral architectures, we suggest (Arkin 1998).

Behavior-based control has a number of advantages with respect to the mainstream approach based on functional decomposition. It results in fast reactions because behavioral modules can immediately map sensory information onto motor actions. It is a robust architecture because if a hardware or software component fails, the robot has a higher probability of retaining at least some behavioral competence. It can accommodate multiple goals that are dealt with asynchronously by multiple behaviors. It is scalable to robots with different degrees of freedom and extensible if new sensors and hardware modules or if new behavioral competences are added to the robot.

The complexity of the robot behavior does not derive from the complexity of the control system or of the world model, as in the functional decomposition approach, but from the interaction among several simple behavioral modules that are continuously interacting with the external world and with each other. Consequently, intelligent behavior results from the dynamics of

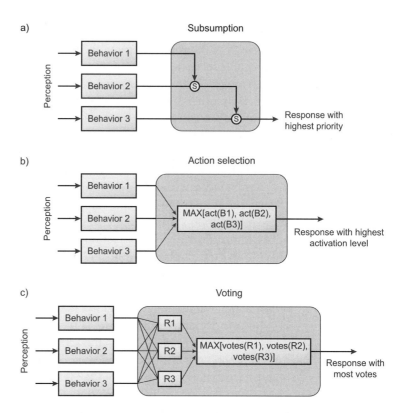

Figure 6.7 A sample of behavior-based architectures: subsumption, action selection, and voting. Adapted from Arkin (1998).

interaction with the world. As Brooks (1991a) put it, intelligence is in the eye of the beholder, not in any particular components or set of components within the machine.

However, it has been argued that the lack of a central world model could restrict the behavioral competences of the robot to those of simple animals; that the lack of planning could result in purely reactive behaviors that cannot anticipate future events, which could have an influence in current decisions; and that the predictability of the robot's behavior may consequently be affected (Arkin 1998). To address these criticisms, Matarić and Brooks (1990) described a robot that is capable of incrementally building a map of the environment without using Cartesian metrics and central representations. The behavioral architecture established a set of interconnected sensorimotor

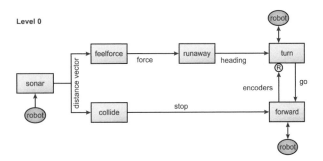

Figure 6.8   Zeroth competence level: runaway from any obstacle. Adapted from Brooks (1986).

nodes that linked perceptual events to motor actions. Brooks et al. (1998) also argued that behavior-based principles could be extended to tackle some aspects of human-like intelligence with the help of a humanoid robot, as we will see in a later section.

Although the theoretical advantages and disadvantages of behavior-based robotics are still widely debated, this approach has found successful applications in several mobile robots that range from commercially available robot toys and domestic assistants all the way to planetary rovers (Brooks 2002).

6.3.1   Example 1: Navigation of a Mobile Robot

Let us start with a simplified version of an example that Brooks (1986) used to illustrate the implementation of a subsumption architecture. The robot consists of a wheeled platform with a circular array of sonar sensors that provide information on distance from objects.

The zeroth competence level (figure 6.8) provides the robot with the ability to steer away from any incoming obstacles or obstacles that it may encounter as it runs away. It consists of several behavioral modules. The `sonar` module receives the sonar readings and produces a vector of distances around the robot. This is sent in parallel to two modules. The `collide` module checks if any of the vector values is below a given threshold, which would signal an obstacle very close to the robot, and if that is the case it outputs a stop signal to the module that is responsible for advancing the robot. Notice that the `collide` module acts independently of what other parts of the architecture are doing. The `feelforce` module instead computes a vector corresponding to the repulsive force that results from the distance vectors and sends it

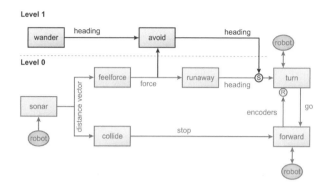

Figure 6.9 First competence level: wander around the environment. Adapted from Brooks (1986).

to the `runaway` module, which outputs a heading direction if the received force is larger than a threshold. The `turn` module rotates the robot to align it with the heading direction, reads the wheel rotation encoders, outputs a go signal, and goes into a busy state. On reception of the go signal, the `forward` module advances the robot by a fixed distance, unless it receives a stop signal, and reads the value of the wheel rotation encoders when the robot is idle. This value is sent back to the `turn` module and acts as a reset signal. The parameters of the modules (thresholds and time constants) are adjusted while the architecture is tested and debugged on the physical robot.

The first level of competence adds to the robot the ability to move around (figure 6.9) and capitalizes on the behavioral competences that the robot already has. It consists of a `wander` module the generates a new heading direction at regular intervals (e.g., every 10 seconds). The `avoid` module combines the new heading direction with the repulsive force, if any, to produce the resulting heading direction, which is connected with a suppression link to the input line of the preexisting `turn` module. The suppression link writes the new heading signal over the heading signal, if any, from the `runaway` module. The first competence level therefore subsumes the zeroth competence level in that it builds on its competences and can take over some of its behavioral modules.

Once the augmented architecture (level 0 + level 1) has been tested, debugged, and its parameters adjusted on the robot, the second level of competence is incrementally developed (figure 6.10). The robot is equipped with a rotating stereo camera that can be used to compute distances. The second

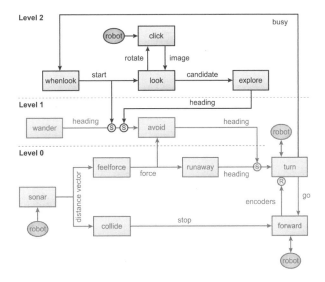

Figure 6.10 Second competence level: vision-based exploration. Adapted and simplified from Brooks (1986).

level of competence is intended to add vision-based exploration to the robot. Whenever the robot has been idle for some time, it looks around for a corridor or an opening. The `whenlook` module monitors the status of the robot from the `turn` module and, if the robot has been idle longer than a predefined time, it sends a start signal that inhibits the output of the `wander` module for a predefined time and activates the `look` module. The `look` module sends a rotation angle to the `click` module, which rotates the camera, takes a snapshot, and passes it back to the `look` module. If the `look` module does not find any candidate corridor or opening, it issues a new rotation angle until a suitable image is found and passed on to the `explore` module, which computes a heading direction. The heading direction is sent through a suppression line to the input of the `avoid` module, which combines it with any repulsive force to generate the resulting heading.

At this stage the robot has a full range of behavioral competences. When somebody approaches, it will move away until it finds a safe open space. If it has been idle for some time, it will search with the camera for an opening and will move in that direction while avoiding obstacles on the way. If no opening is found, it will wander aimlessly in the environment, while avoiding any encountered or approaching obstacle.

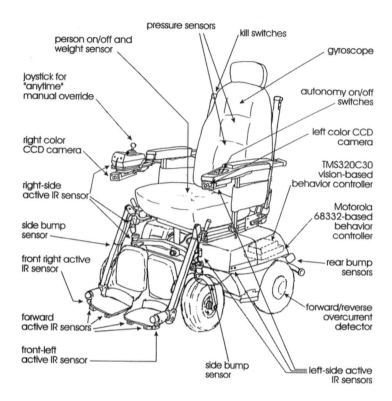

Figure 6.11   A commercially available power wheelchair equipped with sensors for autonomous operation (Gomi and Ide 1996). Image courtesy of Takashi Gomi, Applied A.I. Systems, Inc., Ottawa, Canada.

6.3.2   Example 2: An Intelligent Wheelchair

Behavior-based methodologies have rapidly found applications in several commercially available robots (Brooks 1986). The autonomous wheelchair project described in this section represents probably the best detailed example of a development stage for which details are publicly available. Gomi and Ide (1996) of Applied A.I. Systems Inc., Ottawa, Canada, employed behavior-based control for the development of an intelligent wheelchair that is expected to be fully or semiautonomous.

The wheelchair used in the development stage was a commercially available motorized chair (figure 6.11) with two differentially driven wheels and two free front casters. The front casters add a lot of fluctuations that in mainstream control approaches would require a redesign of the motor platform or specially designed modeling and error-correction routines. The entire mechanical, electrical, and electronic structure of the commercial wheelchair, which included a joystick to manually steer the wheelchair, was used without modifications.

The original platform was extended (figure 6.11) with the addition of several bumper and infrared distance sensors, two color CCD cameras on the armrests, a keypad and a small TV panel for purposes of software development and monitoring, and two processor boards (one for vision-based behaviors and the other for all remaining behaviors).

The behavior-based architecture was incrementally built by adding, testing, and debugging on the wheelchair various levels of behavioral competence (figure 6.12). These levels were organized in terms of safety priorities. Inputs from the joystick had the highest priority, as revealed in the control diagram by the suppression line closest to the motor control of the wheelchair. In other words, the user always had the option to override autonomous behavior and take control of the wheelchair. The second highest priority was given to behaviors that received input signals from the left and right bumpers and infrared distance sensors. These behaviors provided obstacle avoidance competences that prevented uncomfortable or damaging collisions. They were followed by behaviors that used infrared distance sensors in order to "squeeze" the wheelchair through narrow passages. Lower in the priority list were behaviors that handled voice commands and behaviors that produced vision-based corridor following. The last behaviors in the priority list were those that used a vision-based landmark navigation system to take the user to a goal location that has been entered through the keypad.

The autonomous wheelchair was capable of successfully operating autonomously and semiautonomously in several outdoor environments, shopping malls, and other daily situations by physically impaired people who tested it. The reactive and somewhat natural movements produced by the behavior-based architecture induced external observers to believe that the chair was manually operated.

Gomi and Ide (1996) motivated the choice of behavior-based architecture with the ability to operate rapidly in real-world situations and consequently to produce a smooth customer experience. They also indicated that behavior-based control provided an intrinsic fail-safe architecture, whereas mainstream approaches required the addition of special routines for safe standby in case of software or hardware malfunction. Finally, they mentioned that the entire software architecture took only 35 kB for the vision-related behaviors

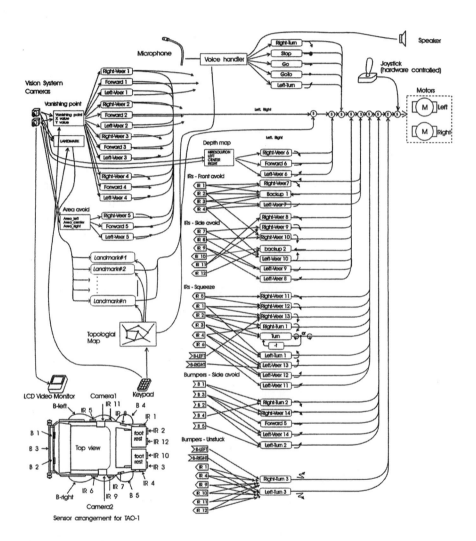

Figure 6.12   Schematic diagram of behavior-based architecture for the autonomous wheelchair (Gomi and Ide 1996). Each behavior may include several other behaviors, which are not shown. Filled arrowheads at the module output indicate the corresponding motion of the wheelchair. Image courtesy of Takashi Gomi, Applied A.I. Systems, Inc., Ottawa, Canada.

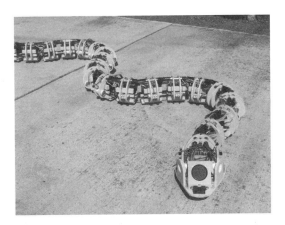

Figure 6.13 A snake robot may provide better agility and flexibility than wheeled or legged robots for search-and-rescue missions in collapsed buildings. Photo courtesy of Gavin Miller, S7 prototype with compass, sonar, and heat sensors.

(out of which 25 kB were dedicate to map generation and low-level image processing) and 32 kB for all other behaviors, including a significant portion to deal with peripheral management, such as keypad, voice input/output, and visual display. This slim code allowed for rapid prototyping and real-time operation.

6.4 Biological Inspiration for Robots

Bioinspired robots are robots that resemble living organisms in some specific characteristics, such as the control system, the morphology, the actuators, or the electronics. Robot engineers take inspiration from biology in order to design robots that have better agility and flexibility, that display a novel functionality, that are more adaptive and intelligent, or that can better operate in the vicinity of humans. For example, snake robots (figure 6.13) are developed for search and rescue in collapsed buildings (G.S.P. Miller 2002) where wheeled and legged robots may not meet the requirements necessary to move over debris, climb high obstacles, and pass through narrow openings.

There are several degrees of bioinspiration, ranging from loose metaphors to high-fidelity replicas of specific parts. The goal is to capture and reproduce the mechanisms that provide better performance for a given engineer-

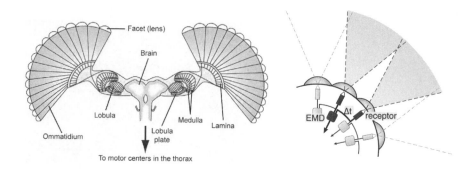

Figure 6.14 *Left*: Top view of the fly visual system. Optical stimulation is collected by several single-lens eyes (omatidia), each containing eight visual receptors, that drive an array of motion-sensitive neurons in the medulla and lobula plate. *Right*: The elementary motion detector (*EMD*) responds to motion in a specific direction by collecting signals from neighboring receptors with temporal delays on the connection lines. Adapted from Franceschini et al. (1992).

ing problem. In the following subsections, we will describe three cases of biological inspiration that provide better performance or a novel functionality to the robots.

6.4.1   Example 1: Vision-based Flying Robots

Robots that can fly autonomously in cluttered or indoor environments could be very useful for search-and-rescue missions, for 3D cartography of urban environments, or for surveillance. In those conditions, GPS signals are often not available, teleoperation is very difficult or impossible, and localization methods used for wheeled robots (such as laser rangefinder or other active sensing) are not applicable because of energy consumption and size.

Vision represents a promising sensing modality because it does not require energy to acquire information and today's cameras are sufficiently small and lightweight to be carried by miniature aircraft. However, it is not trivial to process visual information in real time within low-power, embedded microcontrollers. Insects are marvelous examples of systems that use vision to avoid collisions, take off and land, search, and pursue. For that reason, a number of engineers have taken inspiration from the visual system and behavior of the fly, which has been studied in great neurophysiological detail (Hausen and Egelhaaf 1989).

The fly's primary sensor for flight control consists of two compound eyes (figure 6.14, left) that span an almost omnidirectional field of view with coarse resolution (Land 1997). The optic lobes contain motion-sensitive neurons which respond to retinal image shifts, the so-called optic flow, induced by the motion of the animal relative to the surroundings (Gibson 1950). The fly has several neurons sensitive to optic flow, also known as *elementary motion detectors* (EMDs), that have been linked to specific visually guided behaviors (see Egelhaaf and Kern 2002 for a review).

ELEMENTARY MOTION
DETECTORS (EMDs)

Optic flow is a linear combination of two components, one resulting from rotation and one from translation of the animal. Translational optic flow is the only component whose magnitude depends on distance from an object (Koenderink and van Doorn 1987). If the rotational component is eliminated, the residual optic flow may provide useful information for behaviors that rely on distance from objects, such as collision avoidance, altitude control, or landing.

The need for reducing rotational optic flow may be the reason why flies fly in straight lines separated by rapid turns, known as saccades (Tammero and Dickinson 2002). Straight flight allows flies to experience pure translational optic flow, which may be used to tell how close an obstacle is and decide to initiate a saccadic turn. During saccadic turns, flies seem to ignore visual information, which is dominated by the rotational optic flow.

In pioneering work, Franceschini et al. (1992) built a physical model of the fly's visual system to map visual stimulation into motor commands of a wheeled robot. The authors reproduced a horizontal slice of the fly's visual system with 100 EMDs distributed around the circular body of the robot. The functional diagram of each EMD, shown at the right of figure 6.14, was established on the basis of electrophysiological analyses carried out on the insect and reproduced with analog electronics. An EMD collects signals from neighboring receptors with asymmetric connection lines and a temporal delay. The EMD is active if the sequential stimulation of the receptors occurs in one direction, but not in the other direction. The output of the EMD is proportional to the angular velocity of the object on the receptor surface. If the angle between adjacent EMDs and the translation velocity of the animal is known, the distance to the object can be recovered. The EMD therefore measures distance to objects according to the principles of motion parallax (Whiteside and Samuel 1970).

The behavior of the wheeled robot consisted of a series of translational steps of fixed size during which the EMD signals were collected. At the end of each step, the pattern of EMD activations provided a "snap map" of sur-

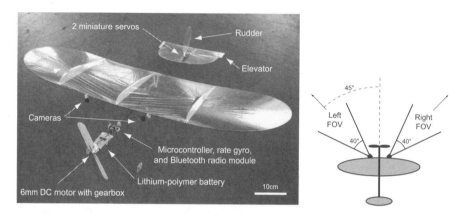

Figure 6.15   A 30 g aircraft for autonomous indoor flight. The signals from two linear cameras, pointing sideways, are used to detect the magnitude of optic flow and decide when to initiate a saccadic turn. An MEMS gyroscope provides information for course stabilization during straight trajectory. In this version altitude was manually controlled.

rounding obstacles in polar coordinates. A steering command was issued from this snap map in order to avoid near obstacles and move in the direction of a target location always visible from a sensor beacon positioned high above the robot center. The 10 kg robot was capable of slaloming through a field of poles at a speed of 50 cm/s toward a distant target location. It did not have a plan or a global map of the environment, but relied entirely on reactive behaviors resulting from a combination of attraction to the target and repulsion from obstacles.

INDOOR FLYING ROBOT   Zufferey and Floreano (2006) developed an indoor flying robot that relied on optic flow to steer away from walls (figure 6.15). In order to cope with the stringent weight constraints of the 30 g platform, they used two linear cameras pointing at $45°$ from the center. The signal of each camera was processed within an onboard microcontroller to extract the optic flow magnitude in each direction by means of an image-interpolation algorithm (Srinivasan 1994). Instead of extracting distance information to decide where to move, as Franceschini's robot did, the flying robot initiated a saccadic turn when the combination of the two optic flow signals exceeded a predefined threshold. The threshold value was chosen to leave sufficient time to the aircraft to avoid a collision. The saccadic turn consists of a deflection of the rudder for 1 s.

The flying platform posed an additional problem with respect to the wheeled robot used by Franceschini: how to maintain straight trajectories and how to recover from a saccadic turn. Since insects face the same problem, it is worth learning how they solve it. Flies possess mechanosensory structures, called halteres, that detect rotations of the body, allowing them to maintain equilibrium in flight (Dickinson 1999). Halteres have also been shown to play an important role in gaze stabilization (Nalbach and Hengstenberg 1994), which may serve to cancel residual components of rotational optic flow due to turbulence while flying on a straight trajectory.

These biological sensors are analogous to microelectromechanical, piezoelectric, rate gyroscopes, which sense Coriolis forces that act on oscillating mechanical parts. Zufferey and Floreano (2006) used the rotation information from a piezoelectric gyroscope to maintain straight trajectory with a proportional feedback controller acting on the rudder and to prevent the onset of additional saccadic turns while the aircraft was recovering from a turn, a strategy that may also be used by insects (Tammero and Dickinson 2002).

Recently, we demonstrated this approach in a 10 g indoor flyer (Zufferey et al. 2006b) and proposed the use of an additional linear camera to maintain altitude by "avoiding" the floor (Beyeler et al. 2007).

6.4.2   Example 2: Wheeled Legs

Wheels allow robots to travel at very high speed, but can operate only on relatively flat terrain. Legs instead allow robots to move on very rugged surfaces, but are slower than wheels on flat terrain. In order to combine the best of both worlds, active wheels have been attached to the extremities of passive articulations (e.g., Estier et al. 2000), but this solution entails a significant cost in mechanical complexity and efficiency of the servomotors embedded in the wheels.

The remarkable mobility and speed displayed by cockroaches in extremely rugged terrains attracted the attention of researchers because it seems that the animal combines a simple, preprogrammed, feedforward control with dynamic stability provided by a compliant leg system (Full and Koditschek 1999). In other words, the animal bounces and collides frequently with the ground and obstacles, but the mechanical properties of its legs coupled with its characteristic gait allow the animal to automatically recover without explicit control. Cockroaches have six legs and display tripod gaits: the front and rear leg on one side move in phase with the central leg on the other side. The remaining three legs display the same pattern shifted by $180°$.

Figure 6.16   RHex is a simple hexapod robot capable of tripod gait and navigation on rough terrains (www.rhex.org).

Several robots with both rigid and elastic legs capable of tripod gait have been developed (Ayers et al. 2002), but most robots use one or two motors per leg (to produce swing and stance), which results in relatively heavy and fragile mechanics. Alternatively, they use pneumatic actuators, which are more robust and agile, but require an external air compressor.

RHEX   RHex (figure 6.16) instead is a hexapod robot that captures the flexibility of the cockroach with simple mechanics and control (Saranli et al. 2001). It has six arched legs rotating around their horizontal axles. Each leg is driven by an electric motor that accelerates the swing phase to produce the tripod gait. The legs are made of elastic material that complies with the ground and the central leg is mounted further out to leave space for the rotation of the other legs. The robot is capable of navigating faster than one body length per second in rough terrains without feedback control and, because of its symmetric design, can keep going even when it rolls over on its back.

Quinn et al. (2002b) designed another hexapod robot that is even simpler, but still captures the control and mechanical properties of cockroach gait. Its locomotion system is based on a combination of wheels and legs (figure 6.17),
WHEGS   and thus has been named *Wheg*. Whegs are rimless wheels with three equally spaced spokes. This configuration allows a spoke to get a foothold on an obstacle that is higher than the length of the spoke. The authors showed that three spokes, as compared to two or four, are optimal in achieving a compro-

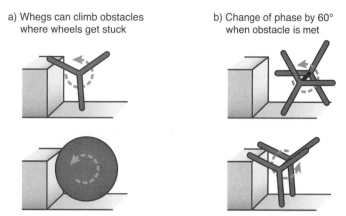

Figure 6.17 Whegs consist of a rimless wheel that can climb obstacles higher than a spoke and of a passive torsional mechanism within the axle that enables the Wheg to a change of phase by as much as 60° when an obstacle is met. A spring within the torsional mechanism brings the wheels out of phase once the obstacle is passed. Figures adapted from Quinn et al. (2002b).

mise between climbing abilities and riding smoothness. For a three-spoke Wheg, the axle travels vertically 13% of the spoke length when the vehicle is moving in a tripod gait on a flat terrain. Smaller vertical displacement can be achieved by manufacturing the spokes out of elastic materials. The position of the front Whegs is as much forward as possible in order to allow the spokes to contact the obstacles before the body of the robot.

When cockroaches meet obstacles, sometimes their front legs move in phase over the top of the obstacle, which facilitates the climbing. Similarly, Whegs include a spring in the drive train enabling them to passively change their relative phase by as much as 60°. When one Wheg hits an obstacle but cannot place its endpoint on top of it, both Whegs will continue to rotate, advancing the robot further until the Wheg on the opposite side reaches the top of the obstacle. At this point, the torsional compliance will keep the spoke in place while the Wheg on the other side will reach the top of the obstacle too. Once the obstacle is passed, the springs drive the two opposing Whegs out of phase for the usual tripod gate.

The authors have designed and tested a variety of Wheg robots. In all cases, a single drive motor is used to rotate all Whegs. This design allows for a more powerful motor and overall lighter weight as compared to robots where each wheel is controlled by a dedicated motor, such as RHex where

Figure 6.18   Miniwhegs feature four Whegs and employ alternating diagonal gait to cover 10 body lengths per second. Images courtesy of Roger Quinn and Andrew Horchler, Case Western Reserve University, Cleveland.

each motor must have sufficient torque to drive the robot out of trouble and is therefore relatively heavy.

Whegs-1 featured six Whegs, measured 50 cm, and moved at three body lengths per second through a thick lawn with tripod gait (Quinn et al. 2002b). The front and rear Whegs could be steered sideways for turning with the aid of two small servomotors, which is equivalent to the way in which cockroaches turn by changing the position of their legs. The robot could keep running even when it rolled over on the dorsal side.

Miniwhegs (figure 6.18) had only four Whegs, measured 9 cm, and moved at 10 body lengths per second with an alternating diagonal gait (Morrey et al. 2003). The size reduction implied a number of modifications. Whegs no longer had a sharp tipped foot, which in larger prototypes provided good traction, because at this smaller size it made the robot jump when the tip got stuck in the ground. Sharp tips were therefore replaced by small feet. The functionality of the passive torsional mechanism within the axle of the larger Whegs was obtained by fabricating the spokes out of compliant material.

6.4.3   Example 3: Wall Climbing

Robots capable of walking on vertical surfaces would in general be very useful for surveillance, inspection, and cleaning. Several robot prototypes have been built that use vacuum suction, magnetic attraction, or grasping. However, all these solutions impose strong constraints. For example, vacuum suction works only on flat surfaces and cannot be used for space applica-

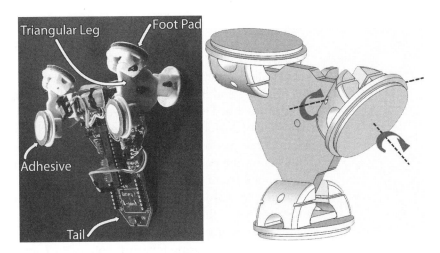

Figure 6.19 Waalbot and detail of the passive joints of a footpad. Images courtesy of Metin Sitti, Carnegie Mellon University, Pittsburgh.

tions; magnetic attraction requires the presence of ferromagnetic surfaces; and grasping does not work on flat surfaces.

Ideally, one would want a clinging mechanism that works on any material without damaging the surface. Geckos seem to have evolved a solution to that problem. Each toe of gecko feet is covered with hundreds of thousands of filaments, known as *setae*, and each seta divides into hundreds of smaller filaments, known as *spatulae*, that have a diameter of approximately 0.2 μm. This structural design allows toes to adapt perfectly to the microstructure of any type of surface and exploit van der Waals forces of weak electrical attraction between molecules in order to stay attached (Autumn et al. 2002). Therefore, they operate both in wet and in dry environments (that's why it is said that they use the principle of *dry adhesion*), do not require energy to attach or cling to any material, and can also work in space where suction-based solutions are not usable. What matters is the size and density of the spatulae, not the material of which they are composed. Autumn et al. (2002) showed almost identical sticking force of microfabricated artificial setae made of two different materials that respected the dimensions and density predicted by mathematical analysis of gecko toes.

While the synthesis of dry adhesives is under way in several academic and industrial labs worldwide, a team led by Metin Sitti at Carnegie Mellon University in Pittsburgh developed a robotic platform that could exploit the

DRY ADHESION

sticking properties of dry adhesives. In particular, the adhesion system requires a method for conforming the adhesive foot to the wall surface and a
WAALBOT method to peel off the foot. Waalbot (figure 6.19) is a 100 g and 13 cm long robot with two three-footed wheels, each driven by a dedicated motor, and a tail (Murphy and Sitti 2007). Although Waalbot's wheels resemble the mechanisms described earlier for RHeX and Whegs, their specificity resides in the twin joint of the foot that makes them particularly suitable for climbing. Each footpad can passively rotate along the axis of the ankle as well as laterally. This allows the pads to conform to the surface as the robot moves forward and enables lateral turns.

During forward motion, the two legs are synchronized. As the motor turns, the tail of the robot presses against the surface and the legs rotate forward. Two footpads, one on each side, are sufficient to hold the robot on vertical surfaces. During forward motion, the compliance of the pads brings four feet temporarily in contact with the surface. As the motor turns, the rear feet are pulled off the surface. The robot turns by advancing one motor at a time while the footpad on the other side rotates around the laterally passive joint. The length of the tail, designed to fit within the rotation circle of the robot, strikes a compromise between the need for long length for reducing peeling force and for short length for reducing weight and the turning angle in narrow passages. The design of the legs allows a smooth walking transition between surfaces at $90°$, e.g., between floor and wall.

A team led by Mark Cutkosky at Stanford University took a different approach that emphasized the morphological and dynamical properties of the gecko's setae. The animal uses two mechanisms to detach: on the microscale it modifies the angle of the setae with the surface and at the macroscale it hyperextends the digits. Although the combination of these two mechanisms may lead one to think that the gecko literally peels off its setae one by one from the surface, recent findings (Autumn et al. 2006) suggest that the the adhesion of the setae is dominated by shear force. The anisotropic design of the spatulae exploits shear forces when the animal is attached vertically on a wall. Instead, when the animal is attached to a ceiling, it must pull together its opposing toes to increase shear forces. Therefore, the hyperextension of the digits is helpful mainly for decreasing shear forces on vertical surfaces.

STICKYBOT Stickybot (figure 6.20) is morphologically similar to a gecko and features four legs and a long tail to decrease the distance between its center of mass and the surface. Each leg has three motors: two for controlling swing and stance phases, and one for hyperextension of the toes. The leg segments are slightly elastic to offer better compliance. The artificial setae are anisotropic

Figure 6.20 Stickybot with detail of the retractable footpads and of the anisotropic artificial setae. Images courtesy of Mark R. Cutkosky, Stanford University, Stanford, CA.

to better exploit shear forces (Park et al. 2007; Santos et al. 2007), and the control system is based on distributed active force control (Kim et al. 2007b). Stickybot is capable of walking on perfectly flat surfaces and moves by lifting and repositioning one leg at a time.

As emphasized in (Murphy and Sitti 2007), robots exploiting dry adhesives will benefit from miniaturization because the mass of the robot is proportional to the cube of its dimension while the adhesive force is proportional to the square of its contact surface. Therefore, as the robot shrinks, the mass decreases faster than the adhesive force.

6.4.4 ● Example 4: Humanoid Robots

For many people, humanoids represent possibly the ultimate goal of bioinspired robotics. Although this view is not shared by all scientists, anthro-

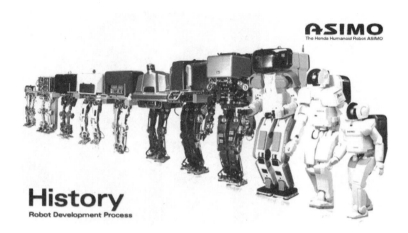

Figure 6.21   Development of Honda humanoid robot from 1987 to 2007. Image courtesy of Honda Motors Co.

pomorphic design may be useful for robots that are expected to operate in environments built for human size and dexterity, such as offices and homes. In addition to that, humans may find it easier to interact with a machine that shares the same physical configuration and gesticularity. Humanoid robots may also be important for investigating biological models that are specific to human-like behavior, such as manipulation and transportation of objects, where computer simulations would be unfeasible or comparatively ineffective.

WABOT   The first humanoid robot, WABOT-1 (*Wa*seda ro*bot* No. 1), was developed in Japan at Waseda University in 1972 (I. Kato et al. 1972). This robot was essentially a biped with a torso and two arms for balancing the body during walk. The second-generation WABOT-2 was capable of playing a piano and performed musical pieces every day during the Tsukuba Science Exhibition in 1985. WABOT-2 was the size of a human, had a single camera mounted on the head, and two arms with articulated fingers (Sugano and Kato 1987). Waseda University continued to produce a range of new humanoid robots over the years, the most recent one being the WABIAN (*Wa*seda *bi*pedal hum*a*noid), a human-size robot with stereo color vision, speaker detection, onboard computation, two arms with detailed hands, and two articulated legs for autonomous walk (Hashimoto et al. 2002).

In 1986 Honda started research on humanoid robotics – although the project was unveiled years later (Hirai et al. 1998) – with the ultimate goal of producing a domestic robot capable of interacting with humans (figure 6.21). The first prototypes of Honda robots were focused on understanding the design and control of bipedal walking. The first unveiled prototype, P2, was a humanoid robot with a height of 182 cm, a weight of 210 kg, two arms, two legs, a head with a camera, and a backpack containing four computers and nickel-zinc batteries for autonomous operation. The robot was capable of opening doors and walking on stairs. The most recent version, Asimo (*a*dvanced *s*tep in *i*nnovative *mo*bility) has the size of a 10-year-old child and perfected mechanics and control. The reduced size makes humans feel more comfortable with interacting with the robot. The robot is sufficiently reliable to allow the company to make it available for rental. Asimo is currently capable of recognizing and autonomously walking toward objects and humans by means of a stereo vision system that emulates human eyes with foveal and peripheral vision (implemented as a double-camera system with narrow and wide fields of view). The robot can also interact to some extent with humans by means of audio processing and speech synthesis. Honda is currently investing research efforts in brain architecture, neural computing, and bioinspired control for adding learning abilities to its robot.

ASIMO

Both the Waseda and Honda robots are made of stiff metallic and plastic structures actuated by electric servomotors. Gait control of biped robots is challenging because the robots tend to tip over when they move (as contrasted to robots with four or more legs that maintain static equilibrium throughout the entire range of motion). Most humanoids, including the Waseda and Honda robots, use active control to maintain the so-called zero-moment point within the support area of the robot. The zero-moment point, first suggested in 1968, is the projection on the ground of the point corresponding to the position where the tipping momentum of the robot is zero (Vukobratovic and Borovac 2004). The control system continuously monitors the error between the desired and the current position of the zero-moment point and generates joint torques that reduce this error by means of a set of differential equations.

ZERO-MOMENT POINT

In recent years, there has been renewed interest in "passive walkers" where the locomotion pattern is generated by gravity and inertia, instead of active control of the joints. McGeer (1990b) provided an analytical study of several passive walkers, ranging from small wooden toys with two rigid legs connected by an axle, to more elaborated structures with articulated knees. When a passive walker is positioned on an inclined surface, it produces a bipedal walking pattern by passively swinging the two legs back and forth in alternation under the combined effects of gravity and inertial forces.

PASSIVE WALKERS

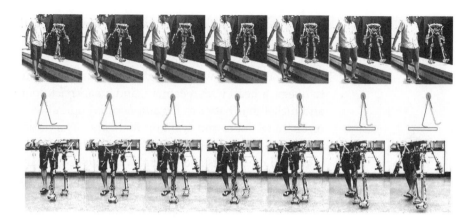

Figure 6.22   Two walk sequences of the Cornell ankle-powered biped on a level surface. Both the robot and the person are walking at about one step per second. From Collins et al. (2005). Reprinted with permission of the American Association for the Advancement of Science..

The problem of foot clearance from the ground when the leg swings forward can be solved in at least two ways. The wooden toys have laterally arched feet that make the toy rock sideways during the swing phase, thus giving sufficient clearance to the other leg. However, the rocking movement dissipates energy and makes the gait less efficient. An alternative solution consists in using articulated knees with a mechanical stop that prevents the hyper-extension of the leg during the forward swing (figure 6.22, center), as in humans. It has been shown that a passive walker with knees can produce efficient walking patterns when started with certain initial positions of the two legs (Mochon and McMahon 1980; McGeer 1990b). Furthermore, the addition of a hip joint with torsional springs, and that of linear springs on the legs can produce passive runners (McGeer 1990a). These developments indicate that the control of dynamic walk can be greatly simplified by taking into account the mechanical design instead of delegating the problem entirely to algorithms that continuously monitor and correct the motion of the joints.

Recently, Collins et al. (2005) extended the principle of passive walk to horizontal surfaces by adding a simple actuation system in place of the gravitational force and designed a series of machines, such as the Cornell biped shown in figure 6.22, which featured two actuated ankles that extended for pushoff when the opposite foot hit the ground, two passive knees with mechanical stops, a hip, and two arms mechanically linked to the opposite leg.

The authors showed that the efficiency (measured as used energy divided by weight of the machine times traveled distance) of these semipassive walkers was comparable to that of humans, whereas the efficiency of the Asimo robot was 10 times worse than that of humans.

The principle of exploiting, rather than controlling, the physical properties of the robot was also used in the design and control of the MIT humanoid

COG   robot COG (Brooks and Stein 1994; Brooks et al. 1998). COG is a human-size torso, with a head and two arms, that has been designed to study embodied cognition and interactions between humanoids and humans (Brooks et al. 1999). The joints of the arms include a novel type of elastic actuator incorporating a spring between the gears and the load point (Pratt and Williamson 1995), which absorbs shocks, makes the arm compliant, and allows stiffness regulation. Therefore, the articulations of COG arms are equivalent to masses connected by springs, just like human arms.

Williamson (1998) showed that this design could be used to perform several natural forms of repetitive motion with minimal control modulating the natural oscillation of the physical arm, instead of attempting to actively control its trajectory. Each joint was controlled by an oscillator built out of two spiking neurons with self- and mutual inhibitory connections where one neuron flexed the joint and the other extended it. Each neuron received the force or position of the joint. Once the joint moves at its natural oscillation frequency, only a small amount of energy is required to maintain it. It seems that also humans exploit natural frequencies of the arms to minimize the metabolic cost of movement (Williamson 1998).

Kuniyoshi et al. (2004) at Tokyo University showed an impressive case where both humans and humanoids partly delegated control to their bodies while performing complex actions. The authors studied the roll-and-rise sequence (figure 6.23) used to move from a lying position to a crouching position without the help of the hands. This motion is very dynamic, depends strongly on the physical properties of the body, and involves several contacts with the ground, all of which make it almost impossible to model and reproduce by means of differential equations. The authors noticed that human subjects asked to perform the movement followed trajectories that varied significantly except at a few points, such as the point where the ground contact

KNACKS   shifts from the hip to the feet. They called these points "knacks" to indicate that they are the critical parts of the behavioral sequence that must be performed precisely, while the other parts of the sequence can be left to the natural dynamics of the body shape, which may vary from person to person or robot. Consequently, the authors modeled and developed control strate-

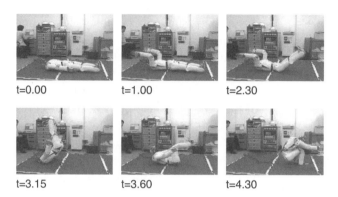

Figure 6.23 Roll-and-rise sequence performed by a humanoid robot that actively controls only a few critical points ("knacks") of the behavioral sequence and leaves the other parts to the natural dynamics of the body (Terada et al. 2003). Reproduced with permission of IEEE Press, Piscataway, NJ.

gies only for those knacks and tested them on an adult-size humanoid robot with padded torso and limbs. The robot performed the entire sequence of actions in less than four seconds with a 100% success rate (figure 6.23). In some cases the robot used its hands to prevent falling forward and to return to the crouching position.

MORPHOLOGICAL
COMPUTATION

Passive walkers, COG arms, and roll-and-rise are examples of what today goes under the name of *morphological computation*, that is, processes performed by a physical body rather than by a brain or a control system. The principle of morphological computation extends well beyond humanoid robotics to encompass the design of any type of embodied behavioral system. Interested readers may also consult (Hara and Pfeifer 2003) for a collection of sample research in this area. Pfeifer and Bongard (2007) suggested that morphologies and materials out of which artificial and biological bodies are made constrain and affect the behavioral and cognitive properties of the organism in important ways. We will come back to this issue, which is very distinctive of embodied and behavioral systems, in later sections of this chapter.

Several research efforts are also being put into principles and technologies necessary to support natural and engaging interactions between humanoids and humans. Although that may seem a daunting task, it turns out that the problem is simplified by our innate tendency to attribute complex emotional and social features even to simple animate objects. For example, Braiten-

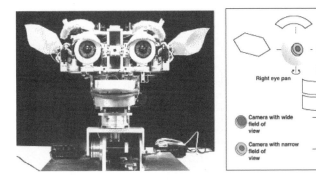

Figure 6.24 Kismet is an anthropomorphic robot head. Three degrees of freedom direct the robot's gaze, another three control the orientation of its head, and the remaining fifteen move its facial features (e.g., eyelids, eyebrows, lips, and ears). Kismet is equipped with four color cameras (two for foveal vision and two for peripheral vision). Reproduced with permission from Breazeal (2003). Courtesy of Elsevier, Amsterdam.

berg's vehicles described earlier in this chapter may express fear, love, or curiosity in the eyes of an observer.

Robots designed for social interaction display several levels of sophistications. *Socially evocative* devices, such as the Furby toy, give the appearance of social states, but do not react to human social expressions. *Androids* are high-fidelity replicas of human faces and body parts with biomimetic skin and realistic expressions (Hara et al. 2001). They may be used to convey speech and emotions from a distant person or in exhibitions and movies, but they do not have internal states and autonomy. However, android technologies have been recently combined with humanoid technologies and artificial intelligence to improve humanoid-human interaction (Hanson 2005; Jun-Ho et al. 2006). *Sociable robots* can express social and emotional states, can perceive and engage in human social interactions, and have internal states and learning that allow them to influence the social interaction with humans. Fong et al. (2003) present a survey of socially interactive robots from the perspective of the technologies and intelligence required by these robots.

Although we do not yet have truly sociable robots, some recent robots, such as Kismet (figure 6.24), can engage a person in natural interactions for several minutes. Kismet is a robotic face that can interact physically, affectively, and socially with humans in order to ultimately learn from them and elicit interactions that improve the learning potential (Breazeal 2002a). It is

SOCIALLY EVOCATIVE
ANDROIDS

SOCIABLE ROBOTS

KISMET

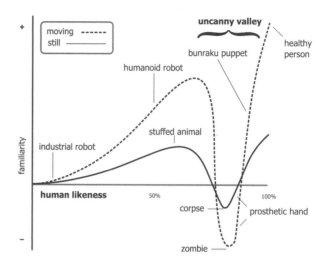

Figure 6.25 The uncanny valley suggested by Mori (1970). Familiarity increases with human likeness until a point is reached at which subtle differences in appearance and behavior create an unpleasant feeling. Image courtesy of Karl F. MacDorman, Indiana University, Bloomington.

equipped with a behavior-based architecture that allows prompt and partly unpredictable, but coherent, responses. The robot can direct its attention to establish a shared reference with a human, give expressive feedback and recognize emotional states from voice features, and take turns to structure learning events (Breazeal 2002a). Kismet takes a proactive role in regulating the interaction with humans so that it is neither overwhelmed nor understimulated. That strategy also facilitates gradual learning.

Although Kismet does not have a full understanding of human language, that does not seem to bother people who engage in conversations that adapt to the robot's feature, e.g., by speaking more slowly, waiting longer for a response, and checking for cues from the robot. As Breazeal (2003: pp. 173-174) puts it, "[S]ocial interaction is not just a scheduled exchange of content, it is a fluid dance between the participants. [...] In short, to offer a high quality (i.e., compelling and engaging) interaction with humans, it is important that the robot not only do the right thing, but also at the right time and in the right manner."

Kismet is not an android in the sense that it does not aim at being indistinguishable from a human face or display exactly the same expressions. This may be one of the reasons why humans like to interact with it. The Japanese robotics researcher Mori (1970) suggested that although a closer resemblance of robots to humans triggers better social interactions, very human-like robots may look "uncanny" and cause unpleasant feelings (figure 6.25). However, positive feelings will appear again if the resemblance is improved. Furthermore, he argued that these reactions, both positive and negative, are amplified if the robot is moving. It has been suggested that the *uncanny valley* is caused by small imperfections that make the android look human but not alive; consequently, the perception of death, or of the living dead if the robot moves, may trigger unpleasant feelings (MacDorman 2005). Although this phenomenon requires further scientific investigation, the film industry seems to be taking it seriously after the negative reactions of some viewers to the animated baby in the movie *Tin Toy*.

UNCANNY VALLEY

6.5 Robots as Biological Models

Robots can also be used as models to investigate biological questions and test hypotheses. As we mentioned in the historical introduction to this chapter, robots have been gradually replacing computers as the preferred tool and metaphor in embodied cognitive science. Robots are becoming increasingly accepted also among experimental biologists and neuroscientists as tools to validate their models.

A robot that operates in the real world constrains the choice of models among all those that could be possibly constructed (Webb 2002). Robotic implementations force scientists to be very concrete in specifying the design of the complete biological system (Franceschini et al. 1992) and help to produce testable hypotheses (Webb 2001). Robot models also allow scientists to study how the interaction with the environment and the operation of the neural system affect each other, whereas disembodied computer studies are necessarily limited in their front end and ability to modify the environment. If the agent-environment interaction is very complex (e.g., flow dynamics or complex sensory input and mechanical interaction), simulations may introduce simplifications that could severely misguide the development of the models (Möller 2000). A realistic interaction with the environment can also provide new data and modification of classic perception-action models (Chiel and Beer 1997; Suzuki et al. 2005). Furthermore, the evolution of robots may allow the testing of evolutionary hypotheses for which there is no fossil record (Nolfi and Floreano 2000). We will describe two such cases where robots are used to test evolutionary models, namely competitive and cooperative coevolution, in chapter 7.

Webb (2001) identified seven dimensions along which robotic models of biology differ: (1) relevance: whether the model tests and generates hypotheses applicable to biology; (2) level: the elemental units of the model in the hierarchy from atoms to societies; (3) generality: the range of biological systems the model can represent; (4) abstraction: the complexity, relative to the target, or amount of detail included in the model; (5) structural accuracy: how well the model represents the actual mechanisms underlying the behavior; (6) performance match: to what extent the model behavior matches the target behavior; (7) medium: the physical basis by which the model is implemented. In particular, she argued that in building robot models biological relevance is more effective than loose biological inspiration and that a physical medium can have significant advantages.

In this section, we will describe three case studies that, according to Webb's classification, have high biological relevance, operate at the level of the individual, explain several biological phenomena, abstract key variables necessary for answering the specific questions, attempt to capture the actual mechanisms that generate corresponding behaviors in biology, match very well the performance of the biological system, and rely on physical robots. Therefore, these three case studies represent high-quality examples of robots as biological models.

6.5.1 Example 1: Song Recognition and Localization

Let us begin with an example drawn from Webb's research where she used a mobile robot to understand how female crickets recognize males' songs and approach them (Webb and Scutt 2000). Crickets have an ear in each foreleg that produces direction-dependent differences in response amplitude. Therefore, one may think that the female simply turns to the side with the strongest response to go toward the source of the male calling song. However, that is not sufficient to explain how females can selectively approach only males of the same species, which emit a characteristic song. A cricket song is composed of short bursts, known as syllables, in a pure tone between 4 and 5 kHz. Syllables are grouped in chirps. Experimental evidence showed that the syllable repetition interval (SRI) is the most important cue for discriminating different songs.

SYLLABLE REPETITION
INTERVAL (SRI)

The biological models available in the literature assume that the cricket nervous system includes both a song recognition system to detect songs with an SRI characteristic of the same species and a localization system to approach the source of the song. Webb made the hypothesis that a simpler model, which does not differentiate between recognition and localization, could explain the selective behavior of female crickets. She used exactly the same experimental method used by biologists for testing the robot model and exposed it to songs recorded from living crickets.

A number of neurons have been identified in the neural circuitry that is responsible for cricket phonotaxis. Some of these include pairs of auditory neurons that receive excitatory input from the ear ipsilateral to their dendritic tree; pairs of omega neurons that in addition to ipsilateral input excitation receive inhibition from the controlateral omega neuron; and brain neurons that receive connections from auditory neurons. It is not known whether the firing rate or the firing time of these neurons matters in producing the phonotactic behavior, but most available models seem to discard firing time. Also, it is not clear how the specific syllable rate (SRI) is recognized because auditory neuron pairs do not display selectivity for temporal patterns.

Webb argued that the firing time of the auditory neurons is sufficient both for producing movement toward the song source and for selecting the appropriate song. Assuming that the relative onset of spiking activity in the left and right auditory neurons is the important factor in deciding in which direction to turn, the female cricket will move only toward songs with a suitable syllable repetition interval because (1) songs with intervals that are too short will excite almost continuously the auditory neurons, making it almost impossible to tell which one fired first; and (2) songs with intervals that are too long may not be sufficient to fire postsynaptic neurons or will cause very few turning behaviors and result in poor tracking of the song source. According to this model, the sensitivity to the species-specific syllable repetition interval depends on specific temporal parameters of the neural circuitry.

The neural model (figure 6.26, left) consisted of only four spiking neurons analogous to the integrate-and-fire neurons with a refractory period described in chapter 3. The four neurons were intended to capture the membrane dynamics of the biological auditory neurons and of the brain neurons, which in the model served as motor neurons issuing motor commands to the robot. The reasons for using such a simple model were that (a) these neurons were considered to be essential to the phonotactic behavior; and (b) it is better to start with a minimal set of elements and add further elements at a later stage only if required to explain more biological phenomena (see also the methodology in behavior-based robotics). The excitatory synaptic weights between auditory and motor neurons halved in value after every signaling

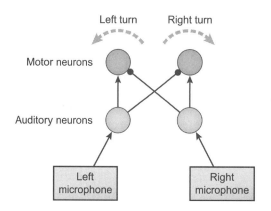

Figure 6.26 *Left*: Neural model comprising two auditory and two motor spiking neurons, dynamic excitatory synapses (arrowhead), and inhibitory synapses (ball-head). *Right*: Khepera robot with electronic auditory circuit comprising two microphones.

event, but recovered exponentially to the original value. The synaptic depression was implemented by means of an inhibitory synapse-to-synapse connection from auditory neurons, but they could have been implemented in a variety of other ways. Consequently, a motor neuron will increase its membrane potential only if the ipsilateral auditory neuron has fired before the contralateral auditory neuron.

The neural network was implemented within the processor of a Khepera robot so that a cycle of neural activation lasted approximately 1 ms. The robot (figure 6.26, right) was equipped with a purpose-built sound-processing circuit designed to mimic the properties of the two cricket ears (Lund et al. 1998). The output of the two artificial ears were scaled and passed to the auditory neurons. Whenever a motor neuron emitted a spike, the robot would turn in place at the same speed and angle measured in crickets. In order to replicate exactly the biological experiments where crickets are put on a treadmill and maintained at a constant distance from sound sources, the robot default movement was set to zero in the absence of turning actions.

The robot was exposed to songs recorded from real crickets exactly in the same conditions used in biological experiments. In all conditions, the robot displayed exactly the same behavior observed in animals. Specifically, it turned toward, and tracked precisely, the source of conspecific songs, but not a continuous tone on the same frequency; it displayed selective sensitivity for a narrow band of syllable repetition interval with a sudden drop in behavioral response for slow songs; it was capable of initiating turning behaviors also when the sound was emitted from above its head at the same distance from both ears (the turning direction being determined by ambient and circuit noise); it avoided a continuous tone when a concurrent conspecific song was played above its head; when presented with two simultaneous songs from different species, it tracked precisely the source of the conspecific song; and when presented with two simultaneous versions of the same conspecific song, it turned toward the song of better quality.

These results suggest that the firing time of auditory neurons coupled with dynamic synaptic connections is sufficient to explain recognition and localization of conspecific male songs as well as a number of other phonotactic phenomena displayed by crickets. Although higher biological realism could be achieved by incorporating other neurons and connections, the model would require the tuning of more parameters and make analysis much more complicated without necessarily better clarifying the original question of how crickets recognize and locate certain songs.

6.5.2   Example 2: Vision-based Homing

Several insects display remarkable abilities to return to a nest or to a food location after traveling for hundreds of meters and for extended periods of time. Social insects, such as ants and honeybees, do so by relying on at least two strategies: path integration and visual piloting (Wehner et al. 1996). In desert ants, path integration seems to be achieved by integration of a direction vector computed from the orientation of polarized light in the sky (Wehner 1997) and of a distance measure estimated by step count (Wittlinger et al. 2006). Path integration is also a frequently used method in robotics for localization and homing. In robots, the polarization filters are often replaced by a magnetic compass and the step-counting mechanism is replaced by wheel odometers. However, path integration is prone to errors for both ants and robots.

When the desert ant *Cataglyphis* approaches the vicinity of the target location, it seems to use visual cues for compensating path integration errors. Since very little is known about the neural circuitry that regulates this behavior, all models of visual homing have been derived from experimental observations of animal behavior. A widely accepted model, known as the *snapshot model*, postulates that the animal memorizes an image (snapshot) of the landscape before leaving the home location and later uses the comparison be-

SNAPSHOT MODEL

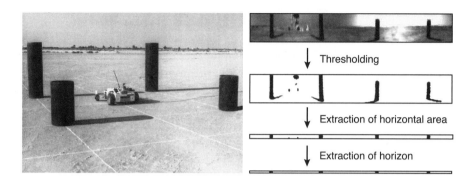

Figure 6.27 *Left*: The Sahabot robot with four sample landmarks in the desert. *Right*: Extraction of snapshot from omnidirectional camera positioned on top of the robot. Images courtesy of Rolf Pfeifer, University of Zurich.

tween this stored image and images taken at its current location to compute a difference vector that points toward the home location (Wehner and Räber 1979; Cartwright and Collett 1983). The snapshot model suggests that ants compare the retinal positions and sizes of distinctive landmarks that appear in the stored snapshot and in the current snapshot. As the animal moves, the retinal size and position of landmarks change. The comparison between the two snapshots requires the alignment of the current snapshot with the stored snapshot, presumably using compass information. Since these insects have almost 360° vision, a snapshot can be considered as a ring centered on the animal that may include several landmarks in its visible surroundings. For each landmark comparison, two unit vectors are generated: a vector tangential to the ring and pointing toward the current retinal position of the landmark, and a centrifugal, or centripetal, vector if the size in the current snapshot is smaller, or larger, respectively, than that in the stored snapshot. The weighted sum of all the pairs of vectors for all landmarks results in a single vector, the homing vector. The animal continuously compares the current view with the stored snapshot to update the homing vector.

Despite its elegance and experimental support, the snapshot model assumes that the nervous system of the insect is capable of memorizing an entire image and performing several complicated comparisons at multiple points during its homing trajectory. Also, the snapshot model does not seem to explain experimental results where the search area of bees was shifted when a portion of a circular array of landmarks was removed between the

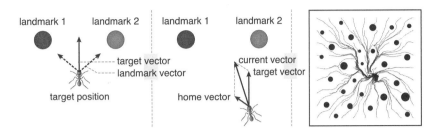

Figure 6.28 *Left*: Average landmark vector model. *Right*: Homing trajectories (image courtesy of Rolf Pfeifer, University of Zurich).

outgoing and return flight, which suggested that such insects rely on the overall configuration of their visual surroundings, rather than on individual landmarks (Anderson 1977). Lambrinos et al. (2000) and Möller (2000) suggested a more parsimonious version of the original snapshot model, known as the *average landmark vector* (ALV) model, and compared its performance to that of the snapshot model in mobile robots (figure 6.27) exposed to situations similar to those faced by insects engaged in visual homing.

In the ALV model (figure 6.28), each visual landmark is associated to a unit vector, known as the landmark vector, that points from the current position of the agent to the landmark. When the animal is at the target location, all landmark vectors are averaged to produce the average landmark vector, which is stored in the memory of the animal ($\vec{v}^t$). When the animal returns to the vicinity of the target location and switches on the visual homing mechanism, it computes in the same way a new average landmark vector at its current location ($\vec{v}^c$) and compares it with the stored average landmark vector. The difference between the two vectors produces the home vector $\vec{v}^h = \vec{v}^c - \vec{v}^t$, which the animal follows with a fixed-length displacement. A new home vector is computed after every displacement until the animal reaches the target location. As in the original snapshot model, the vectors must be aligned with the help of a compass before the comparison. In the robot model, the image is taken from an omnidirectional camera and processed to extract the landmarks (figure 6.27, right).

The ALV model has been tested in real and simulated robots with a varying number of landmarks and compared to the original snapshot model. In all cases, the model led the robot to the target location (figure 6.28, right) and produced trajectories that are comparable, although not exactly similar, to those produced by the original snapshot model. The ALV, as the original

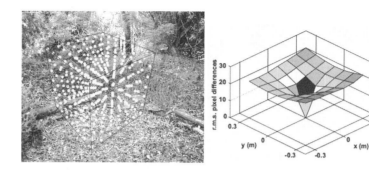

Figure 6.29 Catchment area of panoramic snapshot. *Left*: Position of snapshots. *Right*: Gradient surface of catchment area along x and y for constant z (a similar gradient occurs for z, picture not shown). Images courtesy of Jochen Zeil, Australian National University in Canberra.

snapshot model, can lead the agent toward the target even if some landmarks partially cover each other or if some landmark is missing. Furthermore, the ALV can reproduce the shift of search area when a portion of the landmarks is removed between the outward and inward journeys of the robot, which the original snapshot model could not reproduce (Möller 2000). The ALV makes many fewer assumptions about the computational and memory load of the animal and has also been implemented in analog electronics with operations that may be performed by biological neurons (Möller 2000). Finally, the ALV model resembles another model that has been independently formulated for explaining the visual homing behavior of rats (O'Keefe 1991).

The maximum radius of the homing area around the target where both the ALV and the original snapshot model can operate depends on the resolution of the camera and on the size of the landmarks. One may imagine a simple modification of the model with storage of multiple target vectors that connect multiple areas and operate as route landmarks (Möller 2000). However, it is not clear how the performance of the ALV (and of the original snapshot model) with sparse and highly contrasted landmarks can extend to natural settings, such as a forest, where there are a very large number of potential candidate landmarks and where each of them has complex three-dimensional shapes.

Zeil et al. (2003) studied the information potentially available to a flying insect by means of an omnidirectional camera attached to a robot arm positioned in the middle of a forest to record snapshots at regular positions along the x, y, and z axes (figure 6.29, left). They compared the images by computing the root mean square (RMS) pixel differences between images $\vec{I}_p^c$ taken at positions p in the surroundings of the target location and the image $\vec{I}^t$ taken at the target location in order to produce the RMS surface $\vec{C}$:

$$c_p = \sqrt{\frac{\sum_{x=1}^{X} \sum_{y=1}^{Y} (\vec{I}_p^c(x,y) - \vec{I}^t(x,y))^2}{XY}}$$

where $\vec{I}(x,y)$ is the intensity of the image pixel x, y and X, Y are the dimensions of the image.

RMS ERROR SURFACE   The RMS error surface displays a smooth and monotonic gradient toward the target location. An insect located at any point on this area could move toward the target location by simply descending the gradient of the error surface. Zeil et al. (2003) suggested a simple strategy consisting in moving along the same trajectory as long as the RMS error difference between the current and previous image is negative and turning by $90°$ if the difference is positive. Zampoglou et al. (2006) suggested instead a strategy similar to the chemotaxis behavior of the worm *Caenorhabditis elegans*, for which the biological neural circuit has been elucidated. The size of the area with a gradient leading to the target location, known as the *catchment area*, depends mainly on the resolution of the vision system (in their experiments it was approximately 3 m). The surface of the catchment area displays strong variability in the presence of environmental movements (leaves blown by the wind, other animals, etc.) and changing lighting conditions during the day (shadows, intensities), but the overall gradient information is relatively stable.

Flying insects do not just take a snapshot when they leave a location they wish to return to. Instead, they perform an elaborate flight during which they turn toward the target location and move away from it backward in a series of increasing arcs, while pivoting at the target location (Zeil et al. 1996). Furthermore, when they return to the target location, they do not fly in a straight trajectory, but approach it in a series of sideways movements. Although these flights may be used for other purposes, such as landmark-background segregation and landmark centering, they may also provide the opportunity for memorizing the image of the target location when leaving and descending the gradient of the catchment area when returning.

Although it is not yet clear whether insects rely on feature-based strategies (snapshot or ALV model) or image-based strategies (RMS error surface) for visual homing, the robot models described in these studies have been crucial for the formulation, evaluation, and comparison of the models.

It should be noted that the navigation strategies displayed by these models do not necessarily produce the optimal path toward the target location. A number of other strategies, which are inspired by biological evidence but are not necessarily intended to test biological models, have been developed for the purpose of optimal navigation and used in robotics applications (e.g., Franz et al. 1998; Hong et al. 1992; Möller and Vardy 2006).

SIMULTANEOUS
LOCALIZATION AND
MAPPING (SLAM)

The models described here only tell the animal where to go, but not where it is in space. The knowledge of the agent's position is often requested in robot applications, such as guidance and exploration, and is the focus of a research area named *simultaneous localization and mapping* (SLAM). SLAM methods attempt to build and maintain a map while the robot explores the environment. These methods can generate topological maps or Cartesian maps. Topological maps represent the environment as a list of interconnected nodes and resemble the way in which rats represent space (Arleo et al. 1999). Cartesian maps instead represent the environment in metric coordinates, which, although it may not be biologically plausible, can be useful for human operators that interact with the robot (Dissanayake et al. 2001). Since each map has its advantages and limitations, the combination of the two approaches is currently a topic of interest (Tapus and Siegwart 2006).

6.5.3   Example 3: From Swimming to Walking

The transition from water to land, estimated to have happened 370 million years ago, is one of the most crucial events in the evolution of vertebrates, but the scientific evidence supporting that transition is sparse. Only recently the discovery of a fish fossil with a tetrapod morphology provided a strong link between swimming and walking vertebrates (Long et al. 2006). This fish lived in a semiaquatic environment and its four limbs could provide propulsion both in water and on land, like today's amphibians.

However, the transition between the architectures of the brain tissues responsible for controlling swimming and walking did not leave a fossil record. The salamander is a living amphibian that resembles most closely the first terrestrial tetrapods and is capable of switching between swimming and walking. The swimming pattern is similar to that of the lamprey, a primitive fish that swims by means of fast undulatory movements that travel along the body from head to tail. On the ground, the salamander switches to a slower tetrapod gait where diagonally opposed limbs are moved together while the body undulates with standing waves with nodes at the girdles (Ijspeert et al. 2007). Understanding how the salamander switches between the two modes

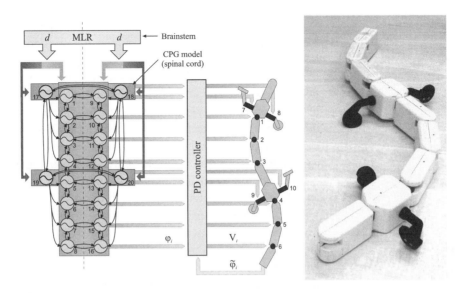

Figure 6.30   Neural control architecture of the robotic salamander. See main text for explanation of symbols. (Ijspeert et al. 2007). Images courtesy of Auke Jan Ijspeert, EPFL, Lausanne, Switzerland.

of locomotion may shed light on the evolutionary transition of locomotion control.

Ijspeert et al. (2007) developed a numerical model of the spinal cord and a robot model of the salamander (figure 6.30) to test the hypothesis that a primitive neural circuit for swimming could be easily extended to walking and to show that such a circuit could rapidly switch between swimming and locomotion as the salamander does. The locomotion of the salamander is controlled by a *central pattern generator* (CPG), which consists of a group of interconnected neurons that display rhythmic oscillations. These neurons oscillate faster for stronger input, but they stop oscillating when the input is larger than a saturation threshold. The salamander CPG is organized into two sub-CPG circuits (figure 6.30, left): the body CPG, which is responsible for the oscillations of the main body; and the limb CPG, which is responsible for the tetrapod walking gait. Both sub-CPG circuits display reciprocal connectivity between nearest-neighbor neurons. The entire CPG circuit, which is distributed along the spinal cord of the animal, receives input from the mesencephalic locomotor region (MLR), which is located in the midbrain.

CENTRAL PATTERN
GENERATOR (CPG)

Low levels of stimulation from the MLR region generate a walking gait that becomes faster as the level of stimulation increases. If the MLR stimulation increases over a certain threshold, the animal switches from walking to swimming, which also becomes faster as the level of stimulation continues to increase. The authors suggested that this phenomenon, which none of the previously developed models could account for, could be explained by assuming that (1) the neurons in the limb CPG oscillate more slowly and saturate at a lower threshold than the neurons in the body CPG; and (2) the connections from limb CPG to body CPG have stronger weights than the connections between body CPG neurons and between body CPG neurons and limb CPG neurons (not drawn in figure 6.30).

The robot (figure 6.30, right) was 85 cm long and designed to approximately match the structure of the salamander. It could produce lateral undulations of the spine with six separately actuated hinges and a tetrapod gait with four rotating limbs. The limb rotation principle was similar to that of the RHex (figure 6.16) and Miniwhegs (figure 6.18) robots described earlier. The outputs of the oscillatory neurons determined the desired angles of the motors that actuated the hinges and the limbs.

The numerical model reproduced the same pattern of neural activity observed in the salamander when increasing the stimulation from the MLR neurons. Low stimulation intensities activated the limb CPG, which dominated the body CPG; as the MLR stimulation increased, the limb oscillations became faster until the entire limb CPG was saturated and shut down. At that point, the body CPG could immediately take over and oscillate increasingly faster as the MRL stimulation increased. The activation of the limb CPG resulted in a tetrapod gait with a standing undulation of the body axis while the activation of the body CPG resulted in a traveling wave from head to tail that would propel the robot in water. Both types of locomotion were similar to those observed in the salamander (figure 6.31).

This model provided a parsimonious explanation of the ability to switch between two locomotion patterns displayed by the salamander. It also suggested that the evolutionary transition from swimming to walking may have happened in a few generations by replicating and adding to the body CPG a set of oscillatory neurons with small variations of their intrinsic and saturation frequencies. In addition, the model provided a number of predictions on the regulation of speed and turning behavior that remain to be tested with the animal. The use of a robot in this study was essential for the development of the model because it required a body and the measurement of the resulting behaviors.

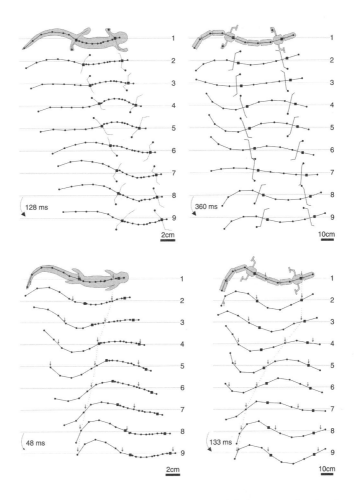

Figure 6.31   Walking (*top*) and swimming (*bottom*) behaviors of animal and robot (Ijspeert et al. 2007). Images courtesy of Auke Jan Ijspeert, EPFL, Lausanne, Switzerland.

6.6   Robot Learning

A key feature of biological organisms is the ability to continuously adapt and develop a range of novel abilities during their lifetime. It would be highly desirable that robots display similar abilities. The area of robotics research that studies the mechanisms of lifelong adaptation that are not entirely determined by genes is known as *epigenetic robotics*. *Developmental robotics* is

EPIGENETIC ROBOTICS

DEVELOPMENTAL
ROBOTICS

another name for indicating approximately the same area of research, but it puts special emphasis on the role of morphological change that occurs during the development of cognitive systems (see Lungarella et al. 2003 for a survey).

A developing biological or artificial organism is capable of finding solutions to multiple problems and executing multiple tasks. Developmental programs are therefore more general and, in principle, simpler than specific hand-designed programs or learning algorithms that take into account the specificities of the problem to be solved (Weng et al. 2001).

Robots represent ideal tools for studying biological development because physical embodiment and behavioral interaction critically affect the formation of the adult organism (Thelen and Smith 1994). At the same time, robotics can benefit from reproducing principles of biological development to endow robots with self-learning and social abilities. This is especially true for humanoid robots where the management of complex sensory inputs, actuators, and social interactions cannot be hand-coded or solved by problem-specific algorithms.

Learning in behavioral systems is a multifaceted phenomenon that includes perceptual self-organization, sensorimotor coordination, memorization, and association, to mention a few. In this section we will mention only three facets of learning in behavioral systems, namely value-based learning, learning with morphological change, and imitation learning. We selected these areas because they strongly rely on the two tenets of behavioral systems: situatedness and embodiment.

6.6.1 Value-based Learning

Robots, like newborn infants, should be able to develop behaviors that satisfy their needs and desires without an external teacher. In artificial intelligence the problem of associating sensory states with suitable actions that lead to the satisfaction of those needs or desires has been framed as the reinforcement learning problem, which has been mentioned in chapter 3.

In its classic and still dominant formulation (Sutton and Barto 1998), reinforcement learning is not easily applicable to real robots because it relies on the mathematical assumption that the problem can be described as that of finding the best reward-generating associations between discrete sensory states and discrete actions among all possible combinations of states and actions. This has two negative consequences for robotics. The first consequence is that learning requires a long and careful phase of exploration of several possible states and actions, which may not always be suitable for a physical robot. The second consequence is that the use of rich analog sensors (vision or touch) and of a real environment, which is essentially analog, implies a rapid combinatorial explosion of the couplings to be explored or some ad hoc solution to reduce the search space to a smaller set of discrete combinations, thus reducing the adaptive potential of the machine.

Classic reinforcement learning theory assumes that the environment provides explicit positive or negative rewards that are directly used by the algorithms to decide what should be learned. However, reward signals in the brain are mediated by neural structures with specific chemical, spatial, and temporal characteristics, whose computational role is not yet fully understood. In addition, humans and some other animals are also capable of internally generating rewards for achieving specific goals.

VALUE SYSTEM   The set of internally mediated and internally generated rewards represents the *value system* of the organism because it discriminates implicitly between good and bad. There is mounting evidence that this value system operates through specific neurons that use neuromodulatory transmitters, which act as global signals that affect the plasticity and response profile of several receiving neurons (see Doya et al. 2002 for a sample of computational models).

NEUROMODULATORY   The neuromodulatory system uses at least four types of neuromodulators
SYSTEM   (dopamine, serotonin, norepinephrine, and acetylcholine), but a single neuromodulatory neuron makes use of only one type of modulator with specific effects. Neuromodulatory neurons are activated by the occurrence of certain sensory events, such as an environmental reward or an unexpected stimulation or other types of salient cues; they display a short-lived activation; they project to large areas of cortical neurons. Furthermore, different neuromodulatory neurons may project to overlapping cortical areas and combine their effects in various ways.

Although it has been suggested (Doya 2002; Suri 2002) that the combined action of the four neuromodulators could provide the neurobiological basis for temporal difference reinforcement learning (see chapter 3), some authors argue that the global and short-lived signal of neuromodulatory neurons serves mainly as a "gating factor" of the onset, duration, and strength of Hebbian learning (e.g., Friston et al. 1994; Pfeifer and Scheier 1999). According to this perspective, the output of the value system V is incorporated in the Hebbian rule as a multiplicative factor so that the weight update of value-based learning takes the general form of

(6.1) $\Delta w_{ij} = v_i \eta x_j y_i$

Figure 6.32 A mobile robot learns to recognize cubes of a specific color and pick them up through a value-based learning architecture. Image courtesy of Olaf Sporns, Indiana University, Bloomington.

where v_i may incorporate a nonlinear function of the raw value signal. However, such a simple formulation of value-based learning cannot readily explain the establishment of associations between events that occurred prior to, and caused the triggering of, a value signal (which instead is explained by temporal difference methods) unless it is selectively used in neural architectures with appropriate modules and dynamics.

For example, Alexander and Sporns (2002) proposed a modular neural architecture for value-based formation of perceptual categories and behavior that approximates the anatomical and functional areas involved in the mammalian dopamine system (Schultz et al. 1997). The architecture included an array of input neurons that unfolded in space the time series of sensory activity (similar to the time delay neural networks described in chapter 3). It also included two value-dependent neurons, one signaling the occurrence of an unpredicted positive reward and the other signaling the lack of predicted reward. The neural model was embedded in a mobile robot with a gripper positioned in an arena with several blue and red cubes (figure 6.32). The robot could perceive the cubes with a color camera and "taste" them by means of resistivity sensors placed within the gripper. Red cubes had resistive surfaces and corresponded to food, which triggered a positive reward signal to the neural architecture. The robot was also equipped with a set of neurally prewired behaviors, such as approach, avoid, and grip.

After 10 minutes of operation, the robot avoided the blue cubes and selectively gripped only the red cubes. During learning the value-dependent neurons developed predictive abilities, that is, they became active in the unpredicted visual perception of a cube that was expected to lead to a positive reward, but not in the presence of the actual reward, as they did at the beginning of learning. Also, they became active when the robot expected a reward, but did not receive it. The development of these predictive properties, which has been documented in the mammalian dopamine system (Schultz 1998), allowed the robot to autonomously adapt if its actions were artificially slowed down, causing a delay between the expected reward from vision of a red cube and the actual reward from tasting it with the gripper. The significance of this work consists in showing that a simple and biologically plausible value-based system embedded in a suitable neural architecture can learn by delayed reward without explicitly resorting to temporal difference methods.

In a later section on evolution and learning, we will describe value-based neural architectures where learning will take place within a value system partly shaped by evolution.

6.6.2   Learning and Morphological Change

Living organisms undergo significant morphological and neural modifications throughout their entire maturation period. In humans, the first neural connections are established within a few weeks after conception while the body undergoes dramatic changes in shape, mass, and elasticity. Morphological and neural development progress in synchrony throughout the entire fetal stage and for several years after birth.

In chapter 4 we described various methods for describing and reproducing morphological growth and we highlighted the challenges of implementing them in physical systems. As a matter of fact, most robots today cannot physically grow, at least not without the help of an engineer who adds or takes away modules. Furthermore, engineers may prefer to design a robot that has the characteristics of a fully functional and mature organism rather than those of a newborn and immature organism. Therefore, at first sight, it may seem awkward to take into account morphological development in the context of robotics.

An important hypothesis of developmental robotics is that physical development is not simply an automatic transition from an imperfect and incomplete state to a fully functional adult state, but rather a crucial process that guides and helps the emergence of complex sensorimotor and cognitive

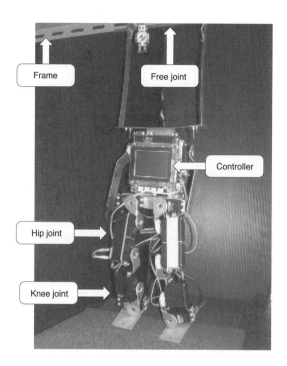

Figure 6.33 The hanging humanoid robot develops a robust swinging behavior through a developmental process that gradually increases the number of degrees of freedom under control. Image courtesy of Max Lungarella, University of Zurich.

skills. The complexity of adult morphology, sensing, actuation, and social environment would be prohibitive for a learning brain to cope with. Physical development provides the opportunity for the nervous system to gradually acquire the skills in simplified conditions that become increasingly more complex. Developmental modifications of body morphology and of neurally mediated control affect each other and proceed through stages that can be more or less discrete according to Piaget (1953) or Thelen and Smith (1994), respectively.

The control of complex robotic forms, such as humanoids, may therefore benefit from a gradual developmental approach. Human movements are an example of how morphological development can guide and help control. The adult human body has a very large number of degrees of freedom that require coordinated control. Bernstein (1967) suggested that the problem of acquiring this complex coordination can be efficiently tackled by a three-stage developmental process. In the first stage, the joints in the periphery of the body are stiffer, thus restricting coordination learning to a much smaller number of joints. In the second stage, the peripheral joints become more mobile and their control is incrementally acquired on top of the previously acquired coordinated movements of the main joints. In the third stage, the organism selects and improves those coordinated movements that are more efficient (e.g., they require less energy and active control because they exploit passive dynamics) or preferred by habit.

Lungarella and Berthouze (2002) translated this hypothesis into a developmental program for a hanging humanoid robot that was expected to develop a swinging behavior (figure 6.33). The robot was equivalent to a pendulum, but swinging was caused by the action of the hip joints and of the knee joints. The robot had more degrees of freedom than actuated joints and the feedback to the learning system was given by the position of the body, not of the legs. Each joint was controlled by a separate neural oscillator that could produce rhythmic activity, as we already saw in a previous section on walking and swimming control. Learning was achieved by simulated annealing (see chapter 1) of the neural parameters where the amount of exploration versus exploitation was inversely proportional to the quality of the swing. In other words, the worse the swing, the larger the parameter change was, and vice versa.

The authors experimentally compared the case where the robot learned simultaneously to control both hip and knee joints with the case where the knee joint was initially frozen and later freed up once the hip-mediated swing had become stable. Although in both cases the robot developed good swinging patterns, in the second –developmental– case the robot developed a more robust control strategy (i.e., less sensitive to initial conditions and external perturbations) where the neural oscillations matched the natural oscillation frequency of the legs. This "physical entrainment" was caused by the incremental fine-tuning of the knee joint controller that improved the performance of, and reduced the sensitivity to, the hip joint controller.

PHYSICAL
ENTRAINMENT

Kuniyoshi and Sangawa (2006) instead investigated the hypothesis that the primitive behaviors displayed by newborn infants may not be genetically determined, but rather emerge from the development of a neural system that becomes physically entrained with its physical body during the fetal and neonate stages. The authors conceived a physics-based simulation of a realistic human baby with 198 muscles that were modeled in detail after measurements of muscle properties and muscle sensors of human babies (finger and face muscles were not simulated). Each muscle was connected bidirectionally (input and output) to weakly coupled neural oscillators capable of generating both periodic and chaotic activity. In addition, they added a simplified somatosensory and motor cortex, simulated as self-organizing maps (see chapter 3) with continuous neuron dynamics, that received inputs from sensors of muscular stretch and tension and provided output to the muscles and to the neural oscillators. The synaptic connections to the somatosensory cortex and from the motor cortex were adapted by means of Hebbian learning so that coherent patterns of motion that persisted for some time were clustered and represented by cortical neurons. Preliminary experiments with the baby model lying in a fenced bed displayed the emergence and transition between coherent motor primitives, such as rolling over from a face-up to a face-down posture and crawling.

At the time of writing this book, the authors were extending the baby model to incorporate realistic morphological change of muscular and skeletal properties as a function of weeks after gestation so that they could simulate physical growth during the fetal and neonatal periods. The fetal-stage baby grew in a simulated uterus filled with liquid while the neonate baby could continue to develop in a playpen. Kuniyoshi et al. (2007) reported preliminary results where the baby acquired coherent movements during the fetal stage that, once connected to a visual system during the neonate stage, could be triggered by recognition of similar movements by other persons. The authors argued that early development of neural control during the fetal stage could serve as a "motor babbling" that chunks movements into sequences that can support early imitation mechanisms displayed by infants (Meltzoff and Moore 1977).

MOTOR BABBLING

6.6.3   Imitation Learning

Imitation is a powerful way of learning because it reduces the number of trials and errors that are required when learning in isolation. In robotics, learning by demonstration has attracted much attention for several years because it can simplify and improve the process of hand-coding a program. For complex robots with several degrees of freedom, such as humanoids, it has been argued that learning by demonstration may actually be the most promising way of achieving complex behaviors (Schaal 1999).

In the classic symbolic approach (Dufay and Latombe 1984), learning by demonstration starts by manually moving the robot through a series of fixed points while recording all angles and forces measured by the robot sensors. Once the entire behavior has been demonstrated, it is decomposed in subbe-

haviors, which, starting from a sensor state corresponding to a fixed point, generate a sequence of control commands to reach the sensor state corresponding to the next fixed point. The strategy for finding a set of suitable control commands that take the robot from an initial state to a desired state
INVERSE MODEL   is known as the *inverse model*. However, this approach does not extend easily to robotic systems characterized by highly nonlinear sensors and actuators that are expected to learn by observation of another agent's behavior, which is ultimately the case with autonomous and socially interactive robots.

Imitation of an actor by an observer entails a number of problems, such as recognition of the movements to be imitated, estimation and tracking of the actor's posture, transformation from external to egocentric coordinates, matching the actor's body to the observer's body, comparing the observed movement with previously learned movements, controlling the degrees of freedom that are not specified by the observed movement, and segmenting and representing the observed behavior in suitable movements, to mention a few (Schaal et al. 2003).

It is also important to distinguish between true imitation and copying (Bil-
TRUE IMITATION   lard 2002). *True imitation* consists in the reproduction of a novel behavior, that is, of a behavior that the observer does not have yet in its behavioral
COPYING   repertoire. *Copying* instead consists in the reproduction of a behavior that the observer already has in its behavioral repertoire. Copying is displayed by several animals and entails all the above-mentioned problems, but does not improve the behavioral repertoire of the animal and may serve mainly as a social response or a facilitator of response. True imitation instead is displayed only by humans and some primates and serves primarily a learning function in the development of the individual.

As we mentioned in the previous section, human neonates display simple imitative skills soon after birth (Meltzoff and Moore 1977). This ability requires neural circuits capable of recognizing specific actions, producing action sequences, and mapping recognized actions onto the production of similar actions. Neurophysiological findings on monkeys seem to provide support for the existence of those circuits. Perrett et al. (1990) found neurons in the superior temporal sulcus that respond selectively to faces, postures, and body movements in viewer-centered coordinates. Rizzolatti et al. (1988) instead found neurons in the area F5 that signal motor actions, such as grasping by hand or holding, and remain active during the generation of the entire motor sequence. It has been suggested that these neurons encode behavioral primitives (Murata et al. 1998). In addition, some of these neurons are active not only when the animal produces the action but also when the animal sees

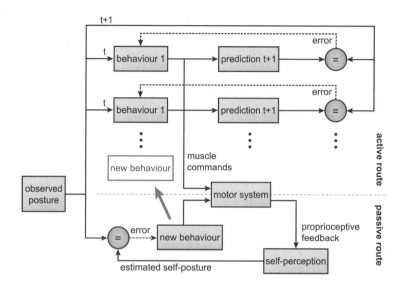

Figure 6.34 The dual-route model of imitation learning consists of a "passive route" that acquires novel behaviors and an "active route" that selects the best-matching behavior for imitation (Demiris and Hayes 2002). The passive route is invoked when none of the existing behaviors in the active route match the behavior to be imitated. See text for explanation.

the same action performed by another person (Rizzolatti et al. 1996). These MIRROR NEURONS neurons were thus called *mirror neurons* to distinguish them from the other "canonical" F5 neurons.

Although there is not yet evidence for a neural path from STS neurons to F5 mirror neurons, the detection of an external action, the mapping onto the same self-generated action, and the production of that action may provide the necessary neural circuitry for imitation. Oztop and Arbib (2002) developed a detailed computational model that suggested how suitable connections from STS to F5 areas could develop through learning.

A number of recent computational models inspired from that line of cognitive and neurophysiological research have been developed and applied to various robotic platforms (see Oztop et al. 2006 for a critical review). Among them, the dual-route model of Demiris and Hayes (2002) is particularly interesting because it integrates both copying and true imitation within the same conceptual framework (figure 6.34). The authors argue that imitation is mediated by two neural circuits or routes: a route that passively acquires observed behaviors as infants do, and a route that recognizes behaviors through the same neural circuitry used for generating those behaviors, as the mirror neurons in the adult brain may do.

PASSIVE ROUTE   The *passive route* is inspired by a detailed model of infant imitation, but it cannot explain all aspects of adult imitation. The neural architecture takes as input the observed posture of an external actor and attempts to produce motor actions that generate the same posture. The architecture relies on a module that matches the posture of the actor expressed in visual coordinates with the posture of the observer expressed in proprioceptive coordinates. The output of this module is used to generate actions that improve the match between the observed and the proprioceptive posture.

ACTIVE ROUTE   The *active route* instead assumes that the observer has already acquired a set of primitive behavioral networks that take as input an observed state and produce as output the motor commands necessary to achieve that state. The motor commands of each behavioral network are sent both to the muscle centers and to a predictive network that outputs the predicted state after execution of those commands. During imitation, the observed state is fed into all behaviors while the output lines to the muscle centers are inhibited and the predicted state resulting from each behavioral network is produced. The internally predicted states are compared to the next observed state of the actor and those behaviors that have produced a close match raise their confidence level while the other behaviors decrease their confidence level. After a few iterations of observation, comparison of internally with externally generated states, and behavior confidence update, only one behavior will reach a sufficiently high confidence to override the inhibition of muscle centers and produce the movements that imitate the observed behavior.

Although the active route is a high-level model, it can account for several aspects of adult imitation, including pathologic conditions where the patient cannot be prevented from executing the demonstrated actions. However, the active route can explain only copying because, if the demonstrated behavior is not in the behavioral repertoire of the agent, all behavioral modules will decrease their confidence and no imitation will take place. Therefore, Demiris and Hayes (2002) suggested that true imitation requires a dual-route architecture where the passive route is invoked if none of the existing behaviors can match the observed behavior via the active route. Once the new behavior has been acquired by the passive route, it is stored in the behavioral repertoire. Both the behavioral modules and the predictive modules can be represented as neural networks. Jordan and Rumelhart (1992) suggest a method for training the predictive modules while the behavioral system executes movements.

Demiris and Hayes (2002) tested the dual-route architecture on a simulated humanoid robot with 13 degrees of freedom that was capable of imitating another humanoid robot controlled by a hand-coded program, even in the case where the dynamics of the joints was different and the visual and proprioceptive inputs were imperfect. Despite these impressive achievements, several open questions remain to be addressed, such as how can a robot know what movements should be imitated, how can a demonstrated complex behavior be decomposed in subbehaviors, how can a robot learn to imitate the goal of an action rather than the action itself, when should imitation be used, and how a more interactive role between the robot and the instructor could facilitate imitation learning (Breazeal 2002b).

A full understanding and synthesis of imitation learning in animals and robots is not yet within reach, but it could have tremendous impact for robots who may use it to engage in social interactions that go beyond mimicking. Rizzolatti and Arbib (1998) suggested that the neural circuitry used for imitation could provide a basis for the emergence of language. As we will see also in the next chapter, a prerequisite of the emergence of communication is the establishment of a common understanding between a signaler and a receiver. The existence of mirror neurons could be instrumental in the establishment of a common ground between signaler and receiver because the reproduction of a behavior (copying) by an observer could be recognized and understood as such by the actor. Starting from the observation that the human equivalent of area F5 is close by, and partially overlapping with, the language area, Rizzolatti and Arbib (1998) developed an intriguing hypothesis of how gesture imitation could become gradually associated to specific sounds and provide the supporting mechanism for the establishment of speech, which gradually expanded and substituted for gesture-mediated communication.

6.7   Evolution of Behavioral Systems

The idea of using artificial evolution for automatically generating control systems dates back to at least the 1950s (Turing 1950) with a more explicit form appearing in the mid 1980s through the ingenious thought experiments of Braitenberg (1984) on neurally driven vehicles. In the early 1990s, the first generation of artificial organisms with a genetic code describing the neural circuitry and morphology of a sensorimotor system began evolving in com-

Figure 6.35 *Left*: The Khepera mobile robot has two wheels and eight infrared sensors that can measure both the amount of infrared light and the distance from objects. *Right*: Bird's-eye view of the looping maze with the Khepera robot.

puter simulations (Beer 1990; Parisi et al. 1990; Floreano et al. 1991; Husbands and Harvey 1992), followed soon after by evolutionary experiments on real robots (Lewis et al. 1992; Floreano and Mondada 1994; Harvey et al. 1994) when the term *evolutionary robotics* was coined (Cliff et al. 1993).

EVOLUTIONARY
ROBOTICS

Evolutionary robotics is the application of artificial evolution to the hardware and software components of a robot where the fitness is given by the behavioral performance of the robot. The field includes almost all the elements of bioinspiration that we have described so far in this book. In this section we will introduce the methodology with the help of examples of evolution of reactive control systems. In later sections we will also cover evolution and learning, evolution and development, coevolution of body and control, and hardware self-replication. In chapter 7, we will extend the description to the evolution of behavioral systems that compete and cooperate.

Most cases of control evolution resort to artificial neural networks because neural networks can easily be mapped to noisy and analog sensors, display intrinsic generalization to novel sensory situations, are amenable to artificial evolution, can be subjected to lifelong learning mechanisms in addition to generational evolution, and can be biologically plausible models of control in living systems, thus facilitating the incorporation of principles extracted from living systems. In relatively few cases, the control system was directly mapped to a field-programmable gate array or represented as a computer program and evolved with genetic programming. However, various forms of genetic programming have been used to evolve the rules of growth of neural networks for robot control, as we will see in a later section.

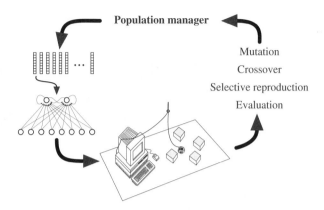

Figure 6.36 Evolutionary experiments on a single robot. Each individual of the population is decoded into a corresponding neurocontroller which reads sensory information and sends motor commands to the robot every 300 ms while its fitness is automatically evaluated and stored away for reproductive selection.

6.7.1   Example 1: Collision-free Navigation

In an early experiment on robot evolution without human intervention carried out at the École Polytechnique Fédérale de Lausanne (EPFL) (Floreano and Mondada 1994), a small wheeled robot was evolved for navigation in a looping maze (figure 6.35, right). The Khepera robot had a diameter of 55 mm and two wheels with controllable velocities in both directions of rotation. It also had eight infrared sensors, six on one side and two on the other side, that can function either in active mode to measure distance from obstacles or in passive mode to measure the amount of (infrared) light in the environment (figure 6.35, left). The robot was connected to a desktop computer through rotating contacts that provided both power supply and data exchange through a serial port (figure 6.36).

KHEPERA

A genetic algorithm with binary encoding, single-point crossover, and rank-based selection (see chapter 1) was used to evolve the synaptic strengths of a neural network composed of eight sensory neurons and two motor neurons. Each sensory unit was assigned to one of the eight active infrared sensors whose value was updated every 300 ms. Each motor unit received weighted signals from the sensory units and from the other motor unit, plus a recurrent connection with itself with a 300 ms delay. The net input of the motor units was offset by a modifiable threshold and passed through a logistic squashing function. The resulting outputs, in the range $[0, 1]$, were used to control the

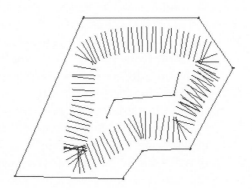

Figure 6.37 Trajectory of the robot with the best neural controller of the last generation. Segments represent the axis between the two wheels. Data were recorded and plotted every 300 ms using an external laser positioning device.

two motors so that an output of 1 generated maximum rotation speed in one direction, an output of 0 generated maximum rotation speed in the opposite direction, and an output of 0.5 did not generate any motion in the corresponding wheel. A population of 80 individuals, each coding the synaptic strengths and threshold values of the neural controllers was initialized with all weights set to small random values centered around zero. Each individual was tested on the physical robot for 80 sensorimotor cycles (approximately 24 seconds) and evaluated at every cycle according to a fitness function with three components measured onboard the robot:

$$(6.2) \qquad \Phi = V \left(1 - \sqrt{\Delta v} \right) (1 - i)$$

where V is the average of the unsigned rotation speeds of the two wheels, Δv is the absolute value of the algebraic difference between the signed speed values of the wheels (positive in one direction, negative in the other), and i is the normalized activation value of the infrared sensor with the highest activity. The first component was maximized by speed, the second by straight motion, and the third by distance from objects. The fitness was accumulated over the entire duration of the test.

Within fewer than 100 generations, the best evolved individuals displayed smooth trajectories around the maze (figure 6.37). Although the fitness function did not specify in what direction the robot should navigate (given that it was perfectly circular and that the wheels could rotate in both directions), after a few generations all the best individuals moved in the direction corresponding to the side with the highest number of sensors. Individuals moving in the other direction had a higher probability of colliding into corners without detecting them and thus disappeared from the population. Furthermore, the cruising speed of the best evolved robots was approximately half of the maximum available speed and did not increase even when the evolutionary experiment was continued up to 200 generations. Further analysis revealed that this self-limitation of the navigation speed had an adaptive function because, considering the sensory and motor refresh rate together with the response profile of the distance sensors, robots that traveled faster had a higher risk of colliding into walls before detecting them and gradually disappeared from the population. Finally, the evolved recurrent connections at the output units allowed the robot to move out of situations where the input signals at two symmetric points around the body of the robot were equal and would cancel each other in a feedforward architecture, such as Braintenberg's vehicles shown in figures 6.2 and 6.3.

Despite its simplicity, this experiment indicated that evolution could automatically discover solutions that matched not only the computational requirements of the task to be solved but also the morphological and mechanical properties of the robot, which are difficult to estimate and incorporate in hand-designed control.

6.7.2 Example 2: Walking

Over the past 15 years or so there has been a growing body of work on evolving locomotion controllers for various kinds of walking robots - a nontrivial sensorimotor coordination task. Early work in this area concentrated on evolving dynamical network controllers for simple (abstract) simulated insects (often inspired by cockroach studies) which were required to walk in simple environments (e.g., de Garis 1990;Beer and Gallagher 1992). Probably the first success in this direction was achieved by Lewis et al. (1992, 1994), who evolved a neural controller for a real hexapod robot using coupled oscillators built from continuous-time, leaky-integrator, artificial neurons. All evaluations were done on the actual robot with each leg connected to its own pair of coupled neurons, leg swing being driven by one neuron and leg elevation by the other. In order to speed up the process they employed *staged evolution* where first an oscillator capable of moving a leg was evolved and

STAGED EVOLUTION

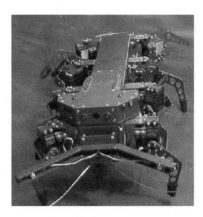

Figure 6.38   The octopod robot developed by Applied AI Systems Inc. was used for various evolutionary projects described in the text.

then an architecture based on these oscillators was further evolved to develop walking. The robot was able to execute an efficient tripod gait on flat surfaces. Gallagher et al. (1996) described experiments where neural networks controlling locomotion in an artificial insect were evolved in simulation and then successfully downloaded onto a real hexapod robot. This machine was more complex than that of Lewis et al., with a greater number of degrees of freedom per leg. In this approach, each leg was controlled by a fully connected network of five continuous-time, leaky-integrator neurons, each receiving a weighted sensory input from that leg's angle sensor.

Galt et al. (1997) used a genetic algorithm to derive the optimal gait parameters for a Robug III robot, an eight-legged, pneumatically powered walking and climbing robot. The individual genotypes represented parameters defining each leg's support period and the timing relationships between leg movements. These parameters were used as inputs to a finite-state machine pattern-generating algorithm that drove the locomotion. Such algorithms, which are often used in conventional walking machines, rely on relatively simple control dynamics and do not have the same potential for the kind of sophisticated multigait coordination that complex dynamical neural network architectures, such as those described above, have been shown to produce. However, controllers were successfully evolved for a wide range of environments and to cope with damage and system failure (although an individual controller had to be tuned to each environment, they were not able to self-adapt across a wide range of conditions).

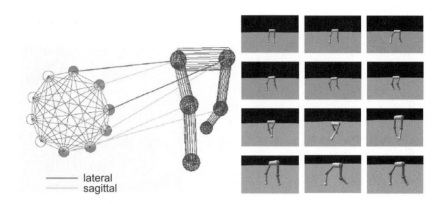

Figure 6.39   *Left*: Fully recurrent, continuous-time neural network with motor neurons controlling the displacement of hip and knee joints. *Right*: Motion sequence of biped controlled by best evolved individual. Frame order is from left to right and top to bottom. Images courtesy of Torsten Reil and Phil Husbands.

Gomi and Ide (1998) evolved the gaits of an eight-legged robot (figure 6.38) using genotypes made of eight similarly organized sets of genes, each gene coding for leg motion characteristics, such as the amount of delay after which the leg begins to move, the direction of the leg's motion, the end positions of both vertical and horizontal swings of the leg, and the vertical and horizontal angular speeds of the leg. After a few dozen generations, in which evaluation was performed on the robot, a mixture of tetrapod and wave gaits were obtained. Using the cellular developmental approach, Gruau and Quatramaran (1997) and Kodjabachian and Meyer (1998a) evolved stable neural controllers for the same eight-legged robot used by Gomi and Ide. This work will be described in more detail in a later section on evolution and development. Jakobi (1998) successfully evolved modular controllers based on Beer's continuous recurrent networks to control the same eight-legged robot as it engaged in walking about its environment avoiding obstacles and seeking out goals. The robot could smoothly change gait, move backward and forward, and even turn on the spot. More recent work has used architectures similar to those explored by the researchers mentioned above to control more mechanically sophisticated robots such as the Sony Aibo (Tellez et al. 2006).

Recently there has been successful work on evolving coupled neural oscillators for the highly unstable dynamic problem of biped walking. Reil and Husbands (2002) used physics-based simulations to evolve controllers able to generate successful bipedal gaits. In a first stage, they evolved the synaptic weights and time constants of the network shown on the left of figure 6.39. The fitness function was the sum of two components, one measuring the distance traveled from the origin and one penalizing individuals that lowered their center of gravity below a certain height. The second component was useful to prevent solutions that displayed grotesque movements. Efficient walkers were evolved in fewer than 60 generations (figure 6.39, right). In a second stage, they provided the walker with two ears and evolved individuals for their ability to approach a sound source. In this case all neurons of the neural network received additional input from a sound sensor whose activation was proportional to the difference between sound strengths measured in the two ears. The previously evolved weights were clamped and only the weights from the sensory unit were genetically encoded and evolved. An additional 60 generations of incremental evolution were sufficient to obtain robots capable of approaching the sound source from any position in their surroundings. Coupled neural oscillators have also been evolved to control the swimming pattern of articulated, snake-like, underwater robots using physics-based simulations (von Haller et al. 2005).

Vaughan used a simulation of a 3D ten-degrees-of-freedom bipedal robot with passive dynamics and compliant tendons to conserve energy while walking on a flat surface (Vaughan et al. 2004a). The parameters of the body and a continuous dynamical neural network controller were under genetic control. The machine started out as a passive dynamic walker (McGeer 1990b) on a slope and then throughout the evolutionary process the slope was gradually lowered to a flat surface. The machine demonstrated resistance to disturbance while retaining passive dynamic features such as a passive swing leg. This machine did not have a torso, but Vaughan has also successfully applied the method to a simplified 2D machine with a torso above the hips. When pushed, this dynamically stable bipedal machine walks either forward or backward just enough to release the pressure placed on it. It is also able to adapt to external and internal perturbations, as well as to variations in body size and mass (Vaughan et al. 2004b).

McHale and Husbands (2004a,b) have compared many forms of evolved neural controllers for bipedal and quadrapedal walking machines. Recurrent dynamical continuous time networks were shown to have advantages in most circumstances. The vast majority of the studies mentioned above were conducted in relatively benign environments. Nevertheless, we can conclude that the more complex dynamical neural network architectures, with their intricate dynamics, generally produce a wider range of gaits and generate smoother, more adaptive locomotion than the more standard use of

Figure 6.40   The gantry robot used in the visual discrimination task. Image courtesy of Phil Husbands, Sussex University, England.

finite-state machine-based systems employing parameterized rules governing the timing and coordination of individual leg movements (e.g., Laszlo et al. 1996).

6.7.3   Example 3: Vision-based Navigation

The experiments described so far used relatively simple distance sensors, such as active infrared or sonar. Pioneering experiments on evolving visually guided behaviors were performed at Sussex University (Harvey et al. 1997) on a specially designed gantry robot (figure 6.40). Discrete-time dynamical recurrent neural networks and visual sampling morphologies were concurrently evolved: the brain was developed in tandem with the visual sensor (Harvey et al. 1994; Husbands et al. 1997; Jakobi 1997a). The robot was designed to allow real-world evolution by having "offboard" power and processing so that the robot could be run for long periods while being monitored by automatic fitness evaluation functions. A CCD camera pointed down toward a mirror angled at $45°$ as shown in figure 6.40. The mirror could rotate around an axis perpendicular to the camera's image plane. The camera was suspended from the gantry allowing motion in the x, y, and z dimensions. This effectively provided an equivalent to a wheeled robot with

GANTRY ROBOT

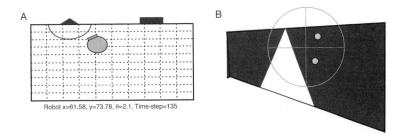

Figure 6.41 The shape discrimination task. *A*, the position of the robot in the arena, showing the target area in front of the triangle. *B*, The robot camera's field of view showing the visual patches selected by evolution for sensory input. Image courtesy of Phil Husbands, Sussex University.

a forward-facing camera when only the x and y dimensions of translation are used.

The apparatus was initially used in a manner similar to the real-world experiments on navigation in the looping maze with the miniature mobile robot described earlier. A population of strings encoding robot controllers and visual sensing morphologies were stored on a computer to be downloaded one at a time onto the robot. The exact position and orientation of the camera head could be accurately tracked and used in the fitness evaluations. A number of visually guided navigation behaviors were successfully achieved, including navigating around obstacles, tracking moving targets and discriminating between different objects (Husbands et al. 1997). In the experiment illustrated in figures 6.40 and 6.41, starting from a random position and orientation the robot had to move to the triangle rather than to the rectangle. This had to be achieved irrespective of the relative positions of the shapes and under very noisy lighting conditions. Recurrent neural network controllers were evolved in conjunction with visual sampling morphologies. Only genetically specified patches from the camera image were used by being connected to input neurons according to the genetic specification. The rest of the image was thrown away. This resulted in extremely minimal systems using only two or three pixels of visual information, yet still able to perform the task reliably under highly variable lighting conditions (Harvey et al. 1994; Husbands et al. 1997).

In another set of experiments, Husbands et al. (1998) evolved vision-based controllers of GasNets, a class of neural networks that incorporate two distinct signaling mechanisms, one "electrical" and one "chemical" (see also the

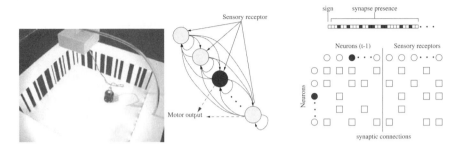

Figure 6.42   The architecture of a network of spiking neurons is evolved to drive the vision-based robot in the arena. The light below the rotating contacts allows continuous evolution also overnight. A binary genetic string encodes the sign of each neuron and the presence/absence of all potential connections to that neuron.

closing remarks in chapter 3). The visual sampling morphology was also under evolutionary control. The original basic GasNet was found to be significantly more evolvable than a variety of other styles of connectionist neural networks. Successful GasNet controllers for this task tended to be rather minimal, in terms of numbers of nodes and connections, while possessing complex dynamics (Husbands et al. 1998).

Floreano and Mattiussi (2001) instead evolved the architecture of a network of spiking neurons for driving a vision-based robot in an arena painted with black stripes of variable size against a white background (figure 6.42). The Khepera robot used in these experiments was equipped with a vision turret composed of one linear array of gray-scale photoreceptors spanning a visual field of $36°$. The output values of a bank of local contrast detection filters (Laplace filters) were converted in spikes –the stronger the contrast, the larger the number of spikes per second (see right side of figure 6.43)– and sent to 10 fully connected spiking neurons implemented according to the spike response model (Gerstner et al. 1996) described in chapter 3. The spike series of a subset of these neurons was translated into motor commands (more spikes per second corresponded to faster rotation of the wheels). The fitness function was the amount of forward translation of the robot measured over two minutes using the wheel rotation encoders. Consequently, robots that turned in place or hit the walls had comparatively lower fitness than robots that could move straight and turn only when it was necessary to avoid walls. The genome of these robots was a bit string that encoded only the sign of the neurons and the presence of synaptic connections. Existing

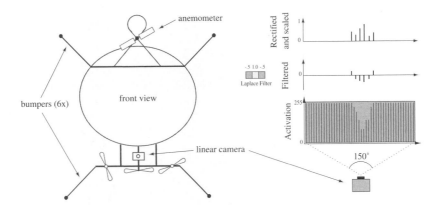

Figure 6.43   *Left*: Front view of the blimp with its main components: the anemometer on top of the envelope, the linear camera pointing forward with a 150° horizontal field of view, the bumpers, and the two propellers. *Right*: Contrast detection is performed by selecting 16 equally spaced photoreceptors and filtering them through a Laplace filter spanning three photoreceptors. Filtered values are then rectified by taking the absolute values and scaling them in the range [0, 1]. These values represent the probability of emitting a spike for each corresponding neuron. The output of the linear camera is the only source of information passed to the evolutionary spiking network. The anemometer is used for fitness computation.

connections were set to 1 and could not change during the lifetime of the robot.

Evolution reliably discovered very robust spiking controllers in approximately 20 generations (approximately 30 hours of evolution on the real robot). Evolved robots could avoid not only walls but any object positioned in front of them. Detailed analysis of the best evolved controllers revealed that neurons did not exploit time differences between spikes, which one would have expected if some type of optic flow information was used to detect distance from walls. Instead, they simply used the number of incoming spikes (firing rate) as an indication of when to turn. When the robot perceived several edges (which generated strong spiking activity in the network) it would go straight, but when the number of perceived edges decreased below a certain level, it started to turn away. Given the low resolution of the camera and its limited field of view, the number of perceived edges is proportional to the distance from the walls.

The same methodology was then applied to a blimp-like robot equipped with a linear camera spanning 150°, two lateral propellers, and an anemome-

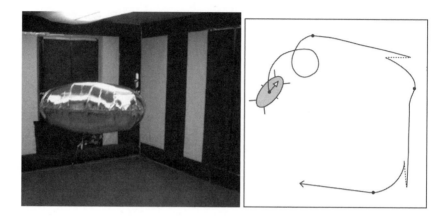

Figure 6.44 Left: The blimp-like robot is evolved in a 5 x 5 x 3 m room with randomly sized black-and-white stripes painted on the walls. The serial data transmission of sensory and motor commands between the blimp and the desktop computer implementing the evolutionary algorithm and spiking neural network is handled by a Bluetooth wireless connection. The blimp has onboard batteries that must be changed every three hours.

ter shown in figure 6.43 (Zufferey et al. 2002). The blimp was equipped with an anemometer whose rotation direction and speed were used as an indication of how fast the blimp moved forward. The fitness function was simply the average rotation speed of the anemometer helix in the direction corresponding to the forward translation of the blimp. The blimp was also equipped with 12 sticks connected to electrical on/off switches whose activation indicated a collision with a wall. Only the visual information was provided as input to the evolutionary neural controller.

The blimp was evolved in a room with randomly sized black-and-white stripes (figure 6.44, left), similar to the arena used for the wheeled Khepera robot described earlier. As for the Khepera robot, 20 generations were sufficient to evolve spiking controllers capable of steering the blimp around the room (figure 6.44, right). The evolved control strategy, however, was different from that evolved in the wheeled robot. Although the steering angle of the evolved blimp was approximately proportional to the amount of contrast (firing rate), the robot was also capable of recognizing collisions with walls, briefly reversing the thrust of the two propellers, and resuming forward motion in a suitable direction. In further work, Zufferey et al. (2006a) devised a physics-based simulation of the blimp that allowed seamless transfer of

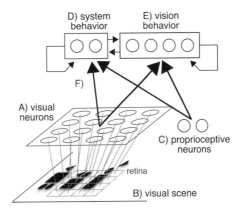

Figure 6.45 The neural architecture of the active vision system is composed of *A*) a grid of visual neurons with nonoverlapping receptive fields whose activation is given by *B*) the gray level of the corresponding pixels in the image; *C*) a set of proprioceptive neurons that provide information about the movement of the vision system; *D*) a set of output neurons that determine the behavior of the system (pattern recognition, car driving, robot navigation); *E*) a set of output neurons that determine the behavior of the vision system; *F*) a set of evolvable synaptic connections. The number of neurons in each subsystem can vary according to the experimental settings.

evolved controllers on the real blimp. This allowed them to start the evaluation of individuals in "difficult" conditions (e.g., against a wall), which resulted in controllers that were capable of maximizing the amount of forward translation by staying very near the walls without ever colliding into them.

In a different line of work, Floreano et al. (2004) studied the coevolution of active vision and of visual feature selection in several types of behavioral systems. *Active vision* is the sequential and interactive process of selecting and analyzing parts of a visual scene (Aloimonos et al. 1987; Bajcsy 1988; Ballard 1991). *Feature selection* instead is the development of sensitivity to relevant features in the visual scene to which the system selectively responds (Hancock et al. 1992). Each of these processes has been investigated and adopted in machine vision, but the codevelopment of active vision and visual feature selection has been largely ignored so far.

ACTIVE VISION

The behavioral systems were equipped with a simple moving camera and a deliberately simple neural architecture (figure 6.45). The neural architecture was composed of an artificial retina and two sets of output units. One set of output units determined the pan, tilt, and zoom values of the cam-

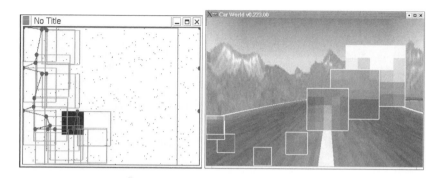

Figure 6.46   *Left*: An evolved individual explores the screen searching for the shape and recognizes it by the presence of a vertical edge. *Right*: Search for the edge of the road at the beginning of a drive over a mountain road.

era and the other set of units determined the behavior of the system. This could vary across experiments; in one case it was the response of a pattern recognition system, in another case it was the actions of a car driver, and in yet another case it was the wheel speeds of a robot. The neural network was embedded in the behavioral system and its input/output values were updated every 300 ms while its fitness was computed. Therefore, the synaptic weights of this network were responsible for both the visual features on which the system based its behavior and for the motor actions necessary to search for those features.

In a first set of experiments, the neural network was embedded within a pan-tilt camera and asked to discriminate between triangles and squares of different size that could appear at any location of a screen (figure 6.46, left), a perceptual task similar to that explored with the gantry robot described earlier. The visual system was free to explore the image for 60 seconds while continuously telling whether the current screen showed a triangle or a square. The fitness was proportional to the amount of correct responses accumulated over the 60 seconds for several screenshots containing various instances of the two shapes. Evolved systems were capable of correctly identifying the type of shape with 100% accuracy after a few seconds of exploration despite the fact that this recognition problem is not linearly separable and that the neural network does not have hidden units, which in theory are necessary to solve nonlinearly separable tasks, as we have seen in chapter 3. Indeed, the same neural network presented with the same set of images and trained with supervised learning, but without the possibility to actively ex-

Figure 6.47 A mobile robot with a pan-tilt camera is asked to move within the walled arena in the office environment.

plore the scene, was not capable of solving the task Kato and Floreano (2001). The evolved active vision system developed sensitivity to vertical edges, oriented edges, and corners and used its movement to search for these features in order to tell whether the shape was a triangle or a square. These features, which are found also in the early visual system of almost all animals, are invariant to size and location.

In a second set of experiments, the neural network was embedded in a simulated car and was asked to drive over several mountain circuits (figure 6.46, right). The simulator was a race car video game. The neural network could move the retina across the scene seen through the windshield from the driver's seat and control the steering, acceleration, and braking of the car. The fitness was inversely proportional to the time taken to complete the circuits without exiting the road. Evolved networks completed all circuits with lap times competitive to those of well-trained students controlling the car with a joystick. Evolved networks started by searching for the edge of the road and tracked its relative position with respect to the edge of the windshield in order to control steering and acceleration. This behavior was supported by the development of sensitivity to oriented edges, as in the previous experiments.

In a third set of experiments, the neural network was embedded in a real mobile robot with a pan-tilt camera that was asked to navigate in a square arena with low walls located in an office (figure 6.47). The fitness was proportional to the amount of straight motion measured over two minutes. Robots that hit the walls because they watched people or other irrelevant features of the office had low fitness. Evolved robots tended to fixate the edge between the floor and the walls of the arena and turned away from the wall when the size of their retinal projection became larger than a threshold. This combination of sensitivity to oriented edges and looming is found also in visual circuits of several insects and birds.

In a further set of experiments (Floreano et al. 2005), the visual pathway of the neural network was augmented by an intermediate set of neurons whose synaptic weights could be modified by Hebbian learning (Sanger rule), which extracts the principal components of the image (see chapter 3) while the robot moved in the environment. All the other synaptic weights were genetically encoded and evolved. The results showed that lifelong development of the receptive fields improved the performance of evolved robots and allowed robust transfer of evolved neural controllers from simulated to real robots because the receptive fields developed a sensitivity to features encountered in the environment where they happened to be born. Furthermore, the results showed that the development of visual receptive fields was significantly and consistently affected by active vision as compared to the development of receptive fields passively exposed to the same set of sample images. In other words, robots evolved with active vision developed sensitivity to a smaller subset of features in the environment and actively tracked those features to maintain a stable behavior.

In conclusion, all these experiments indicate that behavioral systems that are free to explore the environment can solve visually mediated tasks with much simpler architectures and computational resources than those typically advocated in computer vision. This is possible because these systems rely on behavior to self-select the visual stimulation that is most useful for the task to be performed and that matches their computational properties. Artificial evolution plays an important role in the discovery of these vision-based systems because it reduces the bias of the human designer and can explore a space of solutions that capitalize on the interaction between the behavioral system and its environment. Nolfi and Floreano (2000) provide more examples and a detailed explanation of the way in which evolution works from the inside of the system to generate powerful behavioral systems with relatively simple computational strategies.

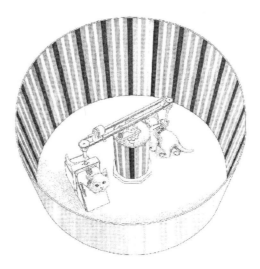

Figure 6.48   The original apparatus in (Held and Hein 1963), where the gross movements of a kitten moving almost freely were transmitted to a second kitten that was carried in a gondola. Both kittens were allowed to move their head. They received essentially the same visual stimulation because of the unvarying pattern on the walls and the center post of the apparatus. Reproduced with permission from Held (1965).

6.7.4   Example 4: Computational Neuro-Ethology

Evolutionary robotics is also used to investigate open questions in neuroscience and cognitive science (D.T. Cliff 1991;Harvey et al. 2005). Although the results should be carefully considered when drawing analogies with biological organisms, evolutionary robotics can generate novel models and test existing hypotheses.

For example, the active vision system with Hebbian plasticity described in the previous section was used to test a hypothesis raised by Held and Hein (1963) who devised the apparatus shown in figure 6.48 where the free movements of a kitten (*active kitten*) were transmitted to a second kitten that was carried in a gondola (*passive kitten*). The second kitten could move its head, but its feet did not touch the ground. Consequently, the two kittens received almost identical visual stimulation, but only one of them received that stimulation as a result of body self-movement. After a few days in that environment, only the active kitten displayed normal behavior in several visually guided tasks. The authors suggested the hypothesis that proprioceptive mo-

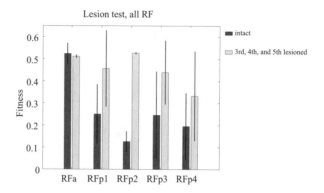

Figure 6.49 Performance of robots with three lesioned visual neurons after development of receptive fields in active condition (RF*a*) or in one of our passive conditions (RF*p*$_{1-4}$). Dark bars show performance levels (average and standard error over 10 trials) of a robot with intact neurons; light bars show performance levels after lesion.

tor information resulting from generation of actions was necessary for the development of normal, visually guided behavior.

The kitten experiments were replicated by using two robots with active vision (figure 6.47). The neural architecture shown in figure 6.45 was augmented with an intermediate layer of five neurons whose synaptic weights could adapt while the robot interacted with the environment according to Sanger's rule. The best evolved network was cloned and the synaptic values of the adaptive visual neurons were randomly initialized in both clones. One cloned robot was then left free to move in the environment, as the active kitten did, while the other cloned robot was forced to move along imposed trajectories, but was free to control its camera position, just like the passive kitten (Suzuki et al. 2005). During these tests the adaptive neurons could change their synaptic strengths (receptive fields). The results indicated that the visual receptive fields and behaviors of robots that developed receptive fields in the passive condition differed significantly from those of robots that developed receptive fields in the active condition. Furthermore, cloned passive robots that were later left free to move could no longer avoid obstacles properly.

A series of lesion studies (figure 6.49) after development of the receptive fields indicated that three of the adaptive visual neurons could be silenced without observing performance loss in active robots. However, when those three neurons were silenced in robots that developed their receptive fields

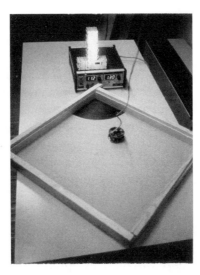

Figure 6.50   Bird's-eye view of the arena with the light tower over the recharging station and the Khepera robot.

in passive conditions, the performance of the robots significantly increased. This and further analysis suggested that passive robots, which could not fully control the series of visual inputs during learning, developed sensitivity to visual features that were not functional for their normal behavior and interfered with other dominant features in the visual field. That explains both why robots that developed in the passive condition could not fully coordinate their behavior when put in normal conditions and why the lesion of some specific neurons restores their behaviors. Whether this explanation holds also for living animals remains to be further investigated, but at least these experiments indicated that motor feedback is not necessary to explain the pattern of pathological behavior observed in animals and robots.

Let us now consider the case of an animal exploring an environment and periodically returning to its nest to feed. It has been speculated that this type of situation requires the formation of spatial representations of the environment that allow the animal to find its way home (Healy 1998). Different neural models with various degrees of complexity and biological detail have been proposed that could provide such functionality (Schmajuk and Blair 1993; Burgess et al. 1997).

Would a robot evolved under similar survival conditions develop a spatial representation of the environment and, if so, would that type of representation resemble that suggested by existing biological models? These questions were explored (Floreano and Mondada 1996a) using a Khepera robot evolved in a square arena with a small patch on the floor in a corner where the robot could instantaneously recharge its (simulated) battery (figure 6.50). The environment was located in a dark room with a small light tower over the "recharging station."

The sensory system of the robot was composed of eight distance sensors, two ambient-light sensors (one on each side), one floor-color sensor, and a sensor for battery charge level. The battery lasted only 20 seconds and had a linear discharge. The evolutionary neural network included five fully connected internal neurons between sensory and motor neurons. The same fitness function described earlier (equation (6.2)) for navigation in the looping maze was used, except for the middle term which had been used to encourage straight navigation in the looping maze. The fitness value was computed every 300 ms and accumulated over the life span of the individual. Therefore, individuals who discovered where the charger was could live longer and accumulate more fitness by exploring the environment (individuals were killed if they survived longer than 60 seconds to limit the experimentation time).

After approximately 200 generations, the robot was capable of navigating around the environment, covering long trajectories while avoiding both walls and the recharging area. When the battery was almost discharged it initiated a straight navigation toward the recharging area and exited immediately after battery recharge to resume navigation. The best evolved individuals always entered the recharging area one or two seconds before full discharge of the battery. That implies that robots must somehow calibrate the timing and trajectory of their homing behavior depending on where they happened to be in the environment.

In order to understand how that behavior could possibly be generated, a set of neuroethological measures were performed. By correlating the robot position and orientation with the activation of the internal neurons in real time while an evolved individual freely moved in the environment, it was found that some neurons specialized in reactive behaviors, such as obstacle avoidance and forward motion. Other neurons instead displayed more complex activation patterns. One of them revealed a pattern of activation levels that depended on whether the robot was oriented facing the light tower or facing the opposite direction (figure 6.51). In the former case, the activation pattern reflected zones of the environment and paths typically followed by

Facing light   Facing opposite corner

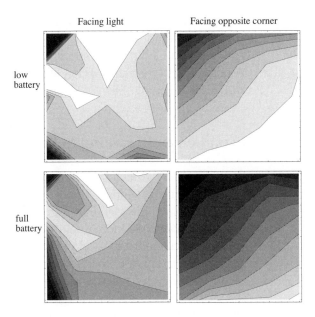

low
battery

full
battery

Figure 6.51   Activation levels (brightness proportional to activation) of an internal neuron plotted over the environment while the robot was positioned at various locations in each of the four conditions (facing recharging area or not, discharged battery or not). The recharging area is located at the top left corner of each map.

the robot during exploration and homing. For example, the robot trajectory toward the recharging area never crossed the two "gate walls" visible in the activation maps around the recharging station. When the robot faced the opposite direction, the same neuron displayed a gradient field orthogonally aligned with the recharging area. This gradient provides an indication of the distance from the recharging area. Interestingly, this pattern of activity is not significantly affected by the charge level of the battery.

The functioning of this neuron recalls the findings on the hippocampus of the rat brain where some neurons (also known as "place cells") selectively PLACE CELLS fire when the rat is in specific areas of the environment (O'Keefe and Nadel 1978). Also, the orientation-specific pattern of neural activation measured on the evolved robot recalls the so-called "head-direction neurons" in the rat HEAD-DIRECTION hippocampus, which are positioned near place cells, whose firing patterns NEURONS depend on the rat heading direction with respect to an environmental landmark (Taube et al. 1990). Although the analogy between brains of evolved robots and of biological organisms should not be taken literally, these results indicate that artificial and biological organisms converge toward a functionally similar computational strategy, which may be more efficient to address this type of situation than a strategy that does not rely on spatial representations, but only on reactive strategies such as random motion, light following, and dead reckoning.

6.8   Evolution and Learning in Behavioral Systems

Evolution and learning are two forms of biological adaptation that differ in space and time. Evolution is based on the existence of a population of individuals displaying variability at the genetic level. Learning, instead, takes place within a single individual during its lifetime. Evolution and learning operate on different time scales. Evolution is a form of adaptation capable of capturing relatively slow environmental changes that might encompass several generations, whereas learning can account for rapid environmental change that occurs within a single generation. Whereas evolution operates on the genotype, learning affects only the phenotype and phenotypic modifications cannot directly modify the genotype. In the introductory section of chapter 3 we described how these two adaptive processes may interact and how evolution tends to assimilate features that are learned during a lifetime when learning implies a cost for the fitness of the individual. In that same chapter we also described several methods to combine artificial evolution and learning for artificial neural networks.

The combination of evolution and learning in behavioral systems serves two purposes (Nolfi 1999): (a) to identify the potential advantage of combining these two methods for developing robust and effective control systems; (b) to understand the role of the interaction between learning and evolution in organisms that interact with their environment.

However, learning can imply evolutionary costs such as (1) a delay in the ability to acquire fitness caused by the time it takes to learn the required characteristics, and (2) an increment of unreliability due to suboptimal learning caused by unpredictable interactions between the agent and its environment (Mayley 1996). Under these circumstances, the Baldwin effect may gradually assimilate learned features into the genotype and reduce the role of learning. This is not necessarily a good effect if one wants to evolve robots that should retain adaptive properties after the evolutionary process is completed.

In the next subsections we will describe examples showing some of the potential advantages of combining evolution and learning. We will also show

Figure 6.52 *Left*: A Khepera robot gains fitness points by finding and staying over a movable target area located on the floor. The walls are covered with white (image) or black paper that changes every generation. The robot has a ground sensor that can detect the target area when it passes over it. *Right*: Self-teaching network. The output values of two teaching neurons are used as teaching values of two motor neurons. The weights that connect the sensory neurons to the teaching neurons do not vary during the robots' lifetime while the weights that connect the sensory neurons to the motor neurons are modified by backpropagation of error.

that evolution can generate robot controllers that display learning-like behaviors without using any form of synaptic plasticity.

6.8.1 Example 1: Evolution of Self-Teaching Controllers

Consider the case of a Khepera robot that should explore an arena surrounded by black or white walls to reach a target placed in a randomly selected location, as shown in figure 6.52 (Nolfi and Parisi 1996). Evolving robots are provided with four sensory neurons that encode the state of four infrared sensors and two motor neurons that control the desired speed of the two wheels. The color of the walls randomly changes between black and white every generation and, since the color significantly affects the intensity of the response of the infrared sensors, evolving robots should develop an ability to understand whether they are currently located in an environment with white or black walls and modify their behavior accordingly.

Robots were provided with a neural controller (figure 6.52) whose output layer included two "teaching neurons" that were used to modify with backpropagation learning the connection weights from the sensory neurons to the motor neurons during the robot's lifetime. This special architecture

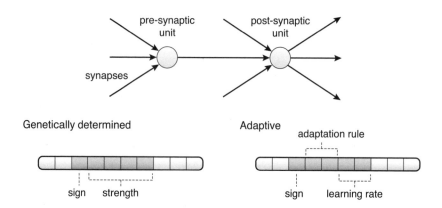

Figure 6.53 Two methods for genetically encoding the properties of a synapse. In the "genetically determined" case, the genetic string encodes the strength of the synapse, which cannot change during the lifetime of the robot. In the "adaptive" case, instead, the genetic string encodes one of four Hebbian learning rules shown on the left and one of four learning rates. In this latter case, the initial synaptic strengths of the initial decoded network are set to small random values and change during the lifetime of the robot according to the learning rules and rates specified in the genetic string.

allows evolving robots to use the sensory information not only to generate behavior but also to generate teaching signals that can modify that behavior.

Analysis of evolved robots revealed that they developed two different behaviors that were adapted to the particular arena where they happened to be "born." Evolving robots did not inherit an ability to behave effectively, but rather a predisposition to learn to behave. This predisposition to learn involved several aspects such as a tendency to move so as to experience useful learning experiences and a tendency to acquire useful adaptive characters through learning (Nolfi and Parisi 1996).

6.8.2   Example 2: Evolution of Learning

In the previous example, the evolutionary neural network learned using an off-the-shelf supervised learning rule that was applied to all synaptic connections. Floreano and collaborators (Floreano and Mondada 1996b) explored the possibility of genetically encoding and evolving the learning rules associated to the different synaptic connections of a neural network embedded in a real robot.

In order to prevent assimilation of learned synaptic weights by evolution (Baldwin effect), which would reduce the role of learning and thus the adaptability of the controller to unforeseen variations, the synaptic weight values were *not* genetically encoded. Each synaptic connection in the network was described by three genes that defined its sign, its learning rule, and its learning rate (figure 6.53). Every time a genome was decoded into a neural network and downloaded onto the robot, the synaptic strengths were initialized to small random values and could change according to the genetically specified learning rules and learning rates while the robot interacted with the environment.

Variations of this methodology included a more compact genetic encoding where all synapses afferent to a neuron were assigned the same properties (sign, learning rule, learning rate), instead of dedicating a set of genes for each synapse. Genes could encode four types of Hebbian learning that were modeled on neurophysiological data and were complementary to each other (Floreano and Urzelai 2000). These rules were (1) the plain Hebb rule (synapse is strengthened if both pre- and postsynaptic units are active); (2) the postsynaptic rule (synapse is modified only if the postsynaptic unit is active: it is strengthened if also the presynaptic unit is active, otherwise it is weakened); (3) the presynaptic rule (synapse is modified only if the presynaptic unit is active: it is strengthened if also the postsynaptic unit is active, otherwise it is weakened); and (4) the covariance rule (synaptic unit modification is proportional to the degree of the correlated activity of pre- and postsynaptic activity).

The methodology was tested both for simple situations, such as navigation in the looping maze described above, and in nontrivial, multitask situations, such as the light-switching task shown in figure 6.54. Results indicated that this methodology had a number of significant advantages with respect to the evolution of synaptic strengths without learning (Urzelai and Floreano 2001). When compared to genetically determined neural controllers without synaptic plasticity, the evolution of learning rules was faster and produced significantly better fitness values in all situations.

Furthermore, evolved behaviors of adaptive individuals were qualitatively different, notably in that they did not exploit minimal solutions tuned to the environment where they evolved. For example, genetically determined individuals (i.e., individuals whose genes encode the synaptic strengths, but not the learning rules) solved the light-switching task by turning in circles tuned to the dimensions of the evolutionary arena which eventually take them over the light switch area and then over the light bulb area. Instead, evolved

Figure 6.54 *Left*: A Khepera robot equipped with a vision module can gain fitness points by staying on the gray area only when the light is on. At the beginning of a trial, the light is off, but it can be switched on if the robot passes over the black area positioned on the other side of the arena. The robot can detect ambient light and wall color with the linear vision module described earlier, but not the color of the floor. The fitness function is simply proportional to the amount of time spent by the robot under the light bulb when the light is on. *Right*: Behavior of an individual evolved in simulation with genetic encoding of learning rules.

adaptive individuals modified their synapses so as to develop in sequence the following set of behaviors: (a) avoid walls; (b) locate and go toward light switch; (c) locate and go toward light bulb; (d) stay under light bulb if it is on. Most importantly, these robots displayed remarkable adaptive properties after evolution. Best evolved individuals (1) transferred perfectly from simulated to physical robots, (2) accomplished the task when the light and reflection properties of the environment were modified, (3) accomplished the task when light bulb and light switch were positioned at different locations, and (4) transferred well across morphologically different robotic platforms. In other words, these robots evolved the ability to solve a partially unknown problem by adapting on the fly, rather than a solution to a specific problem seen during evolution. This result is conceptually similar to the results of the evolution of self-teaching networks described in the previous section.

In further experiments where the genetic code for each synapse of the network included one gene whose value caused its remaining genes to be interpreted as connection strengths (equivalent to the "genetically determined" case) or learning rules and rates (equivalent to the "adaptive" case), 80% of the synapses "made the choice" of using learning, reinforcing the fact that this genetic strategy has a comparatively stronger adaptive power (Floreano and Urzelai 2000). This methodology also has promising applications to the evolution of growing neural networks where synapses are created at run-time and thus their strengths cannot be genetically specified (Floreano and Urzelai 2001).

The adaptive advantages of this methodology were confirmed in the context of evolutionary spiking neurons for robot control. DiPaolo (2003) evolved spiking neural controllers modeled using the spike response model and heterosynaptic variants of the STDP learning rule described in chapter 3 for a robot required to approach a light source, but to avoid it if a sound was detected. The learning rule was described as a polynomial expression of the STDP rule where the components of the rule were weighted by individual constants that were genetically encoded and evolved (similar to the encoding proposed by Chalmers, which is described in chapter 3). The author showed that evolved robots were capable of learning suitable associations between environmental stimuli and behavior.

In the work of both Floreano and of Di Paolo, evolved adaptive controllers exploited the ability to continuously modify the synaptic strengths to regulate the behavior of the robot, instead of converging toward a stable pattern of synaptic strengths that would solve the task as most off-the-shelf neural learning algorithms do. Whether this is the case also for biological organisms or is some artifact of the robotic experiments remains to be investigated. At this stage, it is not yet technically possible to measure and correlate behavioral modification with synaptic modification in living organisms.

Niv et al. (2002) used a similar approach to evolve adaptive controllers in the context of value-dependent learning. The agent was a simulated bee whose fitness was the amount of nectar collected by visiting flowers that contained different quantities of nectar with variable probabilities (figure 6.55, a)). The bee could differentiate flowers through a simple visual system with a small field of view. (The experiments were also reproduced with a mobile Khepera robot equipped with a color camera.)

The neural architecture (figure 6.55, b)) was composed of three modules providing information on the current visual perception, on the difference between the current perception and the perception at the previous time step, and on the amount of nectar collected when the bee landed on a flower. The output of the neural network was the probability (figure 6.55, c)) of choosing a new random flying direction. The architecture contained both normal connections and neuromodulatory connections that modulated synaptic plasticity, as recently discovered in biological neural tissues (e.g., Bailey et al. 2000). Neuromodulation allows a neuron to gate the plasticity of a synaptic connection between two other neurons. In this architecture, the weights of the connections from the three modules toward the output unit (solid lines in

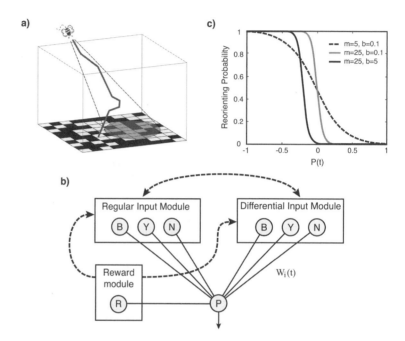

Figure 6.55   *a*) A simulated bee with a 10-degree cone view must collect nectar by flying towards patches of colored flowers. The environment contains two types of flowers, blue (B) and yellow (Y), that provide different quantities of nectar with different probabilities. The area outside the flower patch has no color (N). The bee descends from a random height in steps of one unit that can be taken in any downward direction (360° horizontal and 90° vertical). The nectar is collected when the bee lands on one of the flowers. *b*) The evolutionary neural controller is composed of three modules: the regular input module provides sensory information on the amount of B, Y, and N color currently perceived; the differential input module provides information on the difference between the perception at the current time step and the perception at the previous time step; the activation of the neuron in the reward module acts as a neuromodulatory input (dashed connections) that can enable Hebbian learning in the connections (solid lines) of the other modules. The output of the neural controller is a probability P to randomly reorient the current heading direction according to one of the functions shown in panel *c*). Reproduced from Niv et al. (2002). Image courtesy of Yael Niv, Hebrew University in Jerusalem.

the figure) could be modified if the presynaptic units connected by neuro-modulatory synapses (dashed lines in the figure) were active, according to a heterosynaptic Hebbian learning rule. This corresponded to using a binary value signal in the value-based learning rule of equation (6.1).

The genotype of the agent encoded the values of the constants associated to the components of the heterosynaptic Hebbian rule (as in Chalmers's and Di Paolo's experiments), the presence or absence of the connections, the initial weights of the synapses "at birth," and a global learning rate. The synaptic weight of the reward module was set to one and not evolved.

Bees were evolved in uncertain environments where at each generation one of the flower types was randomly assigned as a constant-yielding high-mean flower (0.7 μl of nectar) and the other as a variable-yielding low-mean flower (1 μl in one-fifth of the flowers, none in the others; mean 0.2 μl of nectar). Furthermore, the amount and mean nectar content was switched in the second or third quarter of the bee's life. Evolved bees learned first to land only on the flower patch and then (in half of the evolutionary runs) to land on flowers that yielded the larger expected amount of nectar. When the nectar distribution was switched, bees learned in a few feeding trials to land on the other type of flower.

Furthermore, evolved bees displayed *risk-aversion behavior* because, when tested in new environments where the two flower types had identical mean nectar content but one was constant and the other was variable, they chose constant-yielding flowers. This risk-aversion behavior was observed also for bees evolved in environments with constant nectar content in the two flowers. Risk aversion, which is observed in several choice scenarios in animals and humans and typically explained by nonlinear properties of the correspondence between the actual and perceived reward (Smallwood 1996), was explained by the authors as a consequence of Hebbian learning in finite time windows where the environmental uncertainty requires high learning rates.

Evolved bees also displayed *probability-matching behavior* whereby, when exposed to flowers containing the same amount of nectar but with different probabilities, they regulated their choice according to the ratios of the two probabilities instead of always choosing the flower type with the higher probability. This behavior, observed also in some animal species (Herrnstein 1997), could be explained by the same mechanism mentioned above. The authors showed that evolved neural controllers implemented (near-) optimal reinforcement learning and were similar to the Actor-Critic model of reinforcement learning described in chapter 3, but changed the synaptic weights

Figure 6.56   A simulated agent (triangle) lives in a 1D world with a visible landmark (rectangle). The agent collects fitness points by reaching a goal point (circle) across several trials where it always starts at the center of the world. The agent is exposed to four sets of trials where the positions of the goal and landmark are changed (*A*, *B*, *C*, and *D*). Within a set of trials, the positions are maintained constant. Adapted from Yamauchi and Beer (1994).

only in the presence of a reward instead of continuously as in the Actor-Critic model.

When compared to the purely epigenetic approach described in the earlier subsection on value-based learning, this approach relies on evolution to shape the weights of the connections that drive the neuromodulatory signals, which corresponds to the application of a genetically wired value system. In both approaches, the neural architecture is handcrafted and only the weights of the connections are allowed to change.

The question of which neurons should be neuromodulatory and how they should be wired to other neurons is not trivial and all computational models presented so far used fairly complicated architectures based on biologically or computationally motivated assumptions. In recent work, Soltoggio et al. (2007) described a set of experiments on artificial evolution of the topology of networks that could include, if needed, modulatory neurons. For the sake of comparison, the new experiments used the same bee problem described in this section. They showed that simulated bees evolved a modulatory network that maximized the total reward using simpler and more effective neural architectures. Evolution operated on genomes with analog genetic encoding (Mattiussi and Floreano 2007; Mattiussi et al. 2008), which was described in chapters 1 and 3. An interesting result of this work was that the temporal difference information, which both reinforcement-learning algorithms and Niv's hand-coded architecture rely on, was not used by evolved neuromodulatory topologies, which found a simpler cue in the sensory input for triggering the release of neuromodulators affecting learning.

6.8.3 Example 3: Evolution of Learning-like Behavior without Synaptic Plasticity

In animal studies it is very difficult to disentangle the contribution of learning to behavioral success (Plotkin 1988). It is therefore worth exploring whether situations that are typically assumed to require some form of learning could be tackled by agents without synaptic plasticity. Yamauchi and Beer (1994) evolved the neural controllers of simulated agents asked to reach for an invisible target area whose location changed during the life of the agent, but was always correlated to the location of a visible landmark for a series of trials (figure 6.56). The correlation between target area and landmark was kept constant across several trials, but could change during the life of an individual. In some trials, the landmark was between the agent's starting location and the target area; in other trials the landmark was at the opposite end of the arena. The fitness was the amount of time spent by the robot on the target area. One would expect that in this scenario, robots should use some sort of reinforcement learning that would associate the position of the landmark with the target location.

Yamauchi and Beer used a modular continuous time recurrent neural network without synaptic plasticity and evolved the synaptic weights and time constants. Evolved agents could solve the problem and switch heading direction depending on the spatial relations between the landmark and the target after a few trials. The authors explained the results by comparing the evolved neural network to a dynamical system that switches between different basins of attraction depending on the sequence of sensory information.

However, the authors could not evolve monolithic controllers for solving this task, raising the question of whether the problem was too difficult to be solved by evolution unless one restricted the search space by using specific, modular architectures. Tuci et al. (2002) addressed this question by evolving a fully connected (monolithic), continuous time recurrent neural network for a mobile Khepera robot in a more realistic version of the landmark-target scenario (figure 6.57). The landmark was a light bulb that could be perceived by the robot anywhere in the environment. The target area was a black stripe that could be perceived by the robot only when it was on it. The fitness function was proportional not only to the time spent by the robot on the target area but also to the proximity of the robot to the target area. In other words, the fitness landscape was smoother than in the case of Yamauchi and Beer, who used a binary fitness function. Evolved neural controllers were capable of detecting the spatial relation between landmark and target and modify

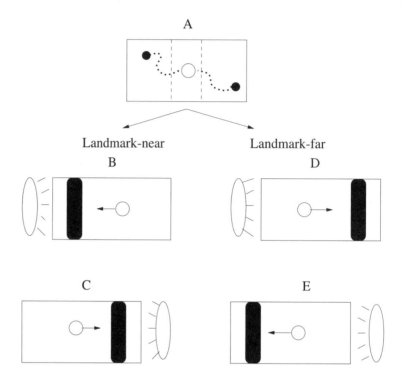

Figure 6.57 A mobile Khepera robot equipped with light sensors and a floor-color sensor lives in an arena with a light source. The fitness of the robot is inversely proportional to its distance to a target area (black stripe painted on the floor) across several trials where it always starts in the center of the environment (dashed area in *A*) at a random orientation. As in the experiments by Yamauchi and Beer (1994), the robot is exposed to four sets of trials where the positions of the target area and of the light source are changed (*B, C, D, E*). Within a set of trials, the positions of the target area and light source are maintained constant. Adapted from Tuci et al. (2002).

their heading direction accordingly. These experiments indicated that the failure with the evolution of monolithic networks by Yamauchi and Beer was not due to architectural constraints, but probably to the fitness function used.

In another series of experiments, Blynel and Floreano (2003) evolved fully connected continuous time recurrent neural networks for a Khepera robot required to find a target location in the T-maze shown in figure 6.58. In this task, which is a classic experimental scenario for studying reinforcement learning in animals (Gallistel 1990), the robot or animal is positioned at the origin of the maze and is let free to explore the maze. If it arrives at the end-

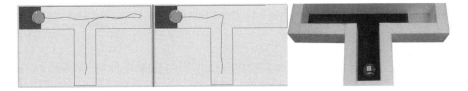

Figure 6.58 The T-maze task. The robot must locate and stay over the colored area across several trials. The location of the colored area is randomly changed every n trials. The robot can perceive the target area when it is on it, using its floor sensor. At the beginning of a trial, the robot starts at the origin of the T-maze. *Left*: First trial of an evolved robot. *Center*: Second trial of the same evolved robot. *Right*: Evolved controllers are successfully transferred to the real robot.

point of the target arm, it receives a reward signal. For animals, the reward is a piece of food or a signal (e.g., sound or light).

The robot instead did not receive a reward, but could perceive the target area with a floor sensor when it was over it and its fitness value was proportional to the amount of time spent over the target in a number of trials. The location of the target area could change during the life of the robot, but it was kept constant for several trials. The states of the neurons were *not* reset at the beginning of a trial. Evolved robots could easily solve the problem by using the first one or two trials to explore the maze while in all remaining trials they correctly went to the target endpoint. If the target endpoint was switched, they required only one trial to move to the other endpoint. Neural controllers evolved in simulation were easily transferred to the real robot. Furthermore, the authors successfully evolved robots capable of solving a double T-maze problem where the endpoint of each of the two arms in the simple T-maze is the origin of another T-maze, giving a total of four possible target areas. Evolved controllers solved the problem by activating one or more neurons when the cue signal was found and maintaining those neurons active in the remaining trials, therefore modifying the internal activation dynamics necessary to produce the correct behavioral trajectories (Blynel 2001).

The results presented in this subsection fit neurophysiological evidence and theories that perception, classification, and memory in animals and humans can be explained by the establishment of chaotic attractors in large assemblies of interconnected neurons (e.g., in the olfactory system) and transition among these attractors (e.g., Freeman 2001). However, the main goal of these experiments with robots was to provoke awareness that synaptic plasticity may not always be necessary to explain behaviors that seem to require learning. There is no doubt that living brains use synaptic plasticity, but that plasticity may not always be required for learning or may be intimately and intricately combined with dynamical properties of neural activation. From an engineering perspective, these results suggest that a promising approach to the evolution of learning agents consists in using genetic encodings that allow for the combination of plastic and nonplastic connections. For example, one could add genes that affect the expression of other genes, which may result in adaptive or in fixed synaptic connections, as mentioned in the previous section (Floreano and Urzelai 2000, 2001).

6.9   Evolution and Neural Development in Behavioral Systems

A developmental genotype-to-phenotype mapping may (a) result in a more compact and evolvable genetic encoding of complex control systems; (b) include the possibility of developing the phenotype while the robot interacts with its environment, akin to a maturation process, thus adding another form of adaptive plasticity to the evolving robot; and (c) incorporate the developmental rules of both the control system and the body morphology. In this section we will describe seminal approaches that address points (a) and (b) and in a later section on coevolution of body and control we will describe approaches that address point (c).

In natural organisms the development of the nervous system begins with the folding in of the ectodermic tissue which forms the neural crest. This structure gives origin to the mature nervous system through three phases: the genesis and proliferation of different classes of neurons by cellular duplication and differentiation, the migration of the neurons toward their final destination, and the growth of terminals (axons, dendrites).

In chapter 4, we described a method for cellular neural encoding developed by Gruau (1994a). He applied this method to evolve neural controllers for a simulated hexapod robot. The fitness function was the distance covered by the robot within a limited amount of time. The controllers were continuous time recurrent neural networks. The model of the robot was a simplified version of a six-legged insect. The genotypes consisted of network-generating programs that could recursively point to subroutines in order to develop complex neural architectures.

Evolved genotypes were more compact and the phenotypes displayed more regular and symmetric architectures, as compared to encoding that

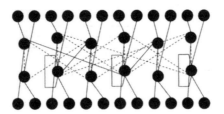 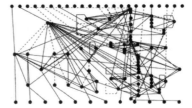

Figure 6.59 *Left*: Neural network architecture resulting from a developmental genetic encoding that can reuse subroutines. *Right*: Neural network architecture resulting from a developmental encoding without reuse of subroutines. Adapted from Gruau (1994a).

could not reuse subroutines (figure 6.59) that matched the spatial distribution of the actuators on the robot body. In the example displayed on the left of figure 6.59, the network included three repeated subnetworks controlling two legs each. In further work, Gruau and Quatramaran (1997) applied this method to the real octopod robot shown in figure 6.38 by including a number of handcrafted constraints that reduced the search space.

SIMPLE
GEOMETRY-ORIENTED
CELLULAR ENCODING
(SGOCE) Kodjabachian and Meyer (1998a) proposed a related approach, known as *simple geometry-oriented cellular encoding* (SGOCE), that was applied to the same insect model. The fitness function was the distance covered during the evaluation period increased by a term encouraging any leg motion. In this approach, neurons grew their connections in a two-dimensional matrix substrate where their positions affected the types of connections that were established. As in the case of Gruau's model, the genotype was formed by a tree of instructions and crossover was accomplished by exchanging subtrees (genetic programming). Genotypes included only six different instructions: (a) a cellular division instruction that takes as parameters the distance and the relative angle of the daughter cell; (b) two instructions that create outgoing or incoming connections and take as parameters the angle and distance of the neuron with which the connection should be made and its connection weight; (c) two instructions that specify the values of the time constant and the bias threshold of the cell; and (d) an instruction that causes a cell to die.

The authors were able to evolve behaviors of increasing complexities in an incremental fashion (Kodjabachian and Meyer 1998b). The evolved neural controller was saved and used as a building block for evolving more complex controllers. In a second evolutionary stage an additional module was evolved. This module received sensory information through additional sensory cells and influenced the walking behavior by establishing connections with the cells of the first module. In one experiment, a gradient-following module was evolved in an additional substrate placed just to the left of the first substrate which was provided with two precursor cells and two photoreceptor sensors. In this second phase, individuals were selected for their ability to approach the light source as quickly as possible. The authors later evolved a third obstacle avoidance module that received information from two additional contact sensors.

When compared to the encoding scheme proposed by Gruau, Kodjabachian and Meyer handcrafted a set of important phenotypical properties: the geometry of the substrate over which the network developed, the positions of the sensory cells and of the motor neurons, and the number and position of six precursor cells. The handcrafting of these features is probably one of the reasons why the authors observed the emergence of walking behaviors in significantly shorter time (after about 100,000 evaluations) than in Gruau's experiments.

Several other authors proposed related approaches that included cell division, migration, and establishment of connections (e.g., Cangelosi et al. 1994; Dellaert and Beer 1996).

Husbands et al. (1994) proposed a developmental approach that did not invoke cellular division. In this approach a number of genetically specified neurons were located in a brain space and connections grew according to a set of differential equations. The genotype encoded the properties of each neuron (the type of neuron, the relative position with respect to the neuron created previously, the initial direction of growth of the dendrites, and the parameters of the equations governing the growth process). During the genotype-to-phenotype process, the genetic string was scanned from left to right until a particular marker was found. When a special marker indicating the beginning of the description of a neuron was encountered, the following bits were read and interpreted as parameters for a new neuron. The presence of an additional marker, however, could indicate that the parameters of the current neuron were specified in a previous portion of the string. This mechanism could potentially allow the emergence of phenotypes with repeated structures formed by reexpression of the same genetic instructions, similar to the approach described by Gruau.

However, in all the approaches described so far development occurred instantaneously before evaluating the fitness of the fully formed phenotype. Nolfi et al. (1994b) used a growing encoding scheme to evolve the architecture and the connection weights of neural networks controlling a Khepera

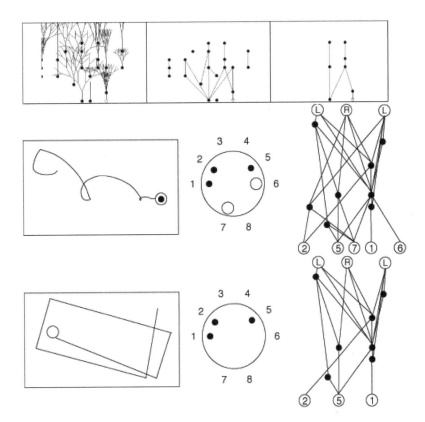

Figure 6.60 *Top*: Growth of axonal connections in the 2D brain space, pruning of axons that do not connect to other neurons, and pruning of all neurons that are not connected to input, output, or other parts of the network. *Center*: Testing an evolved individual in a "light" environment. Trajectory, connected sensors, and neural architecture. *Bottom*: Testing an evolved individual in a "dark" environment. Trajectory, connected sensors, and neural architecture.

robot, but the growth of the neural network occurred *during* the lifetime of the robot while its fitness was evaluated. These controllers were composed of a collection of artificial neurons distributed over a two-dimensional space with growing and branching axons (figure 6.60, top). The genotype specified instructions that control the axonal growth and branching process of a set of neurons. When the axon growing from one neuron reached another neuron, a connection between the two neurons was established. Axons grew and branched only if the neurons displayed an activation variability above a genetically specified threshold. Axons that did not connect to other neurons and neurons that remained unconnected were pruned.

This activity-dependent growth was based on the idea that the sensory information coming from the environment played a critical role in the maturation of the connectivity of the biological nervous system. Indeed, it has been shown that the maturation process is affected by the activity patterns of single neurons (Purves 1994; Quartz and Sejnowski 1997). The developmental process of these individuals was therefore determined by both genetic and environmental factors. The genotype of evolving robots was divided in blocks, each coding for the growth properties of corresponding neurons. The genes of input neurons also specified which sensors they were connected to (out of eight proximity and eight light sensors available on the Khepera robot). Similarly, genes of output neurons specified which of the two wheels they affect.

Robots were evolved in a 60 x 35 cm arena to find and stay on a randomly positioned target. At even generations, the target was illuminated by a small light that could be perceived at a distance by the robots; in odd generations there was no light and the robots could detect the target area with a floor sensor only when they happened to pass by it. Evolved individuals were able to reach the target area most of the time both in dark and in light environments although performances were clearly better in light environments.

Evolved robots displayed different patterns of development, neural architecture, and behavior according to the environment where they were born (figure 6.60, center and bottom). For example, an evolved individual placed in the "light" environment grew an architecture that used both proximity and light sensors in order to avoid walls and approach the light on top of the target area. When this same individual was placed in the "dark" environment, it did not grow connections from light sensors, but displayed an exploratory strategy that increased the chance of finding the target area while avoiding the walls. This experiment clearly indicated that lifetime development represents a form of adaptive plasticity.

Vaario et al. (1997) used a neural growth approach based on diffusion equations guiding the growth of connections (akin to Husbands's approach) that took into account the activity of sensory neurons while the robot interacted with the environment (akin to Nolfi's approach). They successfully evolved neural controllers for mobile robots that were expected to reach light sources while navigating a looping maze similar to that described in an earlier section of this chapter (Floreano and Mondada 1994). However, they did not test the adaptability of the resulting controllers in changing environments, as Nolfi et al. (1994a) did.

To summarize, a developmental mapping from genotype to phenotype can be an advantage if it can generate repeated structures that match the spatial layout of the behavioral system, such as multiple legs. This may result in faster and higher-fitness evolution than the case where each repeated module is separately encoded in the genotype and must be separately evolved. The other advantage of a developmental mapping comes from the adaptive growth of the architecture while the individual interacts with the environment. Developmental costs, such as a retarded appearance of a mature architecture, can be reduced by including the developmental period in the computation of the fitness.

6.10 Coevolution of Body and Control

In the work described so far there has been an overwhelming tendency to evolve control systems for preexisting robots: the brain is constrained to fit a particular body and set of sensors. Of course, in nature the nervous system evolves simultaneously with the rest of the organism. As a result, the nervous system is highly integrated with the sensory apparatus and the rest of the body: the whole operates in a harmonious and balanced way - there are no distinct boundaries between control system, sensors, and body. From the start, work at Sussex University incorporated the evolution of sensor properties, including positions, but other aspects of the physical robot were fixed (Cliff et al. 1993). Although the limitations of not being able to genetically control body morphology were acknowledged at this stage, there were severe technical difficulties in overcoming them, so this issue was somewhat sidelined.

Karl Sims started to unlock the possibilities in his highly imaginative work on evolving simulated 3D "creatures" in an environment with realistic physics (Sims 1994). In this work, the creatures coevolved under a competitive scenario where they were required to gain control of a resource (a cube) placed in the center of an arena (figure 6.61). Both the morphology of the creatures and the neural system controlling their actuators were under evolutionary control. Their bodies were built from rigid 3D primitives with the overall morphology being determined by a developmental process encoded as a directed graph, as described in chapter 4. Various kinds of genetically determined joints were allowed between body parts. A variety of sensors

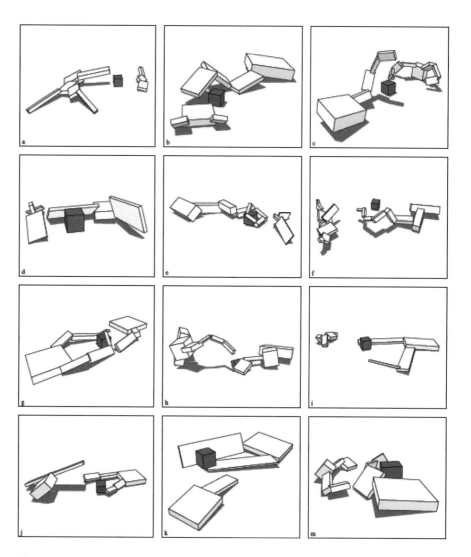

Figure 6.61   Some of the coevolved creatures competing to get hold of the dark cube. Reproduced from Sims (1994).

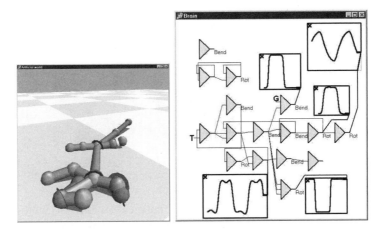

Figure 6.62   A snapshot of the simulator for coevolving body and neural morphologies of stick-like creatures. From Komosinski and Ulatowski (2000).

could be specified for a specific body part. The simulated world included realistic modeling of gravity, friction, collisions, and other dynamics such that behaviors were restricted to be physically plausible. Many different styles of locomotion evolved along with a variety of interesting, and often entertaining, strategies to capture the resource. These included pushing the opponent away and covering up the cube.

With the later developments of sophisticated physics engines for modeling a variety of physical bodies, Sims's work inspired a rush of evolved virtual creatures, including realistic humanoid figures capable of a variety of behaviors (Reil and Husbands 2002). Komosinski and Ulatowski (1999) devised an approach similar to Sims's work, but greatly simplified the genetic representation and the description of the creatures, which consisted of elongated bars joined by muscle-like connections. Each bar, whose shape parameters were genetically encoded, could be equipped with several sensors, such as light, temperature, force, gyroscopes, and food, that were connected to evolvable neural architectures. This work resulted in the software FRAMSTICKS Framsticks (Komosinski and Ulatowski 2006) which can be used to coevolve body and brains of creatures in several environments with user-defined fitness functions (figure 6.62).

Bongard and Pfeifer (2003) expanded this line of work by evolving multicellular organisms (figure 6.63) where individual cells grew and differentiated according to the dynamics of gene regulatory networks. Building

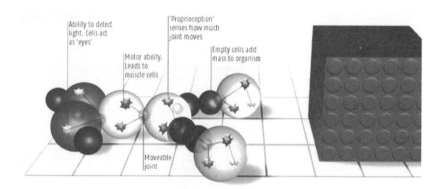

Figure 6.63 Artificial ontogeny of a multicellular organism with sensor, body, and muscle cells (see also figure 4.36). Reproduced from Graham-Rowe (2002). Reprinted with permission of *New Scientist*.

on previous work by Eggenberger (1997b) on morphological differentiation, Bongard and Pfeifer (2003) used a genotype encoding genes that could either affect the phenotype or regulate (promote or inhibit) the activation of other genes (see section 4.8.5). Some of these genes encoded the size and type of cells; other genes encoded the neural structure that could be associated with a cell. Morphological growth and differentiation therefore proceeded along with neural growth and differentiation. However, in contrast to Nolfi's work on neural growth described earlier, in this case the fitness evaluation of the individual started only after development of the organism. The authors showed successful evolution of multicellular organisms capable of growing the appropriate body mass, morphology, sensorimotor apparatus, and control necessary to push boxes of different sizes.

In what might be regarded as a step toward evolving physical robot bodies, Funes and Pollack (1998) explored the use of evolutionary algorithms in the design of physical structures, taking account of stresses and torques. They experimented with evolving structures assembled from elementary components (Lego bricks). Evolution took place in simulation and the designs were verified in the real world. Stable 3D brick structures such as tables, cranes, bridges, and scaffolds were evolved within the restrictions of maximum stress torques at each joint between brick pairs. Each brick was modeled as exerting an external load with a lever arm from its center of mass to the supporting joint, resulting in a network of masses and forces representing the structure. A genetic programming approach was taken us-

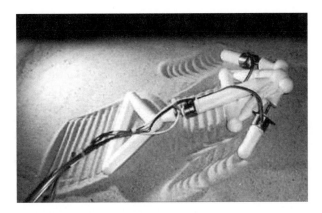

Figure 6.64   A locomoting "creature" evolved by Lipson and Pollack (2000) in research which achieved an autonomous design and fabrication process.

ing tree structures to represent the 3D Lego structures. A mutation operator acted on individual brick parameters while subtree crossover allowed more radical changes to the structure. The fitness function was designed to encourage particular types of structures. It also included a cost factor proportional to the number of bricks that successfully weeded out many of the redundant bricks that inevitably arose. Lego proved to be a predictable building tool with modes of breakage and linkage that could be modeled relatively easily. While this work was successful, producing very strong designs, it focused on static structures, so was limited in terms of its relevance to functional robotic body parts. However, it did demonstrate a viable approach to evolving physical structures.

While various researchers advocated the use of fully evolvable hardware to develop not only a robot's control circuits but also its body plan, which might include the types, numbers, and positions of the sensors; the body size; the wheel radius; actuator properties; and so on (e.g., Lund et al. 1997), this was still largely confined to theoretical discussion until Lipson and Pollack's GOLEM PROJECT work on the golem project (Lipson and Pollack 2000), which was a significant step forward from the earlier Lego work (Funes and Pollack 1998).

Lipson and Pollack, working at Brandeis University, pushed the idea of fully evolvable robot hardware about as far as is reasonably technologically feasible at present. In an important piece of research, directly inspired by Sims's earlier simulation work and Framsticks shapes, autonomous "creatures" were evolved in simulation out of basic building blocks (neurons, plastic bars, actuators) (Lipson and Pollack 2000). The bars could connect together to form arbitrary truss structures with the possibility of both rigid and articulated substructures. Neurons could be connected to each other and to bars whose length they would then control via a linear actuator. Machines defined in this way were required to move as far as possible in a limited period of time. The fittest individuals were then fabricated robotically using rapid manufacturing technology (plastic extrusion 3D printing) to produce results such as that shown in figure 6.64. They thus achieved autonomy of design and construction using evolution in a limited-universe physical simulation coupled to automatic fabrication. The fitness function employed was simply the Euclidean distance moved by the center of mass of a machine over a fixed small number of cycles of its neural controller. A number of different mutation operators acted in concert: small changes to bar or neuron properties, additions and deletions of bars or neurons, changes to connections between neurons and bars, and creation of new vertices. The highly unconventional designs thus realized performed as well in reality as in simulation. The success of this work leads the way to new possibilities in developing energy-efficient, fault-tolerant machines.

In current work, Lipson (2005) is developing novel types of rapid prototyping machines capable of assembling different materials into functional devices, such as batteries and electrical actuators. These machines, also known FABBERS as *fabbers*, position the item to be fabricated and change the deposition material according to computer-generated instructions. Although so far the instructions are specified according to engineering principles, the ultimate goal is to introduce an evolutionary process capable of generating the instructions for robots that will autonomously "walk out" of the machine.

6.11   Toward Self-Reproduction

In the work described so far, robots are either composed of preexisting structures (albeit modifiable), assembled by hand, or built after instructions resulting from software evolution. In order to have a fully autonomous evolutionary robot, robots should be capable of self-assembly and self-reproduction.

The idea of creating self-assembling and self-reproducing physical systems that could evolve into increasingly complex structures was first entertained by John von Neumann in 1948. In a series of informal meetings held at the Institute for Advanced Study at Princeton University, which were reconstructed by A.W. Burks (von Neumann 1966) from the notes and memories of the participants, von Neumann discussed and described the behavioral and physical properties that individual elements of such systems should have. His system was composed of about "one or two dozen parts with simple properties [existing in] the millions." He described eight elements that included a sensory organ, a coincidence organ and an inhibitory organ, a rigid body, a muscle, a fusing organ, a cutting organ, and a stimulus producer (the environment). He also planned to add a source of energy and an element that actively recruited necessary elements from a large collection of other elements, but did not provide further details about them. Von Neumann argued that embedded binary logic in the form of McCulloch-Pitts neurons (McCulloch and Pitts 1943) was sufficient for these systems to self-assemble and self-reproduce, but speculated that inheritable mutations (errors) in the system were necessary to bootstrap the process and lead to the degree of complexity observed in self-reproducing biological systems. He also mentioned that these systems should be put in an environment where they could "float" in order to allow for the necessary patterns of self-connection. Eventually, von Neumann gave up the realization of physical self-assembling and self-reproducing systems because of the technological limitations at that time. Instead, he concentrated on two-dimensional logic systems, which eventually resulted in the conception of cellular automata described in chapter 2 (for a review of self-replication in software systems, see Sipper (1998)).

Other models of physical self-replication were studied a few years later. For example, Penrose concentrated on the shape and possible mechanical interaction of molecules that self-connect and copy themselves by simple vibrations (Penrose 1959, 1962) and demonstrated that with wood blocks. Other attempts include those of Jacobson (1958) and Morowitz (1959) who used a system of railroad cars in a closed-loop circuit, but those attempts were not concerned with the evolution of complexity, which was von Neumann's main concern (McMullin 2000). Eventually, the interest in physically self-reproducing and evolving systems faded out in favor of theoretical models, computer-based simulations, and in vitro experiments with organic molecules.

Von Neumann's ideas on self-assembly developed in the modern field of self-assembling and self-configuring robots, which will be described later in chapter 7. All these robotic systems are composed of simple actuated units (most often consisting of a block and a joint) that can connect to each other in various ways in order to form several shapes. In some cases, these robots can

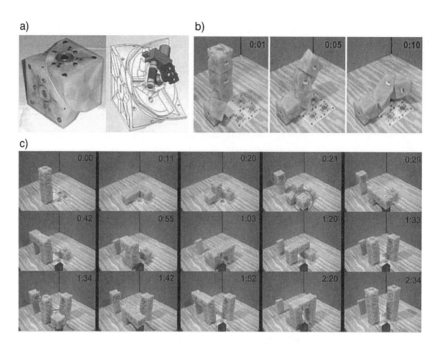

Figure 6.65 Self-reproduction of a four-module robot. *a*) Molecube (basic module), with an illustration of its internal actuation mechanism. *b*) Snapshots from the first 10 seconds showing how a four-module robot transforms when its modules swivel simultaneously. *c*) Sequence of frames showing the self-reproduction process, which spans about 2.5 minutes New molecubes are manually positioned at the two "feeding" locations circled in red. Reprinted by permission of Macmillan Publishers Ltd: *Nature* (Zykov et al. 2005).

even change their shape and behavior without human intervention. However, the basic modules are all equal and are not conceived to self-reproduce.

The idea of robotic self-reproduction was first put forward by Fukuda in his work on cellular robotics (Fukuda and Ueyama 1994) where he described autonomous docking and separation of two wheeled robots as well as a manipulator robot that assembled another robot out of preassembled robot modules. Lipson and colleagues investigated the shape and other physical properties of basic modules for a self-replicating robot (Mytilinaios et al. 2004). Their solution involved a set of identical cubes with magnetic actuators that could switch between repel, attract, and passive modes. In later work, the authors demonstrated physical self-reproduction of multicellular MOLECUBES robots, made out of motorized and articulated cubes (*molecubes*), that could build a copy of themselves (Zykov et al. 2005), as shown in figure 6.65. However, these robotic systems had to be manually fed with other cubes and could only construct a copy of themselves according to prespecified instructions.

At the time of writing this book, the combination of autonomous self-assembly and self-reproduction in physical systems has not yet been achieved. We will return to robots composed of several modular units in chapter 7.

6.12   Simulation and Reality

Few of the experiments described in this chapter were carried out entirely on physical robots because (a) learning or evolution may require several lengthy evaluations, especially if it is carried out on a single robot that incarnates the bodies of all the individuals of an evolving population; (b) the physical robot may be damaged because populations always contain a certain number of poorly performing individuals (for example, colliding against walls) from the effect of random mutations; (c) restoring the environment to initial conditions between trials of different individuals or populations (for example, replenishing the arena with objects) may not always be feasible without human intervention; and (d) evolution of morphologies and evolution of robots that can grow during their lifetime is almost impossible with today's technology without some level of human intervention.

For these reasons, researchers often resort to learning or evolution in simulation and transfer the evolved controllers to the physical robot. However, it is well-known that programs that work well in simulation may not function properly in the real world because of differences in sensing, actuation, and in the dynamic interactions between the robot and the environment (Brooks 1992). This "reality gap" is even more evident in adaptive approaches, such as evolutionary robotics, where the control system and morphology are gradually crafted through the repeated interactions between the robot and the environment. Therefore, robots will evolve to match the specificities of the simulation, which differ from the real world. Although these issues clearly rule out any simulation based on grid worlds or pure kinematics, over the last 10 years simulation techniques have dramatically improved and resulted in software libraries that model reasonably well system dynamics, such as friction, collision, mass, gravity, inertia, etc. (Featherstone and Orin 2000). These software tools allow one to simulate articulated robots of variable morphology as fast as, or faster than, real time on a desktop computer. Today, these

REALITY GAP

physics-based simulations are widely used by most researchers in evolutionary robotics.

Nonetheless, even physics-based simulations include small discrepancies that can accumulate over time and result in behaviors that are very different from reality (for example, a robot may get stuck against a wall in simulation whereas it can free itself in reality, or vice versa). Also, physics-based simulations cannot account for the diversity of response profiles of the individual sensors, motors, and gears of a physical robot. There are at least four methods to cope with these problems and improve the quality of the transfer from simulation to reality.

INDEPENDENT NOISE
- A widely used method consists of adding *independent noise* to the values of the sensors provided by the model and to the end position of the robot computed by the simulator (Jakobi et al. 1995). Some software libraries allow the introduction of noise at several levels of the simulation. This method prevents learning or evolution from finding solutions that rely on the specificities of the simulation model. One could also sample the actual sensor values of the real robot positioned at several angles and distances from objects of different texture. These values are then stored in a lookup table and retrieved with the addition of noise according to the position of the robot in the environment (Miglino et al. 1996). This method proved to be very effective for generating controllers that transfer smoothly from simulation to reality. A drawback of the sampling method is that it does not scale up well to high-dimensional sensors (e.g., vision) or geometrically complicated objects.

MINIMAL SIMULATIONS
- Another method, also known as *minimal simulations*, consists in modeling only those characteristics of the robot and environment that are relevant to the emergence of desired behaviors (Jakobi 1997b). These characteristics, which are referred to as base-set features, should be accurately modeled in simulation. All the other characteristics, which are referred to as implementation aspects, should be randomly varied across several trials of the same individual in order to ensure that evolving individuals do not rely on implementation aspects, but rely on base-set features only. Base-set features must also be varied to some extent across trials in order to ensure some degree of robustness of the individual with respect to base-set features, but this variation should not be so large that reliably fit controllers fail to evolve at all. This method allows for very fast evolution of complex robot-environment situations, as in the example of the hexapod walk described earlier. A drawback of minimal simulations is that it is not always

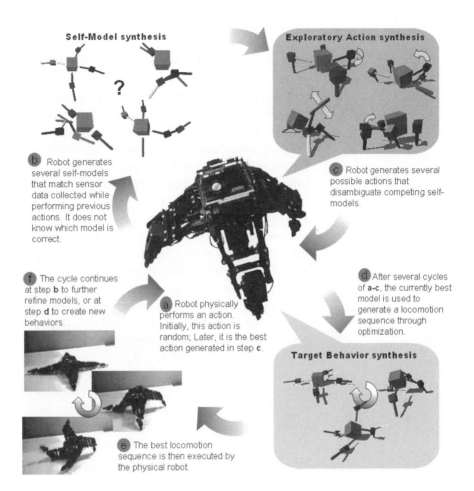

Self-Model synthesis

?

Exploratory Action synthesis

b Robot generates several self-models that match sensor data collected while performing previous actions. It does not know which model is correct.

c Robot generates several possible actions that disambiguate competing self-models.

f The cycle continues at step **b** to further refine models, or at step **d** to create new behaviors.

a Robot physically performs an action. Initially, this action is random; Later, it is the best action generated in step **c**.

d After several cycles of **a-c**, the currently best model is used to generate a locomotion sequence through optimization.

Target Behavior synthesis

e The best locomotion sequence is then executed by the physical robot.

Figure 6.66   Schematic outline of the estimation-exploration algorithm applied to robotics. (*a*) The physical robot begins by performing an action, and then (*b*) uses the resulting sensor data to synthesize a set of self-models that explain the observed data. Then, (*c*) the robot synthesizes a new action that disambiguates between the models generated in step *b*. This cycle continues until the set of models converges or a fixed number of cycles have elapsed. The most accurate model from the set is then passed to the behavior synthesis component (*d*), which creates a behavior for the robot using this model. The best behavior is then executed by the physical robot (*e*). If a new behavior is desired, the experiment resumes at step *d*; if an unexpected motor-sensor pattern is detected, the experiment resumes at step *b*. Reproduced from Bongard et al. (2006). Courtesy of Josh Bongard, University of Vermont, Burlington.

easy to tell in advance which are the base-set features that are relevant to the desired behavior.

- Yet another method consists in the coevolution of a robot's controller and of the simulator parameters that are most likely to differ from the real world and that may affect the quality of the transfer (Bongard and Lipson 2005). This method, also known as the *estimation-exploration algorithm*, consists of coevolving three populations, one encoding possible simulator parameters, one encoding robot controllers used only to find a good simulator, and the third also encoding robot controllers, but these controllers are meant to allow the robot to fulfill its task, such as forward locomotion (figure 6.66). Coevolution happens in several passes through a three-stage process. In stage 1, a randomly generated population of controllers is evolved in the default simulator and the best individual is executed on the real robot while the time series of sensory values is recorded. In stage 2, the population of simulators is evolved for reducing the difference between the time series recorded on the real robot and the time series obtained by running the same controller within the simulator. The best evolved simulator is then used for stage 1 where a new randomly generated population of controllers is evolved and the best individual is tested on the real robot to generate the time series for another pass of simulator evolution. This two-stage coevolution is repeated several times until the error between simulated and real robot behavior is as small as possible. Then, the best simulator found so far is used by a third population (the third stage), in which robot controllers are again evolved, this time to elicit some desired behavior. The best controller from this population is then executed on the physical robot, and, if the evolved simulator was accurate, the real robot will exhibit the same behavior as that seen in the evolved simulator.

  It was shown that 16 passes of the three-stage process were sufficient to evolve a good control system that could be transferred to an articulated robot. In this case, the real robot was used to test only 17 controllers: the 16 controllers used to find an accurate simulator, and the 17th evolved to enable the robot to perform some desired task (Bongard et al. 2006). The authors also showed that this approach could be used to adapt the control system to cope with a damaged robot (for example, loss of a leg, malfunctioning of a sensor, etc.). The advantage of this approach with respect to other methods for damage recovery with evolutionary algorithms (e.g., Grefenstette and Ramsey 1992; Keymeulen et al. 1998; Mahadavi and Bentley 2003) is that it not only incrementally evolves the controller to the changed situation using very few trials on the physical platform but it also reveals in the coevolved simulator the type of damage incurred by the physical robot. This is particularly useful for situations where the physical robot is operating in remote areas, such as other planets, where engineers can only get sensory signals from the robot, but cannot actually see the robot.

EVOLUTION OF
LEARNING

- Finally, another method consists in *evolution of learning*, rather than evolution of control parameters (e.g., connection strengths), as we have seen in a previous section. The parameters of the decoded control system are always initialized to small random values at the beginning of an individual lifetime and must self-organize using the learning rules (Urzelai and Floreano 2001). This method prevents evolution from finding a set of control parameters that fit the specificities of the simulation model and encourages emergence of control systems that remain adaptive to partially unknown environments. When such an evolved individual is transferred to the real robot, it will develop its control parameters online according to the genetically evolved learning rules and taking into account the specificities of the physical world.

6.13   Closing Remarks

Embodiment and situatedness are central aspects of all forms of life on this planet. Organisms exist in a physical form and live by acting on their environment. All living organisms are therefore behavioral systems, but not all bioinspired systems are behavioral systems. For example, the systems that we considered prior to this chapter captured some aspect of biology, but in general they were not behavioral systems.

Behavior introduces unique constraints that make it difficult to reproduce in artificial systems by means of approaches that attempt to hand-code it, reduce noise, and improve controllability. However, difficulties are reduced if the behavioral system is endowed with self-adapting abilities, intrinsic noise is exploited rather than reduced, and control is delegated as much as possible to the intrinsic dynamics of the body and of the environment.

Nonetheless, the design of an artificial behavioral system that is endowed with appropriate embodiment and robust self-adaptation properties is still a major challenge because it requires choosing and balancing multiple components with highly dynamical properties. We believe that artificial evolution of behavioral systems has very promising potentials for the design of behavioral artifacts because it takes a global perspective where the genetically encoded components (morphology, sensing, actuation, wiring, development, adaptation, etc.) can capitalize on each other's interdependencies in endowing the organism with the unique features that are required for its evolutionary success. To this extent, any artificially evolved individual whose fitness does not fluctuate much along several generations is a balanced behavioral system.

However, evolutionary robotics introduces several new challenges that are not yet solved, such as the choice of genetic encodings and mappings that are suitable for complex embodied systems, the choice of fitness functions and methods for assessing behavioral systems, the length of time required for obtaining viable solutions, and the inadequacy of hardware technologies for autonomous evolution, to mention only a few. Although finding solutions to these questions will generate huge rewards, other bioinspired approaches that attempt to capture hardware and control principles of living behavioral systems have had a significant impact and contributed to the dramatic improvement of robot technologies that we have witnessed over the last decade.

It is sometimes argued that artificial evolution and reinforcement learning are similar because in both cases the behavioral system is shaped by a single numerical value (fitness function and reward policy, respectively) that is available only at the end of a sequence of sensorimotor events (lifetime and trial, respectively). However, the two methods are not comparable because (a) evolution involves also the system architecture (both hardware and software), which as we have seen in this chapter affects the characteristics of the computational problem; (b) evolution searches the problem space very differently from reinforcement learning, although no formal comparisons are available yet; and (c) evolution can subsume reinforcement learning or any other type of value-based and non-value-based learning, including the possibility of discovering novel learning structures. It is interesting to notice that robotics experiments with artificial evolution largely outnumber those with reinforcement learning, although this may due to the difficulties inherent in classic reinforcement learning algorithms, as we mentioned in the section on value-based learning.

A major challenge of artificial behavioral systems consists in the development of novel materials that display the flexibility, adaptability, and self-healing properties of the biological materials that make up living behavioral systems. Today's technology relies on metallic and rigid structures, servomotors, and digital computers. These structures are inert, demand control, and naturally lead to the separation of hardware and software design. Recent work in biomimetic design, morphological computation, and self-reproduction described in this chapter represents a first step in the direction of more robust, adaptive, and functional behavioral systems.

6.14 Suggested Readings

The book on ecological perception by Gibson (1979) represents one of the earliest works showing how behavior can simplify the computational operations required of an animal or machine. The book is easy to read, still modern, and useful for robotics researchers alike. The imaginary vehicles described by Braitenberg (1984) represent a stimulating and fun-reading introduction to the importance of taking behavior into account when designing and analyzing both artificial and living control systems. For roboticists, we suggest the annotated collection of papers by Brooks (1999) on the technology, methods, and philosophy of behavior-based robotics. A more comprehensive introduction to new AI, with emphasis on behavioral systems, is provided by Pfeifer and Scheier (1999), who also suggest a series of principles that should be followed in the design of intelligent artifacts. By the same first author, we also recommend (Pfeifer and Bongard 2007) for its emphasis on the role of embodiment and morphology in reducing and affecting computation.

Biorobotics, that is, biological inspiration for robots and robots as biological models, is a rapidly growing field. We recommend the books edited by Webb and Consi (2001) and by Ayers et al. (2002) for a collection of seminal works that include both approaches. We also suggest the book edited by Healy (1998) for a collection of papers on spatial navigation in animals and robots. The book by Nolfi and Floreano (2000) provides a comprehensive introduction to evolutionary robotics with emphasis on the methodology and on the important details of foundational experiments.

Throughout this chapter and in the closing remarks we repeatedly pointed to the importance of novel types of hardware for bioinspired behavioral systems. We definitely recommend the book on comparative biomechanics by Vogel (2003) for an extensive and critical introduction to the methods and technologies used to reproduce biological functionality. We also suggest the book edited by Bar-Cohen (2001) for an overview of recent developments in artificial muscles and other materials with bioinspired functional properties.

Finally, for a more relaxed approach, we recommend the book *Robosapiens* (Menzel and D'Aluiso 2000) for its series of annotated and artistic photographs of robots and their creators, as well as the podcast series *Talking Robots*, directed by Floreano (2006-2008), featuring audio interviews with several of the leading authors mentioned in this chapter on the science, technology, and business of intelligent robotics.

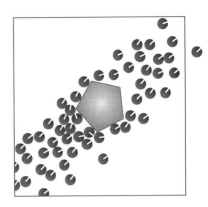

7 *Collective Systems*

In this chapter we describe phenomena that are unique to systems that operate in groups composed of two or more individuals. In particular, we are interested in collective phenomena that provide an adaptive function that is not available to individuals operating in isolation.

Consider, for example, a group of people walking through a shopping mall on a Saturday afternoon or a school of fishes that aggregate in the presence of a predator. In both cases the behavior of an individual is affected by the behaviors of its neighbors. However, while in the former case the trajectories are not instrumental to an adaptive collective behavior, in the latter case they produce a centripetal movement that protects fishes from predators attacking weaker individuals at the periphery of the school. Although the first type of collective movement is worth studying for improving sales or preventing accidents, in this chapter we will focus on collective phenomena of the second type, which provide an adaptive functionality to the participating individuals.

The biological world abounds in collective phenomena that have important adaptive functions, ranging from coordinated movement to nest building, and all the way to communication. In most cases, these phenomena rely on simple rules and local information, do not require a global plan or a central coordinator, and are robust to malfunction or deviations of some individual.

In this chapter we start by reviewing phenomena and models of biological self-organization of collective systems and then describe algorithms and technologies that have been inspired by those models. We will then turn our attention to the evolutionary conditions that lead to the emergence of those collective systems and how they affect the evolutionary dynamics in both biological and artificial systems.

There are some similarities between the models presented in this chapter and some of those presented in chapter 2, especially the agent-based models discussed in section 2.8, since both refer to agents that can interact with their neighbors to produce a nontrivial and self-organizing collective dynamics. However, the models described below are typically based on agents whose interaction with other agents and with the environment are more sophisticated than those defined in a cellular space using a state transition rule. In particular, the agents that will be described in this chapter can be endowed with modalities of perception, action, and embodiment that are generally absent in purely cellular models.

7.1   Biological Self-Organization

The notion of self-organization comes from the world of physical chemistry to indicate a process where local interactions among simple particles generate a structure at a higher level. Examples of self-organization include the appearance of geometric patterns in reaction-diffusion chemical systems, the formation of snow crystals, and the appearance of moving objects within Conway's Game of Life described in chapter 2. The resulting global patterns are also said to be *emergent* because they are more than the sum of the constituent parts.

The principles of self-organization are appealing for explaining biological collective phenomena where the resulting structures and functionalities greatly exceed in complexity the perceptual, physical, and cognitive abilities of the participating organisms. Examples of biological self-organization include the construction of beehives, the foraging strategies of ants, and the regulation of colony life in social insects. In all these cases, the resulting structure emerges from the collective work of individual organisms that execute simple behaviors based on local information and do not possess a global plan of the end result.

Some of those simple behaviors could be genetically programmed, but it is unlikely that long series of actions are encoded in the genotype because that would require large sequences of DNA and would be too inflexible for sudden change that can occur in the environment. A more plausible explanation is that evolution selects for behavioral rules that capitalize on principles of self-organization for producing a collective phenomenon. In other words, most of the resulting behaviors are not genetically encoded, but result from the interactions with other individuals in a self-organizing process.

Self-organization builds on two opponent forces: attraction and repulsion. In biological systems, these forces are often described as positive and negative feedback (Camazine et al. 2001). Feedback occurs when a quantity in the system is fed back into the system to increase or decrease the magnitude of that same quantity. As we will see in the next section, the quantity could be the density of individuals in a group. Small density produces attraction behaviors that increase density, while high densities produce repulsion behaviors that decrease density.

Typically, the equilibrium of a self-organizing system results from the interplay between positive and negative feedback. This equilibrium state is equivalent to an attractor in dynamical systems theory because the system will tend to return to it if perturbed. However, a self-organizing system can display multiple states and chaotic trajectories. Self-organizing systems can be conveniently described by sets of differential equations where the change of system state is a function of the system state at the previous time step (positive feedback) times a limiting factor (negative feedback) whose magnitude is inversely proportional to the magnitude of the system state.

BIFURCATIONS   It has been shown that self-organizing systems with positive and negative feedback can display sudden modifications –or bifurcations– that affect the pattern or functionality at the global level (May 1974). This occurs for particular values of the parameters that describe the system. For example, the logistic equation $N_{t+1} = rN_t(1 - N_t)$, which describes the evolution of the population size N, produces a gradual extinction when the parameter r is smaller than 1, a growth to a constant value for $1 \leq r < 3$, the oscillation between two different sizes for $3 < r \leq 3.4$, and multiple states for $r > 3.4$ that quickly display chaotic trajectories where the population can transit between several unpredictable, but not random, sizes (Crutchfield et al. 1986). Assuming that the parameter r is partly or entirely determined by genetically dependent factors, as the metabolism of a species, evolution may favor parameter values that generate stable behaviors, such as in the regulation of the nest temperature, or parameter values that produce unpredictable behaviors, as in the fleeing behavior of an insect. The genetic description of a single parameter is a very convenient way to produce radically different dynamics instead of the genetic description of all the components that produce the most suitable behavior.

The interaction at a distance between animals occurs by means of cues and signals (Camazine et al. 2001). A cue is an unintentional index that can be picked up by an animal, such as a trail in the snow. The perceiving animal can decide whether to follow (positive feedback) or avoid (negative feedback) the cue. Instead, a signal is an intentional index emitted by an animal that is intended (more or less consciously) to affect the behavior of other receiving animals. For example, the alarm cry of some birds when a predator approaches is an intentional signal.

STIGMERGY   *Stigmergy* is a specific type of social communication through modification of the environment. The result of work by an individual affects the action of another individual. Stigmergy was first advocated by Grassé (1959) to explain the construction of elaborated nest architectures by termites and other insects. It was later extended to explain the foraging strategies of ants that deposit trails of chemicals, known as pheromones. Stigmergy is a form of cue-based communication because it is not intentional.

The intentionality of signals, i.e., signals that can be emitted at will by the animal, can lead to more complex dynamics (Maynard-Smith and Harper 2003; Searcy and Nowicki 2005) and eventually to human language (see Szamado and Szathmary 2006 for a review of theories on the emergence of human language) that involve both evolutionary and cognitive factors.

Let's now consider some examples of self-organization that have inspired computational and robotic models described later in this chapter.

7.1.1   Aggregation

Aggregation is a typical example of self-organization that can be explained by positive and negative feedback. For example, in fish schools large numbers of individuals swim in close formations that can rapidly change direction as well as disperse and reunite. The coordinated, homogeneous, and rapid movement of the school gives the appearance of a single superorganism. Huth and Wissel (1992) suggested a simple model of schooling based on both negative and positive feedback. In its simplified version, a fish displays four behavioral reactions that depend on the position and orientation of other fish: (a) if there is another fish in its immediate neighborhood, the focal individual will move away to avoid collision (negative feedback); (b) if there is another fish at an intermediate distance, the focal individual will tend to align along its orientation; (c) if there is another fish at a greater distance, the focal individual will tend to swim toward it (positive feedback); (d) if there is no fish in sight, the focal individual will perform random search movements. Although the model makes strong assumptions about the perceptual abilities of fish and does not consider other sources of perturbation, it can capture several behaviors observed in animals (Partridge 1982).

7.1.2 Clustering

Several ant species engage in clustering and sorting of objects. For example, corpses of dead ants are organized in large clusters at the periphery, or near walls, of the nest for better circulation (Theraulaz et al. 2002); sand pellets are clustered to form a circular wall that protects the colony (Franks et al. 1992); and eggs are organized in regular patterns where neighboring eggs have similar maturation times for more efficient feeding (Franks and Sendova-Franks 1992). All these types of clustering and sorting behaviors can be explained by variations of a simple behavioral model that combines positive and negative feedback. In its simplest form, the probability that an ant picks up an object is inversely proportional to the number of objects that it has experienced within a short time window. Therefore, the ant will tend to pick up isolated objects, but won't remove objects that occur in clusters. Instead, the probability that an ant deposits an object is directly proportional to the number of perceived objects in a short time window. Therefore, the ant will be more likely to deposit an object near larger clusters of objects (Deneubourg et al. 1991; Theraulaz et al. 2002). Sorting behaviors may be explained by adding different response probabilities for different types of objects in the environment.

7.1.3 Nest Construction

Termites and wasps collectively build nests whose architectural complexities exceed the perceptual and cognitive abilities of single individuals. Even if an individual possessed the full sequence of behaviors necessary to assemble such nests, the question still remains of how that individual could possibly coordinate with other individuals who work in parallel. A number of models based on positive and negative feedback have been advocated to explain how such engineering feats can be realized without a plan or a master architect. All models rely on stigmergic communication (Grassé 1959) whereby the perception of the result of previous work triggers specific construction behaviors genetically encoded as stimulus-response associations. Theraulaz and Bonabeau (1995) proposed a discrete stigmergic model of wasp nest construction. The model operates like a cellular automaton where each cell in a 3D lattice represents a cell of the nest. The wasp decides if and where a new

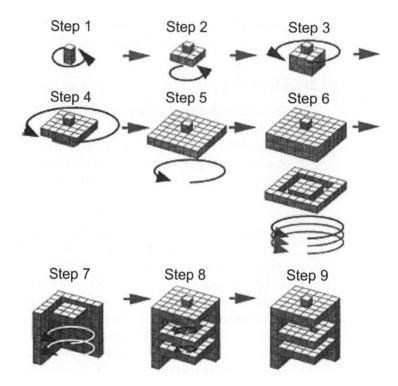

Figure 7.1 Successive steps in the construction of a structured lattice architecture including a pedicel, an external envelope, regularly spaced internal layers or combs, and an entrance opening. This architecture resembles that of the *Epipona* wasp nest. The completion of each step gives rise to stimulating configurations that belong to the next one. In steps 7 to 9, the front and right portions of the external envelope have been cut away. Reproduced with permission from Theraulaz and Bonabeau (1995).

cell is constructed according to a set of predefined rules (figure 7.1). The stigmergic communication is discrete because the animal switches between discrete construction rules depending on discrete perception patterns. The authors have also used an evolutionary approach driven by a complexity fitness measure to investigate the space of construction rules and resulting architectures; some of the resulting nests resembled very closely the wasp nests found in nature.

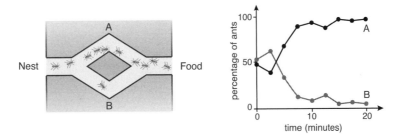

Figure 7.2 Pheromone-depositing ants tend to choose one of two available paths. Adapted from Deneubourg et al. (1990).

7.1.4 Foraging

Stigmergy can also improve the efficiency of collective foraging. Some ant species lay a pheromone trail that is used to select a path, find the shortest one, and establish a link between the food area and the nest. Deneubourg et al. (1990) showed that when pheromone-laying ants are presented with two alternative paths of equal length between the nest and the food area, they distribute equally between the two paths, but after a few minutes most ants will travel only through one of the two paths (figure 7.2). This happens because the path that is initially visited by more ants contains more pheromones, which attracts more ants, which increases the pheromone quantity, and so forth in a positive feedback loop. If the ants are presented with two paths of different length, they will tend to choose the shorter one because ants on the short path return earlier to the nest and thus leave more pheromones on that path (Goss et al. 1989). However, if considerably more ants choose the long rather than the short path in the early stage of exploration, they will tend to stick with the longer path. As we will see later in this chapter, computational models inspired by ant foraging solve this issue by assuming a fast evaporation of the pheromone trail, which favors the choice of short paths; however, it is not clear if in natural systems evaporation is sufficiently rapid to play a role in the choice of the shorter path.

Deneubourg et al. (1989) also showed that positive feedback mediated by pheromone deposition can help blind army ants to raid through a large territory in search of prey while maintaining a path to the nest (figure 7.3). The authors showed that the different raiding patterns displayed by different species of army ants are not generated by different searching rules, but by different dispersion patterns of the prey that the ants chase.

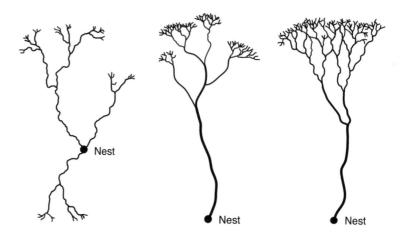

Figure 7.3   Army ants explore the environment in search of prey while maintaining a path to the nest with pheromone deposition. The foraging pattern depends on the scattering of prey. Adapted from Burton and Franks (1985).

7.1.5   Division of Labor

In the examples of self-organization described so far, all individuals have the same set of behavioral rules or perform the same activity. However, several insect societies display division of labor and specialization where different activities are simultaneously performed by specialized individuals (Robinson 1992). Individuals can be specialized in a soft or strong manner. Softly specialized individuals are individuals that can perform several activities, but at every instant tend to perform the activity that is most needed by the group. Strongly specialized individuals instead are individuals that can perform only one or few activities. Examples of division of labor include foraging and nest defense in ants (Detrain and Pasteels 1991), foraging and nursing (Calderone and Page 1996), and nectar and pollen collection (Seeley 1995) in honeybees.

Although genetic factors certainly play a role in some types of strong specialization (Bourke and Franks 1995), such as polyethism (age-dependent specialization) and polymorphism (different body shapes), the unpredictable modifications of the environment and the variability within the colony require additional mechanisms for ensuring dynamic task allocation. For example, Edward O. Wilson (1984) noticed that in some ant species specialists of type A can carry out tasks that they would normally not perform if the number of specialists of type B for those tasks decreases (but not the other way around).

In order to explain these results, Bonabeau et al. (1996) proposed the *response threshold model* whereby an individual performs a task if the stimulus associated to that task exceeds the individual's threshold. For example, the demand for water is a stimulus associated to water collection; similarly, the number of enemies in the surrounding of the nest is a stimulus associated to nest defense. An individual with a low threshold will tend to perform the corresponding task even if the stimulus (need) is very low; conversely, an individual with a high threshold will perform the corresponding task only if the stimulus (need) is very high. The response threshold model can explain the regulation of an essential element, such as food, by allocating individuals to foraging as soon as the food demand increases. Individuals can have as many different thresholds as the many different tasks required for colony survival.

The dynamic task allocation observed by Wilson (1984) can be explained by assuming that specialists of type A have low thresholds for the stimulus associated to task A and moderate thresholds for task B, whereas specialists of type B have low thresholds for the stimulus associated to task B (and high thresholds for task A). As soon as the number of specialists of type B decreases, the stimulus associated to that task will increase until it exceeds the corresponding thresholds of specialists of type A; consequently, some specialists of type A will perform task B until the corresponding stimulus falls below their threshold.

The response threshold model has some similarity with the so-called market-based models of task allocation where agents bid a certain amount of their resources for a good (task) and where the cost of the goods depends on their scarcity and demand (Clearwater 1995). For a comprehensive overview of models of division of labor in social insects, we refer the reader to (Beshers and Fewell 2001).

The examples of collective self-organization and models that we described above inspired the design of novel machine-learning techniques and robotic systems. This area of research, also known as *swarm intelligence* (Beni 2004; Bonabeau et al. 1999), studies large collections of relatively simple agents that can collectively solve problems that are too complex for a single agent or that can display the robustness and adaptability to environmental variation displayed by biological agents. In the following three sections we will describe examples of swarm intelligence applied to computer science and robotics.

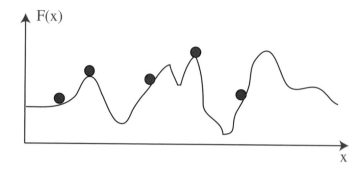

Figure 7.4 A swarm of five particles searching for the global minimum of a one-dimensional function defined over the domain of real values.

7.2 Particle Swarm Optimization

Particle swarm optimization (PSO) is a machine-learning technique loosely inspired by birds flocking in search of food (Kennedy and Eberhart 1995). Imagine a flock where each bird cries at an intensity proportional to the amount of insects that it finds at its current location, can perceive the position of neighboring birds, and can tell which of the neighboring birds emits the loudest cry. There is a good chance that the flock will find a spot with the highest concentration of insects if each bird follows a trajectory that combines three directions: keep flying in the same direction, return to the location where it found the highest concentration of insects so far, and move toward the neighboring bird that cries the loudest.

PSO consists of a number of particles (birds) that collectively move on the search space in search of the global optimum (figure 7.4). Each particle is characterized by its position and performance. For example, in the optimization of a function of one variable, each particle will be characterized by the value of the variable and by the corresponding value of the function.

Initially, particles are randomly distributed on the search space and move according to local information. Each particle communicates its performance to neighboring particles; can remember the position where it recorded the best performance so far; and can tell the position of the neighboring particle with the highest performance. Each particle updates its position by adding up fractions of three displacements: (a) a fraction of the displacement in the same direction that it was following at the previous time step; (b) a fraction of the displacement in the direction toward the position where it recorded

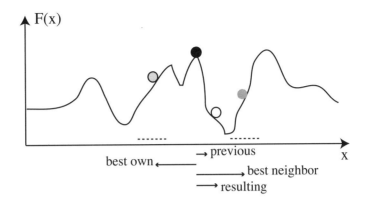

Figure 7.5   The update of the position of a particle (black disk) is given by the sum of three directions: its previous direction, the direction toward its own best position so far (gray disk with black outline), and the direction toward the position of its best neighbor (gray disk without outline) at that moment. Each of the three directions can be separately weighted by the constants a, b, c, which in the figure have been all set to 1. Furthermore, some uncertainty (dotted lines) can be added to the positions of the best own and best neighbor (no uncertainty has been used for the update in the figure).

the highest performance so far; and (c) a fraction of the displacement in the direction toward the position of the neighbor with the highest performance at that moment.

Formally, the new position x_i^{t+1} of a particle i is given by

$$x_i^{t+1} = x_i^t + v_i^{t+1}$$

where v_i^{t+1} is the new velocity of the particle. In this simple case, we consider particles continuously moving on a one-dimensional function of real values.

The velocity of each particle is computed by adding up the three directions (figure 7.5) mentioned above:

$$v_i^{t+1} = av_i^t + b(x_i^p - x_i^t) + c(x_j^t - x_i^t)$$

where a, b, c are constants that separately control the importance of the three directions, $(x_i^p - x_i^t)$ is the difference between the position of the particle where it recorded the best performance p so far and its current position, and $(x_j^t - x_i^t)$ is the difference between the position of the neighboring particle j with the best performance at that moment and the particle's position.

Whenever a particle records a better performance at its new location, the value x_i^p is updated. However, the velocity update requires some element of

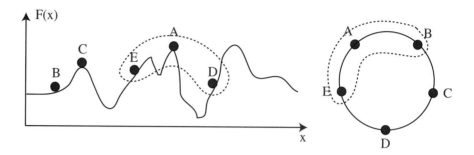

Figure 7.6 Geographical and social neighborhoods. In the geographical neighborhood, a particle speaks and listen to a small number of particles that are nearby in the space of the function to be optimized (dotted area on the left). In social neighborhoods, particles are labeled and defined as neighbors according to some predefined ordering, such as the circle on the right. During search, a particle will always speak and listen to its social neighbors (dotted area on the right), irrespective of where they are in the function space.

randomness in order to allow particles to explore novel areas of the search space and avoid stagnation in local optima. This is obtained by adding a region of uncertainty around the position of the best performance recorded so far and of the neighboring particle with the best performance (dotted regions in figure 7.6), which formally translates into

$$v_i^{t+1} = av_i^t + br_s(x_i^p - x_i^t) + cr_t(x_j^t - x_i^t)$$

where r_s and r_t are random values in the range $[0, s]$ and $[0, t]$, respectively.

The neighborhood of the particles can be geographical, in which case the neighborhood is given by the positions of the particles in the function space (figure 7.6, left), or social, in which case the particles are labeled and defined as neighbors independently of their position. For example, a frequently used social neighborhood consists of ordering particles around a circle (figure 7.6, right). If a social neighborhood is used, a particle will always speak and listen to the same set of particles throughout the search process. The position of the particles in the swarm can be updated synchronously or asynchronously (similarly to the update of neurons in Hopfield neural networks described in chapter 3).

The typical size of the swarm is in the range of $[20, 200]$ particles and the neighborhood is approximately one-tenth of the swarm size. The values of the three constants a, b, c and of the uncertainty regions s and t can be indicatively set in the range $[0.1, 1]$ for low-dimensional functions in the domain of real values. Clerc and Kennedy (2002) provide a formal study of these PSO constants for solving complex functions. The values of the constants, neighborhood, and swarm size can also be adapted during search (Shi and Eberhart 1998).

Particle swarm optimization has been used with success in several domains ranging from regulation of power plants to the class of traveling salesman problems (Kennedy and Eberhart 2001). When compared to other optimization techniques, there is empirical evidence that PSO performs better for optimization of functions of real variables, such as the weights of a neural network. In the case of multidimensional functions and of functions that are defined over discrete or non-Euclidean spaces, one must find the most appropriate ways of computing directions and updating velocities so that particles converge toward the optimum of the function. This choice is very delicate because it transforms the mapping between contiguous points in function space and in particle space, thus affecting the likelihood of optimizing the function.

Just like the evolutionary algorithms (EAs) described in chapter 1, particle swarm optimization searches the problem space in parallel with a population of candidate solutions. However, whereas the search in EA is driven by competition among candidate solutions, the search in PSO is driven by cooperation. The choice of suitable genetic encodings and crossover operators in EA is equivalent to the choice of suitable position and velocity updates in PSO. In both cases, the choices may be easy for real-valued function optimization, but not for more complex and discontinuous problem spaces where different mappings and operators can generate very different results. Langdon and Poli (2007) used genetic programming to find problems which demonstrated the strength and weakness of PSO as compared to other search techniques, including EA.

7.3 Ant Colony Optimization

Ant colony optimization (ACO) is another family of optimization algorithms inspired by pheromone-based strategies of ant foraging. ACO algorithms were originally conceived to find the shortest route in traveling salesman problems. In ACO several ants travel across the edges that connect the nodes of a graph while depositing virtual pheromones.

Ants that travel on the shortest path will be able to make more return trips and deposit more pheromones in a given amount of time. Consequently, that

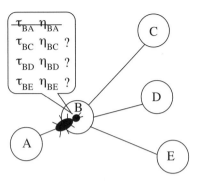

Figure 7.7   A virtual ant arriving from node A considers which edge to choose next on the basis of pheromone levels τ_{ij} and visibilities η_{ij} (inverse of distance). The edge to node A is not considered because that node has already been visited.

path will attract more ants in a positive feedback loop. In nature, however, if more ants choose a longer path during the initial search, that path will become reinforced even if it is not the shortest. To overcome this problem, ACO assumes that virtual pheromones evaporate, thus reducing the probability that long paths are selected.

Several types of ACO algorithms based on those principles have been developed with variations that address the specificities of the problems to be solved. In this chapter we will describe the first ACO algorithm, known as ANT SYSTEM the *ant system* (Dorigo et al. 1996), which provides a foundation for more recent and specific algorithms.

Initially, ants are randomly distributed on the nodes of the graph. Each artificial ant chooses an edge from its location with a probabilistic rule that takes into account the length of the edge and the amount of pheromones on that edge. Edges leading to nodes that have already been visited by that ant are not considered in the probabilistic choice. Once all ants have completed a full tour of the graph, each of them retraces its own route while depositing on the traveled edges an amount of pheromones inversely proportional to the length of the route. Before restarting the ants from random locations for another search, the pheromones on all edges evaporate by a small quantity. The pheromone evaporation, combined with the probabilistic choice of the edge, ensures that ants eventually converge on one of the shortest paths, but some ants continue to travel also on slightly longer paths.

Let us now consider the algorithm more formally. The number of ants M is usually equal to the number of nodes N in the graph. A small amount of virtual pheromones is deposited on all edges at the beginning of the search. The probability p_{ij}^k that ant k chooses the edge from node i to node j (figure 7.7) is given by

$$p_{ij}^k = \frac{\tau_{ij}^a \eta_{ij}^b}{\sum_{h \in J^k}^{H} \tau_{ih}^a \eta_{ih}^b}$$

where τ_{ij} is the amount of virtual pheromones on that edge and η_{ij} is the visibility of the node computed as the inverse of the edge length $1/l_{ij}$. The constants a, b weight the importance of the two factors. If $a = 0$, ants choose solely on the basis of the shortest distance; conversely, if $b = 0$, ants choose solely on the basis of the pheromone amount. The divider in the fraction sums up the pheromone and visibility values for all edges H that are available at the node where the ant sits as long as they belong to the set J^k of nodes that the ant k has not yet visited. As soon as the ant visits a node, this is deleted from the list J^k.

Once all ants have completed a tour of the graph, each ant k retraces its own path and deposits an amount of pheromone $\Delta \tau_{ij}^k$ on traveled edges according to

$$\Delta \tau_{ij}^k = Q/L^k$$

where L^k is the total length of the path found by ant k and Q is a constant, which is set to be the length of the shortest path estimated with a simple heuristic method. The amount of pheromones on each edge after all M ants have retraced their own path is equal to

$$\Delta T_{ij} = \sum_{k}^{M} \Delta \tau_{ij}^k.$$

Before starting all ants again in a new search for the shortest path, pheromone levels evaporate according to

$$\tau_{ij}^{t+1} = (1 - \rho)\tau_{ij}^t + \Delta T_{ij}$$

where $0 \leq \rho < 1$ is the coefficient of pheromone evaporation.

This concludes one iteration of the algorithm. The process is repeated for several hundred iterations until a satisfactorily short path has been found. The ant system is not guaranteed to find the shortest path, but for graphs of moderate size (approximately 30 nodes) it provides equal or better solutions than the best algorithms designed for solving the traveling salesman problems. The performance of the algorithm can be improved by allowing ants

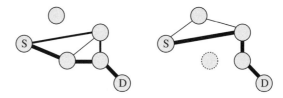

Figure 7.8 *Left*: Virtual ants maintain multiple paths between source and destination nodes. Shorter paths are traversed by more ants (thicker line). *Right*: If a node (or edge) fails, ants immediately use and reinforce the second shortest path available.

that have found the shortest path at each iteration to retrace their own path several times while depositing virtual pheromone.

However, for problems of higher dimensionality or problems that present specific constraints, the ant system is no longer competitive with other algorithms that have been specifically designed for solving those problems. To address this issue, Dorigo and Gambardella (1997) developed the *ant colony system* algorithm which improves on the ant system by introducing local search heuristics in addition to the virtual ants.

Although ACO algorithms are not guaranteed to find the shortest path in complex, high-dimensional graphs, they find satisfactory solutions with relatively little computation. A major advantage of ACO over other algorithms for path finding is that virtual ants discover and maintain several short paths in addition to the best one (figure 7.8, left) because of the probabilistic edge choice. If an edge is no longer traversable or a node is unavailable, ants will quickly use and reinforce the second shortest path (figure 7.8, right), whereas other algorithms must recompute the shortest path on the new graph. Therefore, ACO is particularly suited for dynamic scenarios where alternative solutions must be available immediately.

This is the case of routing in a real communication network where data must be sent between two points in the shortest time and with the best quality despite possible congestion or failure of some nodes. Schoonderwoerd et al. (1996) developed an ACO algorithm for optimizing the load and performance of telephone networks. The algorithm scored significantly better than other shortest-route algorithms in terms of failed calls when tested on a simulation of the 30-node British Telecom network. Di Caro and Dorigo (1998) developed another ACO algorithm for packet routing on the Internet and tested it on realistic simulations of the T1 US backbone network with 14 nodes and 21 bidirectional links and of the Japanese NTT network with

57 nodes and 162 bidirectional links. When compared to a series of industry-standard routing algorithms, ACO scored at least as well as the best standard algorithm in terms of throughput (bits per second) and performed much better than all other algorithms in terms of packet delay time. Furthermore, the algorithm performed equally well on both networks, despite their different topology and properties of the nodes.

ACO is a very powerful method for finding the shortest path in dynamic networks, but it is not straightforward to apply it to other problems, such as function optimization or search on Euclidean spaces. The challenge in that case, similar to PSO and to EA, is to find a suitable representation of the problem so that it can be navigated by virtual ants. Stuetzle and Dorigo (2004) describe several applications of ACO algorithms in a variety of problems with different representations.

7.4   Swarm Robotics

Swarm robotics is the application of the principles of self-organization to collections of simple, autonomous robots (Şahin 2004). By simple robots, we mean robots that do not have sophisticated sensors, electronics, and mechanics and that do not use complex algorithms based on global information or central control. The core idea of swarm robotics is to capitalize on simple interactions among robots in order to solve complex problems by means of emergent behavior, similar to what social insects do.

Swarm robotics is mainly concerned with groups of robots that are larger than those that could be easily controlled with a centralized, top-down approach, but within a reasonable number that could be manufactured at a price that is roughly comparable to that of a few sophisticated robots. Given the constraints of today's technology and manufacturing processes, the typical size of robotic swarms is currently in the range of 10 to 100 units. However, since swarm robotics puts emphasis on hardware and software simplicity, it should benefit from further miniaturization of robotics hardware, which implies simple sensing and computing. The idea of exploiting the potential synergy of several simple robots was first put forward by Fukuda and Ueyama (1994) under the name of *cellular robotics*.

The potential advantages of this approach include the robustness of the swarm to failure of individual robotic units or run-time addition of new units, the scalability of the emergent behaviors to swarms of different sizes, the capitalization of self-organization principles of environmental noise and

individual differences, and the synergetic effect whereby the work of the
swarm is greater than the sum of the work by the individual units, also

SUPERLINEARITY   known as *superlinearity.*

Most developments in swarm robotics focus on populations of homogeneous robotic units, which fall into two distinct classes. One class consists of
groups of relatively inexpensive mobile robots, which are fully autonomous
and can operate also in isolation. Typical examples are the MARS (Fukuda
et al. 1999) and Khepera (Mondada et al. 1993) robots. Within this context,
also known by the name of *collective robotics*, the main goal is devising simple
control rules based on local information that give rise to adaptive functionalities at the level of the swarm, such as coordinated exploration or object
clustering and transportation.

The other class of swarm robotics includes self-reconfigurable and self-
assembling robots, which are composed of several identical modules. Here
the notion of cellularity is taken very literally because the robot is composed
of several, partially independent but interconnected robotic cells. Robotic
cells cannot operate in isolation, but can modify their relative position in order to produce various robotic morphologies with different functionalities.
In this case too, the principles of self-organization, based on simple and local
control, are used to provide emergent functionalities to the robotic superorganism.

Behavior-based control, which was described in chapter 6, is often used in
swarm robotics because it offers a methodology for rapidly adding and fine-
tuning control rules until the desired emergent swarm behavior is obtained,
it maps easily into a distributed and asynchronous collection of physically
distinct robotic units without central control, and it can easily accommodate
the principles of positive and negative feedback that lie at the heart of self-
organization.

7.4.1   Example 1: Coordinated Exploration

A first concern with a swarm of mobile robots is to make sure that individuals remain together as they move through cluttered space.

In a seminal work, Reynolds (1987) suggested three simple rules that individual agents should obey in order to display bird-like flocking or fish

BOIDS   schooling (figure 7.9). Reynolds assumed that agents, known as *boids*, could
perceive the distance and orientation of other agents within a small neighborhood. Each boid followed three rules: (a) if it was too close to other
boids, it moved away from them; (b) if it was too far from other boids, it

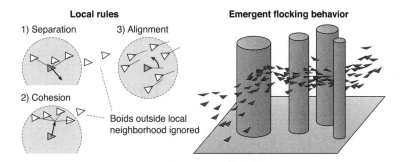

Figure 7.9 Three behavioral rules for *boid* flocking: separation, cohesion, and alignment. When embedded in a flock of agents with local perception, these rules produce coordinated navigation through obstacles without dispersing the agents. Adapted from Reynolds (1987).

moved toward them; and (c) it aligned its orientation along the orientation of neighboring boids. Simulated boids moved together in both empty and cluttered spaces, temporarily diverging and then reaggregating once the obstacles were passed. The film industry adopted this approach because it could rapidly generate computer graphics sequences of natural-looking groups of animals. This not only saved time with respect to drawing each character separately, but also produced sequences that looked more realistic. According to Reynolds's website, which is a rich source of information on the synthesis of group behavior, variations of the boids algorithm were used in the movies *The Lion King*, *Batman Returns*, and *Star Trek*, among others.

The assumption that agents can perceive the distance and orientation of neighboring agents is not easily met in mobile robots, where sensing is still a challenge. Matarić (1992) designed a series of behavior-based modules for mobile robots that, when combined and properly weighted, would produce flocking behavior. Her wheeled robots were equipped with collision sensors (electromechanical switches) and with six infrared distance sensors pointing in different directions around the body of the robot. The sensors could be used to measure the distance to another robot within a small neighborhood and to detect collisions. The robots were equipped with four elementary behaviors. "Collision avoidance" steered the robot away from objects that were closer than a predefined threshold. "Following" steered the robot toward another robot perceived on its left or right, if that was the only perceived object; instead, if robots were perceived on both the left and the right, the robot would move straight for a short time until another robot in front

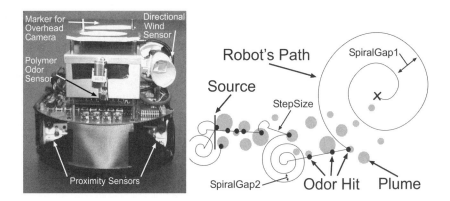

Figure 7.10   *Left*: The 24 cm mobile robot equipped with an anemometer (wind sensing) and conducting polymer for odor detection that was used in experiments on localization of an odor source (Hayes et al. 2002). *Right*: Behavioral strategy of a robot engaged in collective odor localization. Images courtesy of Alcherio Martinoli, EPFL, Lausanne, Switzerland.

was perceived. "Dispersion" and "aggregation" pulled together the distance readings from multiple sensors and steered the robot away and toward, respectively, the computed center of mass. When suitably combined in a subsumption architecture, these behaviors resulted in flocking behavior of five robots through empty spaces. However, flocking through cluttered environments was not tested and may require more sophisticated sensing abilities to align with the orientation of other robots.

Environmental monitoring is a potential application of swarm robotics. For example, Hayes et al. (2002) showed that a swarm of mobile robots with minimal sensors and behaviors could be deployed for localizing an odor source. Although single robots were already suggested for odor localization by moving up the odor gradient in the air and water (Kazadi et al. 2000), the turbulent nature of odor dispersion tends to generate isolated packets of odors rather than plumes with gradient information, which makes the problem more difficult and suitable for collective search. Hayes et al. (2002) devised a simple behavioral controller, rooted in the biological evidence of chemotactic strategies of insects, which combines spiraling and straight trajectories to locate an odor source.

The wheeled robots used in those experiments were equipped with an odor sensor, a wind direction sensor, infrared distance sensors, and a wireless communication chip (figure 7.10, left). Robots autonomously switched among three behaviors. In the absence of other information, a robot performed spirals until it detected an odor packet. At that point, it measured wind direction and moved upwind for a predefined number of steps, which could be adapted to the frequency of odor detection events. At the end of the sequence, if no odor was detected, the robot engaged in another spiraling trajectory whose size depended on the time lag of the previously detected odor events until a new odor packet was detected. During the straight trajectories, the robot kept a memory of the global displacement using the wheel odometers. If no more global displacement was detected after some time, the robot assumed that it reached the source of odor.

The authors compared the performance of a swarm of robots required to localize an odor source with respect to the size of the swarm and to signals exchanged among all robots when the odor was detected. Experimental results indicated that swarms of robots enabled to exchange signals were faster in detecting the odor source than robots operating in isolation and than groups of robots that could not exchange signals.

PHEROMONE ROBOTICS However, global signal exchange may not scale well to a large number of robots. Therefore, Payton et al. (2005) suggested using virtual pheromones in a swarm of robots required to locate a target in cluttered environments and lead a human toward it. The idea of *pheromone robotics* is that robots broadcast optical signals (virtual pheromones) that can be detected by neighboring robots on line of sight and rapidly spread across the swarm indicating a viable path toward the source of the signal. The authors demonstrated this concept with a swarm of 20 wheeled robots equipped with Palm computing devices.

Robots had a circular circuit composed of eight infrared emitters and receivers driven by a microcontroller. The modulated signal allowed robots to send data, such as the signal identity and hop count from the source robot. Received signals (virtual pheromones) were tagged with direction and intensity. Robots received multiple signals, but accepted only signals with hop counts greater than those that were already received earlier. The hop count of accepted signals was decreased and further transmitted along the eight directions. The optical nature of infrared signals, their directionality and intensity, along with the hop-count decrement, rapidly created a gradient through open passages from the source to the periphery of the robotic swarm. A human with a head-mounted camera could detect the directionality and intensity of the infrared signals, thus obtaining a path leading from the periphery of the signal to its source.

The authors tested the method with a swarm of robots equipped with various sensors and behaviors similar to those described by Matarić (1992). The robots quickly located a source of interest, deployed so as to cover the open space, and indicated the path to a human equipped with the head-mounted camera and display. A potential problem with this method is the mismatch between the size of robots and of humans because objects that obstruct optical signals between the small robots may not necessarily be obstacles for humans, and small openings where robots can exchange optical signals may not be traversed by humans.

CHAINING   Target search and path formation were also implemented by means of chaining methods (Goss and Deneubourg 1992; Drogoul and Ferber 1992; Werger and Matarić 1996) where robots establish long head-to-tail chains linking a target with a source. A limitation with chaining is that robots must continuously touch each other to maintain contact, or pass messages to establish directionality of the chain, as in the pheromone approach described previously. Nouyan and Dorigo (2006) suggested an alternative chaining strategy where robots could light up in one of four different colors to signal their position in the chain. Each robot was equipped with three behaviors: searching for a target, chaining by setting up a color according to the color of the neighboring robot (if any), and exploring by moving up the chain to the target location while following the color order. Robots switched among these behaviors with predefined probabilities until a link between target and starting locations was established. The preprogrammed color rule allowed robots to form repeated sequences of colors that gave a directionality to the chain.

7.4.2   Example 2: Transportation and Clustering

Swarming has also been studied in the context of physical work. As we mentioned earlier, social insects excel in carrying out physical work that a single animal could not perform.

In one of the earliest studies of collective physical work, Kube and Zhang (1993) studied the problem of coordinated box pushing by a team of robots without explicit communication. The box was brightly lighted and too heavy for a single robot. Robots were equipped with directional light sensors for detecting the box and directional infrared sensors for detecting obstacles and other robots. When the box and the robots were positioned in an enclosed arena, only three behaviors (box pushing, robot following, and ob-

Figure 7.11   Nests constructed by one, two, and four robots using a blind bulldozing strategy (Parker et al. 2003). Image courtesy of Chris Parker, University of Alberta, Canada.

stacle avoidance) organized in a subsumption architecture were sufficient to push the box toward a wall.

However, the swarm was stuck if several robots happened to push the box from opposite directions. In later work, Kube and Bonabeau (2000) realized that ants solve this problem during the transportation of heavy prey by randomly realigning their pulling and pushing directions when when they do not detect any displacement for some time. The authors added a similar control rule to the robots and observed that even large swarms of robots could reliably push the box all the way to a wall. Box pushing has been studied also by other authors who used teams of robots with different sensors, shape, and locomotion mechanisms (Parker 2000).

Parker et al. (2003) showed that collective pushing can also result in primitive forms of construction. The authors took inspiration from nest construction strategies, known as *blind bulldozing*, displayed by ants that build circular nests in rock crevices where light cannot enter (Franks et al. 1992). Each ant pushes granules encountered on its path until the piled-up material offers too much resistance. At that point, the ant turns in a different direction and resumes the pushing behavior. The combined work of the ants results in a clear area surrounded by walls that is sufficiently large for the ant colony to live within.

The authors programmed a number of mobile robots with three behaviors that were sequentially activated according to predefined transition rules. The robots moved in straight trajectories pushing whatever dirt was found. If a robot encountered resistance above a predefined threshold, it reoriented itself in a new random direction, and resumed pushing. The robot reoriented itself also if it encountered another robot detected with dedicated sensors.

BLIND BULLDOZING

This behavior resulted in a clean, circular area surrounded by walls of collected material. The shape of the nest-like area was not affected by the number of robots in the swarm (figure 7.11). It should be noticed that although nest construction resulted from stigmergy (the result of piled-up material affects the robot's actions), in this case only a single robot could be sufficient for construction of the nest.

CLUSTERING   Beckers et al. (1994) reported that a similar set of pushing behaviors, when embedded in a swarm of robots, can result in clustering and sorting of spatially dispersed objects, as observed in the biological ants described earlier. Holland and Melhuish (1999) systematically studied the behavioral, morphological, and environmental conditions that lead to the formation of clusters collected by a swarm of mobile robots. The objects were disks of two colors distributed in an arena. The robots were equipped with a disk-grasping mechanism and three sensors: a grasp sensor, a color sensor that reported the color of the grasped disk, and an obstacle detection sensor. Robots moved according to three behavioral rules: (1) If the robot held a disk and detected an obstacle (wall or other robot), it chose with a predefined probability either to turn away and keep the disk, or to release the disk and turn away; (2) if the robot held a disk and detected another disk in its path, it released the disk and turned away in a random direction; (3) in all other cases, the robot moved forward. A swarm of 10 robots was put in a large arena with 44 spatially dispersed disks.

The authors noticed that the work of the swarm produced two different results for different values of the probability of the first rule: when the probability of turning away from obstacles while retaining the disk was set to 1, the robots reliably brought all disks into a single cluster at the center of the arena; instead, when the probability was below 0.88, the robots started to form clusters along the walls of the arena; when the probability value was 0, the robots lined up all objects at the boundaries of the arena, but no clusters were formed. As we have seen earlier, this sudden bifurcation between stable and different outcomes is a typical behavior of self-organizing systems and has also been observed in the behavior of ants clearing up dead corpses. Holland and Melhuish (1999) also noticed that by adding to the second rule a pull-back and release behavior if the disk held in the gripper was of a certain color, the robots sorted out the disks into a cluster with all the disks of that color surrounding all the disks of the other color. This result indicated that sorting and clustering may not be the result of different behaviors, but variations of the same set of rules.

Figure 7.12   A team of robots cooperate to pull long sticks out of the ground (Ijspeert et al. 2001). Image courtesy of Alcherio Martinoli, EPFL, Lausanne, Switzerland.

The identification of critical parameters, such as the disk release probability described above, whose variation can affect the emergence of collective behavior, represents a major research goal in the study of self-organization. Along this line of investigation, Ijspeert et al. (2001) analyzed the conditions under which a swarm of robots can cooperatively pull sticks out of the ground. Groups of two to six Khepera robots were placed in a circular arena that contained four thin sticks partially buried in the ground (figure 7.12). Robots were equipped with grippers that could be used to lift sticks, but the maximum elevation of the gripper was shorter than the portion of the stick in the ground. Collaboration was thus required, with one robot lifting the stick half out of the ground until another robot approached the stick from the opposite direction and lifted the stick completely out. Robots used infrared sensors to discriminate sticks from walls and from other robots because sticks were much thinner; robots could also feel the effect on their own gripper of another robot attempting to extract the stick. The control system was composed of few behaviors that allowed the robot to wander in the arena while avoiding walls and other robots. If a stick was encountered, the robot gripped and pulled it up. If it did not feel any other robot holding the stick, it waited some time with the partially elevated stick for another robot to help. Instead, if a robot was already holding the stick and another robot arrived to help, it released the grip, allowing the full extraction of the stick by the second robot. If no robot came to help within a predefined amount of time, the first robot released the stick and resumed the wandering behavior.

   The authors studied the optimal waiting time relative to the number of cooperative events (successfully extracted sticks) within a predefined time window as a function of the number of robots. When all robots adopted the same waiting time, two conditions clearly emerged. If the number of robots was larger than the number of sticks, the optimal waiting time was very large because there was always at least one free robot that sooner or later would arrive to help. Instead, if the number of robots was smaller than the number of sticks, the authors found an optimum waiting time that maximized the number of extracted sticks. Furthermore, they discovered a *superlinear effect* of cooperation (cooperation increased not only the total number of successful events but also the number of successful events per robot) for groups of up to six robots, independently of how the waiting time was set. For larger groups, the cooperation effect decreased because of overcrowding and interferences among robots. The authors also realized that when the number of robots was smaller than the number of sticks, i.e., the condition when the setting of the waiting time was important, the number of cooperative events increased if the robots adopted different waiting times. However, this heterogeneity did not have any effect on swarms of more than six robots.

Although it would be interesting to check whether similar findings on the interactions between group size, relatedness, and criticality hold also in other experimental settings, this experiment revealed several properties of collective self-organization, namely synergetic cooperation to perform tasks that a single individual cannot tackle, complex behavior emerging from interaction of simple behaviors, and superlinear performance of swarms.

7.4.3   Example 3: Reconfiguration

Reconfigurable robots are composed of several robotic units and can change shape to perform different functions. As we already anticipated in chapter 6, these robots are rooted in early suggestions by von Neumann and others in the 1960s. The first prototype of a reconfigurable robot was the CEBOT (*ce*llular ro*bot*), whose history is described in (Fukuda and Ueyama 1994).

The robotic units that compose reconfigurable robots are sometimes assimilated to the cells that compose an organism. In other cases the robotic units are assimilated to the individuals that compose a society. The most important functionality of a reconfigurable robot is the ability to change shape, possibly autonomously, in order to execute a variety of different tasks. Therefore, their measure of success is generality and flexibility of operation rather than optimality in performing a specific task.

Most reconfigurable robots are composed of several homogeneous modules. Homogeneity of the constituent modules brings several advantages,

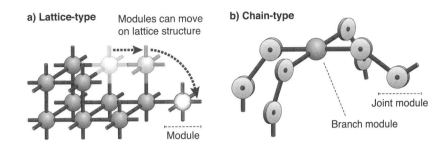

Figure 7.13 Lattice-type (*a*) and chain-type (*b*) reconfigurable robots. Adapted from Murata and Kurokawa (2007).

such as reduced development and manufacturing costs, but also requires greater effort in the design of the module, which affects the shape and functionality that the assembled robot can display. Homogeneity also brings the possibility of self-repair with identical elements being shifted to replace malfunctioning elements.

Reconfigurable robots fall into two classes (figure 7.13), according to their connectivity modes (see Murata and Kurokawa 2007 for a recent review). CHAIN-TYPE ROBOTS Chain-type robots are composed of modules with few connecting points that can be assembled in chains and, where connections on the sides of the modules exist, in branched structures. These robots typically take the form of snakes, loops, and legged robots, such as the salamander robot described LATTICE-TYPE ROBOTS in chapter 6. Lattice-type robots instead are composed of modules with more connecting points and can be assembled in 2D and 3D lattice structures. These robots can change shape by displacement of constituent modules throughout the structure of the robot. The regularity of the lattice structure allows better self-reconfigurability and offers the possibility of generating more diverse shapes than chain structures where the extremities hang in the void and are thus difficult to guide into precise docking to other elements of the robot. However, chain robots offer better mobility, strength, and speed for functions that require locomotion.

CONRO (*con*figurable *ro*bot) is an example of a chain-type reconfigurable robot composed of a set of homogeneous modules (Castano et al. 2000). Compared to similar robots, such as the Polybot (Yim et al. 2002), the CONRO design puts emphasis on the energy and motion self-sufficiency of each module. A module is an articulated segment composed of three parts (figure 7.14, top): an active connector that can attach to other modules, a passive connector that accepts connections from other modules, and

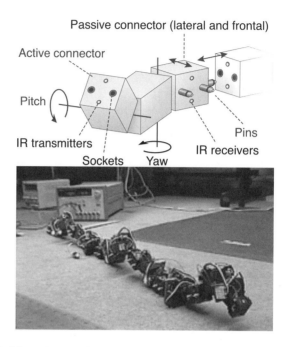

Figure 7.14 A CONRO module and a fully assembled CONRO robot (Castano et al. 2000). Photograph courtesy of Andres Castano, NASA JPL, Pasadena, CA.

a central body with two servomotors that provide pitch motion to the active connector and yaw motion to the passive connector. The passive connector has three identical sides with two pins and two infrared receivers where the single-sided active connector can dock with the help of two sockets and two infrared transmitters. Infrared communication is used for aligning the faces of the modules prior to attachment and for exchanging simple messages between attached modules. The active connector incorporates an electromechanical device for locking in the pins after attachment and for releasing them when required. Each module incorporates a battery and a microcontroller that can control the movement of the connectors relative to the main body. Additional elements, such as antennas or cameras, must be piggybacked on one of the modules.

CONRO modules can operate either in master-slave mode, where the instructions for motion patterns are computed by an external computer and propagated to the individual modules, or in distributed mode, where each module receives the same set of rules and executes a particular movement according to the state or messages received from neighboring modules. CONRO successfully demonstrated rapid wave-like motion in snake formation (figure 7.14, bottom) as well as walking in quadruped and hexapod formation. Thanks to the self-sufficiency of modules, the robot can disassemble into multiple functional robots and vice versa. However, autonomous self-assembly is rather difficult, as in other chain-type robots, because it requires precise guidance of the extremities of the robot.

The M-TRAN (*m*odular *tran*sformer) instead is an example of a lattice-type reconfigurable robot (Murata et al. 2002), although it can also reconfigure itself as chain type. M-TRAN modules are designed to fill the spaces of a regular 3D lattice and can self-reconfigure in a larger variety of shapes than CONRO by precisely shifting individual modules along other modules in the structure. An M-TRAN module is composed of two interlinked cubes that can independently rotate around their parallel joints (figure 7.15, top). As in CONRO, one cube is an active connector and the other is a passive connector. Although each cube has connecting devices only on one side, the connection can occur in any of the four possible orientations. Cubes rotate along the linked axis in steps of 90° precisely connecting to the face of another cube in the lattice. When connected in series, M-TRAN modules can transform into a chain-type robot. M-TRAN robots demonstrated self-reconfiguration between several different shapes and the ability to move in snake, circle, and quadruped formation. However, compared to CONRO where modules can rotate around two perpendicular axes, M-TRAN modules can rotate only around a parallel axis and thus require more hardware or reconfiguration steps to perform turns during locomotion. Another recent lattice-type robot, known as ATRON, addresses this limitation by using quasi-spherical modules composed of two rotating hemispheres with multiple attachment points on the surface (Jorgensen et al. 2004).

The control of lattice-type robots is still challenging because shape reconfiguration requires not only complicated displacements of the units but also complex calculations to prevent intermediate configurations where the structure can no longer transform into the desired shape. Therefore, almost all demonstrated robots rely on precomputed displacements of individual modules.

Cellular automata have been recently considered as a promising approach to autonomous self-reconfiguration of lattice-type robots (figure 7.16). In this approach, which, incidentally, reconnects the field of reconfigurable robots to its roots conceived by von Neumann, each module incorporates the same set of transition rules that are executed according to the presence and messages

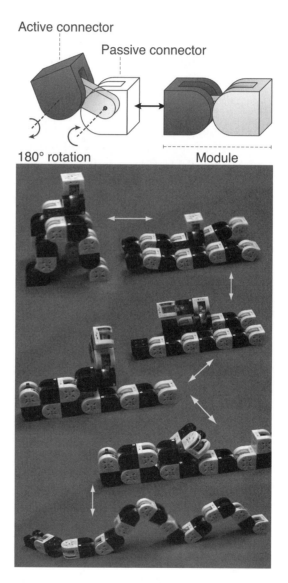

Figure 7.15   An M-TRAN module and a fully assembled M-TRAN robot (Murata et al. 2002). Images courtesy of Haruhisa Kurokawa, AIST, Tsukuba, Japan.

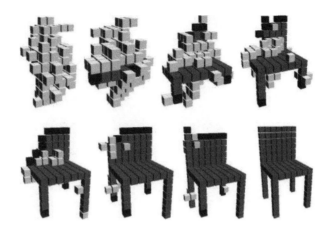

Figure 7.16 A simulated self-reconfigurable robot that transforms from a random assembly into a chair-like structure (Stoy 2006). Image courtesy of Kasper Stoy, University of Southern Denmark, Odense, Denmark.

of neighboring modules. Cellular automata were used to produce relative movements of modules that generate locomotion of a preassembled robot (Butler et al. 2001) and to self-reconfigure into desired 3D shapes (Stoy 2006). In this latter case, the transition starts with a randomly selected module acting as a seed and broadcasting the transition rules to all other elements. In order to attract modules that are far away from the seed module, a digital signal is diffused throughout the structure providing a gradient that individual modules follow. When new modules are in place, they act as seeds, until all modules reach their final destination. Stoy (2006) showed that the combination of cellular automata and gradient diffusion guarantees convergence to a desired shape if certain precautions are taken in the choice of the transition rules and in preventing the structure from obstructing the trajectories of modules to their destination. However, the modules used in the simulations rely on motion abilities that have not yet been met by hardware prototypes of lattice-type robots.

Finally, Swarm-bots (figure 7.17) are robots that cut across the dichotomy between collective and self-reconfigurable robots because they are composed of several autonomous units, known as s-bots, which can operate either 'in isolation or in physically connected assemblies (Mondada et al. 2004). S-bots can coordinate their behavior by exchanging light signals and can physically self-assemble in the most appropriate shape for solving problems that a sin-

Figure 7.17 Swarm-bots are composed of several s-bot units that can operate in isolation or in physically connected assemblies (Mondada et al. 2004). *a*, A group of s-bots are deployed for coordinated search. *b*, A Swarm-bot of five s-bots is passing over a step. *c*, A Swarm-bot of five s-bots transports a heavy objects towards a target region. *d*, A Swarm-bot of 18 s-bots pulls a child to a safe location.

gle s-bot could not, such as passing over large gaps, climbing stairs, or transporting heavy objects. S-bots can connect to each other with a strong gripper and can feel the torque exerted by other connected robots. They also have an omnidirectional color camera, several infrared distance sensors pointing around and under the body of the robot, directional microphones, inclinometers, and temperature sensors. The combinations of wheels and tracks allows an s-bot to move with high precision on flat surfaces as well as on rough and stony terrain. S-bots can also rotate their body with respect to the wheel base in order to grasp other robots or objects and can light up their translucent belt in different colors, patterns, and frequencies. Swarm-bots can operate as a coordinated and physically disjoint swarm, or in physically connected 2D and 3D assemblies, but do not offer the rich 3D reconfiguration abilities of other chain and lattice robots. Swarm-bots were used for coordinated search, navigation through gaps, and obstacles larger than a single s-bot, and for transportation of heavy objects (figure 7.17).

The field of reconfigurable robots is still in its infancy and there are several electromechanical and control challenges that remain to be solved, but it offers enormous potential both in the context of swarm intelligence and of evolutionary robotics to approximate the self-reproducing and self-organizing properties of living organisms.

7.5 Coevolutionary Dynamics: Biological Models

In the previous sections we have described how collections of agents can interact with each other during their lifetime in order to produce emergent functionalities. In the rest of this chapter we will describe how interacting agents can affect evolutionary dynamics and what evolutionary conditions are necessary for the emergence of cooperation.

7.5.1 Predator-Prey Competition

As we have seen in the first chapter, evolution occurs in populations where the struggle for limited resources implies a competition with other individuals in the environment. There are situations where the competition between individuals sharing the same environment becomes so dominant over other environmental factors that evolutionary dynamics can be significantly affected.

Let us consider the extreme case of two species that coevolve in competition with each other, such as prey and predators or hosts and parasites. In this case of competitive coevolution, the reproduction probability of individuals of one species depends on the abilities of individuals in the opponent species. It has been suggested that competing species may reciprocally drive themselves to increasing levels of performance by producing an evolutionary arms race (Dawkins and Krebs 1979).

Formal models of competitive coevolution are based on a set of differential equations, first developed by Lotka and Volterra in the mid 1920s (Lotka 1925; Volterra 1926), that describe the variations in size of the two populations, assuming a certain number of characters in the two species:

LOTKA-VOLTERRA
MODEL

$$\frac{dN_1}{dt} \;=\; N_1(r_1 - b_1 N_2)$$

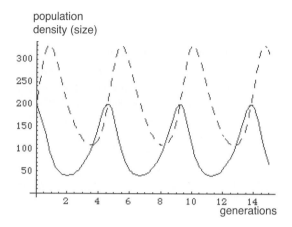

Figure 7.18   Variation of population size of two competing species according to the Lotka-Volterra model.

$$\frac{dN_2}{dt} = N_2(-r_2 + b_2 N_1)$$

where N_1 and N_2 are the population sizes of the two species (number of prey and number of predators, respectively), r_1 is the growth rate of prey in the absence of predators, r_2 is the death rate of predators in the absence of prey, b_1 is the death rate of prey caused by predators, and b_2 is the effect of prey capture on the reproduction rate of predators.

The model predicts that increments of population size in one species cause decrements of population size in the opponent species, with cyclical dynamics as long as the two populations remain viable (figure 7.18). Later experimental results with coevolving parasites and hosts in controlled laboratory conditions displayed similar dynamics (Utida 1957). However, the Lotka-Volterra model does not allow for change in the characteristics of the individuals, which may affect the reproduction rates of the two species and consequently the evolutionary dynamics. Therefore, the Lotka-Volterra models cannot tell whether competitive coevolution may lead to incremental progress in the characteristics and behaviors of the opponent species. As we will show in a later section, the application of artificial evolution to competing agents can provide an answer to this question, at least in the context of artificial agents.

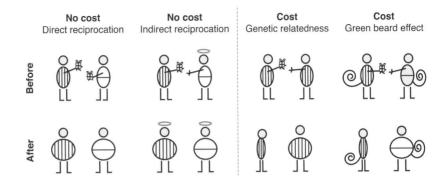

Figure 7.19 Conditions for evolution of cooperation (after Lehmann and Keller 2006). When there is no cost for the cooperator, cooperation can evolve if there is direct reciprocation or indirect reciprocation (in the latter case, reputation may help). When there is a cost to the cooperator, cooperation can evolve if individuals have a high level of genetic relatedness or if they both have green beard genes. The pattern indicates the genetic similarity between individuals. The size change after cooperation indicates the cost or benefit of cooperation. Adapted from E.W. Wilson (2000).

7.5.2   Cooperation

Competition for survival among individuals of the same species seems at odds with the observation that some organisms display cooperative behaviors. In order to better understand the conditions where cooperation can evolve, Lehmann and Keller (2006) suggested distinguishing between two types of cooperation (figure 7.19), namely the situations where a cooperator does not pay a fitness cost for helping other individuals and the situations where a cooperator must pay a fitness cost for helping other individuals. Let us remember from the first chapter that in biology fitness benefits and costs translate into the number of genetic copies that an individual can produce or lose with respect to its baseline reproduction rate.

The situation where cooperation generates a fitness benefit without any cost to the cooperator is relatively common in nature. This situation can be further divided into two cases: when the benefit is immediate or direct and when the benefit is indirect. Examples of cooperation with direct benefits include nest building and group hunting. Whenever a cooperator obtains an immediate and direct benefit from helping another individual, cooperation will always evolve and remain stable, no matter whether the receiving individuals belong to another species or have never been seen before.

Instead, if the benefit is indirect, i.e., the act of helping is not immediately reciprocated or the benefit appears only in the long term, cooperation evolves only if individuals have an initial tendency to cooperate, interact together several times, and can both recognize the partner and remember the outcome of previous interactions. If these conditions are satisfied, cooperation will always evolve and remain stable even if the cooperating individual belongs to a different species (see box 7.1).

It has also been shown that recognition of other individuals and memorization of the outcomes of the interactions is not necessary if there is a reputation system that informs how cooperative an individual is (Nowak and Sigmund 1998). The way in which animals and people decide to cooperate has been studied also in game theory (see box 7.1).

ALTRUISM
Instead, the situation where cooperation implies a fitness cost to the cooperator is less common. Cooperation with a cost is also known as *altruism* because the cooperator helps other individuals at its own expense. Parental care is an instance of altruism directed toward offspring of the individual because it implies an energetic cost to the parent. The alarm call emitted when a predator is approaching a group is another example of altruism because the emitter is more likely to attract the attention of the predator and be killed while the other members of the group can escape. The specialization of ant colonies into large numbers of sterile workers (for food collection, nest defense, rearing of the pupae, etc.) is yet another instance of altruistic cooperation where the helping workers incur the highest fitness cost because they cannot reproduce.

Building on earlier intuitions by Haldane (1955), Hamilton (1964) suggested that altruism can evolve if the cooperator is genetically related to the recipient of help. In this case, even if the cooperator cannot propagate its own genes to the next generation, its altruistic act will increase the probability that a large portion of those genes will be propagated through the reproduction of the recipient of the altruistic act. Hamilton (1964) proposed the notion of
INCLUSIVE FITNESS
inclusive fitness, which is the sum of the individual fitness and of the fitness effects caused by its own act on the portion of genes shared with other individuals. The portion of shared genes between two individuals is known as *genetic relatedness*. Hamilton (1964) predicted that altruistic cooperation will evolve if the inclusive fitness of the helper is larger than zero:

(7.1) $rb - c > 0$

where r is the coefficient of genetic relatedness, b is the fitness benefit of the recipient(s) of help, and c is the fitness cost of the helper. To use an example suggested by Haldane, in the case of brothers, where $r = 1/2$, an individual may be willing to sacrifice its own life and thus pay the maximum cost $c = 1$ if its act increases more than twice $b > 2$ the fitness of the brother. For cousins, where $r = 1/8$, an individual may be willing to pay the maximum cost if its act increases the fitness of the cousin more than eight times.

Hamilton's inequality applies to average genetic relatedness over the entire genotype and population, i.e. it is not restricted to the sharing of a specific set of genes. It also applies to the case where the act of cooperation benefits multiple individuals with various degrees of relatedness. The theory of *kin selection* (Maynard-Smith 1964), which developed from Hamilton's model, predicts that the ratio of altruistic individuals in a population is related to the degree of kinship, or genetic relatedness, among individuals. Although the theory is widely accepted, its quantitative validation in nature has not yet been done because it is difficult to precisely measure the values of the three variables in equation (7.1).

KIN SELECTION

For evolution of altruism to occur, helping should be directed toward related individuals. This is more likely to happen when individuals share the same geographical space, such as a nest, for social activities. Indeed, most cases of altruistic cooperation are found in families of social insects (Keller and Chapuisat 2002). Kin selection does not require that individuals recognize kin individuals or know their degree of genetic relatedness. As long as the act of altruism benefits genetically related individuals, altruism will spread throughout the population and remain stable.

A particular case of altruism occurs when individuals share few specific genes that favor cooperating behaviors only between individuals having a specific phenotypic character, such as a green beard (Dawkins 1976), and that express the same phenotypic character. However, altruism due to *green beard effects* can be disrupted if the linkage between the genes responsible for the green beard and the genes responsible for altruistic behavior is disrupted. For example, a mutant individual with a green beard but without the altruistic behavior will have larger inclusive fitness than individuals who have both types of genes; consequently, it will spread in the population and destroy altruistic cooperation (Lehmann and Keller 2006).

The four conditions for the evolution of cooperation, direct or indirect reciprocity, genetic relatedness, and green beard genes, which can all be included within a single model (Lehmann and Keller 2006), hold only if cooperation brings a net fitness advantage to the individuals. In some societies, the actual values of benefits and costs are distorted by means of coercion and punishment to ensure maintenance of cooperative behavior.

Box 7.1: The prisoner's dilemma

The prisoner's dilemma game is a general framework for studying decision strategies in a situation where the pursuit of self-interest by each player leads to a poor outcome for all. The game takes place between two players who are assumed to make decisions to maximize their own self-interest. The name of the game comes from the metaphor of two imprisoned suspects, *a* and *b*, being questioned separately by the police. The police offers each of the two suspects the same deal: "If you testify against your partner (and your partner refuses to testify against you), you are free and your partner receives a 10-year sentence; if both you and your partner refuse to testify, you will both get half a year in prison; if you both testify against each other, each of you gets a 5-year sentence." The dilemma comes from the fact that both prisoners must make a decision without knowing what the partner decides. If they knew, the best decision for both of them would be to refuse to testify. However, since they don't know what the partner decides, the best decision in their self-interest is to betray the partner. In doing so, they both end up in a worse situation than if they had cooperated.

	Cooperate (b)	Defect (b)
Cooperate (a)	R(a)=3, R(b)=3	S(a)=0, T(b)=5
Defect (a)	T(a)=5, S(b)=0	P(a)=1, P(b)=1

The game is often reformulated as a cooperation-defection game where two players, a and b, receive a specific payoff associated to the outcomes of the game (see table): a temptation payoff to defect when the other cooperates, a reward payoff for reciprocal cooperation, a punishment payoff for reciprocal defection, and a sucker's payoff for deciding to cooperate when the other defects. Both players know the payoff matrix, but must decide whether to cooperate or defect without knowing the decision of the other player. The actual payoff values are not important as long as they are identical for both players; they are ordered so that $T > R > P > S$; and R is larger than the average of T and S to prevent alternation of reciprocal exploitations.

Axelrod (1989) used the prisoner's dilemma to study the efficiency of various decision strategies and understand under what conditions both players would engage in reciprocal cooperation. When the game is played only once, both players choose defection.

If the game is played several times by the same players (also known as the *iterated prisoner's dilemma*), but for a finite number of times, reciprocal cooperation will never emerge because both players know   *(cont.)*

Box 7.1: continued

that the other player will defect in the last game. However, if the game is played for an *indefinite* number of times, reciprocal cooperation can emerge and remain stable. This latter situation is common in natural settings where an animal does not know when the last interaction with another animal will occur.

When comparing several play strategies, Axelrod discovered that the most efficient one was the so-called *tit for tat*. This strategy started by cooperating and then did what the other player had done at the previous move. Axelrod highlighted the four properties of tit for tat that contributed to its efficiency: (1) start with cooperation; (2) retaliate if the other player does not cooperate; (3) forgive if the other player reverts to cooperation; and (4) be simple and clear so that the other player can understand your rules and comply with them.

Although tit for tat has shown high performance in maximizing self-interest, it does not guarantee the establishment of mutual cooperation (and thus even better self-interest). For cooperation to emerge, the other player must start with a cooperative move. When the prisoner's dilemma is played between several players in a community, cooperation can emerge if the players have the chance to have more than one interaction (many interactions are not necessary, but it is important that the number is unknown to the players) and if there is a certain number of players that start with a cooperative move. This latter condition raises the question of how those cooperators could have been there in the first place, which can be answered in an evolutionary context.

Despite its level of abstraction, the prisoner's dilemma captures a large number of situations in the animal and human kingdom, ranging from small fishes cleaning the mouth of large predators to stopping at road junctions when the red light is on, all the way to wars between countries.

GROUP SELECTION   Yet another explanation for the evolution of altruistic cooperation is provided by the theory of *group selection*, which argues that altruistic cooperation may also evolve in groups of genetically unrelated individuals that are selected and reproduced together at a higher rate than the single individuals composing the group (Wynne-Edwards 1986). This could happen in situations where the synergetic effect of cooperation by different individuals provides a higher fitness to the group with respect to other competing groups. In those situations, cooperating individuals are assimilated by a superorganism that becomes a unit of selection. It has been suggested that group selection may be a driving force behind the transition from unicellular to multicellular organisms (Michod 1999).

However, the theory of group selection has been criticized because genetic mutations at the level of the individual are more likely and frequent than mutations at the level of the group, thus creating stronger competition among individuals than among groups. It has also been argued that the transition from unicellular to multicellular organisms can be explained by kin selection because all cells share the same genotype (Wolpert and Szathmary 2002). Although proponents of group selection respond to these criticisms by pointing to evidence for the evolution of group-level features that decrease individual conflict (such as a reduced mutation rate of individual organisms or cells that compose the group), the theory of group selection is still widely debated. Furthermore, group selection may eventually lead to high genetic relatedness, thus making even more difficult the disambiguation between the original driving forces that led to altruistic cooperation.

7.6 Artificial Evolution of Competing Systems

In the previous section, we described a biological model of competitive coevolution between predators and prey. In this section we will describe the application of artificial evolution to competing agents. We will start by reviewing an artificial ecological system evolving in the digital memory of a computer where parasites and hosts emerge autonomously. We will then describe an application of competitive coevolution to computer programs and suggest ways to enforce coevolutionary progress. Finally, we will describe the application of competitive coevolution to behavioral robots.

7.6.1 Tierra

The dynamics of an artificial coevolutionary system were first investigated by Ray (1992) with the aim of understanding the origins of biological diversity that occurred during the Cambrian explosion 600 million years ago by replicating those conditions in an artificial world, named *Tierra*, which existed within the memory of a computer. Ray started by creating a primordial artificial organism composed of a list of machine code instructions whose only ability was to self-replicate by allocating space in the memory and making a copy of itself. This self-replication ability did not require any external piece of code for evaluating fitness and for selecting individuals for reproduction as in the evolutionary algorithms described earlier. Individuals could replicate within Tierra as long as their own machine code and environment allowed them to do so. Therefore, Tierra was a potentially open-ended evolutionary system when compared to traditional evolutionary algorithms, which are limited by an externally imposed fitness function. According to Ray, self-replication, as opposed to fitness-driven selective reproduction, is a critical feature for emergence of complexity and creativity in artificial evolution because individuals are free to invent their own implicit fitness functions in unexpected ways.

Since machine languages are not designed to be evolvable because single mutations have a very high probability of crashing the entire system, Ray devised a virtual computer within the memory of a regular computer. Each creature had a central processing unit (CPU) with registers and pointers within the virtual computer. A CPU could execute simple arithmetic operations on the registers and move instruction pointers around the memory (RAM) of the computer where all the creatures lived. The genetic code of a creature was its own machine code.

A creature had a membrane defined by `start` and `end` bytes that define the size of the creature in the memory space. Each creature had write privileges within its own membrane and could also read and execute instructions of other creatures, but could not write within the membrane of other creatures. A creature could allocate an additional block of memory, which could later be used to spin off a daughter creature.

Tierra had various sources of mutation that flipped the bits of the digital substrate with very low probability. When the memory of the computer was filled at more than 80%, creatures with the largest number of errors died by deallocating their memory, but their instructions were not erased from the environment and could be executed and incorporated by other creatures.

The Tierran ecosystem was started by inoculating a single creature consisting of 80 instructions and letting it free to replicate. Soon the computer memory was filled by daughters of the ancestor creature (figure 7.20). During prolonged runs of Tierra, Ray observed several phenomena found in biological life. Initially, creatures were small and their turnover rate was very rapid. As time progressed, there was an increasing diversity of the population with respect to size, longevity, and replication rate. At this point parasite creatures emerged that did not harm their hosts, but generated competition for memory space. For example, some parasites did not have a `copy` instruction, but could reproduce by executing the `copy` instruction of another creature. As

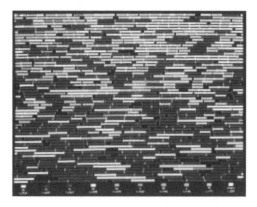

Figure 7.20 A snapshot of the Tierra ecosystem in the artificial life monitor (ALmond), a program developed by Marc Cygnus. Shaded blocks correspond to bodies of creatures in the memory space.

parasites started to occupy larger portions of the Tierran memory space, host creatures became rarer. As a consequence, parasite density decreased and host density rebounded, in cycles predicted by the Lotka-Volterra dynamics explained earlier. In addition, Ray observed the emergence of social communities with a high degree of genetic relatedness that could self-reproduce if they lived close enough in the memory space. In very long runs, the pattern of evolutionary change was characterized by long periods of stasis followed by rapid evolutionary change, similar to the pattern of *punctuated equilibria* observed in natural evolution (Gould and Elredge 1977).

Tierra was conceived as a tool to re-create and study evolution of life in silico. A more recent and publicly available system to investigate evolutionary and biological phenomena in silico, known as *Avida*, has been created through a joint collaboration between physicist Chris Adami at Caltech and biologist Richard Lenski at the University of Michigan (Adami 1998). The Avida program is based on concepts similar to those employed by Tierra, but in Avida the population can also evolve though a combination of self-replication and of an externally imposed fitness function provided by the researcher. Avida has been used to study several biological phenomena, including the role of gene interactions (Lenski et al. 1999), increment of complexity (Adami et al. 2000), and effects of mutation rates (Wilke et al. 2001), to mention a few.

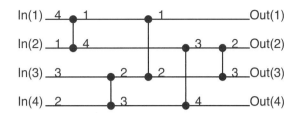

Figure 7.21   A sorting network for arranging four arbitrary numbers in nondecreasing order. Vertical connectors represent comparators that exchange the position of two incoming elements if the lower is smaller than the higher. Redrawn from (Knuth 1998, p. 221).

7.6.2   Competitive Coevolution of Programs

The potential advantage offered by coevolving parasite programs was explored by Hillis (1992) to generate efficient sorting networks. A sorting network is an algorithm that sorts a series of n characters according to a predefined order. Such networks can be represented as graphs made of horizontal input-output lines and vertical connectors that represent comparators (figure 7.21). In order to reduce computation time, it is desirable to execute as few comparisons as possible. Although it can be demonstrated that the four-input sorting network of figure 7.21 can sort any series of four numbers, it is not obvious that it is demonstrable for networks of arbitrary size (Knuth 1998, p. 223). However, it has been shown that if a network with n input lines can sort all 2^n sequences of 0s and 1s into nondecreasing order, it will also sort arbitrary sequences of any n numbers into nondecreasing order (Knuth 1998, pp. 223-225). This result greatly reduces the number of tests that must be made in order to check the efficiency of large sorting networks.

Capitalizing on this result, Hillis attempted to evolve sorting networks of 0s and 1s without and with coevolving parasite programs. He focused his experiments on a 16-input network because this was a well-studied case for which there existed a number of previously suggested algorithms requiring between 65 and 60 comparisons (the latter being the best found so far). The sorting network was represented by genetically encoding pairs of numbers that point to the elements to be compared and exchanged. Individuals could include between 60 and 120 pairs (comparators), but they all started with the same pattern of 32 comparators defined in the first half of the genotype because this was known to be an efficient subsolution for networks of 16 elements. The remaining part of the genotype was randomly initialized.

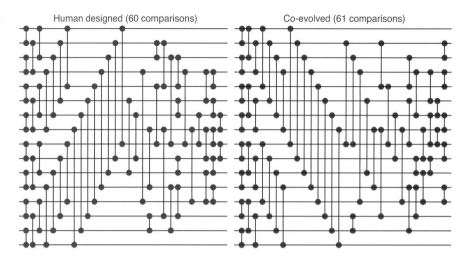

Figure 7.22 Shortest 16-element sorting network developed so far by humans (Green, cited by Knuth 1998) and best coevolved network (redrawn from Hillis 1992).

In the case of evolution without parasites, the fitness was the percentage of test input sequences for which the network produced the correct order. The test cases were randomly generated out of all 2^{16} possible sequences. The best sorting networks discovered by evolution required 65 comparisons, which, although not optimal, ranked among the best found by humans. However, Hillis noticed that most evolutionary runs tended to stagnate into local minima and would often score well on the test cases used during evolution, but not so well on new test cases.

Hillis then introduced a coevolving population of parasites that encoded the test sequences and used them to evaluate the quality of the sorting networks. The fitness of the parasites was inversely proportional to the fitness of the sorting networks. In other words, while the goal of the sorting networks was to sort out the test cases in the best possible order, the goal of the parasites was to generate test cases that would be very hard to sort. Although the original article does not give much detail on the procedure, Hillis mentioned that both parasites and sorting networks evolved at the same speed, meaning that the generational turnover happened at the same time for the two competing populations. In other words, the fitness of the two species was computed by pairwise association of test cases and sorting networks.

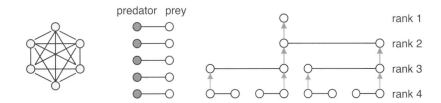

Figure 7.23 Strategies for organizing tournaments in competitive coevolution. *Left*: All against all. *Center*: Pairwise (separate species). *Right*: Tournament-based ranking. Adapted from Angeline and Pollack (1993).

The best coevolved sorting network (figure 7.22, right) displayed 61 comparisons, only one more that the best sorting network found so far by humans. Hillis suggested that these results were obtained because coevolving parasites generated continuously changing test sequences that effectively probed weak parts of sorting networks and at the same time maintained a higher diversity in the population, thus preventing premature convergence.

7.6.3 Progress in Competitive Coevolution

In competitive coevolution solutions coevolve and provide a continuously changing set of challenges rather than a predefined and immutable set as in standard evolutionary algorithms. This has two potential advantages. Evolution of a new strategy in one species may represent a new challenge for the competing species, which in turn would develop a new and more efficient strategy. From the perspective of a single species, this situation corresponds to a *pedagogical series* of challenges that requires gradually more complex solutions (Rosin and Belew 1997). As a whole, the continuation of this reciprocal process of coadaptation may drive both species toward increasingly better solutions.

The other advantage of competitive coevolution is a lower probability of stagnation in local minima thanks to the preservation of higher genetic diversity, which is caused by the continuously changing fitness landscape. In other words, the combinations of genes that correspond to high fitness may change as the opponents change strategies. Consequently, individuals with relatively low fitness in the current generation may still score higher fitness in later generations before a suboptimal genotype spreads too far within the population.

Angeline and Pollack (1993) extended the framework of competitive co-evolution to the more general case of a single population with tournament selection. They pointed out that the choice of opponents for the tournaments of a competitive coevolutionary run is crucial to both the success of the experiment and to the number of tournaments (figure 7.23). Axelrod (1989), for example, considered an exhaustive approach where each individual is tested against all other individuals in the current generation, which, for a population of n individuals, requires $(n(n-1))/2$ tournaments. In the work by Hillis on sorting networks, each parasite is coupled to one sorting network of the same generation, which corresponds to only $n/2$ tournaments (assuming that n includes both the parasite test cases and the sorting networks), but it may not always be practical for other problems to devise competing populations and fitness functions as in Hillis's work.

Angeline and Pollack (1993) therefore proposed a solution where members of the population undergo a series of binary tournaments to determine their relative ranking. Two individuals are randomly selected (without replacement) and the one with highest fitness is selected to enter the second level of tournaments. Once the first round of tournaments is completed, individuals are randomly paired again and only the best ones move on to the next round of tournaments. This is repeated until there is one winner. The fitness of an individual is its rank in the tournament stages. Individuals with higher fitness have higher reproduction probability, but it is important to ensure that individuals within the same rank are randomly selected for reproduction in order to ensure sufficient genetic variability. This method requires $n-1$ tournaments and does not necessarily require two separate populations and different fitness functions.

However, there are at least two factors that may hamper continuous progress in coevolution of competing populations. Coevolving populations may drive each other into twisting pathways where new solutions are good enough to defeat current strategies used by the coevolving population, but are not necessarily better than solutions discovered in earlier generations in the sense that they may not be able to defeat earlier competitors. Furthermore, competing populations may enter cycling dynamics where previously evolved strategies are rediscovered over and over again (figure 7.24).

RED QUEEN EFFECT

Another potential problem is caused by the continuously changing fitness landscape (see chapter 1). This problem is known as the *Red Queen* effect, from the name of the chess character in Lewis Carroll's *Through the Looking Glass* who was always running without making any progress because the landscape was moving in the same direction. In competitive coevolution,

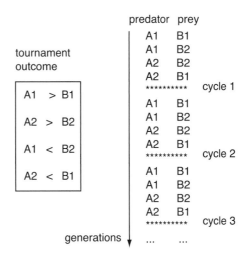

Figure 7.24   Cycling dynamics of two coevolving populations with pairwise tournaments. Each population develops two types of strategies: A1 and A2, and B1 and B2, respectively. The same strategies are rediscovered over and over again.

the reproductive value (fitness) of a specific combination of genes can change as the opponents coevolve (figure 7.25, top right). Although this may have beneficial effects to avoid stagnation into local minima and encourage population diversity, there is the danger that a particular combination of features evolved over some generations may become ineffective in later generations.

Furthermore, the instantaneous fitness (that is, the fitness measured at each generation) is no longer an indicator of progress because a decline may be caused either by less effective strategies than those used in earlier generations or by a more effective strategy discovered by the opposing species (figure 7.25, bottom right). The same ambiguity holds in the case where the instantaneous fitness displays an increment over generations.

Rosin and Belew (1997) suggested a method for preventing the lack of "generational memory" and the Red Queen effect. The method consists of HALL OF FAME setting up a *Hall of Fame* that records a copy of all the best individuals generated over generations. Each individual of the evolving population is then tested against all the best opponents recorded so far in the Hall of Fame. This method was used to coevolve play strategies for a 3D version of the tictactoe game and was shown to produce continuous progress in the sense that evolved players were capable of defeating both current and previous

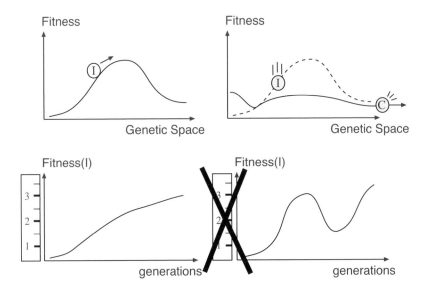

Figure 7.25   *Top*: The Red Queen effect. Individuals evolved in isolation (*left*) tend to converge toward combinations of genes with higher fitness. When a coevolving competitor is introduced (*right*), the fitness landscape is modified so that the fitness of certain combinations of genes is lowered by the adaptive modifications of the competitor. *Bottom*: Instantaneous fitness of a coevolving population (*right*) is no longer a reliable indicator of progress, as is the case for populations evolved in isolation (*left*), because it is not possible to disambiguate between fitness variations due to change in the measured population and fitness variations due to change in the competing population.

opponents. As the Hall of Fame grows over generations, the number of tournaments that each individual must face increases too. However, Rosin and Belew (1997) also showed that to ensure progress it is sufficient to test individuals against a fixed number of opponents randomly chosen from the Hall of Fame.

7.6.4   Competitive Coevolution of Behavioral Systems

Cliff and Miller realized the potential of coevolution of pursuit-evasion tactics for evolutionary robotics (see also chapter 6). In the first of a series of papers (G.F. Miller and Cliff 1994), they introduced a 2D simulation of predator and prey robots. Later, they proposed a set of visualization techniques, CIAO PLOTS   known as CIAO plots (*c*urrent *i*ndividual vs. *a*ncestral *o*pponent), in order

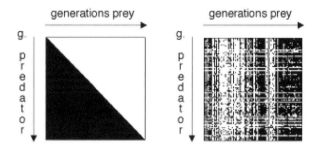

Figure 7.26   CIAO plots show the result of tournaments where the best individual of one generation is tested against the best individual of another generation. Black: predator wins; white: prey wins. *Left*: CIAO plot of an ideal situation where competitive coevolution produces increasingly better solutions in both species. *Right*: CIAO plot of a situation where competitive coevolution does not produce increasingly better solutions.

to detect evolutionary progress which could not be tracked otherwise due to the Red Queen effect (D. Cliff and Miller 1995).

A CIAO plot (figure 7.26) is a matrix whose rows and columns indicate the generations of predators and prey. Each cell in the matrix shows in white or black whether the best predator or the best prey, respectively, of the corresponding generations won the tournament. If competitive coevolution generated increasingly better solutions in the two competing species, the CIAO plot should display two distinct black and white areas (figure 7.26, left). If instead, competitive coevolution did not display continuous progress, the CIAO plot may display patches of black and white areas distributed across the entire surface (figure 7.26, right).

In a later paper (D. Cliff and Miller 1996), the authors described experiments where simulated prey and predator robots were coevolved for pursuit and evasion. The fitness of the prey f_{py} was proportional to its distance from the predator scaled in the range $[0, 1]$ and the fitness of the predator f_{pr} was its complement $1 - f_{py}$. In the first generation individuals were tested against a set of randomly chosen individuals from the opponent species. In all subsequent generations, individuals were tested against the best opponent of the previous generation to improve coevolutionary stability, as suggested by Rosin and Belew (1997). The genetic code of the robots included not only the parameters of the neural controllers but also the layout of the neurons in 2D space and the morphologies of their eyes. The CIAO analysis did not display a clear pattern of continuous progress although one could detect temporally

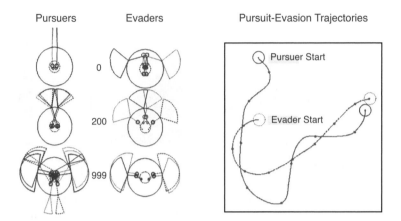

Figure 7.27 *Right*: Pursuit-evasion trajectories resulting from testing an evader from generation 200 against a pursuer from generation 999. *Left*: Plot of the sensorimotor morphologies of best pursuers and evaders at generations 0, 200, and 999. Each plot is a top-down view of the animat where the large circle is the extent of the animat body and the inner dotted circle is the area where terminations of synaptic connections (smallest circles) contribute to the movement of the animat. Other circles are neurons. The genetic specification of visual neurons includes the orientation and angle of acceptance of the visual receptor (shown as solid and dotted sectors). Reproduced from D. Cliff and Miller (1996).

limited improvements. Some of the best evolved predators used four wide-field lateral visual sensors and two narrow-field rear visual sensors, whereas some of the best evolved prey used only four wide-field lateral visual sensors (figure 7.27, left). Cliff and Miller (1996) reported successful pursuit or evasion strategies only when competitors that were far apart in evolutionary time were tested together (figure 7.27, right).

Despite these promising achievements, if one looks carefully at the results described in the literature, focusing on competitive coevolution of pursuit-evasion behaviors, it is easy to notice that coevolutionary benefits often come at the cost of several thousand individuals per population (C.W. Reynolds 1994), several hundred generations with little progress (D. Cliff and Miller 1996), or repeated trials of evolutionary runs with alternating success (Sims 1994). Moreover, since all the experiments were conducted in simulation, often the results cannot be directly applied to real robots, either because agent descriptions are too abstract or technically unfeasible, or because the fitness

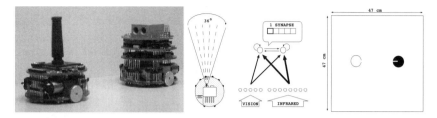

Figure 7.28 *Left*: The predator and prey robots. The predator (to the right) is equipped with a linear vision module. The prey (to the left) does not have vision and has a black hat that can be detected by the predator, but its maximum speed is twice that of the predator. Both predator and prey are equipped with eight infrared proximity sensors (maximum detection range was 2 cm in the environment used for this experiment). *Center*: Details of the visual field of view, divided into five sectors, of the neural architecture, and of the genetic encoding. The prey differs from the predator in that it does not have five input units for vision. Each synapse in the network is coded by five bits, the first bit determining the sign of the synapse and the remaining four bits its strength. *Right*: Initial starting position for prey (*left*, empty disk with small opening corresponding to frontal direction) and predator (*right*, black disk with line corresponding to frontal direction) when tested after coevolution.

function is the distance between the competing agents (see also discussion of fitness space in chapter 1).

Floreano and Nolfi (1997b) and Floreano et al. (2001) explored competitive coevolution of predator and prey systems in real mobile robots. The authors speculated that the difficulties found in previous work were due to one or more of the following reasons: (a) the genetic encodings and neural controllers were exceedingly complex for the situation under study and may have introduced unnecessary complications in the fitness landscape and thus the evolvability of the system; (b) the two agents were equipped with the same set of sensorimotor abilities (even if in some cases these could evolve), thus creating a symmetric situation that may have hampered exploitation by each species of the weaknesses of the opponent species; (c) the fitness function was based on distance between the opponents, thus forcing coevolution to explore a very limited set of potential behaviors that the agents could have developed.

In order to address these issues, they used simple feedforward neural networks without hidden units, direct genetic encoding of the connection weights in bit strings, two robots with different sensorimotor abilities, and an internal fitness function based on time to contact measured by the robot's internal clock.

Two Khepera robots were coevolved in a square arena with white walls (figure 7.28). The predator robot was equipped with a linear vision module and eight short-range (2 cm) proximity sensors. The prey robot had only the short-range proximity sensors, but had a maximum available speed that was twice the available speed of the predator. The prey had a black hat that could be detected by the predator with the vision module. Both robots were connected to a desktop computer through a triple set of rotating contacts that provided power supply and data communication while preventing twisting of the cables as the two robots engaged in pursuit-evasion strategies. Coevolutionary experiments were carried out both with the real robots and with simulated robots that allowed the exploration of long evolutionary runs. The overall pattern of results was not significantly different in physical and simulated evolutionary runs.

Two populations of 100 individuals each were coevolved for 30 generations in the physical experiments and for 100 generations in the simulation experiments. Each individual was tested against the best competitors of the 10 previous generations in order to improve coevolutionary stability. At the beginning of each tournament, the two robots were positioned at random locations and orientations in the arena. A tournament ended either when the predator touched the prey or after 50 seconds. The fitness of the prey was the time to collision T scaled by the maximum tournament duration, while the fitness of the predator was $1 - T$.

In all evolutionary runs, the fitness graphs of the two species displayed counterphase oscillations, as one would expect from competitive coevolutionary dynamics. When the fitness of the predator increased, the fitness of the prey decreased, and vice versa. The authors observed spontaneous evolution of obstacle avoidance, visual tracking, object discrimination (prey vs. wall), following, and a variety of other behaviors that appeared and disappeared over generations. However, after approximately 20 generations, a small set of behavioral strategies were cyclically adopted by the two opponent robots.

MASTER
TOURNAMENTS

In order to measure coevolutionary progress, if any, the authors devised what they called *master tournaments* where the best individual of a species recorded at each generation was tested 10 times against all the best competitors of all generations. The master fitness values resulting from these tournaments provide an absolute measure of progress. If a species developed increasingly more powerful solutions, that is, solutions capable of defeating a

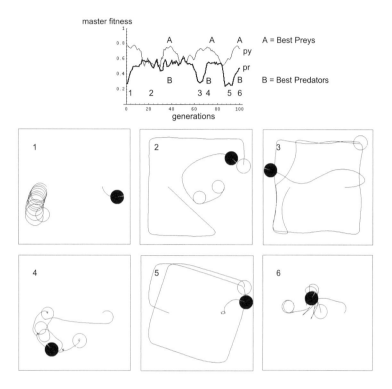

Figure 7.29 *Top*: Fitness of best individuals in master tournaments. Letters indicate position of best prey and best predators. Numbers indicate position of individuals whose tournaments are displayed below. *Bottom*: Behaviors recorded at interesting points of coevolution, representing typical strategies. Black disk is predator, white is the prey. See text for explanation.

larger number of opponent strategies, its master fitness would increase over generations. However, this was not the case in the experiments with robots (figure 7.29, top), where the master fitness of the two species remained almost constant over generations. Indeed, a CIAO analysis revealed patterns of local dominance by one species followed by patterns of local dominance by the other species. Master tournaments tell two additional things: at which generation one can find the best prey and the best predator, and at which generation one should observe the most complex tournaments. The best individuals are those corresponding to generations with the highest fitness when the competitor also reports the highest fitness (marked by the letters A and B in the graph). Instead, the most complex tournaments are those that take place between individuals that report the same fitness level, because those are the situations where both species have the same level of ability to win over the competitor.

Despite the lack of increasing progress, evolved behaviors displayed a very rich variety of relatively complex and surprising strategies. In the lower part of figure 7.29, behaviors of the best competitors at critical stages of co-evolution, as indicated by master tournament data, give a more intuitive idea of how pursuit-evasion strategies coevolved. Initially, the predator tended to stop in front of walls while the prey moved in circles (panel 1). Later, the prey moved fast with straight trajectories, avoiding walls, while the predator tracked it from the center and quickly attacked when the prey was closer (panel 2). Interestingly, predators developed the ability to know how distant the prey was by using information on how fast they moved on their visual field. Decrement of predator performance around generation 65 was due to a temporary loss of the ability to discriminate between walls and prey, probably because most of the time predators were successful at catching the prey without experiencing collisions with walls. The decrement of predator performance was most likely due to a more efficient prey strategy. As shown in panel 3, the predator intercepted the prey, but missed it and ended up against the wall. Around generation 75 (panel 4), the prey moved in circles and, when the predator got closer, it rapidly avoided it. In fact, prey that moved too fast around the environment sometimes could not avoid an approaching predator because they detected it too late (proximity sensors have lower sensitivity for an object as small as a robot, which reflects less infrared light than a white flat wall). Therefore, it paid off for prey to wait for the slower predator and accurately avoid it. However, some predators became smart enough to perform a small circle once they had missed the target, and to attack again until, by chance, the prey offered the side with the wheels where there are no proximity sensors. Consequently, prey rediscovered the fast-moving trajectories around the environment. At that point, predators developed a "spider strategy" (panel 5): instead of running after the prey, they slowly moved backward until they hit a wall where they waited for the fast-approaching prey, which collided into the "spider robot" before detecting it. However, this strategy did not pay off when the prey stayed in the same place. Finally, at generation 99 a new interesting strategy emerged (panel 6): the predator quickly tracked and reached the prey, which had been quietly rotating in small circles. As soon as the prey sensed the predator, it backed off and then approached the predator (without touching it) from the

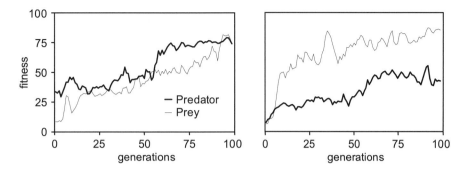

Figure 7.30   *Left*: Master fitness of predator and prey coevolved with Hall of Fame strategy (Rosin and Belew 1997). *Right*: Master fitness of predator and prey with standard coevolution where the prey is equipped with a vision module spanning 240° (much larger than that of the predator).

side where it could not be seen; consequently, the predator quickly turned in an attempt to visualize the prey which rotated around it, producing an entertaining dance.

Nolfi and Floreano (1999a) also showed that real coevolutionary progress (as revealed by master tournaments) could be obtained either by using the Hall of Fame strategy (figure 7.30, left) or by providing the prey with a vision module spanning a much larger field of view than that of the predator (figure 7.30, right). They also showed that predators evolved to catch some of the best (previously) coevolved prey could not develop efficient strategies, as predators coevolved with prey, indicating that competitive coevolution was more effective than evolution of individual agents. Finally, the authors measured the fitness of all coevolved prey and predators using the distance metrics used by Cliff and Miller and noticed that while prey robots did obtain a very high score, predator robots obtained a very low score, indicating that predators used other strategies to maximize the probability of hitting the prey (e.g., visual tracking and interception or spider strategies).

The results of this set of experiments suggest that the difficulties encountered by authors of previous work on competing agents may have been caused either by complicated genetic and neural structures or by a constraining fitness function that did not allow more exploration by coevolution.

7.6.5   Ontogenetic Plasticity in Competitive Coevolution

In the experiments described above, predator and prey robots could not modify the synaptic strengths during their lifetime. Considering that the environment faced by those robots is very dynamic, lifetime synaptic plasticity may affect the outcome of coevolutionary dynamics.

In order to study this issue, Floreano and Nolfi (1997a) carried out two additional sets of coevolutionary experiments under exactly the same conditions described earlier. In the first set of experiments, two bits of each synaptic gene (see figure 7.28) were used to determine the strength of the synapse and the two remaining bits were translated into a quantity of uniform noise (centered around zero) added to the synaptic strength at each sensorimotor cycle. In this "noise" condition, synaptic plasticity was not directional (i.e., it was independent of the behavior of the competing robot) and served as a control situation. In a second set of experiments, instead, the four bits encoded one of the four Hebbian rules and four possible learning rates, exactly as in the experiments on the evolution of learning described in chapter 6 (figure 6.53). In this "adaptive" condition, synaptic plasticity could be potentially exploited to develop abilities aimed at improving the individual's performance according to the competitor's behavior.

Floreano and Nolfi (1997a) then measured the relative performance of the two species' master fitness in the genetically determined condition (previous section), in the noise condition, and in the adaptive condition. The relative performance consisted in counting how often in the master fitness graph the fitness of one species was higher than that of the other species. For a coevolutionary run of 100 generations, the relative performance could be in the range $[-100, 100]$. The results of this analysis indicated that the genetically determined condition and the noise condition were not statistically different and that both predators and prey prevailed in an almost equal number of generations. However, the adaptive condition was significantly different because predators prevailed in more than 70% of the generations.

Behavioral strategies of coevolved adaptive robots were similar to those of coevolved genetically determined robots. However, compared to the genetically determined condition where predators tracked and attacked the prey only in one direction (figure 7.29), in the adaptive condition predators efficiently tracked and attacked in both directions by adjusting their trajectories according to the observed behaviors of the prey (figure 7.31). Whereas activity-dependent synaptic change was exploited by the far-sighted predator, the same did not happen for the prey whose short-range proximity sen-

 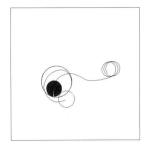 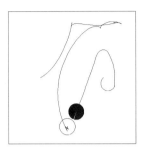

Figure 7.31   Behavioral strategies of predator and prey with adaptive synapses. Black disk is predator, white disk is prey. *Left*. Generation 20. *Center*. Generation 70. *Right*. Generation 95.

sors did not provide sufficient information to develop an effective counter-strategy.

In order to check whether such improvements were due to a real advantage of the predator provided by Hebbian learning, rather than to some difficulty of the prey to cope with Hebbian learning, the authors organized master tournaments between predators coevolved in adaptive conditions and prey coevolved in genetically determined conditions. Also in this case, predators won most of the tournaments, suggesting that Hebbian learning did not penalize prey in the adaptive condition, but rather provided an advantage to predators.

Finally, Floreano et al. (2001) used a genetic representation with two additional bits per synapse whose values indicated whether the remaining bits were interpreted as synaptic strength (genetically determined condition), synaptic strength plus uniform noise (noise condition), or Hebbian rules plus learning rate (adaptive condition). The results indicated that all best coevolved predators used Hebbian learning in most synapses, whereas best coevolved prey oscillated between noise and Hebbian learning across generations in a significantly smaller set of synapses. Relative performance measured in master tournaments showed that predators won most of the tournaments in this case too.

The results described in this subsection carry two important messages. The first is that ontogenetic plasticity can affect the outcome of coevolutionary dynamics. The second is that the morphology and physical properties of the agent can affect the efficiency of neural plasticity and produce quite different behavioral outcomes. This specific and opportunistic view of learning reinforces the criticism of learning as a general-purpose mechanism that we mentioned in the closing remarks in chapter 3.

7.7   Artificial Evolution of Cooperation

Cooperation describes the situation where a group of individuals obtain a net benefit by working together. A first major issue in the artificial evolution of cooperation is whether robots should be genetically identical or different and whether the fitness used for selection should take into account the performance of the entire group or only that of single individuals. These two choices are analogous to the issues of genetic relatedness and of group selection that were discussed earlier in the context of the biological literature.

If we consider only the extreme cases, individuals in a team can be genetically homogeneous (clones) or heterogeneous (they differ from each other); and the fitness can be computed at the level of the team (in which case, the entire team of individuals is reproduced) or at the level of the individual (in which case, only individuals of the team are selected for reproduction).

The majority of current approaches to the evolution of multiagent systems use genetically homogeneous teams evolved with team-level selection (a comparative survey can be found in Waibel et al. 2008). Where the reasons for the choice of genetically homogeneous teams are made explicit, it is argued that homogeneous teams are easy to use (Baray 1997; Trianni et al. 2006), require fewer evaluations (Luke et al. 1997; Richards et al. 2005), scale more easily (Bryant and Miikkulainen 2003), and are more robust against the failure of team members (Bryant and Miikkulainen 2003; Quinn et al. 2002a) than heterogeneous teams.

Many other approaches use genetically heterogeneous teams evolved with individual-level selection. Genetically heterogeneous teams are sometimes seen as providing more behavioral flexibility (Luke et al. 1997) and as allowing specialization (Baldassarre et al. 2003b; J.C. Bongard 2000; Luke et al. 1997; Quinn et al. 2002a). However, the reasons for the choice of team-level or individual-level selection are rarely made explicit.

The terms "homogeneous team" and "heterogeneous team" used in the current literature cover many different aspects. It is important to note that while all agents in genetically homogeneous teams share the same genes, agents can nevertheless be behaviorally heterogeneous. This can happen when agents differentiate during their lifetime, for example, due to varying initial conditions (Quinn et al. 2003), or due to developmental processes or learning. This can also happen when agents "activate" different parts of their genome, for example, when each agent's behavior is controlled by a different section of a single team genome (J.C. Bongard 2000; Haynes and Sen 1997; Miconi 2003; Robinson and Spector 2002). In this case, agents can specialize in different functions, yet be genetically identical, just like specialized cells in a biological organism.

In some cases, teams consist of clonal subteams (Luke et al. 1997; Luke 1998) or of agents that share only part of their genome. Teams with agents that are, on average, genetically more similar (but not identical) to members of their team than to members of the rest of the population are termed "partially heterogeneous." The effects of partial genetic heterogeneity on the evolution of multiagent teams are not yet fully explored in evolutionary computation (Mirolli and Parisi 2005), but have been thoroughly studied in biology (Hamilton 1964; Lehmann and Keller 2006).

The choice of level of selection is rarely discussed explicitly. Some research has addressed the related issue of credit assignment for the evolution of multiagent systems (Agogino and Tumer 2004). In the context of multiagent systems, credit assignment is concerned with distributing fitness rewards among individual agents. Fitness distribution leads to credit assignment problems (Grefenstette 1988; Minsky 1961) in many cooperative multiagent tasks because individual contributions to team performance are often difficult to estimate or difficult to monitor (Panait and Luke 2005). Selection is usually performed on the basis of accumulated individual or team fitness, which may be the result of many fitness rewards with different types of credit assignment. Therefore an optimal choice of level of selection is not only influenced by the type of task but also by the types of credit assignment used.

7.7.1   Evolutionary Conditions and Task Demands

Waibel et al. (2008) systematically compared the performance of robot teams evolved in four evolutionary conditions (figure 7.32): genetically homogeneous teams evolved with team-level selection; genetically homogeneous teams evolved with individual-level selection; genetically heterogeneous teams evolved with team-level selection; and genetically heterogeneous teams evolved with individual-level selection. Team-level selection (akin to group selection) consisted in computing the fitness of the team and reproducing the robots in the best teams to create a new population of robot teams. Individual-level selection instead consisted in computing the fitness of individual robots (notice that even robots with identical genomes can obtain

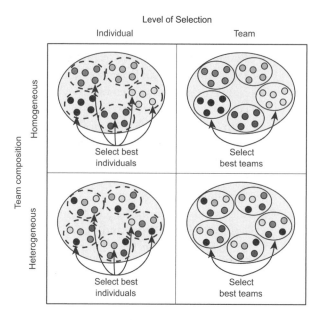

Figure 7.32 Four methods for evolution of cooperation. A population (large oval) was composed of several teams (medium-sized ovals), each of which was composed of several robots (small circles). Genetic team composition was varied by either composing teams of robots with identical genomes (homogeneous, identical shading), or different genomes (heterogeneous, different shading). The level of selection was varied by either selecting teams (team selection) or selecting individuals independently of their team affiliation (individual selection).

different fitness because they are exposed to different situations) and reproducing the best ones independently of their team affiliation to re-create new teams.

Since the relative performance of the four conditions may vary according to the type of cooperative task, the authors evaluated the performance of robot teams in three classes of multirobot tasks (figure 7.33, left): a collective task that did not require cooperation; a task that required cooperation but did not imply a cost to cooperators; and a task that required altruistic cooperation, i.e., a task that implied an individual fitness cost to cooperators. The experiments were carried out in physics-based simulations of 10 microrobots that had to find and bring to a foraging area small and large food tokens (figure 7.33, right). A small token could be pushed by a single robot, which gained one fitness point; instead, a large token required two or more robots

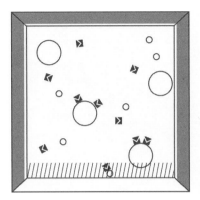

Figure 7.33 *Left*: The experimental setup for the altruistic foraging task. Ten microrobots (black squares with arrows) searched for small and large tokens and transported them to the target area (hatched area at bottom) under the white wall (the other three walls were black). An identical setup was used in the other two experimental conditions, except that the arena contained either only small tokens in the collective task (no cooperation required), or only large tokens in the cooperative task (cooperation required, but no cost to the individual). *Right*: Three microrobots in the altruistic cooperative foraging task. The robot in the background could transport the small token by itself. The robot at the left could not transport the large token by itself and needed to wait for the arrival of a second robot. From Waibel et al. (2008).

to be pushed, but all robots in the team received one fitness point independently of whether they helped to bring the token to the foraging area.

When only small tokens were placed in the arena, robots were not required to cooperate because each one could forage in isolation. When only large tokens were placed in the arena, robots had to cooperate, but there was no cost because that was the only way robots could obtain fitness. When both large and small food tokens were present, robots that cooperated to push large tokens paid a cost with respect to robots that pushed small tokens because it took a longer time to carry a large token and it was more difficult to bring it all the way to the foraging area. However, teams of robots that displayed altruistic cooperation obtained higher team fitness than teams of robots that pushed only small objects.

The results indicated that when no cooperation was required, heterogeneous robots with individual-level selection obtained the highest performance. Instead, when some degree of cooperation was required, either with or without cost to the cooperators, homogeneous teams evolved either with team-level selection or with individual-level selection obtained consistently

Figure 7.34   A Swarm-bot composed of four interconnected s-bots in chain formation.

the highest performance. In the case of altruistic foraging, heterogeneous teams were more efficient at bringing objects to the foraging area, but those objects were mainly small tokens, whereas homogeneous teams carried fewer objects, but those were mainly large tokens.

These results carry two messages. The first is that if some degree of cooperation is required, it is advisable to use genetically related controllers and team-level selection. The second is that the choice of evolutionary method depends on the type of cooperative task. Tasks that do not require specialization are carried out most efficiently by homogeneous teams; however, if a task benefits from some degree of specialization, then heterogeneous teams may obtain higher performance (J.C. Bongard 2000). Armed with the insights from these experiments, we can now approach the literature on evolution of cooperative behavior and better appreciate the choice of evolutionary methods.

Let us consider the case of evolution of Swarm-bots (Mondada et al. 2004) described earlier in this chapter. From a morphological point of view, s-bots are not specialized because they have been created mainly with the purpose of cooperating by dynamically self-assembling when required by the task at hand. Indeed, almost all experiments with evolution of Swarm-bots resorted to populations of homogeneous robots with team-level selection (Dorigo et al. 2004). The evolutionary tasks that were studied included heavy object transportation, collective exploration, communication, and navigation in swarm formation.

In a simple case, Swarm-bots of four s-bots assembled in chain formation were evolved for the ability to move coordinately on a flat terrain (figure 7.34). Each s-bot was provided with a simple neural controller where sensory neurons were directly connected to the motor neurons that controlled the desired speed of the tracks and whether or not a sound signal was produced. Evolved neural controllers were also capable of producing coordinated movement when the Swarm-bot was augmented by additional s-bots and reorganized in different shapes. Swarm-bots also dynamically rearranged their shape so as to effectively negotiate narrow passages and were capable of moving on rough terrains by negotiating situations that could not be handled by a single robot. Such robots also collectively avoided obstacles and coordinated to transport heavy objects (Baldassarre et al. 2003a,b; Trianni et al. 2006).

Let us instead consider the case of evolving a team of agents that must accomplish a foraging task typically faced by honeybees. The animals (and agents) must collect both pollen and nectar, but must ensure that the quantity of nectar in the hive is within a certain small quantity in order to ensure the viability of the colony. Too little or too much nectar is deleterious to colony survival. This task typically requires division of labor and specialization between agents collecting pollen or nectar. Tarapore et al. (2006) used a response threshold model whereby agents had two different behaviors, nectar and pollen collection, characterized by genetically encoded thresholds. The question in this case was whether the best colony performance, defined as the amount of pollen collected for a regulated nectar quantity, was obtained by genetically related or genetically different teams. The results of the computational experiments predicted that colonies characterized by high genetic diversity resulting from the queen mating with several different males performed much better than genetically related colonies. These predictions were later verified by experimental evidence collected with the biological animals (Mattila and Seeley 2007).

7.7.2   Evolution of Communication

In social species communication plays a pivotal organizing role, allowing the transfer of vital information among colony members, for example, to detect predators and find food sources (E.O. Wilson 2000). While much is known concerning the neurophysiological processes by which signals are produced, conducted, perceived, and interpreted, the conditions conductive to the evolution of communication and the paths by which reliable systems of commu-

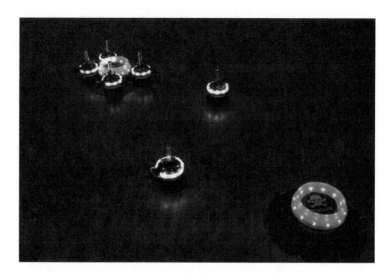

Figure 7.35   A team of s-bots engaged in cooperative communication (Floreano et al. 2007). A group of four s-bots feed on the food objects while they are lighted up in blue color. Two s-bots in white color are attracted by the blue signal and move away from the poison object.

nication become established remain largely unknown because communication does not leave a fossil record. This is a particularly challenging problem because efficient communication requires tight coevolution between the signal emitted and the response elicited (Maynard-Smith and Harper 2003). Under natural conditions, most communication systems are also costly because of the energy required for signal production and/or increased competition for resources resulting from information transfer about food location.

Werner and Dyer (1992) studied the emergence of communication in a population of simulated agents evolving in grid worlds. Female agents had the ability to see males and to emit sounds. Male agents were blind, but could hear signals from females. Thus, the environment was designed to favor organisms that evolved to generate and interpret meaningful signals. Starting with random neural networks, the simulation resulted in a progression of generations that exhibit increasingly effective mate-finding strategies. In addition, a number of distinct subspecies, i.e., groups with different signaling protocols or dialects, evolved and competed. These protocols became a behavioral barrier to mating that supported the formation of distinct subspecies.

However, that study did not clarify whether communication involved a cost, to what extent the population was genetically related, and did not leave agents the possibility of surviving without communication. To address these issues, Floreano et al. (2007) used teams of 10 s-bots that could forage in an environment containing a food and a poison source that both emitted a red light and could only be discriminated at close range (figure 7.35). Under such circumstances, foraging efficiency could potentially be increased if robots transmitted information on food and poison location. However, such communication also incurred direct costs to the signaler because it resulted in higher robot density and increased competition and interference near the food (i.e., spatial constraints around the food source allowed a maximum of 8 robots out of 10 to feed simultaneously and resulted in robots sometimes pushing each other away from the food). Thus, while beneficial to other team members, signaling of a food location effectively constituted a costly act (Hamilton 1964; Lehmann and Keller 2006) because it decreased the food intake of signaling robots. This setting thus mimics the natural situation where communicating almost invariably incurs costs in terms of signal production or increased competition for resources (Zahavi and Zahavi 1997).

The authors evolved teams of robots under the four evolutionary conditions of genetic relatedness and level of selection by using physics-based simulations that precisely model the dynamical properties of real robots. At the end of the experiments evolved genomes were transferred in real robots that displayed the same behavior observed in simulation. Robots could signal by switching on and off their light ring in blue light. Robots could see the red food and poison items as well as the blue-lighted robots with the omni-directional color camera. In a control situation, the evolutionary experiments were repeated with the light ring disabled.

In evolving teams where robots could produce blue light, foraging efficiency greatly increased over generations and was significantly greater compared to control experiments for all evolutionary conditions, except for the condition of heterogeneous teams under individual-level selection. An analysis of the robot behavior revealed that this performance increment in the three conditions of genetic relatedness or team-level selection was associated with the evolution of effective systems of communication.

In teams of genetically related robots with team-level selection, two distinct communication strategies evolved. In 12 of the 20 evolutionary replicates, robots preferentially produced light in the vicinity of the food and were attracted by blue light (figure 7.36, left). Instead, in the other 8 evolutionary replicates, robots tended to emit light near the poison and were

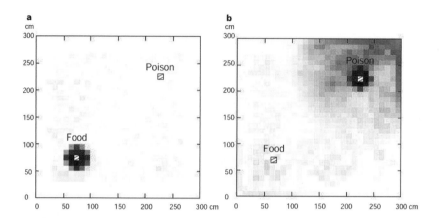

Figure 7.36 Signaling frequency measured in each area of the arena for robots from two different evolved teams. *a*, This team was one where robots signal the presence of food. *b*, In this team robots signal the presence of poison. The darkness of each square is proportional to the amount of signaling in that area of the arena. From Floreano et al. (2007).

repulsed by blue light (figure 7.36, right). Teams of robots that signaled food resulted in higher team performance. Interestingly, once one type of communication was well established, there was no transition to the alternate and more efficient strategy. This was because a change in either the signaling or response strategy would completely destroy the communication system and result in a performance decrease. Thus, each communication strategy effectively constituted an adaptive peak separated by a valley with lower performance values.

These results bring two messages. The first message is that cooperative communication is expected to occur principally among genetically related individuals or when selection takes place at the team rather than the individual level. Consistent with this view, most sophisticated systems of communication indeed occur in animals forming kin groups, as exemplified by pheromone communication in social insects (E.O. Wilson 1971; Bourke and Franks 1995) and quorum sensing in clonal colonies of bacteria (Keller and Surette 2006). Humans are a notable exception, but other selective forces such as reciprocal altruism and reputation-based systems of reciprocity may operate to favor altruism (Nowak and Sigmund 2005) and costly communication.

The second message is that, once a given system of communication has evolved, it may constrain the evolution of more efficient communication systems because it would require going through a stage where communication between signalers and receivers is perturbed. This finding supports the idea of the possible arbitrariness and imperfection of communication systems, which can be maintained despite their suboptimal nature. Similar observations have been made about evolved biological systems (Jacob 1981), which are formed by the randomness of the evolutionary selection process, leading, for example, to different dialects in the honeybee dance language (von Frisch 1967).

The study of the evolution of embodied communication is a growing field of research. For example, Cangelosi (2001) showed that agents provided with suitable neural architectures can evolve increasingly complex communication patterns from signals to symbols and all the way to syntax. However, a better understanding of the evolution of human-like language will require capturing the evolutionary, neural, physical, and environmental conditions that have led to the emergence of this unique phenomenon in the history of our planet.

7.8   Closing Remarks

In this chapter we reviewed theories, methods, and technologies developed to understand and replicate the functional organization of living collective systems. The distinction between self-organizing and evolutionary processes adopted here is deeper than a necessity of organizing the material to be presented. We witness a similar gap between researchers that pursue either one of the two approaches and, interestingly, the same occurs both in the biological and engineering sciences.

In our daily investigations of collective systems and throughout the research of background material for the preparation of this chapter, we noticed that, on the one hand, biologists who study the self-organization of insect societies tend to ignore the evolutionary reasons for the observed state of affairs, and vice versa; and that, on the other hand, computer scientists and engineers who develop collective systems capable of self-organization tend to ignore how alternative organization principles could be brought about by artificial evolution, and vice versa.

This dichotomy is certainly a gross approximation because there are documented exceptions in both scientific areas. Nonetheless, we never came across a deep reflection or experimental study of the extent to which functional phenomena observed in collective systems should be explained with the language of self-organization or of evolution. We may have neglected important papers, but we are not alone in our conclusions. On the biological side, Camazine et al. (2001, pp. 88-89) raise a similar issue and conclude that

> There is no contradiction or competition between self-organization and natural selection. Instead, it is a *cooperative "marriage"* in which self-organization allows tremendous economy in the amount of information that natural selection needs to encode in the genome.

But they also reveal their interest in self-organization by closing the paragraph with a defensive note

> In this way, the study of self-organization in biological systems promotes orthodox evolutionary explanation, not heresy.

On the engineering side, a certain pragmatism favors the adoption of the methodology that produces the best results or best matches the available technology without much preoccupation with explanatory principles. This approach is certainly justified, but a more critical reflection may be useful because both self-organization and artificial evolution present significant shortcomings as they stand today. For example, within a self-organization approach it is not clear how to design simple rules that, once embedded in collections of interacting agents, will generate the desired functionality. Furthermore, since a self-organizing system develops over time, it is not guaranteed that satisfactory behavior observed now will remain stable in the future or will not drastically bifurcate due to external perturbations. Winfield et al. (2004) have explicitly raised these issues and suggested a methodology to engineer self-organizing systems for dependability. Self-organization relies mainly on collection of simple and reactive agents. As we have seen in the section on swarm robotics, these agents may stagnate in suboptimal solutions because they cannot forecast the consequence of their actions at a global level, unless they result from an evolutionary process where behavioral rules that produce inferior outcomes are selected out of the population. Furthermore, some aspects of collaboration or competition demand consideration of evolutionary factors. For example, although cooperative behaviors that do not imply a cost for the cooperating individual may result from a variety of conditions, altruistic cooperation with a cost to the cooperator demands specific design principles that are rooted in evolutionary conditions, such as team relatedness or team selection. Similarly, a purely evolutionary approach may not be sufficiently justified considering the amount of resources it takes (populations of agents must be continuously tested for several generations), which may not even be feasible in hardware systems. Furthermore, in the absence of other self-organizing or adaptive mechanisms, artificial evolution may result in solutions that work only in the environmental conditions used during generational selection.

Finally, some reader may be surprised that in this chapter we did not cover recent work on the topology and dynamics of biological networks, which was contributed mainly by the community of physicists. Indeed, there is mounting evidence that several living systems and phenomena are not randomly or uniformly interconnected and that some observed network topologies, such as small-world networks, may offer advantageous adaptive functionalities, such as robustness and efficient spread of information (e.g., Sole and Goodwin 2000; Barabási 2002; Watts 2003). The accounts reported in technical and popular writings, ranging from the organization of brains to the organization of societies and languages, are extremely fascinating and intriguing. However, in our opinion it is not yet sufficiently clear to what extent those network topologies and dynamics are a side effect of other mechanisms and processes or a cause of the adaptive functionalities that they are associated with. At the same time, it is not yet clear at this stage if, and how, constraints on network topologies and dynamics should, and could, be incorporated in engineering practice to produce more efficient systems.

7.9   Suggested Readings

For an introduction to self-organization in animal societies we recommend the book by Camazine et al. (2001), which was a major source of inspiration and information for the first section of this chapter. The book is organized in two parts, the first part providing a general overview of physical and behavioral phenomena that are common to many species of animals while the second part delves into experimental and modeling aspects of specific animals and behaviors previously published as separate articles by the authors. However, this book does not cover the evolution of self-organization in biological systems.

For an introduction to self-organization in artificial systems, instead, we recommend the book on swarm intelligence by Bonabeau et al. (1999), which describes both ant colony optimization and early experiments in collective and reconfigurable robotics. Although parts of the book are very technical and the area of robotics has progressed significantly since its publication, the book represents a well-structured and lucid guide to algorithms that are currently witnessing a proliferation of applications and modifications. Unfortunately, this book does not cover particle swarm optimization, whose treatment can be found in the book *Swarm Intelligence* by Kennedy and Eberhart (2001), who are among the first proposers of this approach. Although both books mention artificial evolution in several chapters, neither addresses the evolution of swarm intelligence.

The imbalance between self-organization and evolution can be easily corrected by reading *Sociobiology* by Edward Wilson (2000), which in its 25th anniversary edition remains a classic treatment of evolutionary factors that can lead to cooperation in animal and human societies. The *Evolution of Cooperation* by Axelrod (1989), despite its somewhat misleading title, is an extremely clear treatment of the conditions that can lead to cooperation within the framework of game theory. The author describes with great clarity the prisoner dilemma problem and its variations, showing the effects of different strategies and establishing links with both animal and human behavior. One of the chapters provides also an interpretation of the game in the context of evolutionary theory. Important aspects, puzzles, and suggested solutions of the evolution of communication can be found in *Animal Signals* by Maynard-Smith and Harper (2003).

Finally, the science fiction novel *Prey* by Michael Crichton (2002), which was a number 1 *New York Times* bestseller, is an entertaining story based on the nanotechnology, swarm robotics, and competitive coevolution described in this chapter.

Conclusion

A careful reader may have noticed that we have not yet defined what intelligence is. This was done on purpose because intelligence has different meanings for different persons and in different situations. For example, some believe that intelligence is the ability to be creative; others think that it is the ability to make predictions; and others believe that intelligence exists only in the eye of the observer. In this book we have shown that biological and artificial intelligence manifests itself through multiple processes and mechanisms that interact at different spatial and temporal scales to produce emergent and functional behavior. The most important implication of the approaches presented here is that understanding and engineering intelligence does not reduce to replicating a mammalian brain in a computer but requires also capturing multiple types and levels of interactions, such as those between brains and bodies, individuals and societies, learning and behavior, evolution and development, self-protection and self-repair, to mention a few.

The approaches and examples described in this book show that biology is a bewildering source of inspiration for the design of intelligent artifacts capable of efficient and autonomous operation in unknown and changing environments. It is difficult to resist the fascination of creating artifacts that display elements of lifelike intelligence, but we should keep in mind two caveats. The first caveat is that copying a mechanism from biology does not necessarily bring an advantage either because the technology may not match the biology or because the desired functionality may be different from that of the biological mechanisms. In this case, the artifact is no more than a curiosity or, at best, a piece of art with no function. The second caveat consists of resorting to biological inspiration (for example by evolving a complicated neural control system) without understanding how the resulting artifact works. In this case, one cannot characterize the performance of the artifact and predict under which conditions it will fail, which is definitely not good engineering practice.

Proper practice of bio-inspired artificial intelligence requires a *scientific effort* to extract the principles of biological intelligence from the data and theories provided by biologists, and an *engineering effort* to translate those principles into functional artifacts and technologies. The rejoining of science and engineering in bio-inspired artificial intelligence is witnessed by the numerous examples described in this book where intelligent artifacts, such as evolutionary software or mobile robots, are used to test biological hypotheses and make new predictions about biological organisms and processes.

We believe that the theories, methods, and technologies presented in this book have strong potentials for the engineering of intelligent artifacts capable of self-organization and self-adaptation. We also believe that since these artifacts originate from the same processes that operate in nature, they are more likely to capture the very essence of biological intelligence. Even if intelligence is only a subjective phenomenon, we are more likely to attribute intelligence to artifacts whose behaviors emerge from self-organization and self-adaptation than to artifacts that behave according to a predefined plan.

References

Abeles, M. (1991). *Corticonics*. Cambridge University Press, Cambridge, UK.

Abelson, H. and diSessa, A. A. (1981). *Turtle Geometry : The Computer as a Medium for Exploring Mathematics*. MIT Press, Cambridge, MA.

Ackley, D. H., Hinton, G. E., and Sejnowski, T. J. (1985). A learning algorithm for Boltzmann machines. *Cognitive Science*, 9:147–169.

Ackley, D. H. and Littman, M. L. (1992). Interactions between learning and evolution. In Langton, C., Farmer, J., Rasmussen, S., and Taylor, C., editors, *Artificial Life II: Proceedings Volume of Santa Fe Conference*, volume 11, pages 487–510. Addison Wesley, Redwood City, CA.

Adami, C. (1998). *Introduction to Artificial Life*. Springer-Verlag, New York.

Adami, C., Ofria, C., and Collier, T. C. (2000). Evolution of biological complexity. *Proceedings of the National Academy of Sciences USA*, 97:4463.

Adleman, L. M. (1994). Molecular computation of solutions to combinatorial problems. *Science*, 226:1021–1024.

Adrian, E. D. (1928). *The Basis of Sensation: The Action of the Sense Organs*. Norton, New York.

Agogino, A. and Tumer, K. (2004). Efficient evaluation functions for multi-rover systems. In Deb, K., Poli, R., Banzhaf, W., Beyer, H.-G., Burke, E. K., Darwen, P. J., Dasgupta, D., Floreano, D., Foster, J. A., Harman, M., Holland, O., Lanzi, P. L., Spector, L., Tettamanzi, A., Thierens, D., and Tyrrell, A. M., editors, *Proceedings of the 2004 Genetic and Evolutionary Computation Conference (GECCO04)*, volume 3102 of *Lecture Notes in Computer Science*, pages 1–11. Springer-Verlag, Berlin.

Aickelin, U. and Cayzer, S. (2002). The danger theory and its application to artificial immune systems. In *Proceedings of the First International Conference on Artificial Immune Systems (ICARIS-2002), Canterbury, UK*, pages 141–148.

Alberts, B., Johnson, A., Lewis, J., Raff, M., Roberts, K., and Walter, P. (2002). *Molecular Biology of the Cell*, 4th edition. Garland, New York.

Alexander, W. H. and Sporns, O. (2002). An embodied model of learning, plasticity, and reward. *Adaptive Behavior*, 10:143–159.

Allen, P. E. and Holberg, D. R. (2002). *CMOS Analog Circuit Design*, 2nd edition. Oxford University Press, Oxford.

Aloimonos, J., Weiss, I., and Bandopadhay, A. (1987). Active vision. *International Journal of Computer Vision*, 1(4):333–356.

Alpaydin, G., Balkir, S., and Dundar, G. (2003). An evolutionary approach to automatic synthesis of high-performance analog integrated circuits. *IEEE Transactions on Evolutionary Computation*, 7(3):240–252.

Amit, D., Gutfreund, H., and Sompolinsky, H. (1985). Spin-glass models of neural networks. *Physical Review A*, 32:1007–1018.

Anderson, A. M. (1977). A model of landmark learning in the honey-bee. *Journal of Comparative Physiology A*, 114:335–355.

Anderson, J. A. and Rosenfeld, E., editors (1988). *Neurocomputing: Foundations of Research*. MIT Press, Cambridge, MA.

Anderson, J. A. and Rosenfeld, E. (1998). *Talking Nets: An Oral History of Neural Networks*. MIT Press, Cambridge, MA.

Angeline, P. J. and Pollack, J. B. (1993). Competitive environments evolve better solutions for complex tasks. In Forrest, S., editor, *Proceedings of the Fifth International Conference on Genetic Algorithms*, pages 264–270. Morgan Kaufmann, San Mateo, CA.

Arbib, M., editor (1995). *The Handbook of Brain Theory and Neural Networks*. MIT Press, Cambridge, MA.

Arena, P., Caponetto, R., Fortuna, L., and Manganaro, G. (1997). Cellular neural networks to explore complexity. *Soft Computing - A Fusion of Foundations, Methodologies and Applications*, 1(3):120–136.

Arkin, R. C. (1990). Integrating behavioral, perceptual and world knowledge in reactive navigation. *Robotics and Autonomous Systems*, 6:105–122.

Arkin, R. C., editor (1998). *Behavior-Based Robotics*. MIT Press, Cambridge, MA.

Arleo, A., Millan, J., and Floreano, D. (1999). Learning of variable-resolution cognitive maps for autonomous indoor navigation. *IEEE Transactions on Robotics and Automation*, 15(6):990–1000.

Ashby, W. R. (1960). *Design for a Brain*, 2nd revised edition. Chapman and Hall, London.

Astor, J. and Adami, C. (2000). A developmental model for the evolution of artificial neural networks. *Artificial Life*, 6(3):189–218.

Autumn, K., Dittmore, A., Santos, D., Spenko, M., and Cutkosky, M. (2006). Frictional adhesion: A new angle on gecko attachment. *Journal of Experimental Biology*, 209:3569–3579.

Autumn, K., Sitti, M., Liang, Y. A., Peattie, A. M., Hansen, W. R., Sponberg, S., Kenny, T. W., Fearing, R., Israelachvili, J. N., and Full, R. J. (2002). Evidence for van der Waals adhesion in gecko setae. *Proceedings of the National Academy of Sciences USA*, 99(19):12252–12256.

Avnir, D., Biham, O., Lidar, D., and Malcai, O. (1998). Is the geometry of nature fractal? *Science*, 279(5347):39–40.

Axelrod, R. (1989). *The Evolution of Cooperation*. Basic Books, New York.

Ayers, J., Davis, J. L., and Rudolph, A., editors (2002). *Neurotechnology for Biomimetic Robots*. MIT Press, Cambridge, MA.

Bäck, T. (1992). Self-adaptation in genetic algorithms. In Varela, F. and Bourgine, P., editors, *Proceedings of the First European Conference on Artificial Life*, pages 263–271. MIT Press, Cambridge, MA.

Bäck, T. (1996). *Evolutionary Algorithms in Theory and Practice*. Oxford University Press, Oxford.

Bäck, T., Fogel, D. B., and Michalewicz, Z. (2000). *Evolutionary Computation 2: Advanced Algorithms and Operators*. Institute of Physics, Bristol, UK.

Baddeley, R. J. and Hancock, P. J. (1991). A statistical analysis of natural images matches psychophysically derived orientation tuning curves. *Proceedings of the Royal Society of London B*, 246:219–223.

Bailey, C. H., Giustetto, M., Huang, Y.-Y., Hawkins, R. D., and Kandel, E. R. (2000). Is heterosynaptic modulation essential for stabilizing Hebbian plasticity and memory? *Nature Reviews Neuroscience*, 1(1):11–20.

Bajcsy, R. (1988). Active perception. *Proceedings of the IEEE*, 76:996–1005.

Baker, R. W. and Herman, G. T. (1972a). Simulation of organisms using a developmental model. Part 1: Basic description. *International Journal of Bio-Medical Computing*, 3(3):201–215.

Baker, R. W. and Herman, G. T. (1972b). Simulation of organisms using a developmental model. Part 2: The heterocyst formation problem in blue-green algae. *International Journal of Bio-Medical Computing*, 3(4):251–267.

Baldassarre, G., Nolfi, S., and Parisi, D. (2003a). Evolving mobile robots able to display collective behaviour. *Artificial Life*, 9:255–267.

Baldassarre, G., Parisi, D., and Nolfi, S. (2003b). Coordination and behavior integration in cooperating simulated robots. In Schaal, S., Ijspeert, A., Billard, A., Vijayakumar, S., Hallam, J., and Meyer, J.-A., editors, *From Animals to Animats 8: Proceedings of the Eight International Conference on Simulation of Adaptive Behavior*. MIT Press, Cambridge, MA.

Baldwin, J. M. (1896). A new factor in evolution. *American Naturalist*, 30:441–451.

Ball, P. (1999). *The Self-Made Tapestry: Pattern Formation in Nature*. Oxford University Press, Oxford.

Ball, P. (2004). *Critical Mass*. Arrow Books, London.

Ballard, D. H. (1991). Animate vision. *Artificial Intelligence*, 48(1):57–86.

Baluja, S. (1996). Evolution of an artificial neural network-based autonomous land vehicle controller. *IEEE Transactions on Systems, Man, and Cybernetics-Part B*, 26:450–463.

Baluja, S. (1997). Genetic algorithms and explicit search statistics. In Mozer, M. C., Jordan, M. I., and Petsche, T., editors, *Advances in Neural Information Processing Systems 9*, pages 319–325. MIT Press, Cambridge, MA.

Baluja, S. and Caruana, R. (1995). Removing the genetics from the standard genetic algorithm. In Prieditis, A. and Russel, S., editors, *Proceedings of the Twelfth International Conference on Machine Learning*, pages 38–46. Morgan Kaufmann, San Mateo, CA.

Bandini, S. and Mauri, G. (1999). Multilayered cellular automata. *Theoretical Computer Science*, 217(1):99–113.

Banzhaf, W., Beslon, G., Christensen, S., Foster, J. A., Képés, F., Lefort, V., Miller, J. F., Radman, M., and Ramsden, J. J. (2006). From artificial evolution to computational evolution: A research agenda. *Nature Reviews Genetics*, 7:729–735.

Bar-Cohen, Y., editor (2001). *Electroactive Polymer (EAP) Actuators as Artificial Muscles*. SPIE Press, Bellingham, WA.

Barabási, A.-L. (2002). *Linked. The New Science of Networks*. Perseus, Cambridge, MA.

Baray, C. (1997). Evolving cooperation via communication in homogeneous multi-agent systems. *Intelligent Information Systems, 1997. IIS'97. Proceedings*, pages 204–208.

Barlow, H. (1989). Unsupervised learning. *Neural Computation*, 1:295–311.

Barto, A. G. (1995). Adaptive critic and the basal ganglia. In Houk, J. C., Davis, J. L., and Beiser, D. G., editors, *Models of Information Processing in the Basal Ganglia*, pages 215–232. MIT Press, Cambridge, MA.

Baxter, J. (1992). The evolution of learning algorithms for artificial neural networks. In Green, D. and Bossomaier, T., editors, *Complex Systems*. IOS Press, Amsterdam.

Beckers, R., Holland, O., and Deneubourg, J.-L. (1994). From local actions to global tasks: Stigmergy and collective robotics. In Brooks, R. and Maes, P., editors, *Proceedings of the Fourth Workshop on Artificial Life*. MIT Press, Cambridge, MA.

Beer, R. D. (1990). *Intelligence as Adaptive Behavior: An Experiment in Computational Neuroethology*. Academic Press, San Diego.

Beer, R. D. and Gallagher, J. C. (1992). Evolving dynamical neural networks for adaptive behavior. *Adaptive Behavior*, 1:91–122.

Behera, N. and Nanjundiah, V. (1995). An investigation into the role of phenotypic plasticity in evolution. *Journal of Theoretical Biology*, 172:225–234.

Bejan, A. (1997). Constructal-theory network of conducting paths for cooling a heat generating volume. *International Journal of Heat and Mass Transfer*, 40(4):799–811.

Bejan, A. (2000). *Shape and Structure, from Engineering to Nature*. Cambridge University Press, Cambridge, UK.

Belew, R. K., McInerney, J., and Schraudolph, N. N. (1992). Evolving networks: Using the genetic algorithm with connectionistic learning. In Langton, C. G., Taylor, C., Farmer, J. D., and Rasmussen, S., editors, *Proceedings of the Second Conference on Artificial Life*, pages 511–548. Addison-Wesley, Reading, MA.

Belew, R. K. and Mitchell, M., editors (1996). *Adaptive Individuals in Evolving Populations: Models and Algorithms*. Addison-Wesley, Redwood City, CA.

Bellman, R. (1961). *Adaptive Control Processes: A Guided Tour*. Princeton University Press, Princeton, NJ.

Beni, G. (2004). From swarm intelligence to swarm robotics. In Şahin, E. and Spears, W. M., editors, *Proceedings of the Swarm Robotics Workshop*, pages 1–9. Springer-Verlag, Heidelberg, Germany.

Bentley, P. J. (2004). Fractal proteins. *Genetic Programming and Evolvable Machines*, 5(1):71–101.

Bentley, P. J., Greensmith, J., and Ujjin, S. (2005). Two ways to grow tissue for artificial immune systems. In *Proceedings of the Fourth International Conference on Artificial Immune Systems (ICARIS 2005)*, volume 3627 of *Lecture Notes in Computer Science*, pages 139–152. Springer-Verlag, Berlin.

Berek, C. and Ziegner, M. (1993). The maturation of the immune response. *Immunology Today*, 14(8):400–404.

Berlekamp, E., Conway, J. H., and Guy, R. (2004). *Winning Ways for Your Mathematical Plays*, 2nd edition. A K Peters, Wellesley, MA.

Bernstein, N. (1967). *The Coordination and Regulation of Movement*. Pergamon Press, London.

Beshers, S. N. and Fewell, J. H. (2001). Models of division of labor in social insects. *Annual Review of Entomology*, 46:413–440.

Betsch, B. Y., Einhäuser, W., Körding, K. P., and König, P. (2004). The world from a cat's perspective – statistics of natural videos. *Biological Cybernetics*, 90(1):41–50.

Beyeler, A., Zufferey, J.-C., and Floreano, D. (2007). 3D vision-based navigation for indoor microflyers. In *IEEE International Conference on Robotics and Automation (ICRA'07)*.

Bi, G.-Q. and Poo, M.-M. (2001). Synaptic modification by correlated activity: Hebb's postulate revisited. *Annual Review of Neuroscience*, 24:139–166.

Billard, A. (2002). Imitation. In Arbib, M. A., editor, *Handbook of Brain Theory and Neural Networks*, pages 566–569. MIT Press, Cambridge, MA.

Bishop, C. M. (1995). *Neural Networks for Pattern Recognition*. Oxford University Press, New York.

Bithell, M. and Macmillan, W. (2007). Escape from the cell: Spatially explicit modelling with and without grids. *Ecological Modelling*, 200(1-2):59–78.

Blynel, J. (2001). Evolving reinforcement learning-like abilities for robots. In Tyrrell, A., Haddow, P., and Torresen, J., editors, *Fifth International Conference on Evolvable Systems (ICES'03)*.

Blynel, J. and Floreano, D. (2003). Exploring the T-maze: Evolving learning-like robot behaviors using CTRNNs. In Raidl, G., editor, *Second European Workshop on Evolutionary Robotics*.

Bochev, P. and Hyman, J. (2006). Principles of mimetic discretizations of differential operators. In Arnold, D. N., Bochev, P. B., Lehoucq, R. B., Nicolaides, R. A., and Shashkov, M., editors, *Compatible Spatial Discretizations*, volume 142 of *The IMA Volumes in Mathematics and Its Applications*, pages 89–119. Springer-Verlag, New York.

Boers, E. and Sprinkhuizen-Kuyper, I. (2001). Combined biological metaphors. In Patel, M., Honavar, V., and Balakrishnan, K., editors, *Advances in the Evolutionary Synthesis of Intelligent Agents*, pages 153–183. MIT Press, Cambridge, MA.

Bonabeau, E., Dorigo, M., and Theraulaz, G. (1999). *Swarm Intelligence: From Natural to Artificial Systems*. Oxford University Press, New York.

Bonabeau, E., Theraulaz, G., and Deneubourg, J.-L. (1996). Quantitative study of the fixed threshold model for the regulation of division of labour in insect societies. *Proceedings of the Royal Society of London B*, 263:1565–1569.

Bongard, J. and Lipson, H. (2005). Nonlinear system identification using coevolution of models and tests. *IEEE Transactions on Evolutionary Computation*, 9(4):361–384.

Bongard, J. and Pfeifer, R. (2001). Repeated structure and dissociation of genotypic and phenotypic complexity in artificial ontogeny. In Spector, L., Goodman, E. D., Wu, A., Langdon, W. B., Voigt, H.-M., Gen, M., Sen, S., Dorigo, M., Pezeshk, S., Garzon, M. H., and Burke, E., editors, *GECCO 2001*, pages 829–836. Morgan Kaufmann, San Francisco.

Bongard, J. and Pfeifer, R. (2003). Evolving complete agents using artificial ontogeny. In Hara, F. and Pfeifer, R., editors, *Morpho-Functional Machines: The New Species (Designing Embodied Intelligence)*, pages 237–258. Springer-Verlag, Berlin.

Bongard, J., Zykov, V., and Lipson, H. (2006). Resilient machines through continuous self-modeling. *Science*, 5802:1118–1121.

Bongard, J. C. (2000). The legion system: A novel approach to evolving heterogeneity for collective problem solving. In Poli, R., Banzhaf, W., Langdon, W., Miller, J., Nordin, P., and Fogarty, T., editors, *Genetic Programming*, pages 16–28. Springer-Verlag, Berlin.

Bourke, A. F. G. and Franks, N. R. (1995). *Social Evolution in Ants*. Princeton University Press, Princeton, NJ.

Bradley, D. W. and Tyrrell, A. M. (2002). Immunotronics - novel finite-state-machine architectures with built-in self-test using self-nonself differentiation. *IEEE Transactions on Evolutionary Computation*, 6(3):227–238.

Braitenberg, V. (1984). *Vehicles. Experiments in Synthetic Psychology*. MIT Press, Cambridge, MA.

Breazeal, C. (2002a). *Designing Sociable Robots*. MIT Press, Cambridge, MA.

Breazeal, C. (2002b). Robots that imitate humans. *Trends in Cognitive Sciences*, 6(11):481–487.

Breazeal, C. (2003). Toward sociable robots. *Robotics and Autonomous Systems*, 42(3-4):167–175.

Brooks, R. A. (1986). A robust layered control system for a mobile robot. *IEEE Journal of Robotics and Automation*, RA-2(1):14–23.

Brooks, R. A. (1989). A robot that walks: Emergent behavior from a carefully evolved network. *Neural Computation*, 1(2):253–262.

Brooks, R. A. (1991a). Intelligence without reason. In *Proceedings of the International Joint Conference on Artificial Intelligence*, pages 569–595. Morgan Kaufmann, San Mateo, CA.

Brooks, R. A. (1991b). Intelligence without representation. *Artificial Intelligence Journal*, 47:139–160.

Brooks, R. A. (1991c). New approaches to robotics. *Science*, 253:1227–1232.

Brooks, R. A. (1992). Artificial life and real robots. In Varela, F. J. and Bourgine, P., editors, *Toward a Practice of Autonomous Systems: Proceedings of the First European Conference on Artificial Life*, pages 3–10. MIT Press, Cambridge, MA.

Brooks, R. A. (1999). *Cambrian Intelligence. The Early History of the New AI*. MIT Press, Cambridge, MA.

Brooks, R. A. (2002). *Flesh and Machines: How Robots Will Change Us*. Pantheon, New York.

Brooks, R. A., Breazeal, C., Marjanovic, M., Scassellati, B., and Williamson, M. (1998). The cog project: Building a humanoid robot. In Nehaniv, C., editor, *Computation for Metaphors, Analogy and Agents*. Springer-Verlag, Berlin.

Brooks, R. A., Breazeal, C., Scassellati, B., and O'Reilly, U. (1999). Technologies for human/humanoid natural interactions. In Nehaniv, C., editor, *The Second International Symposium on Humanoid Robots (HURO99)*. Tokyo, Japan,.

Brooks, R. A. and Stein, L. A. (1994). Building brains for bodies. *Autonomous Robots*, 1:7–25.

Bryant, B. and Miikkulainen, R. (2003). Neuroevolution for adaptive teams. In *Proceedings of the 2003 Congress on Evolutionary Computation, CEC '03*, volume 3, pages 2194–2201.

Bryson, A. E. and Ho, Y.-C. (1969). *Applied Optimal Control*. Blaisdell, New York.

Bull, L. and Kovacs, T., editors (2005). *Foundations of Learning Classifier Systems*. Springer-Verlag, Berlin.

Burgess, N., Donnett, J. G., Jeffery, K. J., and O'Keefe, J. (1997). Robotic and neuronal simulation of the hippocampus and and rat navigation. *Philosophical Transactions of the Royal Society B*, 352:1535–1543.

Burke, R., Gustafson, S., and Kendall, G. (2002). A survey and analysis of diversity measures in genetic programming. In Langdon, W. B., Cantú-Paz, E., Mathias, K., Roy, R., Davis, D., Poli, R., Balakrishnan, K., Honavar, V., Rudolph, G., Wegener, J., Bull, L., Potter, M. A., Schultz, A. C., Miller, J. F., Burke, E., and Jonoska, N., editors, *Proceedings of Genetic and Evolutionary Computation Conference - GECCO 2002*, pages 716–723. Morgan Kaufmann, San Francisco.

Burks, A., editor (1970). *Essays on Cellular Automata*. University of Illinois Press, Urbana, IL.

Burton, J. L. and Franks, N. R. (1985). The foraging ecology of the army ant *Eciton rapax*: An ergonomic enigma? *Ecological Entomology*, 10:131–141.

Butler, Z., Kotay, K., Rus, D., and Tomita, K. (2001). Cellular automata for decentralized control of self-reconfigurable robots. In *Proceedings of the Workshop on Modular Self-Reconfigurable Robots at the International Conference on Robotics and Automation, Seoul, Korea*. IEEE Press,, Piscataway, NJ.

Calderone, N. W. and Page, R. E. (1996). Temporal polyethism and behavioural canalization in the honey bee, *Apis mellifera*. *Animal Behavior*, 51:631–643.

Camazine, S., Deneubourg, J.-L., Franks, N. R., Sneyd, J., Theraulaz, G., and Bonabeau, E. (2001). *Self-Organization in Biological Systems*. Princeton University Press, Princeton, NJ.

Cangelosi, A. (2001). Evolution of communication and language using signals, symbols, and words. *IEEE Transactions on Evolutionary Computation*, 5:93–101.

Cangelosi, A., Parisi, D., and Nolfi, S. (1994). Cell division and migration in a genotype for neural networks. *Network*, 5:497–515.

Caporale, L. H. (1984). Is there a higher level genetic code that directs evolution? *Molecular and Cellular Biochemistry*, 64:5–13.

Caporale, L. H. (2003). *Darwin in the Genome: Molecular Strategies in Biological Evolution*. McGraw-Hill, New York.

Caporale, L. H. (2004). Genomes don't play dice. *New Scientist*, 181(2437):42–45.

Carpenter, G. A. and Grossberg, S. (1987). A massively parallel architecture for self-organizing neural pattern recognition machines. *Computer Vision, Graphics, and Image Processing*, 37:54–115.

Carr, C. E. and Konishi, M. (1988). Axonal delay lines for time measurement in the owl's brainstem. *Proceedings of the National Academy of Sciences USA*, 85:8311–8315.

Cartwright, B. A. and Collett, T. S. (1983). Landmark learning in bees. *Journal of Comparative Physiology A*, 151:521–543.

Casasent, D. (1992). Optical processing in neural networks. *IEEE Expert*, October:55–61.

Castano, A., Shen, W., and Will, P. (2000). CONRO: Towards deployable robots with inter-robot metamorphic capabilities. *Autonomous Robots*, 8:309–324.

Cavalier-Smith, T. (1978). Nuclear volume control by nucleoskeletal DNA, selection for cell volume and cell growth rate and the solution to the DNA C-value paradox. *Journal of Cell Science*, 34:247–278.

Chalmers, D. J. (1990). The evolution of learning: An experiment in genetic connectionism. In Touretzky, D. S., Elman, J. L., Sejnowski, T., and Hinton, G. E., editors, *Proceedings of the 1990 Connectionist Models Summer School*, pages 81–90. Morgan Kaufmann, San Mateo, CA.

Chauvin, Y. (1989). A back-propagation algorithm with optimal use of hidden units. In Touretzky, D., editor, *Advances in Neural Information Processing Systems 1*. Morgan Kaufmann, San Francisco.

Chiel, H. and Beer, R. (1997). The brain has a body: Adaptive behaviour emerges from interactions of nervous system, body and environment. *Trends in Neurosciences*, 20:553–557.

Chomsky, N. (1957). *Syntactic Structures*. Mouton, The Hague.

Chopard, B. and Droz, M. (1998). *Cellular Automata Modeling of Physical Systems*. Cambridge University Press, Cambridge, UK.

Chua, L. and Yang, L. (1988a). Cellular neural networks: Applications. *IEEE Transactions on Circuits and Systems*, 35(10):1273–1290.

Chua, L. and Yang, L. (1988b). Cellular neural networks: Theory. *IEEE Transactions on Circuits and Systems*, 35(10):1257–1272.

Chua, L. O., Sbitnev, V. I., and Yoon, S. (2003). A nonlinear dynamics perspective of Wolfram's new kind of science. Part 2: Universal neuron. *International Journal of Bifurcation and Chaos*, 13(9):2377–2491.

Chua, L. O., Sbitnev, V. I., and Yoon, S. (2004). A nonlinear dynamics perspective of Wolfram's new kind of science. Part 3: Predicting the unpredictable. *International Journal of Bifurcation and Chaos*, 14(11):3689–3820.

Chua, L. O., Sbitnev, V. I., and Yoon, S. (2005). A nonlinear dynamics perspective of Wolfram's new kind of science. Part 4: From Bernoulli shift to $1/f$ spectrum. *International Journal of Bifurcation and Chaos*, 15(4):1045–1183.

Chua, L. O., Yoon, S., and Dogaru, R. (2002). A nonlinear dynamics perspective of Wolfram's new kind of science. Part 1: Threshold of complexity. *International Journal of Bifurcation and Chaos*, 12(12):2655–2766.

Churchland, P. M. and Sejnowski, T. J. (1992). *The Computational Brain*. MIT Press, Cambridge, MA.

Clark, A. (1989). *Microcognition: Philosophy, Cognitive Science, and Parallel Distributed Processing*. MIT Press, Cambridge, MA.

Clark, A. (1997). *Being There: Putting Brain, Body and World Together Again*. MIT Press, Cambridge, MA.

Clearwater, S. H. (1995). *Market-Based Control: A Paradigm for Distributed Resource Allocation*. World Scientific, Singapore.

Clerc, M. and Kennedy, J. (2002). The particle swarm-explosion, stability, and convergence in a multidimensional complex space. *IEEE Transactions on Evolutionary Computation*, 6:58–73.

Cliff, D., Harvey, I., and Husbands, P. (1993). Explorations in evolutionary robotics. *Adaptive Behavior*, 2:73–110.

Cliff, D. and Miller, G. F. (1995). Tracking the Red Queen: Measurements of adaptive progress in co-evolutionary simulations. In Morán, F., Moreno, A., Merelo, J. J., and Chacón, P., editors, *Advances in Artificial Life: Proceedings of the Third European Conference on Artificial Life*, pages 200–218. Springer-Verlag, Berlin.

Cliff, D. and Miller, G. F. (1996). Co-evolution of pursuit and evasion II: Simulation methods and results. In Maes, P., Matarić, M., Meyer, J., Pollack, J., Roitblat, H., and Wilson, S., editors, *From Animals to Animats IV: Proceedings of the Fourth International Conference on Simulation of Adaptive Behavior*. MIT Press, Cambridge, MA.

Cliff, D. T. (1991). Computational neuroethology: A provisional manifesto. In Meyer, J. A. and Wilson, S. W., editors, *From Animals to Animats: Proceedings of the First International Conference on Simulation of Adaptive Behavior*. MIT Press, Cambridge, MA.

Codd, E. (1968). *Cellular Automata*. Academic Press, New York.

Coen, E. (1999). *The Art of Genes.* Oxford University Press, Oxford.

Coico, R., Sunshine, G., and Benjamini, E. (2003). *Immunology: A Short Course*, 5th edition. Wiley-Liss, Hoboken, NJ.

Collins, S., Ruina, A., Tedrake, R., and Wisse, M. (2005). Efficient bipedal robots based on passive-dynamic walkers. *Science*, 307:1082–1085.

Conrad, M. (1988). The price of programmability. In Herken, R., editor, *The Universal Turing Machine: A Fifty Year Survey*, pages 285–307. Oxford University Press, Oxford.

Conrad, M. (1990). The geometry of evolution. *Biosystems*, 24:61–81.

Cook, N. D. (1995). Artefact or network evolution? *Nature*, 374:313–314.

Cook, N. D., Fruh, H., and Landis, T. (1995). The cerebral hemispheres and neural network simulations: Design considerations. *Journal of Experimental Psychology: Human Perception and Performance*, 95 (21):410–422.

Crichton, M. (2002). *Prey.* Harper Collins (Avon Books), New York.

Crutchfield, J. P. (2003). When evolution is revolution–origins of innovation. In Crutchfield, J. P. and Schuster, P., editors, *Evolutionary Dynamics: Exploring the Interplay of Selection, Accident, Neutrality, and Function*, SFI Studies in the Sciences of Complexity, pages 101–133. Oxford University Press, Oxford.

Crutchfield, J. P., Farmer, J., Packard, N. H., and Shaw, R. S. (1986). Chaos. *Scientific American*, 255:46–57.

Culik, K. and Yu, S. (1988). Undecidability of CA classification schemes. *Complex Systems*, 2(2):177–190.

Damiani, E., Tettamanzi, A., and Liberali, V. (1999). On-line evolution of FPGA-based circuits: A case study on hash functions. In Stoica, A., Keymeulen, D., and Lohn, J., editors, *First NASA/DoD Workshop on Evolvable Hardware (EH '99), July 19-21, 1999, Pasadena, CA*, pages 26–33. IEEE Computer Society, Los Alamitos, CA.

Darwin, C. (1859). *On The Origin of Species by Means of Natural Selection, or The Preservation of Favoured Races in the Struggle for Life.* Murray, London.

Dasdan, A. and Oflazer, K. (1993). Genetic synthesis of unsupervised learning algorithms. Technical report, Department of Computer Engineering and Information Science, Bilkent University, Ankara, Turkey.

Dasgupta, D. (1997). Artificial neural networks and artificial immune systems: similarities and differences. In *1997 IEEE International Conference on Systems, Man, and Cybernetics, 1997, Orlando, FL, 12-15 Oct. 1997*, volume 1, pages 873–878. IEEE Press, Piscataway, NJ.

Dasgupta, D. and Michalewicz, Z., editors (1997). *Evolutionary Algorithms in Engineering Applications.* Springer-Verlag, New York.

Dastidar, T. R., Chakrabarti, P. P., and Ray, P. (2005). A synthesis system for analog circuits based on evolutionary search and topological reuse. *IEEE Transactions on Evolutionary Computation*, 9(2):211–234.

Davis, D. M. (2006). Intrigue at the immune synapse. *Scientific American*, 294(2):48–55.

Davis, D. M. and Dustin, M. L. (2004). What is the importance of the immunological synapse? *Trends in Immunology*, 25(6):323–327.

Davis, L., editor (1989). *Genetic Algorithms and Simulated Annealing*. Morgan Kaufmann, San Mateo, CA.

Davis, L., editor (1991). *Handbook of Genetic Algorithms*. Van Nostrand Reinhold, New York.

Dawkins, R. (1976). *The Selfish Gene*. Oxford University Press, Oxford.

Dawkins, R. (1986). *The Blind Watchmaker*. Longman, Essex, UK.

Dawkins, R. and Krebs, J. R. (1979). Arms races between and within species. *Proceedings of the Royal Society of London B*, 205:489–511.

de Castro, L. N. and Timmis, J. I. (2002). *Artificial Immune Systems: A New Computational Intelligence Approach*. Springer-Verlag, London.

de Castro, L. N. and Von Zuben, F. J. (2002). Learning and optimization using the clonal selection principle. *IEEE Transactions on Evolutionary Computation*, 6(3):239–251.

de Garis, H. (1990). Genetic programming: Evolution of time dependent neural network modules which teach a pair of stick legs to walk. In *Proceedings of the Ninth European Conference on Artificial Intelligence*, pages 204–206. Stockholm.

de Garis, H. (1993). Growing an artificial brain with a million neural net modules inside a trillion cell cellular automaton machine. In *Fourth International Symposium on Micro Machine and Human Science*, pages 211–214. Nagoya, Japan.

De Lillo, C., Floreano, D., and Antinucci, F. (2001). Transitive choices by a simple, fully connected, backpropagation neural network: Implications for the comparative study of transitive inference. *Animal Cognition*, 4:61–66.

Deb, K. (2001). *Multi-Objective Optimization Using Evolutionary Algorithms*. Wiley, Chichester, UK.

Deiss, S. R., Douglas, R. J., and Whatley, A. M. (1999). A pulse-coded communications infrastructure for neuromorphic systems. In Maass, W. and Bishop, C. M., editors, *Pulsed Neural Networks*. MIT Press, Cambridge, MA.

Dellaert, F. and Beer, R. (1996). A developmental model for the evolution of complete autonomous agents. In Maes, P., Mataric, M., Meyer, J.-A., Pollack, J., and Wilson, S., editors, *From Animals to Animats IV*, pages 393–401. MIT Press, Cambridge, MA.

DeMarse, T. B., Wagenaar, D. A., Blau, A. W., and Potter, S. M. (2001). The neurally controlled animat: Biological brains acting with simulated bodies. *Autonomous Robots*, 11:305–310.

Demiris, Y. and Hayes, G. (2002). Imitation as a dual-route process featuring predictive and learning components: A biologically-plausible computational model. In Dautenhahn, K. and Nehaniv, C., editors, *Imitation in Animals and Artifacts*. MIT Press, Cambridge, MA.

Deneubourg, J.-L., Goss, S., Franks, N. R., and Pasteels, J. M. (1989). The blind leading the blind: Modelling chemically mediated army ant raid patterns. *Journal of Insect Behavior*, 2:719–725.

Deneubourg, J.-L., Goss, S., Franks, N. R., Sendova-Franks, A., Detrain, C., and Chretien, L. (1991). The dynamics of collective sorting: Robot-like ant and ant-like robot. In Meyer, J.-A. and Wilson, S. W., editors, *Proceedings of the First International Conference on Simulation of Adaptive Behavior: From Animals to Animats*, pages 356–365. MIT Press, Cambridge, MA.

Deneubourg, J.-L., Gregoire, J. C., and Le Fort, E. (1990). Kinetics of the larval gregarious behaviour in the bark beetle *Dendroctonus micans*. *Journal of Insect Behavior*, 3:169–182.

Detrain, C. and Pasteels, J. M. (1991). Caste differences in behavioral thresholds as a basis for polyethism during food recruitment in the ant *Pheidole pallidula* (nyl.) (Hymenoptera: Myrmicinae). *Journal of Insect Behavior*, 4:157–176.

Deussen, O. and Lintermann, B. (2005). *Digital Design of Nature: Computer Generated Plants and Organics*. Springer-Verlag, Berlin.

Deutsch, A. and Dormann, S. (2005). *Cellular Automaton Modeling of Biological Pattern Formation*. Birkhäuser, Boston.

Dewdney, A. K. (1993). Misled by metaphors: Two tools that don't always work. In Haken, H., Karlqvist, A., and Svedin, U., editors, *The Machine as Metaphor and Tool*. Springer-Verlag, Berlin.

DeYoe, E. A. and van Essen, D. C. (1988). Concurrent processing streams in monkey visual cortex. *Trends in Neuroscience*, 11:219–226.

D'haeseleer, P., Forrest, S., and Helman, P. (1996). An immunological approach to change detection: Algorithms, analysis and implications. In *Proceedings of the 1996 IEEE Symposium on Security and Privacy*, pages 110–119.

Di Caro, G. and Dorigo, M. (1998). Antnet: Distributed stigmergetic control for communications networks. *Journal of Artificial Intelligence Research*, 9:317–365.

Diamantaras, K. I. and Kung, S. Y., editors (1996). *Principal Component Analysis*. Wiley, New York.

DiCarlo, A., Milicchio, F., Paoluzzi, A., and Shapiro, V. (2007). Solid and physical modeling with chain complexes. In *Proceedings of the 2007 ACM Symposium on Solid and Physical Modeling*, Beijing, pages 73–84. ACM Press, New York.

Dickinson, M. (1999). Haltere-mediated equilibrium reflexes of the fruit fly, *Drosophila melanogaster*. *Philosophical Transactions: Biological Sciences*, 354 (1385):903–916.

Dickinson, W. J. and Seger, J. (1999). Cause and effect in evolution. *Nature*, 399(6731):30.

Diermeier, D. (2007). Arguing for computational power. *Science*, 318(5852):918–919.

DiPaolo, E. (2003). Evolving spike-timing-dependent plasticity for single-trial learning in robots. *Philosophical Transactions of the Royal Society of London A*, 361:2299–2319.

DiPaolo, E. A. (2000). Homeostatic adaptation to inversion of the visual field and other sensorimotor disruptions. In Meyer, J., Berthoz, A., Floreano, D., Roitblat, H., and Wilson, S., editors, *From Animals to Animats VI: Proceedings of the Fifth International Conference on Simulation of Adaptive Behavior*, pages 440–449. MIT Press, Cambridge, MA.

Dissanayake, M., Newman, P., Clark, S., Durrant-Whyte, H., and Csorba, M. (2001). A solution to the simultaneous localization and map building (SLAM) problem. *IEEE Transactions on Robotics and Automation*, 17(3):229–241.

Dobzhansky, T. (1973). Nothing in biology makes sense except in the light of evolution. *The American Biology Teacher*, 35:125–129.

Dorigo, M. and Gambardella, L. (1997). Ant colony system: A cooperative learning approach to the traveling salesman problem. *IEEE Transactions on Evolutionary Computation*, 1:53–66.

Dorigo, M., Maniezzo, V., and Colorni, A. (1996). The ant system: Optimization by a colony of cooperating agents. *IEEE Transactions on Systems, Man, and Cybernetics B*, 26:29–41.

Dorigo, M., Trianni, V., Scedilahin, E., Gross, R., Labella, T. H., Baldassarre, G., Nolfi, S., Deneubourg, J. L., Mondada, F., Floreano, D., and Gambardella, L. M. (2004). Evolving self-organizing behaviors for a Swarm-bot. *Autonomous Robots*, 17(2-3):223–245.

Douglas, R. J. and Martin, K. A. C. (1990). Neocortex. In Shepherd, G. M., editor, *The Synaptic Organization of the Brain*. Oxford University Press, Oxford.

Doya, K. (2002). Metalearning and neuromodulation. *Neural Networks*, 15:495–506.

Doya, K., Dayan, P., and Hasselmo, M. E. (2002). Special issue on Computational Models of Neuromodulation. *Neural Networks*, 15:475–774.

Doya, K., Kimura, H., and M., K. (2001). Neural mechanisms of learning and control. *IEEE Control Systems Magazine*, 21:42–54.

Dreyfus, H. L. (1992). *What Computers Still Can't Do: A Critique of Artificial Reason.* MIT Press, Cambridge, MA.

Drogoul, A. and Ferber, J. (1992). From Tom Thumb to the dockers: Some experiments with foraging robots. In Meyer, J., Roitblat, H. L., and Wilson, S. W., editors, *From Animals to Animats II: Proceedings of the Second International Conference on Simulation of Adaptive Behavior*, pages 451–459. MIT Press, Cambridge, MA.

Drossel, B. and Schwabl, F. (1992). Self-organized critical forest-fire model. *Physical Review Letters*, 69(11):1629–1632.

Drossel, B. and Schwabl, F. (1994). Formation of space-time structure in a forest-fire model. *Physica A: Statistical and Theoretical Physics*, 204(1-4):212–229.

Dufay, B. and Latombe, J. C. (1984). An approach to automatic robot programming based on inductive learning. *International Journal of Robotics Research*, 3:3–20.

Durlauf, S. N. (1997). Insights for socioeconomic modeling. *Complexity*, 2(3):47–49.

Dürr, P., Mattiussi, C., and Floreano, D. (2006). Neuroevolution with analog genetic encoding. In Runarsson, T. P., Beyer, H.-G., Burke, E. K., Guervós, J. J. M., Whitley, L. D., and Yao, X., editors, *Parallel Problem Solving from Nature - PPSN IX*, volume 9, pages 671–680. Springer-Verlag, Berlin.

Edelman, G. M. (1988). *Neural Darwinism: The Theory of Neuronal Group Selection.* Basic Books, New York.

Egelhaaf, M. and Kern, R. (2002). Vision in flying insects. *Current Opinion in Neurobiology*, 12(6):699–706.

Eggenberger, P. (1997a). Creation of neural networks based on developmental and evolutionary principles. In Gerstner, W., Germond, A., Hasler, M., and Nicoud, J.-D., editors, *Proceedings of the International Conference on Artificial Neural Networks, ICANN'97*, Lausanne, Switzerland, October 8-10, 1997, volume 1327 of *Lecture Notes in Computer Science*, pages 337–342. Springer-Verlag, Berlin.

Eggenberger, P. (1997b). Evolving morphologies of simulated 3D organisms based on differential gene expression. In Husbands, P. and Harvey, I., editors, *Proceedings of the Fourth European Conference on Artificial Life, ECAL97*, pages 205–213. MIT Press, Cambridge, MA.

Eggenberger, P. (2003). Genome-physics interaction as a new concept to reduce the number of genetic parameters in artificial evolution. In *Proceedings of the 2003 Congress on Evolutionary Computation. CEC03.*, volume 1, pages 191–198. IEEE Press, Piscataway, NJ.

Eggenberger, P. (2004a). Asymmetric cell division and its integration with other developmental processes for artificial evolutionary systems. In Pollack, J., Bedau, M., Husbands, P., Ikegami, T., and R., W., editors, *Proceedings of the Ninth International Conference on Artificial Life, ALIFE IX*, Boston, September 12–15, pages 387–392. MIT Press, Cambridge, MA.

Eggenberger, P. (2004b). Comparing direct and developmental encoding schemes in artificial evolution: A case study in evolving lens shapes. In Maes, P., Mataric, M., Meyer, J.-A., Pollack, J., and Wilson, S., editors, *Proceedings of the 2004 Congress on Evolutionary Computation, CEC 2004*. IEEE Press, Piscataway, NJ.

Eiben, A. E. and Smith, J. E., editors (2003). *Introduction to Evolutionary Computing*. Springer-Verlag, Berlin.

Eilon, S. (1972). Goals and constraints in decision-making. *Operational Research Quarterly*, 23(1):3–15.

Eldredge, J. G. and Hutchings, B. L. (1994). Density enhancement of a neural network using FPGAs and run-time reconfiguration. In Buell, D. A. and Pocek, K. L., editors, *IEEE Workshop on FPGAs for Custom Computing Machines*, pages 180–188. IEEE Computer Society Press, Los Alamitos, CA.

Eliasmith, C. and Anderson, C. H. (2003). *Neural Engineering: Computation, Representation, and Dynamics in Neurobiological Systems*. MIT Press, Cambridge, MA.

Elman, J. L. (1990). Finding structure in time. *Cognitive Science*, 14:179–211.

Engelhard, V. H. (1994). How cells process antigens. *Scientific American*, 271(2):54–61.

Enquist, M. and Arak, A. (1994). Symmetry, beauty, and evolution. *Nature*, 372:169–172.

Epstein, J. M. (2006). *Generative Social Science: Studies in Agent-based Computational Modeling*. Princeton University Press, Princeton, NJ.

Epstein, J. M. and Axtell, R. (1996). *Growing Artificial Societies: Social Science from the Bottom Up*. Complex Adaptive Systems. Brookings Institution Press, Washington, DC.

Escuela, G., Ochoa, G., and Krasnogor, N. (2005). Evolving L-systems to capture protein structure native conformations. In *Proceedings of the Eigth European Conference, EuroGP 2005*, Lausanne, Switzerland, March 30 - April 1, 2005, volume 3447 of *Lecture Notes in Computer Science*, pages 74–84. Springer-Verlag, Berlin.

Estier, T., Crausaz, Y., Merminod, B., Lauria, M., R.Piguet, and Siegwart, R. (2000). An innovative space rover with extended climbing alilities. In *Proceedings of Space and Robotics 2000*. Albuquerque, NM.

Fahlman, S. E. (1989). Fast-learning variations on back-propagation: An empirical study. In Touretzky, D., Hinton, G., and Sejnowski, T., editors, *Proceedings of the 1988 Connectionist Models Summer School*, pages 38–51. Morgan Kaufmann, San Francisco.

Fahlman, S. E. and Lebiere, C. (1990). The cascade-correlation learning architecture. In Touretzky, D., editor, *Advances in Neural Information Processing Systems 2*. Morgan Kaufmann, San Francisco.

Farah, M. J. and McClelland, J. L. (1991). A computational model of semantic memory impairment: Modality specificity and emergent category specificity. *Journal of Experimental Psychology: General*, 120:339–357.

Farge, M. (2007). Numerical experimentation: A third way to study nature. In Kaneda, Y., Kawamura, H., and Sasai, M., editors, *Frontiers of Computational Science. Proceedings of the International Symposium on Frontiers of Computational Science 2005*, pages 15–30. Springer-Verlag, Berlin.

Fatès, N. and Morvan, M. (2004). Perturbing the topology of the Game of Life increases its robustness to asynchrony. In *Cellular Automata: Sixth International Conference on Cellular Automata for Research and Industry, ACRI 2004*, Amsterdam, October 25-28, 2004, volume 3305 of *Lecture Notes in Computer Science*, pages 111–120. Springer-Verlag, Berlin.

Featherstone, R. and Orin, D. E. (2000). Robot dynamics: Equations and algorithms. In *Proceedings of the IEEE International Conference on Robotics and Automation*. IEEE Press, Piscataway, NJ.

Fellous, J.-M. and Linster, C. (1998). Computational models of neuromodulation. *Neural Computation*, 10(4):771–805.

Field, D. J. (1994). What is the goal of sensory coding? *Neural Computation*, 4:559–601.

Fisher, R. A. (1930). *The Genetical Theory of Natural Selection*. Oxford University Press, Oxford.

Fleischer, K. and Barr, A. (1994). A simulation testbed for the study of multicellular development: The multiple mechanisms of morphogenesis. In Langton, C., editor, *Artificial Life III* , *Proceedings of the Third International Workshop on the Synthesis and Simulation of Living Systems*, pages 246–257. Addison-Wesley, Reading, MA.

Fleming, K. M., Reger, B. D., Sanguineti, V., Alford, S., and Mussa-Ivaldi, F. A. (2000). Connecting brains to robots: An artificial animal for the study of learning in vertebrate nervous systems. In Meyer, J., Berthoz, A., Floreano, D., Roitblat, H., and Wilson, S., editors, *From Animals to Animats V: Proceedings of the Sixth International Conference on Simulation of Adaptive Behavior*. MIT Press, Cambridge, MA.

Floreano, D. (1992). Emergence of home-based foraging strategies in ecosystems of neural networks. In Meyer, J., Roitblat, H. L., and Wilson, S. W., editors, *From Animals to Animats II: Proceedings of the Second International Conference on Simulation of Adaptive Behavior*. MIT Press, Cambridge, MA.

Floreano, D. (2006-2008). Talking robots. http://lis.epfl.ch/podcast.

Floreano, D., Kato, T., Marocco, D., and Sauser, E. (2004). Coevolution of active vision and feature selection. *Biological Cybernetics*, 90(3):218–228.

Floreano, D. and Mattiussi, C. (2001). Evolution of spiking neural controllers for autonomous vision-based robots. In Gomi, T., editor, *Evolutionary Robotics. From Intelligent Robotics to Artificial Life*. Springer-Verlag, Tokyo.

Floreano, D., Miglino, O., and Parisi, D. (1991). Emergent complex behaviours in ecosystems of neural networks. In Caianiello, E., editor, *Parallel Architectures and Neural Networks*. World Scientific Press, Singapore.

Floreano, D., Mitri, S., Magnenat, S., and Keller, L. (2007). Evolutionary conditions for the emergence of communication in robots. *Current Biology*, 17:514–519.

Floreano, D. and Mondada, F. (1994). Automatic creation of an autonomous agent: Genetic evolution of a neural-network driven robot. In Cliff, D., Husbands, P., Meyer, J., and Wilson, S. W., editors, *From Animals to Animats III: Proceedings of the Third International Conference on Simulation of Adaptive Behavior*, pages 421–430. MIT Press, Cambridge, MA.

Floreano, D. and Mondada, F. (1996a). Evolution of homing navigation in a real mobile robot. *IEEE Transactions on Systems, Man and Cybernetics, Part B*, 26(3):396–407.

Floreano, D. and Mondada, F. (1996b). Evolution of plastic neurocontrollers for situated agents. In Maes, P., Matarić, M., Meyer, J., Pollack, J., Roitblat, H., and Wilson, S., editors, *From Animals to Animats IV: Proceedings of the Fourth International Conference on Simulation of Adaptive Behavior*, pages 402–410. MIT Press, Cambridge, MA.

Floreano, D. and Nolfi, S. (1997a). Adaptive behavior in competing co-evolving species. In Husbands, P. and Harvey, I., editors, *Proceedings of the Fourth European Conference on Artificial Life*, pages 378–387. MIT Press, Cambridge, MA.

Floreano, D. and Nolfi, S. (1997b). God save the Red Queen! Competition in co-evolutionary robotics. In Koza, J., Deb, K., Dorigo, M., Fogel, D., Garzon, M., Iba, H., and Riolo, R. L., editors, *Proceedings of the Second International Conference on Genetic Programming*. Morgan Kaufmann, San Mateo, CA.

Floreano, D., Nolfi, S., and Mondada, F. (2001). Co-evolution and ontogenetic change in competing robots. In Patel, M., Honavar, V., and Balakrishnan, K., editors, *Advances in the Evolutionary Synthesis of Intelligent Agents*. MIT Press, Cambridge, MA.

Floreano, D., Schoeni, C., Caprari, G., and Blynel, J. (2002). Evolutionary bits'n'spikes. In Standish, R. K., Bedau, M. A., and Abbass, H. A., editors, *Artificial Life VIII. Proceedings of the Eighth International Conference on Artificial Life*. MIT Press, Cambridge, MA.

Floreano, D., Suzuki, M., and Mattiussi, C. (2005). Active vision and receptive field development in evolutionary robots. *Evolutionary Computation*, 13(4):527–544.

Floreano, D. and Urzelai, J. (2000). Evolutionary robots with online self-organization and behavioral fitness. *Neural Networks*, 13:431–443.

Floreano, D. and Urzelai, J. (2001). Neural morphogenesis, synaptic plasticity, and evolution. *Theory in Biosciences*, 120(3-4):225–240.

Fogel, D. (1998). *Evolutionary Computation: The Fossil Record*. Wiley-IEEE Press, New York.

Fogel, L. J., Owens, A. J., and Walsh, M. J. (1966). *Artificial Intelligence through Simulated Evolution*. Wiley, New York.

Fong, T., Nourbakhsh, I., and Dautenhahn, K. (2003). A survey of socially interactive robots. *Robotics and Autonomous Systems*, 42:143–166.

Fonseca, C. M. and Fleming, P. J. (2002). Multiobjective optimization. In Bäck, T., Fogel, D. B., and Michalewicz, Z., editors, *Evolutionary Computation 2: Advanced Algorithms and Operators*, pages 25–37. Institute of Physics, Bristol, UK.

Fontanari, J. F. and Meir, R. (1991). Evolving a learning algorithm for the binary perceptron. *Network*, 2:353–359.

Forrest, S. and Hofmeyr, S. A. (2001). Immunology as information processing. In Segel, L. and Cohen, I., editors, *Design Principles for the Immune System and Other Distributed Autonomous Systems*, Santa Fe Institute Studies in the Sciences of Complexity, chapter 17, pages 361–387. Oxford University Press, Oxford.

Forrest, S., Perelson, A. S., Allen, L., and Cherukuri, R. (1994). Self-nonself discrimination in a computer. In *Proceedings of the 1994 IEEE Symposium on Security and Privacy*, pages 202–212.

Fortuna, L., Arena, P., Balya, D., and Zarandy, A. (2001). Cellular neural networks: A paradigm for nonlinear spatio-temporal processing. *Circuits and Systems Magazine, IEEE*, 1(4):6–21.

Franceschini, N., Pichon, J. M., and Blanes, C. (1992). From insect vision to robot vision. *Philosophical Transactions of the Royal Society B*, 337:283–294.

Franco, S. (2001). *Design with Operational Amplifiers and Analog Integrated Circuits*. McGraw-Hill, New York.

Franks, N. R. and Sendova-Franks, A. B. (1992). Brood sorting in ants: Distributing the workload over the work-surface. *Behavioural Ecology and Sociobiology*, 30:109–123.

Franks, N. R., Wilby, A., Silverman, B. W., and Tofts, C. (1992). Self-organizing nest construction in ants: Sophisticated building by blind bulldozing. *Animal Behaviour*, 44:357–375.

Franz, M. O., Schölkopf, B., Mallot, H. A., and Bülthoff, H. H. (1998). Where did I take that snapshot? Scene-based homing by image matching. *Biological Cybernetics*, 79:191–202.

Frean, M. R. (1990). The upstart algorithm: A method for constructing and training feedforward neural networks. *Neural Computation*, 2:190–209.

Fredkin, E. (1992). A new cosmogony. Department of Physics, Boston University, Boston.

Freeman, W. J., editor (2001). *How Brains Make Up Their Minds*. Columbia University Press, New York.

Freeman, W. J. (2003). W. Grey Walter: Biographical essay. In Nadel, L., editor, *McMillan Encyclopedia of Cognitive Science*, volume 4, pages 537–539. McMillan, London.

Friedl, P., DenBoer, A. T., and Gunzer, M. (2005). Tuning immune responses: Diversity and adaptation of the immunological synapse. *Nature Reviews Immunology*, 5(7):532–545.

Frisch, U., Hasslacher, B., and Pomeau, Y. (1986). Lattice-gas automata for the Navier-Stokes equation. *Physical Review Letters*, 56(14):1505–1508.

Friston, K., Tononi, G., Reeke, G., Sporns, O., and Edelman, G. (1994). Value-dependent selection in the brain: Simulation in a synthetic neural model. *Neuroscience*, 59:229–243.

Fromherz, P. (2003). Neuroelectronic interfacing: Semiconductor chips with ion channels, nerve cells, and brain. In Waser, R., editor, *Nanoelectronics and Information Technology*. Wiley-VCH, Berlin.

Fuks, H. (1997). Solution of the density classification problem with two cellular automata rules. *Physical Review E*, 55(3):R2081–R2084.

Fukuda, T., Mizoguchi, H., Sekiyama, K., and Arai, F. (1999). Group behavior control for MARS. In *Proceedings of the IEEE International Conference on Robotics and Automation*, pages 1550–1555. IEEE Press, Piscataway, NJ.

Fukuda, T. and Ueyama, T. (1994). *Cellular Robotics and Micro Robotic Systems*. World Scientific, Singapore.

Full, R. J. and Koditschek, D. E. (1999). Templates and anchors: Neuromechanical hypotheses of legged locomotion on land. *Journal of Experimental Biology*, 202:3325–3332.

Funes, P. and Pollack, J. (1998). Evolutionary body building: Adaptive physical designs for robots. *Artificial Life*, 4(4):337–357.

Gallagher, J., Beer, R., Espenschiel, M., and Quinn, R. (1996). Application of evolved locomotion controllers to a hexapod robot. *Robotics and Autonomous Systems*, 19(1):95–103.

Gallistel, C. R., editor (1990). *The Organization of Learning*. MIT Press, Cambridge, MA.

Galt, S., Luk, B., and Collie, A. (1997). Evolution of smooth and efficient walking motions for an 8-legged robot. In *Proceedings of the Sixth European Workshop on Learning Robots*. Brighton,UK.

Gardner, M. (1970). The fantastic combinations of John Conway's new solitaire game "life". *Scientific American*, 223(4):120–123.

Gardner, M. (1971). On cellular automata, self-reproduction, the Garden of Eden and the game "life". *Scientific American*, 224(2):112–117.

Garrett, S. M. (2005). How do we evaluate artificial immune systems? *Evolutionary Computation*, 13(2):145–178.

Garthwaite, J. (1991). Glutamate, nitric oxide and cell-cell signalling in the nervous system. *Trends in Neuroscience*, 87:3547–3551.

Gaylord, R. J. and D'Andria, L. J. (1998). *Simulating Society : A Mathematica Toolkit for Modeling Socioeconomic Behavior*. Springer-Verlag, New York.

Gaylord, R. J. and Nishidate, K. (1996). *Modeling Nature: Cellular Automata Simulations with Mathematica*. Springer-Verlag, New York.

Gaylord, R. J. and Wellin, P. R. (1995). *Computer Simulations with Mathematica: Explorations in Complex Physical and Biological Systems*. Springer-Verlag, New York.

Gerhart, J. and Kirschner, M. (1997). *Cells, Embryos, and Evolution*. Blackwell, London.

Gerstner, W. (1999). Spiking neurons. In Maass, W. and Bishop, C. M., editors, *Pulsed Neural Networks*. MIT Press, Cambridge, MA.

Gerstner, W. and Kistler, W. (2002). *Spiking Neuron Models*. Cambridge University Press, Cambridge, UK.

Gerstner, W., van Hemmen, J. L., and Cowan, J. D. (1996). What matters in neuronal locking? *Neural Computation*, 8:1653–1676.

Ghodoossi, L. (2004). Conceptual study on constructal theory. *Energy Conversion and Management*, 45(9-10):1379–1395.

Giacobini, M., Tomassini, M., Tettamanzi, A., and Alba, E. (2005). Selection intensity in cellular evolutionary algorithms for regular lattices. *IEEE Transactions on Evolutionary Computation*, 9(5):489–505.

Gibson, J. J. (1950). *The Perception of the Visual World*. Houghton Mifflin, Boston.

Gibson, J. J. (1979). *The Ecological Approach to Visual Perception*. Houghton Mifflin, Boston.

Gilbert, B. (1991). Where do little circuits come from? In Williams, J., editor, *Analog Circuit Design: Art, Science, and Personalities*, pages 177–186. Butterworth-Heinemann, Boston.

Gilbert, B. (2002). Design for manufacture. In Toumazou, C., Moschytz, G., and Gilbert, B., editors, *Trade-Offs in Analog Circuit Design*, pages 7–74. Kluwer, Boston.

Gilboa, E. (2004). The promise of cancer vaccines. *Nature Reviews Cancer*, 4(5):401–411.

Glickman, M., Balthrop, J., and Forrest, S. (2005). A machine learning evaluation of an artificial immune system. *Evolutionary Computation*, 13(2):179–212.

Gluck, M. (1991). Stimulus sampling and distributed representations in adaptive network theories of learning. In Healy, A., Kosslyn, S., and Shiffrin, R., editors, *Festschrift for W. K. Estes*. Erlbaum, Hillsdale, NJ.

Glusman, G., Yanai, I., Rubin, I., and Lancet, D. (2001). The complete human olfactory subgenome. *Genome Research*, 11:685–702.

Goldberg, D. E. (1989). *Genetic Algorithms in Search, Optimization and Machine Learning*. Addison-Wesley, Redwood City, CA.

Gomi, T. and Ide, K. (1996). The Tao project: Intelligent wheelchairs for the handicapped. In V. Mittal, H. Yanco, J. A. and Simpson, R., editors, *AAAI Fall Symposium*. MIT Press, Cambridge, MA.

Gomi, T. and Ide, K. (1998). Emergence of gaits of a legged robot by collaboration through evolution. In *IEEE World Congress on Computational Intelligence*. IEEE Press, New York.

Gordon, T. G. W. and Bentley, P. J. (2002). On evolvable hardware. In Ovaska, S. and Sztandera, L., editors, *Soft Computing in Industrial Electronics*, pages 279–323. Physica-Verlag, Heidelberg, Germany.

Goss, S., Aron, S., and Deneubourg, J.-L. (1989). Self-organized shortcuts in the argentine ant. *Naturwissenschaften*, 76:579–581.

Goss, S. and Deneubourg, J.-L. (1992). Harvesting by a group of robots. In Varela, F. J. and Bourgine, P., editors, *Toward a Practice of Autonomous Systems: Proceedings of the First European Conference on Artificial Life*, pages 195–204. MIT Press, Cambridge, MA.

Gould, S. J. (1977). *Ontogeny and Phylogeny*. Harvard University Press, Cambridge, MA.

Gould, S. J. (1997). *Full House: The Spread of Excellence from Plato to Darwin*, 2nd edition. Random House, New York.

Gould, S. J. and Elredge, N. (1977). Punctuated equilibria: The tempo and mode of evolution reconsidered. *Paleobiology*, 3:115–151.

Gouyet, J.-F. (1996). *Physics and Fractal Structures*. Masson, Paris.

Graham, P. (1993). *On LISP*. Prentice Hall, Englewood Cliffs, NJ.

Graham-Rowe, D. (2002). Animals grown from an artificial embryo. *New Scientist*.

Grassé, P.-P. (1959). La reconstruction du nid et les coordinations interindividuelles chez *Bellicositermes natalensis* et *Cubitermes* sp. la théorie de la stigmergie: Essai d'interprétation du comportement des termites constructeurs. *Insectes Sociaux*, 6:41–83.

Graur, D. and Wen-Hsiung, L. (1999). *Fundamentals of Molecular Evolution*. Sinauer Associates, Sunderland, MA.

Gray, P. R., Hurst, P. J., Lewis, S. H., and Meyer, R. G. (2001). *Analysis and Design of Analog Integrated Circuits*, 4th edition. Wiley, New York.

Greenwood, G. W. and Tyrrell, A. M. (2007). *Introduction to evolvable hardware: A practical guide for designing self-adaptive systems*. Wiley, Hoboken, NJ.

Grefenstette, J. (1988). Credit assignment in rule discovery systems based on genetic algorithms. *Machine Learning*, 3(2):225–245.

Grefenstette, J. J. and Ramsey, C. L. (1992). An approach to anytime learning. In *Proceedings of the Ninth International Workshop on Machine Learning*, pages 189–195. Morgan Kauffmann, San Mateo, CA.

Grimbleby, J. B. (2000). Automatic analogue circuit synthesis using genetic algorithms. *IEE Proceedings - Circuits, Devices & Systems*, 147(6):319–323.

Grossberg, S. (1980). How does the brain build a cognitive code? *Psychological Review*, 87:1–51.

Grossberg, S. (1987). Competitive learning: From interactive activation to adaptive resonance. *Cognitive Science*, 11:121–134.

Gruau, F. (1994a). Automatic definition of modular neural networks. *Adaptive Behavior*, 3(2):151–183.

Gruau, F. (1994b). Genetic microprogramming of neural networks. In Kinnear, K., editor, *Advances in Genetic Programming*, pages 495–518. MIT Press, Cambridge, MA.

Gruau, F. and Quatramaran, K. (1997). Cellular encoding for interactive evolutionary robotics. In Husbands, P. and Harvey, I., editors, *Proceeding of the Fourth European Conference on Artificial Life*, pages 368–377. MIT Press,, Cambridge, MA.

Haldane, J. B. S. (1955). Population genetics. *New Biology*, 18:34–51.

Hamilton, W. D. (1964). The genetical evolution of social behavior, I and II. *Journal of Theoretical Biology*, 7:1–52.

Hamilton, W. D., Axelrod, R., and Tanese, R. (1990). Sexual reproduction as an adaptation to resist parasites (a review). *Proceedings of the National Academy of Sciences USA*, 87(9):3566–3573.

Hammerstein, P., Hagen, E. H., Herz, A. V. M., and Herzel, H. (2006). Robustness: A key to evolutionary design. *Biological Theory*, 1(1):90–93.

Hampton, A. and Adami, C. (2004). Evolution of robust developmental neural networks. In Pollack, J., Bedau, M., Husbands, P., Ikegami, T., and R., W., editors, *Proceedings of the Ninth International Conference on Artificial Life, ALIFE IX*, Boston, September 12–15, pages 438–443. MIT Press, Cambridge, MA.

Hancock, P. J. (1992). Genetic algorithms and permutations problems: A comparison of recombination operators for neural structure specification. In Whitley, D., editor, *Proceedings of the International Workshop on the Combination of Genetic Algorithms and Neural Networks*, Baltimore. IEEE Press, Piscataway, NJ.

Hancock, P. J., Baddeley, R. J., and Smith, L. S. (1992). The principal components of natural images. *Network*, 3:61–70.

Hanson, D. (2005). Expanding the aesthetics possibilities for humanlike robots. In *Proceedings of the IEEE Humanoid Robotics Conference*, Tsukuba, Japan. IEEE Press, Piscataway, NJ.

Hanssen, S. A., Hasselquist, D., Folstad, I., and Erikstad, K. E. (2005). Costs of immunity: Immune responsiveness reduces survival in a vertebrate. *Proceedings of the Royal Society of London B*, 271(1542):925–930.

Hara, F., Akazawa, H., and Kobayashi, H. (2001). Realistic facial expressions by SMA driven face robot. In *Proceedings of the Tenth IEEE International Workshop on Robot and Human Interactive Communication*. Bordeaux, France.

Hara, F. and Pfeifer, R., editors (2003). *Morpho-Functional Machines: The New Species*. Springer-Verlag, Tokyo.

Hardy, J., de Pazzis, O., and Pomeau, Y. (1976). Molecular dynamics of a classical lattice gas: Transport properties and time correlation functions. *Physical Review A*, 13(5):1949–1961.

Harold, F. (2001). *The Way of the Cell*. Oxford University Press, Oxford.

Harp, S. A., Samad, T., and Guha, A. (1989). Toward the genetic synthesis of neural networks. In Schaffer, J. D., editor, *Proceedings of the Third International Conference on Genetic Algorithms*. Morgan Kaufmann, San Mateo, CA.

Hartl, D. L. (2000). Molecular melodies in high and low C. *Nature Reviews Genetics*, 1:145–149.

Harvey, I. (1992). Species adaptation genetic algorithms: A basis for a continuing SAGA. In Varela, F. J. and Bourgine, P., editors, *Toward a Practice of Autonomous Systems: Proceedings of the First European Conference on Artificial Life*, pages 346–354. MIT Press, Cambridge, MA.

Harvey, I. (1997). Is there another new factor in evolution? *Evolutionary Computation*, 4(3):313–329.

Harvey, I., Di Paolo, E., Wood, R., Quinn, M., and Tuci, E. (2005). Evolutionary robotics: A new scientific tool for studying cognition. *Artificial Life*, 11(1-2):79–98.

Harvey, I., Husbands, P., and Cliff, D. (1994). Seeing the light: Artificial evolution, real vision. In Cliff, D., Husbands, P., Meyer, J.-A., and Wilson, S., editors, *From Animals to Animats 3: Proceedings of the Third International Conference on Simulation of Adaptive Behaviour, SAB94*, pages 392–401. MIT Press, Cambridge, MA.

Harvey, I., Husbands, P., Cliff, D., Thompson, A., and Jakobi, N. (1997). Evolutionary robotics: The Sussex approach. *Robotics and Autonomous Systems*, 20:205–224.

Harvey, I. and Thompson, A. (1996). Through the labyrinth, evolution finds a way: A silicon ridge. In Higuchi, T., Iwata, M., and Liu, W., editors, *Proceedings of the First International Conference on Evolvable Systems: From Biology to Hardware*. Springer-Verlag, Tokyo.

Hashimoto, S., Narita, S., Kasahara, H., Shirai, K., Kobayashi, T., Takanishi, A., Sugano, S., Yamaguchi, J., Sawada, H., Takanobu, H., Shibuya, K., Morita, T., Kurata, T., Onoe, N., Ouchi, K., Noguchi, T., Niwa, Y., Nagayama, S., Tabayashi, H., Matsui,

I., Obata, M., Matsuzaki, H., Murasugi, A., Kobayashi, T., Haruyama, S., Okada, T., Hidaki, Y., Taguchi, Y., Hoashi, K., Morikawa, E., Iwano, Y., Araki, D., Suzuki, J., Yokoyama, M., Dawa, I., Nishino, D., Inoue, S., Hirano, T., Soga, E., Gen, S., Yanada, T., Kato, K., Sakamoto, S., Ishii, Y., Matsuo, S., Yamamoto, Y., Sato, K., Hagiwara, T., Ueda, T., Honda, N., Hashimoto, K., Hanamoto, T., Kayaba, S., Kojima, T., Iwata, H., Kubodera, H., Matsuki, R., Nakajima, T., Nitto, K., Yamamoto, D., Kamizaki, Y., Nagaike, S., Kunitake, Y., and Morita, S. (2002). Humanoid robots in Waseda University – Hadaly-2 and WABIAN. *Autonomous Robots*, 12:25–38.

Haugeland, J., editor (1985). *Artificial Intelligence: The Very Idea*. MIT Press, Cambridge, MA.

Hausen, K. and Egelhaaf, M. (1989). Neural mechanisms of visual course control in insects. In Stavenga, D. G. and Hardie, R. C., editors, *Facets of Vision*. Springer-Verlag, Berlin.

Hayes, A. T., Martinoli, A., and Goodman, R. M. (2002). Distributed odor source localization. *IEEE Sensors Journal*, 2:260–271.

Haykin, S. (2007). *Neural Networks. A Comprehensive Foundation*, 3rd edition. Prentice Hall, Upper Saddle River, NJ.

Haynes, T. and Sen, S. (1997). Crossover operators for evolving a team. In Koza, J. R., Deb, K., Dorigo, M., Fogel, D. B., M. Garzon, H. I., and Riolo, R. L., editors, *Proceedings of the Second Annual Conference on Genetic Programming*, Stanford, CA, pages 162–167. Morgan Kaufmann„ San Mateo, CA.

Healy, S., editor (1998). *Spatial Representations in Animals*. Oxford University Press, Oxford.

Hebb, D. O. (1949). *The Organisation of Behavior*. Wiley, New York.

Hegselmann, R. and Flache, A. (1998). Understanding complex social dynamics: A plea for cellular automata based modelling. *Journal of Artificial Societies and Social Simulation*, 1(3).

Held, R. (1965). Plasticity in sensory-motor systems. *Scientific American*, 213(5):84–94.

Held, R. and Hein, A. (1963). Movement-produced stimulation in the development of visually guided behavior. *Journal of Comparative and Physiological Psychology*, 56(5):872–876.

Herrnstein, R. J. (1997). *The Matching Law: Papers in Psychology and Economy*. Harvard University Press, Cambridge, MA.

Hertz, J., Krogh, A., and Palmer, R. G. (1991). *Introduction to the Theory of Neural Computation*. Addison Wesley, Redwood City, CA.

Herz, A. V. M., Gollisch, T., Machens, C. K., and Jaeger, D. (2006). Modeling single-neuron dynamics and computations: A balance of detail and abstraction. *Science*, 314:80–85.

Higuchi, T., Iwata, M., Keymeulen, D., Sakanashi, H., Murakawa, M., Kajitani, I., Takahashi, E., Toda, K., Salami, N., Kajihara, N., and Otsu, N. (1999). Real-world applications of analog and digital evolvable hardware. *IEEE Transactions on Evolutionary Computation*, 3(3):220–235.

Hillis, W. D. (1987). The Connection Machine. *Scientific American*, 256:108–115.

Hillis, W. D. (1992). Co-evolving parasites improve simulated evolution as an optimization procedure. In Langton, C., Farmer, J., Rasmussen, S., and Taylor, C., editors, *Artificial Life II: Proceedings Volume of Santa Fe Conference*, volume 11. Series of the Santa Fe Institute Studies in the Sciences of Complexities, Addison Wesley, Redwood City, CA.

Hinton, G. E. and Nowlan, S. J. (1987). How learning can guide evolution. *Complex Systems*, 1:495–502.

Hinton, G. E. and Sejnowski, T. J., editors (1999). *Unsupervised Learning*. MIT Press, Cambridge, MA.

Hiptmair, R. (2001). Discrete Hodge operators. *Numerische Mathematik*, 90(2):265–289.

Hirai, K., Hirose, M., and Takenaka, T. (1998). The development of the Honda humanoid robot. In *Proceedings of the IEEE International Conference on Robotics and Automation*, Leuven, Belgium. IEEE Press, Piscataway, NJ.

Hirotsune, S., Yoshida, N., Chen, A., Garrett, L., Fumihiro, S., Takahashi, S., Yagami, K., Wynshaw-Boris, A., and Yoshiki, A. (2003). An expressed pseudogene regulates the messenger-RNA stability of its homologous coding gene. *Nature*, 423:91–96.

Hodgkin, A. L. and Huxley, A. F. (1952). A quantitative description of membrane current and its application to conduction and excitation in nerve. *Journal of Physiology (London)*, 108:500–544.

Hofmeyr, S. A. (2001). An interpretative introduction to the immune system. In Segel, L. and Cohen, I., editors, *Design Principles for the Immune System and Other Distributed Autonomous Systems*, Santa Fe Institute Studies in the Sciences of Complexity, chapter 1, pages 3–26. Oxford University Press, Oxford.

Hofmeyr, S. A. and Forrest, S. (2000). Architecture for an artificial immune system. *Evolutionary Computation*, 8(4):443–473.

Holland, J. H. (1975). *Adaptation in Natural and Artificial Systems*. University of Michigan Press, Ann Arbor.

Holland, J. H. (1976). Adaptation. In Rosen, R. and Snell, F. M., editors, *Progress in Theoretical Biology IV*. Academic Press, New York.

Holland, O. (2003). Exploration and high adventure: The legacy of Grey Walter. *Philosophical Transactions of the Royal Society*, 361(1811):2085–2121.

Holland, O. E. and Melhuish, C. (1999). Stigmergy, self-organisation, and sorting in collective robotics. *Artificial Life*, 5:173–202.

Hong, J., Tan, X., Pinette, B., Weiss, R., and Riseman, E. (1992). Image-based homing. *Control Systems Magazine, IEEE*, 12(1):38–45.

Hopcroft, J. E., Motwani, R., and Ullman, J. D. (2006). *Introduction to Automata Theory, Languages, and Computation*, 3rd edition. Addison Wesley, Menlo Park, CA.

Hopfield, J. J. (1982). Neural networks and physical systems with emergent collective computational abilities. *Proceedings of the National Academy of Sciences USA*, 79:3088–3092.

Hopfield, J. J. (1995). Pattern recognition computation using action potential timing for stimulus representation. *Nature*, 376:33–36.

Hornby, G., Lipson, H., and Pollack, J. (2003). Generative representations for the automated design of modular physical robots. *IEEE Transactions on Robotics and Automation*, 19(4):703–719.

Hornby, G. and Pollack, J. (2001a). The advantages of generative grammatical encodings for physical design. In *Proceedings of the 2001 Congress on Evolutionary Computation*, volume 1, pages 600–607.

Hornby, G. S. and Pollack, J. B. (2001b). Evolving L-systems to generate virtual creatures. *Computers & Graphics*, 25(6):1041–1048.

Hubel, D. H. and Wiesel, T. N. (1963). Receptive fields of cells in striate cortex of very young, visually inexperienced kittens. *Journal of Neurophysiology*, 26:994–1002.

Hubel, D. H. and Wiesel, T. N. (1977). Functional architecture of a macaque monkey visual cortex. *Proceedings of the Royal Society of London B*, 198:1–59.

Humphrey, N. (2002). *Great Expectations: The Evolutionary Psychology of Faith Healing and the Placebo Effect*, chapter 19, pages 255–285. Oxford University Press, Oxford.

Husbands, P. and Harvey, I. (1992). Evolution versus design: Controlling autonomous robots. In *Integrating Perception, Planning and Action, Proceedings of Third IEEE Annual Conference on Artificial Intelligence, Simulation and Planning*, pages 139–146. IEEE Press, Piscataway, NJ.

Husbands, P., Harvey, I., Cliff, D., and Miller, G. (1994). The use of genetic algorithms for the development of sensorimotor control systems. In Gaussier, P. and Nicoud, J.-D., editors, *From Perceptin to Action*. IEEE Press, Los Alamitos, CA.

Husbands, P., Harvey, I., Cliff, D., and Miller, G. (1997). Artificial evolution: A new path for artificial intelligence? *Brain and Cognition*, 34:130–159.

Husbands, P., Smith, T., Jakobi, N., and O'Shea, M. (1998). Better living through chemistry: Evolving gasnets for robot control. *Connection Science*, 10:185–210.

Huth, A. and Wissel, C. (1992). The simulation of the movement of fish schools. *Journal of Theoretical Biology*, 156:365–385.

Huynen, M. A., Stadler, P. F., and Fontana, W. (1996). Smoothness within ruggedness: The role of neutrality in adaptation. *Proceedings of the National Academy of Sciences USA*, 93:397–401.

Hyvärinen, A., Karhunen, J., and Oja, E. (2001). *Independent Component Analysis*. Wiley, Indianapolis, IN.

Ienne, P., Cornu, T., and Kuhn, G. (1996). Special-purpose digital hardware for neural networks: An architectural survey. *Journal of VLSI Signal Processing*, 13:5–25.

Ijspeert, A. J., Crespi, A., Ryczko, D., and Cabelguen, J.-M. (2007). From swimming to walking with a salamander robot driven by a spinal cord model. *Science*, 315:1416–1420.

Ijspeert, A. J., Martinoli, A., Billard, A., and Gambardella, L. M. (2001). Collaboration through the exploitation of local interactions in autonomous collective robotics: The stick pulling experiment. *Autonomous Robots*, 11:149–171.

Ilachinski, A. (2001). *Cellular Automata: A Discrete Universe*. World Scientific, Singapore.

Indiveri, G. and Douglas, R. (2000). Robotic vision: Neuromorphic vision sensors. *Science*, 288:1189–1190.

International Human Genome Sequencing Consortium (1995). An estimation of minimal genome size required for life. *FEBS Letters*, 362:257–260.

International Human Genome Sequencing Consortium (2001). Initial sequencing and analysis of the human genome. *Nature*, 409:860–921.

Ishida, Y. (2004). *Immunity-Based Systems: A Design Perspective*. Advanced Information Processing. Springer-Verlag, Berlin.

Jacob, C. (2001). *Illustrating Evolutionary Computation with Mathematica*. Morgan Kaufmann, San Diego.

Jacob, F. (1981). *Le jeu des possibles*. Librairie Arthéme Fayard, Paris.

Jacobs, R. A. and Kosslyn, S. M. (1994). Encoding shape and spatial relations: The role of receptive field size in coordinating complementary representations. *Cognitive Science*, 18:361–386.

Jacobson, H. (1958). On models of reproduction. *American Scientist*, 46:255–284.

Jaeger, H. and Haas, H. (2004). Harnessing nonlinearity: Predicting chaotic systems and saving energy in wireless communication. *Science*, 304:78–80.

Jakobi, N. (1997a). Evolutionary robotics and the radical envelope of noise hypothesis. *Adaptive Behavior*, 6:131–174.

Jakobi, N. (1997b). Half-baked, ad-hoc and noisy: Minimal simulations for evolutionary robotics. In Husbands, P. and Harvey, I., editors, *Proceedings of the Fourth European Conference on Artificial Life*, pages 348–357. MIT Press, Cambridge, MA.

Jakobi, N. (1998). Running across the reality gap: Octopod locomotion evolved in a minimal simulation. In Husbands, P. and Meyer, J.-A., editors, *Evolutionary Robotics: First European Workshop, EvoRobot98*, pages 39–58. Springer-Verlag, London.

Jakobi, N. (2003). Harnessing morphogenesis. In Kumar, S. and Bentley, P., editors, *On Growth, Form and Computers*, pages 392–404. Academic Press, London.

Jakobi, N., Husbands, P., and Harvey, I. (1995). Noise and the reality gap: The use of simulation in evolutionary robotics. In Moran, F., Moreno, A., Merelo, J., and Chacon, P., editors, *Advances in Artificial Life: Proceedings of the Third European Conference on Artificial Life*, volume 929 of *Lecture Notes in Artificial Intelligence*, pages 704–720. Springer-Verlag, London.

James, W. (1890). *The Principles of Psychology*. Dover, New York. Reprint 1950.

Jaynes, E. T. (2003). *Probability Theory: The Logic of Science*. Cambridge University Press, Cambridge, UK.

Jennings, H. S. (1906). *Behavior of the Lower Organisms*. Indiana University Press, Bloomington. Reprint 1976.

Jerne, N. K. (1974). Towards a network theory of the immune system. *Annales de l'Institut Pasteur. Immunologie*, 125(C):373–389.

Ji, Z. and Dasgupta, D. (2007). Revisiting negative selection algorithms. *Evolutionary Computation*, 15(2):223–251.

Jordan, M. and Rumelhart, D. (1992). Forward models: Supervised learning with a distal teacher. *Cognitive Science*, 16:307–354.

Jordan, M. I. (1989). Serial order: A parallel, distributed processing approach. In Elman, J. and Rumelhart, D., editors, *Advances in Connectionist Theory: Speech*. Erlbaum, Hillsdale, NJ.

Jorgensen, M. W., Ostergaard, E. H., and Lund, H. H. (2004). Modular ATRON: Modules for a self-reconfigurable robot. In *Proceedings of the IEEE/RSJ International Conference on Robotics and Intelligent Systems (IROS), Sendai, Japan*. IEEE Press, Piscataway NJ.

Jun-Ho, O., Hanson, D., Han, I. Y., Kim, J. K., Kim, W. S., and Park, I. W. (2006). Design of android type humanoid robot Albert HUBO. In *Proceedings of the International IEEE/RJS Conference on Intelligent Robotics and Systems*. Beijing.

Kaelbling, L. P., Littman, M. L., and Moore, A. W. (1996). Reinforcement learning: A survey. *Journal of Artificial Intelligence Research*, 4:237–285.

Kandel, E. R., Schwartz, J. H., and Jessell, T. M. (2000). *Principles of Neural Science*, 4th edition. McGraw-Hill, New York.

Kaneko, K. (1992). Overview of coupled map lattices. *Chaos: An Interdisciplinary Journal of Nonlinear Science*, 2(3):279–282.

Karolyi, A. and Kertesz, J. (1999). Granular medium lattice gas model: The algorithm. *Computer Physics Communications*, 121-122:290–293.

Kato, I., Mori, Y., and Masuda, T. (1972). Pneumatically powered artificial legs walking automatically under various circumstances. In *Proceedings of the Fourth International Symposium on External Control of Human Extremities*, pages 458–470.

Kato, T. and Floreano, D. (2001). An evolutionary active-vision system. In *Proceedings of the Congress on Evolutionary Computation (CEC01)*. IEEE Press, Piscataway, NJ.

Katz, R. H. and Borriello, G. (2004). *Contemporary Logic Design*, 2nd edition. Prentice Hall, Upper Saddle River, NJ.

Kazadi, S., Goodman, R., Tsikate, D., and Lin, H. (2000). An autonomous water vapor plume tracking robot using passive resistive polymers sensors. *Autonomous Robots*, 9:175–188.

Keller, L. and Chapuisat, M. (2002). Eusociality and cooperation. In *Encyclopedia of Life Sciences*, pages 1–8. MacMillan, London.

Keller, L. and Surette, M. G. (2006). Communication in bacteria. *Nature Reviews Microbiology*, 4:249–258.

Keller, P. E., Kouzes, R. T., and Kangas, L. J. (1994). Three neural network based sensor systems for environmental monitoring. In *IEEE Electro/94 International Combined Conference Proceedings*, pages 378–382. IEEE Press, Piscataway, NJ.

Kelso, S., Ganong, A., and Brown, T. (1986). Hebbian synapses in hippocampus. *Proceedings of the National Academy of Sciences USA*, 83:5326–5330.

Kennedy, J. and Eberhart, R. C. (1995). Particle swarm optimization. In *Proceedings of the IEEE International Conference on Neural Networks*, pages 1942–1948. IEEE Press, Piscataway, NJ.

Kennedy, J. and Eberhart, R. C. (2001). *Swarm Intelligence*. Morgan Kaufmann, San Francisco.

Kennedy, P. and Osborn, T. (2001). A model of gene expression and regulation in an artificial cellular organism. *Complex Systems*, 13(1):33–59.

Keymeulen, D., Iwata, M., Kuniyoshi, Y., and Higuchi, T. (1998). Online evolution for a self-adapting robotic navigation system using evolvable hardware. *Artificial Life*, 4(4):359–393.

Keymeulen, D., Konaka, K., Iwata, M., Kuniyoshi, Y., and Higuchi, T. (1997). Off-line evolution for a robot navigation system based on a gate-level evolvable hardware. Technical report, Electrotechnical Laboratory, Ibaraki, Japan.

Kim, J., Bentley, P., Aickelin, U., Greensmith, J., Tedesco, G., and Twycross, J. (2007a). Immune system approaches to intrusion detection – a review. *Natural Computing*, 6(4):413–466.

Kim, J. and Bentley, P. J. (1999). Negative selection and niching by an artificial immune system for network intrusion detection. In Banzhaf, W., Daida, J., Eiben, A. E., Garzon, M. H., Honavar, V., Jakiela, M., and Smith, R. E., editors, *Proceedings of the 1999 Conference on Genetic and Evolutionary Computation (GECCO-99)*, Orlando, FL July 13-17, 1999, pages 149–158. Morgan Kaufmann, San Francisco.

Kim, J., Greensmith, J., Twycross, J., and Aickelin, U. (2005). Malicious code execution detection and response immune system inspired by the danger theory. In *Proceedings of the Adaptive and Resilient Computing Security Workshop (ARCS-05)*, Santa Fe, NM.

Kim, S., Spenko, M., Trujillo, S., Heyneman, B., Mattoli, V., and Cutkosky, M. R. (2007b). Whole body adhesion: Hierarchical, directional and distributed control of adhesive forces for a climbing robot. In *Proceedings of the IEEE International Conference on Robotics and Automation*. Rome.

Kimura, M. (1983). *The Neutral Theory of Molecular Evolution*. Cambridge University Press, Cambridge, UK.

Kirkland, K. L. (2002). High-tech brains: A history of technology and models of nerve and brain function. *Perspectives in Biology and Medicine*, 45 (2):212–223.

Kirkpatrick, S., Gelatt, C. D., and Vecchi, M. P. (1983). Optimization by simulated annealing. *Science*, 220:671–680.

Kirschner, M. and Gerhart, J. (1998). Evolvability. *Proceedings of the National Academy of Sciences USA*, 95(15):8420–8427.

Kirschner, M. and Gerhart, J. C. (2005). *The Plausibility of Life: Resolving Darwin's Dilemma*. Yale University Press, New Haven, CT.

Kitano, H. (1990). Designing neural networks using genetic algorithms with graph generation system. *Complex Systems*, 4(4):461–476.

Kitano, H. (1995). A simple model of neurogenesis and cell differentiation based on evolutionary large-scale chaos. *Artificial Life*, 2(1):79–99.

Kleinfeld, D. and Sompolinski, H. (1989). Associative network models for central pattern generators. In Koch, C. and Segev, I., editors, *Methods in Neuronal Modeling: From Synapses to Networks*, pages 195–246. MIT Press, Cambridge, MA.

Klimovich, V. B. (2002). Actual problems of evolutionary immunology. *Journal of Evolutionary Biochemistry and Physiology*, 38(5):562 – 574.

Knudsen, E. I., du Lac, S., and Esterly, S. D. (1987). Computational maps in the brain. *Annual Review of Neuroscience*, 10:41–65.

Knuth, D. E. (1998). *The Art of Computer Programming*. Volume 3: *Sorting and Searching*, 2nd edition. Addison-Wesley, Boston.

Kodjabachian, J. and Meyer, J.-A. (1998a). Evolution and development of modular control architectures for 1-D locomotion in six-legged animats. *Connection Science*, 10(3-4):211–254.

Kodjabachian, J. and Meyer, J.-A. (1998b). Evolution and development of neural networks controlling locomotion, gradient following and obstacle avoidance in artificial insects. *IEEE Transactions on Neural Networks*, 9:796–812.

Koenderink, J. and van Doorn, A. (1987). Facts on optic flow. *Biological Cybernetics*, 56:247–254.

Kohonen, T. (1982). Self-organized formation of topologically correct feature maps. *Biological Cybernetics*, 43:59–69.

Kohonen, T. (1989). *Self-Organization and Associative Memory*. Springer Verlag, Berlin.

Kolen, J. F. and Pollack, J. B. (1991). Back propagation is sensitive to initial conditions. In Lippmann, R. P., Moody, J. E., and Touretzky, D. S., editors, *Advances in Neural Information Processing Systems*, volume 3, pages 860–867. Morgan Kaufmann, San Francisco.

Komosinski, M. and Ulatowski, S. (1999). Framsticks: Towards a simulation of a nature-like world, creatures and evolution. In Floreano, D., Nicoud, J.-D., and Mondada, F., editors, *Advances in Artificial Life - ECAL99*. Springer-Verlag, Berlin.

Komosinski, M. and Ulatowski, S. (2000). The world of Framsticks: Simulation, evolution, interaction. In *Proceedings of Second International Conference on Virtual Worlds*, pages 214–224. Springer-Verlag, Berlin.

Komosinski, M. and Ulatowski, S. (2006). *Framsticks Manual*. Available at http://www.framsticks.com.

Kondrashov, A. S. and Crow, J. F. (1993). A molecular approach to estimating human deleterious mutation rate. *Human Mutation*, 2:229–234.

Koonin, E. and Mushegian, A. (1996). Complete genome sequences of cellular life forms: Glimpses of theoretical evolutionary genomics. *Current Opinion in Genetics and Development*, 6:757–762.

Korneev, S. A., Park, J., and O'Shea, M. (1999). Neuronal expression of neural nitric oxide synthase (nNOS) protein is suppressed by an antisense RNA transcribed from an NOS pseudogene. *Journal of Neuroscience*, 19:7711–7720.

Kosslyn, S. M., Chabris, C. F., Marsolek, C. J., and Koenig, O. (1992). Categorical versus coordinate spatial relations: Computational analyses and computer simulations. *Journal of Experimental Psychology: Human Perception and Performance*, 18:562–577.

Koza, J. R. (1992). *Genetic Programming: On the Programming of Computers by Means of Natural Selection*. MIT Press, Cambridge, MA.

Koza, J. R. (1993). Discovery of rewrite rules in Lindenmayer systems and state transition rules in cellular automata via genetic programming. In *Symposium on Pattern Formation (SPF-93)*, Claremont, CA.

Koza, J. R. (1994). *Genetic programming II: Automatic Discovery of Reusable Programs*. MIT Press, Cambridge, MA.

Koza, J. R., Bennett III, F. H., Andre, D., and Keane, M. A. (1999). *Genetic Programming III: Darwinian Invention and Problem Solving*. Morgan Kaufmann, San Francisco.

Koza, J. R., Jones, L. W., Keane, M. A., Streeter, M. J., and Al-Sakran, S. H. (2005). Toward automated design of industrial-strength analog circuits by means of genetic programming. In O'Reilly, U.-M., Yu, T., Riolo, R., and Worze, B., editors, *Genetic Programming Theory and Practice II*, chapter 8, pages 121–142. Springer-Verlag, New York.

Koza, J. R., Keane, M. A., Streeter, M. J., Mydlowec, W., Yu, J., and Lanza, G. (2003). *Genetic Programming IV: Routine Human-Competitive Machine Intelligence*. Kluwer, Norwell, MA.

Kruiskamp, W. and Leenaerts, D. (1995). DARWIN: CMOS opamp synthesis by means of a genetic algorithm. In *Proceedings of the 32nd ACM/IEEE Design Automation Conference*, pages 433–438.

Kube, C. R. and Bonabeau, E. (2000). Cooperative transport by ants and robots. *Robotics and Autonomous Systems*, 30:85–101.

Kube, C. R. and Zhang, H. (1993). Collective robotics: From social insects to robots. *Adaptive Behavior*, 2:189–218.

Kuddusi, L. and Eğrican, N. (2008). A critical review of constructal theory. *Energy Conversion and Management*, 49(5):1283–1294.

Kumar, S. and Bentley, P. J., editors (2003). *On Growth, Form and Computers*. Academic Press, Orlando, FL.

Kuniyoshi, Y., Ohmura, Y., Terada, K., and Nagakubo, A. (2004). Dynamic roll-and-rise motion by an adult-size humanoid robot. *International Journal of Humanoid Robotics*, 1:497–516.

Kuniyoshi, Y. and Sangawa, S. (2006). Early motor development from partially ordered neural-body dynamics: Experiments with a cortico-spinal-musculo-skeletal model. *Biological Cybernetics*, 95:589–605.

Kuniyoshi, Y., Yorozu, Y., Suzuki, S., Sangawa, S., Ohmura, Y., Terada, K., and Nagakubo, A. (2007). Emergence and development of embodied cognition – a constructivist approach using robots. In von Hofsten, C. and Rosander, K., editors, *From Action to Cognition*. Elsevier, Amsterdam.

Lamarck, J. B. (1914). *Zoological Philosophy*. Macmillan, London. Relevant parts reprinted in Belew, R. K. and Mitchell, M., editors, (1996).

Lambrinos, D., Moöller, R., Labhart, T., Pfeifer, R., and Wehner, R. (2000). A mobile robot employing insect strategies for navigation. *Robotics and Autonomous Systems*, 30:39–64.

Land, M. (1997). Visual acuity in insects. *Annual Review of Entomology*, 42:147–177.

Land, M. F. and Fernald, R. D. (1992). The evolution of eyes. *Annual Review of Neuroscience*, 15(1):1–29.

Langdon, W. B. and Poli, R. (2007). Evolving problems to learn about particle swarm optimizers and other search algorithms. *IEEE Transactions on Evolutionary Computation*, 11:561–578.

Langton, C. G. (1984). Self-reproduction in cellular automata. *Physica D: Nonlinear Phenomena*, 10(1-2):135–144.

Langton, C. G. (1990). Computation at the edge of chaos: Phase transitions and emergent computation. *Physica D: Nonlinear Phenomena*, 42(1-3):12–37.

Langton, C. G. (1996). Artificial life. In Boden, M. A., editor, *The Philosophy of Artificial Life*, pages 39–94. Oxford University Press, Oxford.

Lanzi, P. L., Stolzmann, W., and Wilson, S. W., editors (2000). *Learning Classifier Systems: From Foundations to Applications*. Springer-Verlag, Berlin.

Laszlo, J. F., van de Panne, M., and Fiume, E. (1996). Limit cycle control and its application to the animation of balancing and walking. In *Proceedings of the 23rd Annual Conference on Computer Graphics and Interactive Techniques (SIGGRAPH'96)*, pages 155–162. ACM Press, New York.

Lawrence, P. A. and Levine, M. (2006). Mosaic and regulative development: Two faces of one coin. *Current Biology*, 16(7):R236–R239.

Layzell, P. (1998). A new research tool for intrinsic hardware evolution. In *ICES '98: Proceedings of the Second International Conference on Evolvable Systems*, volume 1478 of *Lecture Notes in Computer Science*, pages 47–56.

Lazzaro, J. P., Wawrzynek, J., and Kramer, A. (1994). Systems technologies for silicon auditory models. *IEEE Micro*, 14:7–15.

Le Cun, Y., Boser, B., Denker, J. S., Henderson, R. E., Howard, R. E., Hubbard, W., and Jackel, L. D. (1990). Handwritten digit recognition with a back-propagation network. In Touretzky, D., editor, *Advances in Neural Information Processing Systems 2*, pages 396–404. Morgan Kaufmann, San Francisco.

Le Masson, G., Renaud-Le Masson, S., Debay, D., and Bal, T. (2000). Feedback inhibition controls spike transfer in hybrid thalamic circuits. *Nature*, 417:854–858.

Lehmann, L. and Keller, L. (2006). The evolution of cooperation and altruism – a general framework and a classification of models. *Journal of Evolutionary Biology*, 19:1365–1376.

Lenski, R. E., Ofria, C., Collier, T. C., and Adami, C. (1999). Genomic complexity, robustness, and genetic interactions in digital organisms. *Nature*, 400:661–664.

Lewis, M., Fagg, A., and Bekey, G. (1994). Genetic algorithms for gait synthesis in a hexapod robot. In Zheng, Y., editor, *Recent Trends in Mobile Robots*, pages 317–331. World Scientific, Singapore.

Lewis, M. A., Fagg, A. H., and Solidum, A. (1992). Genetic programming approach to the construction of a neural network for a walking robot. In *Proceedings of the IEEE*

International Conference on Robotics and Automation, pages 2618–2623. IEEE Press, Piscataway, NJ.

Lewontin, R. C. (1978). Adaptation. *Scientific American*, 293(3):157–169.

Lewontin, R. C. (1996). Evolution as engineering. In Collado-Vides, J., Smith, T., and Magasanik, B., editors, *Integrative Approaches to Molecular Biology*, pages 1–10. MIT Press, Cambridge, MA.

Li, W., Luo, C., and Wu, C. (1985). Evolution of DNA sequences. In MacIntyre, R. J., editor, *Molecular Evolutionary Genetics*. Plenum, New York.

Lindenmayer, A. (1968). Mathematical models for cellular interactions in development I. Filaments with one-sided inputs. *Journal of Theoretical Biology*, 18(3):280–299.

Lindenmayer, A. (1974). Adding continuous components to L-systems. In Rozenberg, G. and Salomaa, A., editors, *L-Systems*, volume 15 of *Lecture Notes in Computer Science*, pages 53–68. Springer-Verlag, Berlin.

Lindenmayer, A. (1975). Developmental systems and languages in their biological context. In Herman, G. T. and Rozenberg, G., editors, *Developmental Systems and Languages*, pages 1–40. North-Holland, Amsterdam.

Lindsay, P. N. and Norman, D. A. (1972). *Human Information Processing*. Academic Press, New York.

Linsker, R. (1986). From basic network principles to neural architecture (series). *Proceedings of the National Academy of Sciences USA*, 83:7508–7512, 8390–8394, 8779–8783.

Linsker, R. (1988). Self-organization in a perceptual network. *Computer*, 3:105–117.

Lipson, H. (2005). Homemade: The future of functional rapid prototyping. *IEEE Spectrum*, 42(5):24–31.

Lipson, H. and Pollack, J. B. (2000). Automatic design and manufacture of robotic lifeforms. *Nature*, 406:974–978.

Liu, S.-C., Kramer, J., Indiveri, G., Delbrueck, T., and Douglas, R. (2002). *Analog VLSI: Circuits and Principles*. MIT Press, Cambridge, MA.

Lohn, J. D. and Colombano, S. P. (1999). A circuit representation technique for automated circuit design. *IEEE Transactions on Evolutionary Computation*, 3(3):205–219.

Lohn, J. D. and Hornby, G. S. (2006). Evolvable hardware: Using evolutionary computation to design and optimize hardware systems. *IEEE Computational Intelligence Magazine*, 1(1):19–27.

Lohn, J. D., Hornby, G. S., and Linden, D. S. (2004). An evolved antenna for deployment on NASA's space technology 5 mission. In O'Reilly, U.-M., Riolo, R. L., Yu, T., and Worzel, B., editors, *Genetic Programming Theory and Practice II*. Kluwer, Amsterdam.

Loker, E. S., Adema, C. M., Zhang, S. M., and Kepler, T. B. (2004). Invertebrate immune systems – not homogeneous, not simple, not well understood. *Immunological Reviews*, 198(1):10–24.

Long, J. A., Young, G. C., Holland, T., Senden, T. J., and Fitzgerald, E. M. G. (2006). An exceptional Devonian fish from Australia sheds light on tetrapod origins. *Nature*, 444:199–202.

Lotka, A. J. (1925). *Elements of Physical Biology*. Williams and Wilkins, Baltimore.

Luke, S. (1998). Genetic programming produced competitive soccer softbot teams for RoboCup 97. In Banzhaf, W., Chellapilla, K., Deb, K., Dorigo, M., Fogel, D., Garzon, M., Goldberg, D., H., I., Koza, J., and Riolo, R., editors, *Genetic Programming 1998: Proceedings of the Third Annual Conference*, pages 214–222. Morgan Kaufmann, San Francisco.

Luke, S., Hohn, C., Farris, J., Jackson, G., and Hendler, J. (1997). Co-evolving soccer softbot team coordination with genetic programming. In *Proceedings of the First International Workshop on RoboCup, (IJCAI-97)*, Nagoya, Japan, pages 214–222. Springer-Verlag, London.

Lund, H. H., Hallam, J., and Lee, W. (1997). Evolving robot morphology. In *Proceedings of IEEE Fourth International Conference on Evolutionary Computation*, pages 197–202. IEEE Press, Piscataway, NJ.

Lund, H. H., Webb, B., and Hallam, J. (1998). Physical and temporal scaling considerations in a robot model of cricket calling song preference. *Artificial LIfe*, 4:95–107.

Lungarella, M. and Berthouze, L. (2002). On the interplay between morphological, neural and environmental dynamics: A robotic case-study. *Adaptive Behavior*, 10:223–241.

Lungarella, M., Metta, G., Pfeifer, R., and Sandini, G. (2003). Developmental robotics: A survey. *Connection Science*, 15:151–190.

Lydyard, P. M., Whelan, A., and Fanger, M. W. (2000). *Immunology*, 2nd edition. BIOS Scientific Publishers, London.

Maass, W. and Bishop, C. M., editors (1999). *Pulsed Neural Networks*. MIT Press, Cambridge, MA.

Maass, W., Natschlager, T., and Markram, T. (2002). Real-time computing without stable state: A new framework for neural computation based on perturbations. *Neural Computation*, 14(11):2531–2560.

MacDorman, K. F. (2005). Androids as an experimental apparatus: Why is there an uncanny valley and can we exploit it? In *Proceedings of the CogSci 2005 Workshop*, pages 106–118. Cognitive Science Society, Wheat Ridge, CO.

MacKay, D. J. C. (2003). *Information Theory, Inference, and Learning Algorithms*. Cambridge University Press, Cambridge, UK.

Maddox, J. (1992). Forest-fires, sandpiles and the like. *Nature*, 359(6394):359–359.

Maerivoet, S. and De Moor, B. (2005). Cellular automata models of road traffic. *Physics Reports*, 419(1):1–64.

Maes, P. (1989). The dynamics of action selection. In Sridharan, N. S., editor, *Proceedings of the Eleventh International Joint Conference on Artificial Intelligence (IJCAI-89)*, Detroit, pages 51–58. Morgan Kaufmann, San Francisco.

Mahadavi, S. H. and Bentley, P. J. (2003). An evolutionary approach to damage recovery of robot motion with muscles. In Banzhaf, W., Christaller, T., Dittrich, P., Kim, J. T., and Ziegler, J., editors, *Proceedings of the Seventh European Conference on Artificial Life*, pages 248–255. Springer-Verlag, Berlin.

Mahmood, A. and McCluskey, E. J. (1988). Concurrent error detection using watchdog processors–a survey. *IEEE Transactions on Computers*, 37(2):160–174.

Mahner, M. and Bunge, M. (1997). *Foundations of Biophilosophy*. Springer-Verlag, Berlin.

Mahovald, M., editor (1994). *An Analog VLSI System for Stereoscopic Vision*. Kluwer, Boston.

Makowski, L., Casper, D. L. D., and Philips, W. C. (1977). Gap junction structure. II. Analysis of x-ray diffraction data. *Journal of Cellular Biology*, 74:629–645.

Mandelbrot, B. B. (1982). *The Fractal Geometry of Nature*. Freeman, San Francisco.

Mange, D., Sipper, M., Stauffer, A., and Tempesti, G. (2000). Toward robust integrated circuits: The embryonics approach. *Proceedings of the IEEE*, 88(4):516–543.

Marijuán, P. (1996). Gloom in the society of enzymes: On the nature of biological information. *BioSystems*, 38(2-3):163–171.

Mark I Perceptron (1960). *MARK I Perceptron Press Conference Records (CBI 48)*, Charles Babbage Institute, University of Minnesota, Minneapolis. Cornell Aeronautical Laboratory, Inc. Material archived by K. D. Corbitt in March 1991.

Marr, D. (1982). *Vision*. Freeman, New York.

Matarić, M. (1992). Designing emergent behaviors: From local interactions to collective intelligence. In Meyer, J., Roitblat, H. L., and Wilson, S. W., editors, *From Animals to Animats II: Proceedings of the Second International Conference on Simulation of Adaptive Behavior*, pages 432–441. MIT Press, Cambridge, MA.

Matarić, M. and Brooks, R. A. (1990). Learning a distributed map representation based on navigation behaviors. In *Proceedings of the 1990 USA-Japan Symposium on Flexible Automation*. Kyoto, Japan.

Mattila, H. R. and Seeley, T. D. (2007). Genetic diversity in honeybee colonies enhances productivity and fitness. *Science*, 317:317–364.

Mattiussi, C. (1997). An analysis of finite volume, finite element, and finite difference methods using some concepts from algebraic topology. *Journal of Computational Physics*, 133(2):289–309.

Mattiussi, C. (2000). The finite volume, finite difference, and finite elements methods as numerical methods for physical field problems. *Advances in Imaging and Electron Physics*, 113:1–146.

Mattiussi, C. (2002). A reference discretization strategy for the numerical solution of physical field problems. *Advances in Imaging and Electron Physics*, 121:143–279.

Mattiussi, C. (2005). *Evolutionary Synthesis of Analog Networks*. PhD thesis, École Polytechnique Fédérale de Lausanne, Lausanne.

Mattiussi, C. and Floreano, D. (2004). Evolution of analog networks using local string alignment on highly reorganizable genomes. In Zebulum, R. S., Gwaltney, D., Keymeulen, D., Lohn, J., and Stoica, A., editors, *Proceedings of the 2004 NASA/DoD Conference on Evolvable Hardware,* Seattle, June 24-26, 2004, pages 30–37. IEEE Computer Society, Los Alamitos, CA.

Mattiussi, C. and Floreano, D. (2005). Viability evolution. Technical report, Laboratory of Intelligent Systems, École Polytechnique Fédérale de Lausanne (EPFL), Lausanne, Switzerland.

Mattiussi, C. and Floreano, D. (2007). Analog genetic encoding for the evolution of circuits and networks. *IEEE Transaction on Evolutionary Computation*, 11(5):596–607.

Mattiussi, C., Marbach, D., Dürr, P., and Floreano, D. (2008). The age of analog networks. *AI Magazine*. In press.

Mattiussi, C., Waibel, M., and Floreano, D. (2004). Measures of diversity for populations and distances between individuals with highly reorganizable genomes. *Evolutionary Computation*, 12:495–515.

Matzinger, P. (1994). Tolerance, danger, and the extended family. *Annual Review of Immunology*, 12:991–1045.

Matzinger, P. (1998). An innate sense of danger. *Seminars in Immunology*, 10(5):399–415.

Matzinger, P. (2002). The danger model: A renewed sense of self. *Science*, 296(5566):301–305.

Matzinger, P. (2007). Friendly and dangerous signals: Is the tissue in control? *Nature Immunology*, 8(1):11–13.

Maupertuis, P. M. d. (1753). *The Earthly Venus*. Translated by S. Boas, 1966. Johnson Reprint, New York.

May, R. M. (1974). Biological populations with nonoverlapping generations: Stable points, stable cycles, and chaos. *Science*, 186:645–647.

Mayley, G. (1996). Landscapes, learning costs and genetic assimilation. *Evolutionary Computation*, 4(3):213–234.

Maynard-Smith, J. (1964). Group selection and kin selection. *Nature*, 201:1145–1147.

Maynard-Smith, J. and Harper, D. (2003). *Animal Signals*. Oxford University Press, Oxford.

Maynard-Smith, J. and Szathmáry, E. (1995). *The Major Transitions in Evolution*. Oxford University Press, Oxford.

Maynard-Smith, J. and Szathmáry, E. (1999). *The Origins of Life*. Oxford University Press, Oxford.

Mayr, E. (2001). *What Evolution Is*. Basic Books, New York.

McClelland, J., Rumelhart, D. E., and the PDP Research Group (1986). *Parallel Distributed Processing: Explorations in the Microstructure of Cognition*. Volume 2: *Psychological and Biological Models*. MIT Press, Cambridge, MA.

McCulloch, W. and Pitts, W. (1943). A logical calculus of the ideas immanent in nervous activity. *Bulletin of Mathematical Biophysics*, 5:115–133.

McFarland, D. J. and Boesser, T. (1993). *Intelligent Behavior in Animals and Robots*. MIT Press, Cambridge, MA.

McGeer, T. (1990a). Passive bipedal running. *Proceedings of the Royal Society of London. B*, 240:107–134.

McGeer, T. (1990b). Passive dynamic walking. *International Journal of Robotics Research*, 9:62–82.

McHale, G. and Husbands, P. (2004a). Gasnets and other evolvable neural networks applied to bipedal locomotion. In Schaal, S., editor, *From Animals to Animats 8: Proceedings of the Eighth International Conference on Simulation of Adaptive Behaviour (SAB'2004)*, pages 163–172. MIT Press, Cambridge, MA.

McHale, G. and Husbands, P. (2004b). Quadrupedal locomotion: Gasnets, ctrnns and hybrid ctrnn/pnns compared. In Pollack, J., Bedau, M., Husbands, P., Ikegami, T., and R., W., editors, *Proceedings of the Ninth International Conference on Artificial Life, ALIFE IX*, Boston, September 12–15, pages 106–112. MIT Press, Cambridge, MA.

McMullin, B. (2000). John von Neumann and the evolutionary growth of complexity: Looking backward, looking forward. *Artificial Life*, 6(4):347–361.

McVean, G. A. T., Myers, S. R., Hunt, S., Deloukas, P., Bentley, D. R., and Donnelly, P. (2004). The fine-scale structure of recombination rate variation in the human genome. *Science*, 304(5670):581–584.

Mead, C. (1989). *Analog VLSI and Neural Systems*. Addison-Wesley, Reading, MA.

Mead, C. (1990). Neuromorphic electronic systems. *Proceedings of the IEEE*, 78(10):1629–1636.

Meltzoff, A. N. and Moore, M. K. (1977). Imitation of facial and manual gestures by human neonates. *Science*, 198:74–78.

Menzel, P. and D'Aluiso, F., editors (2000). *Robosapiens*. MIT Press, Cambridge, MA.

Merleau-Ponty, M. (1962). *Phenomenology of Perception*. Routledge and Kegan Paul, London. Translated by C. Smith from *Phénoménologie de la perception*, Gallimard, Paris, 1945.

Merzenich, M. M. and Kaas, J. H. (1980). Principles of organization of sensory-perceptual systems in mammals. In *Progress in Psychobiology and Physiological Psychology*, volume 9, pages 1–42. Academic Press, London.

Metzgar, D. and Wills, C. (2000). Evidence for the adaptive evolution of mutation rates. *Cell*, 101:581–584.

Michalewicz, Z. (1996). *Genetic Algorithms + Data Structures = Evolution Programs*, 3rd edition. Springer-Verlag, Berlin.

Michalewicz, Z. and Fogel, D. B. (2004). *How to Solve It: Modern Heuristics*, 2nd edition. Springer-Verlag, Berlin.

Michod, R. E. (1999). *Darwinian Dynamics. Evolutionary Transitions in Fitness and Individuality*. Princeton University Press, Princeton, NJ.

Miconi, T. (2003). When evolving populations is better than coevolving individuals: The blind mice problem. In Gottlob, G. and Walsh, T., editors, *Proceedings of the Eighteenth International Joint Conference on Artificial Intelligence*, pages 647–652. Morgan Kaufmann, San Francisco.

Miglino, O., Lund, H. H., and Nolfi, S. (1996). Evolving mobile robots in simulated and real environments. *Artificial Life*, 2:417–434.

Miller, G. A. (1956). The magical number seven, plus or minus two: Some limits on our capacity for processing information. *Psychological Review*, 63(2):81–97.

Miller, G. F. and Cliff, D. (1994). Protean behavior in dynamic games: Arguments for the co-evolution of pursuit-evasion tactics. In Cliff, D., Husbands, P., Meyer, J., and Wilson, S. W., editors, *From Animals to Animats III: Proceedings of the Third International Conference on Simulation of Adaptive Behavior*. MIT Press, Cambridge, MA.

Miller, G. S. P. (2002). Snake robots for search and rescue. In *Neurotechnology for Biomimetic Robots*, pages 271–284. MIT Press, Cambridge, MA.

Miller, J. and Banzhaf, W. (2003). Evolving the program for a cell: From French flags to Boolean circuits. In Kumar, S. and Bentley, P., editors, *On Growth, Form and Computers*, pages 278–302. Academic Press, London.

Miller, J. F., Job, D., and Vassilev, V. K. (2000). Principles in the evolutionary design of digital circuits – Part I. *Genetic Programming and Evolvable Machines*, 1(1-2):7–35.

Minsky, M. (1961). Steps toward artificial intelligence. *Proceedings of the Institute of Radio Engineers*, 49:8–30.

Minsky, M. (1982). Cellular vacuum. *International Journal of Theoretical Physics*, 21(6-7):537–551.

Mirolli, M. and Parisi, D. (2005). How can we explain the emergence of a language that benefits the hearer but not the speaker? *Connection Science*, 17:307–324.

Mitchell, M. (1996). *An Introduction to Genetic Algorithms*. MIT Press, Cambridge, MA.

Mitchell, M. (1998). Computation in cellular automata: A selected review. In Gramss, T., Bornholdt, S., Gross, M., Mitchell, M., and Pellizzari, T., editors, *Nonstandard Computation: Molecular Computation, Cellular Automata, Evolutionary Algorithms, Quantum Computers*, pages 95–140. Wiley-VCH, Weinheim, Germany.

Mitchell, M., Crutchfield, J. P., and Hraber, P. T. (1994). Evolving cellular automata to perform computations: Mechanisms and impediments. *Physica D: Nonlinear Phenomena*, 75(1-3):361–391.

Mitchell, M., Hraber, P. T., and Crutchfield, J. P. (1993). Revisiting the edge of chaos: Evolving cellular automata to perform computations. Technical report, Santa Fe Institute, Working Paper 93-03-014, Santa Fe, NM.

Mitchison, A. (1993). Will we survive? *Scientific American*, 269(3):136–144.

Mizutani, E. and Dreyfus, S. E. (1998). Totally model-free reinforcement learning by actor-critic Elman networks in non-Markovian domains. In *Proceedings of the IEEE World Congress on Computational Intelligence*. IEEE Press, Piscataway, NJ.

Mochon, S. and McMahon, T. (1980). Ballistic walking: An improved model. *Mathematical Biosciences*, 52:241–260.

Möller, R. (2000). Insect visual homing strategies in a robot with analog processing. *Biological Cybernetics*, 83:231–243.

Möller, R. and Vardy, A. (2006). Local visual homing by matched-filter descent in image distances. *Biological Cybernetics*, 95:413–430.

Mondada, F., Franzi, E., and Ienne, P. (1993). Mobile robot miniaturization: A tool for investigation in control algorithms. In Yoshikawa, T. and Miyazaki, F., editors, *Proceedings of the Third International Symposium on Experimental Robotics, Tokyo*, pages 501–513. Springer-Verlag, Berlin.

Mondada, F., Pettinaro, G., Guignard, A., Kwee, I., Floreano, D., Deneubourg, J.-L., Nolfi, S., Gambardella, L., and Dorigo, M. (2004). Swarm-bot: A new distributed robotic concept. *Autonomous Robots*, 17:193–221.

Montague, P., Dayan, P., and Sejnowski, T. (1996). A framework for mesencephalic dopamine systems based on predictive Hebbian learning. *Journal of Neuroscience*, 16(5):1936–1947.

Montana, D. and Davis, L. (1989). Training feedforward neural networks using genetic algorithms. In *Proceedings of the Eleventh International Joint Conference on Artificial Intelligence*, pages 529–538. Morgan Kaufmann, San Mateo, CA.

Moody, D. B., Zajonc, D. M., and Wilson, I. A. (2005). Anatomy of CD1-lipid antigen complexes. *Nature Reviews Immunology*, 5(5):387–399.

Morgan, C. L. (1896). *Habit and Instinct*. Edward Arnold, London.

Mori, M. (1970). Bukimi no tani [the uncanny valley]. *Energy*, 7(4):33–35. In Japanese. English translation in MacDorman (2005).

Morowitz, H. J. (1959). A model of reproduction. *American Scientist*, 47:261–263.

Morowitz, H. J. (1984). The completeness of molecular biology. *Israel Journal of Medical Sciences*, 20:750–753.

Morrey, J. M., Lambrecht, B., Horchler, A. D., Ritzmann, R. E., and Quinn, R. D. (2003). Highly mobile and robust small quadruped robots. In *IEEE International Conference on Intelligent Robots and Systems*. Las Vegas.

Mountcastle, V. B., Poggio, G. F., and Werner, G. (1963). The relation of thalamic cell response to peripheral stimuli varied over an intensive continuum. *Journal of Neurophysiology*, 26:807–834.

Mukherjee, B., Heberlein, L. T., and Levitt, K. N. (1994). Network intrusion detection. *IEEE Network*, 8(3):26–41.

Murata, A., Fadiga, L., Fogassi, L., Gallese, V., Raos, V., and Rizzolatti, G. (1998). Object representation in the ventral premotor cortex (area F5) of the monkey. *Journal of Neurophysiology*, 78:2226–2230.

Murata, S. and Kurokawa, H. (2007). Self-reconfigurable robots. *IEEE Robotics and Automation Magazine*, March:71–78.

Murata, S., Yoshida, E., Kamimura, A., Kurokawa, H., Tomita, K., and Kokaji, S. (2002). M-TRAN: Self-reconfigurable modular robotic system. *IEEE/ASME Transactions on Mechatronics*, 7:431–441.

Murphy, M. and Sitti, M. (2007). Waalbot: An agile small-scale wall climbing robot utilizing pressure sensitive adhesives. *IEEE/ASME Transactions on Mechatronics*, 12(3):330–338.

Mytilinaios, E., Desnoyer, M., Marcus, D., and Lipson, H. (2004). Designed and evolved blueprints for physical self-replicating machines. In Pollack, J., Bedeau, M., Husbands, P., Ikegami, T., and Watson, R. A., editors, *Artificial Life IX. Proceedings of the Ninth International Conference on the Simulation and Synthesis of Living Systems*. MIT Press, Cambridge, MA.

Nachenberg, C. (1997). Computer virus-antivirus coevolution. *Communications of the ACM*, 40(1):46–51.

Nagel, K. and Schreckenberg, M. (1992). A cellular automaton model for freeway traffic. *Journal de Physique I*, 2:2221–2229.

Nägeli, C. (1845). Wachsthumsgeschichte der Laub- und Lebermoose. *Zeitschrift für wissenschaftliche Botanik*, 1:138–210.

Nalbach, G. and Hengstenberg, R. (1994). The halteres of the blowfly *Calliphora*. Three-dimensional organization of compensatory reactions to real and simulated rotations. *Journal of Comparative Physiology A*, 175:695–708.

Nam, D., Seo, Y. D., Park, L. J., Park, C. H., and Kim, B. (2001). Parameter optimization of an on-chip voltage reference circuit using evolutionary programming. *IEEE Transactions on Evolutionary Computation*, 5(4):414–421.

Nayfeh, B. A. (1993). Cellular automata for solving mazes. *Dr. Dobb's Journal*, 18(2):32–38.

Neal, R. M. (1996). *Bayesian Learning for Neural Networks*. volume 118 of *Lecture Notes in Statistics*. Springer-Verlag, Secaucus, NJ.

Neisser, U. (1967). *Cognitive Psychology*. Appleton-Century-Crofts, New York.

Neisser, U. (1976). *Cognition and Reality. Principles and Implications of Cognitive Psychology*. Freeman, San Francisco.

Newell, A. and Simon, H. (1972). *Human Problem Solving*. Prentice-Hall, Englewood Cliffs, NJ.

Nijhout, H. F. (1997). Pattern formation in biological systems. In Nijhout, H. F., Nadel, L., and Stein, D., editors, *Pattern Formation in the Physical and Biological Sciences*, volume 5 of *Studies in the Sciences of Complexity*, pages 269–297. Addison-Wesley, Reading, MA.

Niv, Y., Joel, D., Meilijson, I., and Ruppin, E. (2002). Evolution of reinforcement learning in uncertain environments: A simple explanation for complex foraging behaviors. *Adaptive Behavior*, 10(1):5–24.

Nobili, R. and Pesavento, U. (1996). Generalised von Neumann's automata I: A revisitation. In Besussi, E. and Cecchini, A., editors, *Artificial Worlds and Urban Studies*. DAEST, Venezia, Italy.

Noë, A. (2004). *Action in Perception*. MIT Press, Cambridge, MA.

Nolfi, S. (1999). How learning and evolution interact: The case of a learning task which differs from the evolutionary task. *Adaptive Behavior*, 7(2):231–236.

Nolfi, S., Elman, J. L., and Parisi, D. (1994a). Learning and evolution in neural networks. *Adaptive Behavior*, 3:5–28.

Nolfi, S. and Floreano, D. (1999a). Co-evolving predator and prey robots: Do "arms races" arise in artificial evolution? *Artificial Life*, 4:311–335.

Nolfi, S. and Floreano, D. (1999b). Learning and evolution. *Autonomous Robots*, 7(1):89–113.

Nolfi, S. and Floreano, D. (2000). *Evolutionary Robotics: Biology, Intelligence, and Technology of Self-Organizing Machines*. MIT Press, Cambridge, MA.

Nolfi, S., Miglino, O., and Parisi, D. (1994b). Phenotypic plasticity in evolving neural networks. In Gaussier, D. and Nicoud, J.-D., editors, *From Perception to Action: Proceedings of the International Conference*, pages 146–157. IEEE Computer Society Press, Los Alamitos, CA.

Nolfi, S. and Parisi, D. (1996). Learning to adapt to changing environments in evolving neural networks. *Adaptive Behavior*, 5(1):75–98.

Nouyan, S. and Dorigo, M. (2006). Chain based path formation in swarms of robots. In Dorigo, M., editor, *ANTS 2006*. Springer-Verlag, Heidelberg, Germany.

Nowak, M. and Sigmund, K. (1998). Evolution of indirect reciprocity by image scoring. *Nature*, 393:573–577.

Nowak, M. and Sigmund, K. (2005). Evolution of indirect reciprocity. *Nature*, 437:1291–1298.

Nowak, M. A. (2006). *Evolutionary Dynamics: Explorng the Equations of Life*. Harvard University Press, Cambridge, MA.

Nowak, M. A. and McMichael, A. J. (1995). How HIV defeats the immune system. *Scientific American*, 273(2):58–65.

Nüsslein-Volhard, C. (2006). *Coming to Life: How Genes Drive Development*. Kales Press, San Diego.

Ochoa, G. (1998). On genetic algorithms and Lindenmayer systems. In Eiben, A. E., Bäck, T., Schoenauer, M., and Schwefel, H.-P., editors, *Proceedings of Parallel Problem Solving from Nature – PPSN V*, volume 1498 of *Lecture Notes in Computer Science*, pages 335–344. Springer-Verlag, Berlin.

Ohno, S. (1970). *Evolution by Gene Duplication*. Springer-Verlag, Berlin.

Oja, E. (1982). A simplified neuron model as a principal component analyzer. *Journal of Mathematical Biology*, 15:267–273.

Oja, E. (1989). Neural networks, principal components, and subspaces. *International Journal of Neural Systems*, 1:61–68.

O'Keefe, J. (1991). The hippocampal cognitive map and navigational strategies. In Paillard, J., editor, *Brain and Space*, pages 273–295. Oxford University Press, Oxford.

O'Keefe, J. and Nadel, L. (1978). *The Hippocampus as a Cognitive Map*. Clarendon Press, Oxford.

O'Regan, J. K. and Noë, A. (2001). A sensorimotor approach to vision and visual consciousness. *Behavioral and Brain Sciences*, 24 (5):939–973.

Orgel, L. E. and Crick, F. H. C. (1980). Selfish DNA: The ultimate parasite. *Nature*, 284:604–607.

Osborn, H. F. (1896). Ontogenetic and phylogenetic variation. *Science*, 4:786–789.

Oztop, E. and Arbib, M. A. (2002). Schema design and implementation of the grasp-related mirror neuron system. *Biological Cybernetics*, 87(2):116–140.

Oztop, E., Kawato, M., and Arbib, M. (2006). Mirror neurons and imitation: A computationally guided review. *Neural Networks*, 19(3):254–271.

Packard, N. H. (1988). Adaptation toward the edge of chaos. In Kelso, J. A. S., Mandell, A. J., and Shlesinger, M. F., editors, *Dynamic Patterns in Complex Systems*, pages 293–301. World Scientific, Singapore.

Palmer, E. (2003). Negative selection – clearing out the bad apples from the T-cell repertoire. *Nature Reviews Immunology*, 3(5):383–391.

Panait, L. and Luke, S. (2005). Cooperative multi-agent learning: The state of the art. *Autonomous Agents and Multi-Agent Systems*, 11:387–434.

Pardoll, D. (1998). Cancer vaccines. *Nature Medicine*, 4(5):525–531.

Pardoll, D. (2003). Does the immune system see tumors as foreign or self? *Annual Review of Immunology*, 21:807–839.

Parisi, D., Cecconi, F., and Nolfi, S. (1990). Econets: Neural networks that learn in an environment. *Network*, 1:149–168.

Park, Y.-L., Chau, K., Black, R. J., and Cutkosky, M. R. (2007). Force sensing smart robot fingers using embedded fiber Bragg grating sensors and shape deposition manufacturing. In *Proceedings of the IEEE International Conference on Robotics and Automation*. Rome.

Parker, C. A. C., Zhang, H., and Kube, C. R. (2003). Blind bulldozing: Multiple robot nest construction. In *Proceedings of the IEEE/RSJ International Conference on Robotics and Intelligent Systems (IROS), Las Vegas*. IEEE Press, Piscataway, NJ.

Parker, D. B. (1985). Learning logic. Technical report 47, Center for Computational Research in Economics and Management Science. MIT Press, Cambridge, MA.

Parker, G. H., editor (1919). *The Elementary Nervous System*. Lippincott, Philadelphia.

Parker, L. E. (2000). Lifelong adaptation in heterogeneous multi-robot teams: Response to continual variation in individual robot performance. *Autonomous Robots*, 8:239–267.

Partridge, B. L. (1982). The structure and function of fish schools. *Scientific American*, 246:90–99.

Payton, D., Estkowski, R., and Howard, M. (2005). Pheromone robotics and the logic of virtual pheromones. In Şahin, E. and Spears, W. M., editors, *Proceedings of the Swarm Robotics Workshop*. Springer-Verlag, Heidelberg, Germany.

Payton, D. W. (1986). An architecture for reflexive autonomous vehicle control. In *Proceedings of the IEEE International Conference on Robotics and Automation*. San Francisco.

Pease, B. (1991). The story of the P2 - the first successful solid-state operational amplifier with picoampere input currents. In Williams, J., editor, *Analog Circuit Design: Art, Science, and Personalities*, pages 67–78. Butterworth-Heinemann, Boston.

Peitgen, H.-O., Jürgens, H., and Saupe, D. (1992). *Fractals for the Classroom.* Part 2: *Complex Systems and Mandelbrot Set.* Springer-Verlag, New York.

Peitgen, H.-O. and Saupe, D., editors (1988). *The Science of Fractal Images.* Springer-Verlag, New York.

Penrose, L. S. (1959). Self-reproducing machines. *Scientific American*, 200 (6):105–113.

Penrose, L. S. (1962). On living matter and self-replication. In Good, I. J., Mayne, A. J., and Maynard-Smith, J., editors, *The Scientist Speculates: An Anthology of Partly-Baked Ideas.* Heinemann, London.

Percus, J. K., Percus, O. E., and Perelson, A. S. (1993). Predicting the size of the T-cell receptor and antibody combining region from consideration of efficient self-nonself discrimination. *Proceedings of the National Academy of Sciences USA*, 90(5):1691–1695.

Perelson, A. S. and Oster, G. F. (1979). Theoretical studies of clonal selection: Minimal antibody repertoire size and reliability of self-non-self discriminationt. *Journal of Theoretical Biology*, 81(4):645–670.

Perez-Uribe, A. and Sanchez, E. (1996). FPGA implementation of an adaptable-size neural network. In von der Malsburg, C., von Seelen, W., Vorbrüggen, J. C., and Sendhoff, B., editors, *Proceedings of the International Conference on Artificial Neural Networks*, pages 383–388. Springer-Verlag, Berlin.

Perrett, D. I., Harries, M. H., Mistlin, A. J., Hietanen, J. K., Benson, P. J., Bevan, R., Thomas, S., Oram, M. W., Ortega, J., and Brierley, K. (1990). Social signals analyzed at the single cell level: Someone is looking at me, something touched me, something moved! *International Journal of Comparative Psychology*, 4:25–55.

Pesavento, U. (1995). An implementation of von Neumann's self-reproducing machine. *Artificial Life*, 2(4):337–354.

Petrov, D. A., Lozovskaya, E. R., and Hartl, D. L. (1996). High intrinsic rates of DNA loss in *Drosophila. Nature*, 384:346–349.

Pfeifer, R. and Bongard, J. C. (2007). *How the body shapes the way we think.* MIT Press, Cambridge, MA.

Pfeifer, R. and Scheier, C. (1999). *Understanding Intelligence.* MIT Press, Cambridge, MA.

Piaget, J. (1953). *The Origins of Intelligence.* Routledge, New York.

Piel, J., editor (1993). *Life, Death, and the Immune System.* 269 (3), *Scientific American.*

Pine, J. (1980). Recording action potentials from cultured neurons with extracellular microcircuit electrodes. *Journal of Neuroscience Methods*, 2:19–31.

Plaut, D. C. and Shallice, T. (1993). Deep dyslexia: A case study of connectionist neuropsychology. *Cognitive Neuropsychology*, 10:377–500.

Plotkin, H. C. (1988). Learning and evolution. In Plotkin, H. C., editor, *The Role of Behavior in Evolution*, pages 133–164. MIT Press, Cambridge, MA.

Potter, D. (1973). *Computational Physics*. Wiley, London.

Potter, S. M. and DeMarse, T. B. (2001). A new approach to neural cell culture for long-term studies. *Journal of Neuroscience Methods*, 110:17–24.

Poundstone, W. (1985). *The Recursive Universe*. Oxford University Press, Oxford.

Pratt, G. A. and Williamson, M. M. (1995). Series elastic actuators. In *Proceedings of the IEEE/RSJ International Conference on Intelligent Robots and Systems*, pages 399–406. Pittsburgh.

Prusinkiewicz, P. (1986). Graphical applications of L-systems. In *Proceedings on Graphics Interface '86/Vision Interface '86*, pages 247–253. Canadian Information Processing Society, Toronto.

Prusinkiewicz, P. and Lindenmayer, A. (1990). *The Algorithmic Beauty of Plants*. Springer-Verlag, Berlin.

Psaltis, D., Brady, D., Gu, X.-G., and Lin, S. (1990). Holography in artificial neural networks. *Nature*, 343:325–330.

Purves, D. (1994). *Neural Activity in the Growth of the Brain*. Cambridge University Press, Cambridge, UK.

Quartz, S. and Sejnowski, T. J. (1997). The neural basis of cognitive development: A constructivist manifesto. *Behavioral and Brain Science*, 4:537–555.

Quinn, M., Smith, L., Mayley, G., and Husbands, P. (2002a). Evolving teamwork and role-allocation with real robots. In *Artificial Life VIII: Proceedings of the Eighth International Conference*, pages 302–311. MIT Press, Cambridge, MA.

Quinn, M., Smith, L., Mayley, G., and Husbands, P. (2003). Evolving controllers for a homogeneous system of physical robots: Structured cooperation with minimal sensors. *Philosophical Transactions of the Royal Society of London A: Mathematical, Physical and Engineering Sciences*, 361:2321–2344.

Quinn, R. D., Offi, J. T., Kingsley, D. A., and Ritzmann, R. E. (2002b). Improved mobility through abstracted biological principles. In *Proceedings of the International Conference on Intelligent Robots and Systems*. Lausanne, Switzerland.

Radcliffe, N. J. (1991). Forma analysis and random respectful recombination. In Belew, R. K. and Booker, L. B., editors, *Proceedings of the Fourth International Conference on Genetic Algorithms*. Morgan Kaufmann, San Mateo, CA.

Ramón y Cajal, S., editor (1909, 1911). *Histologie du système nerveux de l'homme et des vertébrés*. 2 volumes. Maloine, Paris.

Ray, T. S. (1992). An approach to the synthesis of life. In Langton, C., Farmer, J., Rasmussen, S., and Taylor, C., editors, *Artificial Life II: Proceedings Volume of Santa Fe Conference*, volume 11 of *Series of the Santa Fe Institute Studies in the Sciences of Complexities*. Addison Wesley, Redwood City, CA.

Rechenberg, I. (1965). *Cybernetic Solution Path of an Experimental Problem*. Royal Aircraft Establishment, Ministry of Aviation, Farnborough Hants, UK.

Rechenberg, I. (1973). *Evolutionstrategie: Optimierung technischer Systeme nach Prinzipien der biologischen Evolution*. Friedrich Fromann Verlag, Stuttgart.

Rédei, M., editor (2005). *John von Neumann : Selected letters*. American Mathematical Society, Providence, RI.

Redish, A. D. (1999). *Beyond the Cognitive Map. From Place Cells to Episodic Memory*. MIT Press, Cambridge, MA.

Reed, R. D. and Marks, R. J., II. (1999). *Neural Smithing: Supervised Learning in Feedforward Artificial Neural Networks*. MIT Press, Cambridge, MA.

Reil, T. and Husbands, P. (2002). Evolution of central pattern generators for bipedal walking in real-time physics environments. *IEEE Transactions on Evolutionary Computation*, 6(2):10–21.

Reis, A. H., Miguel, A. F., and Aydin, V. (2004). Constructal theory of flow architecture of the lungs. *Medical Physics*, 31(5):1135–1140.

Reisberg, D. (1999). Learning. In Wilson, R. A. and Keil, F. C., editors, *The MIT Encyclopedia of the Cognitive Sciences*, pages 460–461. MIT Press, Cambridge, MA.

Reynolds, C. (1987). Flocks, herds, and schools: A distributed behavioral model. *Computer Graphics*, 21(4):25–34.

Reynolds, C. W. (1994). Competition, coevolution and the game of tag. In Brooks, R. and Maes, P., editors, *Proceedings of the Fourth Workshop on Artificial Life*, pages 59–69. MIT Press, Cambridge, MA.

Richards, F. C., Meyer, T. P., and Packard, N. H. (1990). Extracting cellular automaton rules directly from experimental data. *Physica D: Nonlinear Phenomena*, 45(1-3):189–202.

Richards, M., Whitley, D., Beveridge, J., Mytkowicz, T., Nguyen, D., and Rome, D. (2005). Evolving cooperative strategies for UAV teams. In *Proceedings of the 2005 Conference on Genetic and Evolutionary Computation*, pages 1721–1728.

Ridley, M. (2004). *Evolution*, 3rd edition. Blackwell Publishing, Oxford.

Rieke, F., Warland, D., van Steveninck, R., and Bialek, W. (1997). *Spikes. Exploring the neural code*. MIT Press, Cambridge, MA.

Rizzolatti, G. and Arbib, M. A. (1998). Language within our grasp. *Trends in Neurosciences*, 21(5):188–194.

Rizzolatti, G., Camarda, R., Fogassi, L., Gentilucci, M., Luppino, G., and Matelli, M. (1988). Functional organization of inferior area 6 in the macaque monkey. II. Area F5 and the control of distal movements. *Experimental Brain Research*, 71:491–507.

Rizzolatti, G., Fadiga, L., Gallese, V., and Fogassi, L. (1996). Premotor cortex and the recognition of motor actions. *Cognitive Brain Research*, 3:131–141.

Roach, S. (1995). Signal conditioning in oscilloscopes and the spirit of invention. In Williams, J., editor, *The Art and Science of Analog Circuit Design*, pages 65–84. Butterworth-Heinemann, Boston.

Robinson, A. and Spector, L. (2002). Using genetic programming with multiple data types and automatic modularization to evolve decentralized and coordinated navigation in multi-agent systems. In *Late-Breaking Papers of the Genetic and Evolutionary Computation Conference (GECCO-2002)*. The International Society for Genetic and Evolutionary Computation, New York.

Robinson, G. E. (1992). Regulation of division of labor in insect societies. *Annual Review of Entomology*, 37:637–65.

Roggen, D., Federici, D., and Floreano, D. (2007). Evolutionary morphogenesis for multi-cellular systems. *Genetic Programming and Evolvable Machines*, 8(1):61–96.

Roggen, D., Floreano, D., and Mattiussi, C. (2003a). A morphogenetic evolutionary system: Phylogenesis of the POEtic circuit. In *Proceedings of the Fifth International Conference on Evolvable Systems (ICES'2003)*, pages 153–164.

Roggen, D., Hofmann, S., Thoma, Y., and Floreano, D. (2003b). Hardware spiking neural network with run-time reconfigurable conectivity in an autonomous robot. In Lohn, J., Zebulum, R., Steincamp, J., Keymeulen, D., Stoica, A., and Ferguson, M. I., editors, *NASA/DoD Conference on Evolvable Hardware*, pages 189–198. IEEE Computer Society Press, Los Alamitos, CA.

Rosenblatt, F. (1962). *Principles of Neurodynamics*. Spartan Books, New York.

Rosin, C. and Belew, R. (1997). New methods for competitive co-evolution. *Evolutionary Computation*, 5(1):1–29.

Rumelhart, D. E., Hinton, G. E., and Williams, R. J. (1986a). Learning representations by back-propagation of errors. *Nature*, 323:533–536.

Rumelhart, D. E., McClelland, J., and the PDP Research Group (1986b). *Parallel Distributed Processing: Explorations in the Microstructure of Cognition*. Volume 1: *Foundations*. MIT Press, Cambridge, MA.

Russo, L. (2004). *The Forgotten Revolution*. Springer-Verlag, Berlin.

Şahin, E. (2004). Swarm robotics: From sources of inspiration to domains of application. In Şahin, E. and Spears, W. M., editors, *Proceedings of the Swarm Robotics Workshop*, pages 10–20. Springer-Verlag, Heidelberg, Germany.

Sakoda, J. M. (1971). The checkerboard model of social interaction. *Journal of Mathematical Sociology*, 1(2):119–132.

Salzberg, C., Antony, A., and Sayama, H. (2004). Evolutionary dynamics of cellular automata-based self-replicators in hostile environments. *Biosystems*, 78(1-3):119–134.

Samad, T. and Harp, S. A. (1989). Self-organization with partial data. *Network*, 3:205–212.

Sanger, T. D. (1989). Optimal unsupervised learning in a single-layer feedforward neural network. *Neural Networks*, 2:459–473.

Santos, D., Kim, S., Spenko, M., Parness, A., and Cutkosky, M. R. (2007). Directional adhesive structures for controlled climbing on smooth vertical surfaces. In *Proceedings of the IEEE International Conference on Robotics and Automation*. Rome.

Saranli, U., Buehler, M., and Koditschek, D. E. (2001). RHex - a simple and highly mobile hexapod robot. *International Journal of Robotics Research*, 20(7):616–631.

Sarpeshkar, R. (1998). Analog versus digital: Extrapolating from electronics to neurobiology. *Neural Computation*, 10(7):1601–1638.

Sarpeshkar, R. (2006). Brain power. *IEEE Spectrum*, 43(5):24–29.

Sasaki, T. and Tokoro, M. (1997). Adaptation toward changing environments: Why Darwinian in nature? In Husbands, P. and Harvey, I., editors, *Proceedings of the Fourth European Conference on Artificial Life*. MIT Press, Cambridge, MA.

Scalettar, R. and Zee, A. (1988). Emergence of grandmother memory in feed forward networks: Learning with noise and forgetfulness. In Waltz, D. and Feldman, J. A., editors, *Connectionist Models and Their Implications: Readings from Cognitive Science*. Ablex, Norwood, MA.

Schaal, S. (1999). Is imitation learning the route to humanoid robots? *Trends in Cognitive Sciences*, 3:233–242.

Schaal, S., Ijspeert, A., and Billard, A. (2003). Computational approaches to motor learning by imitation. *Philosophical Transaction of the Royal Society of London B*, 358:537–547.

Schaffer, J. D., Whitley, D., and Eshelman, L. J. (1992). Combinations of genetic algorithms and neural networks: A survey of the state of the art. In Whitley, D. and Schaffer, J. D., editors, *Proceedings of an International Workshop on the Combinations of Genetic Algorithms and Neural Networks (COGANN-92)*. IEEE Press, Piscataway, NJ.

Schmajuk, N. A. and Blair, H. T. (1993). Place learning and the dynamics of spatial navigation: A neural network approach. *Adaptive Behavior*, 1:353–385.

Schmid-Hempel, P. (2003). Variation in immune defence as a question of evolutionary ecology. *Proceedings of the Royal Society of London B*, 270(1513):357–366.

Schonfisch, B. and de Roos, A. (1999). Synchronous and asynchronous updating in cellular automata. *Biosystems*, 51(3):123–143.

Schoonderwoerd, R., Holland, O., Bruten, J., and Rothkrantz, L. (1996). Ant-based load balancing in telecommunication networks. *Adaptive Behavior*, 5:169–207.

Schraudolph, N. N. and Belew, R. K. (1992). Dynamic parameter encoding for genetic algorithms. *Machine Learning*, 9:9–21.

Schultz, W. (1998). Predictive reward signal of dopamine neurons. *Journal of Neurophysiology*, 80:1–27.

Schultz, W., Dayan, P., and Montague, P. R. (1997). A neural substrate of prediction and reward. *Science*, 275(5306):1593–1599.

Searcy, W. A. and Nowicki, S., editors (2005). *The Evolution of Animal Communication: Reliability and Deception in Signaling Systems*. Princeton University Press, Princeton, NJ.

Seeley, T. D. (1995). *The Wisdom of the Hive*. Harvard University Press, Cambridge, MA.

Segel, L. A. and Cohen, I. R. (2001). *Design Principles for the Immune System and Other Distributed Autonomous Systems*. Santa Fe Institute Studies in the Sciences of Complexity. Oxford University Press, Oxford.

Seidenberg, M. S. and McClelland, J. L. (1989). A distributed, developmental model of word recognition and naming. *Psychological Review*, 96:523–568.

Sejnowski, T. J. and Rosenberg, C. R. (1987). Parallel networks that learn to pronounce English text. *Complex Systems*, 1:145–168.

Sekanina, L. (2004). *Evolvable components: From theory to hardware implementations*. Springer-Verlag, Berlin.

Seong, S.-Y. and Matzinger, P. (2004). Hydrophobicity: An ancient damage-associated molecular pattern that initiates innate immune responses. *Nature Reviews Immunology*, 4(6):469–478.

Shahaf, G. and Marom, S. (2001). Learning in networks of cortical neurons. *Journal of Neuroscience*, 21:8782–8788.

Shannon, C. E. (1949). Communication in the presence of noise. *Proceedings of the Institute of Radio Engineers*, 37(1):10–21.

Shapiro, J. (2005). A 21st century view of evolution: Genome system architecture, repetitive DNA, and natural genetic engineering. *Gene*, 345(1):91–100.

Shashkov, M. and Steinberg, S. (1995). Support-operator finite-difference algorithms for general elliptic problems. *Journal of Computational Physics*, 118(1):131–151.

Shepherd, G. M., editor (1990). *The Synaptic Organization of the Brain*. Oxford University Press, Oxford.

Sherrington, C. S., editor (1906). *Integrative Action of the Nervous System*. Yale University Press, New Haven, CT.

Shi, Y. and Eberhart, R. C. (1998). A modified particle swarm optimizer. In *Proceedings of the IEEE Congress on Evolutionary Computation*, pages 69–73. IEEE Press, Piscataway, NJ.

Siddiqi, A. and Lucas, S. (1998). A comparison of matrix rewriting versus direct encoding for evolving neural networks. In *Proceedings of the 1998 IEEE World Congress on Computational Intelligence*, pages 392–397.

Simon, H. A. (1996). *The Science of the Artificial*, 3rd edition. MIT Press, Cambridge, MA.

Sims, K. (1994). Evolving 3D morphology and behavior by competition. In Brooks, R. and Maes, P., editors, *Proceedings of Artificial Life IV*, pages 28–39. MIT Press, Cambridge, MA.

Singer, W. (1987). Activity-dependant self-organisation of synaptic connections as a substrate of learning. In Changeux, J. P. and Konishi, M., editors, *The Neural and Molecular Bases of Learning*. Wiley, London.

Singer, W. (1990). Search for coherence: A basic principle of cortical self-organization. *Concepts in Neuroscience*, 1:1–26.

Singer, W. and Gray, C. M. (1995). Visual feature integration and the temporal correlation hypothesis. *Annual Review of Neuroscience*, 18:555–586.

Sipper, M. (1996). Co-evolving non-uniform cellular automata to perform computations. *Physica D: Nonlinear Phenomena*, 92(3-4):193–208.

Sipper, M. (1998). Fifty years of research on self-replication: An overview. *Artificial Life*, 4(3):237–257.

Sipper, M., Tomassini, M., and Capcarrere, M. S. (1997). Designing cellular automata using a parallel evolutionary algorithm. *Nuclear Instruments and Methods in Physics Research Section A: Accelerators, Spectrometers, Detectors and Associated Equipment*, 389(1-2):278–283.

Sivia, D. S. (2006). *Data Analysis: A Bayesian Tutorial*, 2nd edition. Oxford University Press, Oxford.

Skinner, B. F. (1938). *The Behavior of Organisms*. Appleton-Century-Crofts, New York.

Smallwood, P. D. (1996). An introduction to risk sensitivity: The use of Jensen's inequality to clarify evolutionary arguments of adaptation and constraint. *American Zoologist*, 36:392–401.

Smith, D. J., Forrest, S., Ackley, D. H., and Perelson, A. S. (1999). Variable efficacy of repeated annual influenza vaccination. *Proceedings of the National Academy of Sciences USA*, 96(24):14001–14006.

Smith, T. M. C., Husbands, P., and O'Shea, M. (2001). *Not* measuring evolvability: Initial investigation of an evolutionary robotics search space. In Angeline, P., Michaelewicz, M., Schonauer, G., Yao, X., and Zalzala, Z., editors, *Proceedings of the 1999 Congress on Evolutionary Computation*. IEEE Press, Piscataway, NJ.

Sokal, A. D. (1996). Transgressing the boundaries: Towards a transformative hermeneutics of quantum gravity. *Social Text*, 46/47:217–252.

Sole, R. V. and Goodwin, B. (2000). *Signs of Life: How Complexity Pervades Biology*. Basic Books, New York.

Soltoggio, A., Duerr, P., Mattiussi, C., and Floreano, D. (2007). Evolving neuro-modulatory topologies for reinforcement learning-like problems. In Angeline, P., Michaelewicz, M., Schonauer, G., Yao, X., and Zalzala, Z., editors, *Proceedings of the 2007 Congress on Evolutionary Computation*. IEEE Press, Piscataway, NJ.

Sompayrac, L. (2003). *How the Immune System Works*, 2nd edition. Blackwell, Malden, MA.

Song, S. and Abbott, L. F. (2000). Temporally asymmetric Hebbian learning, spike timing and neuronal response variability. In Kearns, M. S., Solla, S. A., and Cohn, D. A., editors, *Advances in Neural Information Processing Systems 11*. MIT Press, Cambridge, MA.

Srinivasan, M. (1994). An image-interpolation technique for the computation of optic flow and egomotion. *Biological Cybernetics*, 71:401–416.

Stanley, K. and Miikkulainen, R. (2002). Evolving neural networks through augmenting topologies. *Evolutionary Computation*, 10(2):99–127.

Stanley, K. and Miikkulainen, R. (2003). A taxonomy for artificial embryogeny. *Artificial Life*, 9(2):93–130.

Stanton, P. K. and Sejnowski, T. J. (1989). Associative long-term depression in the hippocampus induced by Hebbian covariance. *Nature*, 339:215–218.

Steinberg, S. (2004). A discrete calculus with applications of high-order discretizations to boundary-value problems. *Computational Methods in Applied Mathematics*, 4(2):228–261.

Stent, G. (1973). A physiological mechanism for Hebb's postulate of learning. *Proceedings of the National Academy of Sciences USA*, 70:997–1001.

Stoica, A. (1999). Toward evolvable hardware chips: Experiments with a programmable transistor array. In *Proceedings of the Seventh International Conference on Microelectronics for Neural, Fuzzy and Bio-Inspired Systems, MicroNeuro99*, Granada, Spain, pages 156–162.

Stoica, A., Arslan, T., Keymeulen, D., Duong, V., Zebulum, R., Ferguson, I., and Daud, T. (2004). Evolutionary recovery from radiation induced faults on reconfigurable devices. In *Proceedings 2004 IEEE Aerospace Conference*, March 6-13, 2004, pages 2449–2457.

Stoica, A., Keymeulen, D., and Zebulum, R. (2001a). Evolvable hardware solutions for extreme temperature electronics. In *Proceedings of the Third NASA/DoD Workshop on Evolvable Hardware, 2001*, pages 93–97.

Stoica, A., Zebulum, R., Keymeulen, D., Tawel, R., Daud, T., , and Thakoor, A. (2001b). Reconfigurable VLSI architectures for evolvable hardware: From experimental field programmable transistor arrays to evolution-oriented chips. *IEEE Transactions on VLSI Systems*, 9(1):227–232.

Stoy, K. (2006). Using cellular automata and gradients to control self-reconfiguration. *Robotics and Autonomous Systems*, 54:135–141.

Stuetzle, T. and Dorigo, M., editors (2004). *Ant Colony Optimization*. MIT Press, Cambridge, MA.

Sugano, S. and Kato, I. (1987). WABOT-2: Autonomous robot with dexterous finger-arm. In *Proceedings of the IEEE International Conference on Robotics and Automation*, Raleigh, NC, pages 90–97. IEEE Press, Piscataway, NJ.

Suri, R. E. (2002). TD models of reward predictive responses in dopamine neurons. *Neural Networks*, 15:523–534.

Sutton, R. S. (1988). Learning to predict by the method of temporal difference. *Machine Learning*, 3:9–44.

Sutton, R. S. and Barto, A. G. (1998). *Reinforcement Learning. An Introduction*. MIT Press, Cambridge, MA.

Suzuki, M., Floreano, D., and Di Paolo, E. A. (2005). The contribution of active body movement to visual development in evolutionary robots. *Neural Networks*, 18(5/6):656–665.

Svensson, E., Raberg, L., and Koch, C .and Hasselquist, D. (1998). Energetic stress, immunosuppression and the costs of an antibody response. *Functional Ecology*, 12(6):912–919.

Swanson, L. (2003). *Brain Architecture. Understanding the Basic Plan*. Oxford University Press, New York.

Syswerda, G. (1989). Uniform crossover in genetic algorithms. In *Proceedings of the Third International Conference on Genetic Algorithms*, pages 2–9. Morgan Kaufmann, San Mateo, CA.

Szamado, S. and Szathmary, E. (2006). Selective scenarios for the emergence of natural language. *Trends in Ecology and Evolution*, 21:555–561.

Takagi, H. (2001). Interactive evolutionary computation: Fusion of the capacities of EC optimization and human evaluation. *Proceedings of the IEEE*, 89:1275–1296.

Talia, D. (2000). Cellular processing tools for high-performance simulation. *Computer*, 33(9):44–52.

Tammero, L. and Dickinson, M. (2002). The influence of visual landscape on the free flight behavior of the fruit fly *Drosophila melanogaster*. *Journal of Experimental Biology*, 205:327–343.

Tan, K. C., Lee, T. H., and Khor, E. F. (2002). Evolutionary algorithms for multi-objective optimization: Performance assessments and comparisons. *Artificial Intelligence Review*, 17(4):251–290.

Tapus, A. and Siegwart, R. (2006). A cognitive modeling of space using fingerprints of places for mobile robot navigation. In *Proceedings of the IEEE International Conference on Robotics and Automation*.

Tarapore, D., Floreano, D., and Keller, L. (2006). Influence of the level of polyandry and genetic architecture on division of labour. In Rocha, L. M., Yaeger, L. S., Bedeau, M. A., Floreano, D., Goldstone, R. L., and Vespignani, A., editors, *The Tenth International Conference on the Simulation and Synthesis of Living Systems*, pages 358–364. MIT Press, Cambridge, MA.

Taube, J. S., Muller, R. U., and Ranck, J. B. J. (1990). Head-direction cells recorded from the postsubiculum in freely moving rats. I. Description and quantitative analysis. *Journal of Neuroscience*, 10:420–435.

Taubes, G. A. (1995). The rise and fall of thinking machines. *Inc. Magazine*, September.

Taylor, C. P. and Dudek, F. E. (1984). Excitation of hippocampal pyramidal cells by electrical field effect. *Trends in Neuroscience*, 11:126–142.

Teixeira, F. and Chew, W. (1999). Lattice electromagnetic theory from a topological viewpoint. *Journal of Mathematical Physics*, 40(1):169–187.

Tellez, R., Angulo, C., and Pardo, D. (2006). Evolving the walking behaviour of a 12 DOF quadruped using a distributed neural architecture. In *Second International Workshop on Biologically Inspired Approaches to Advanced Information Technology (BioADIT'2006), LNCS vol. 3853*, pages 5–19. Springer-Verlag, Berlin.

Tempesti, G., Roggen, D., Sanchez, E., and Thoma, Y. (2002). A POEtic architecture for bio-inspired hardware. In Standish, R. K., Bedau, M. A., and Abbas, H. A., editors, *Proceedings of the Eighth International Conference on Artificial Life, ALIFE VIII*, pages 111–115. MIT Press, Cambridge, MA.

Tempesti, G., Roggen, D., Sanchez, E., Thoma, Y., Canham, R., and Tyrrell, A. M. (2003). Ontogenetic development and fault tolerance in the poetic tissue. In *Proceedings of the Fifth International Conference on Evolvable Systems: From Biology to Hardware*, pages 335–363. Springer-Verlag, Berlin.

Tenaillon, O., Taddei, F., Radman, M., and Matic, I. (2001). Second-order selection in bacterial evolution: Selection acting on mutation and recombination rates in the course of adaptation. *Research in Microbiology*, 152(1):11–16.

Terada, K., Ohmura, Y., and Kuniyoshi, Y. (2003). Analysis and control of whole body dynamic humanoid motion. Towards experiments on a roll-and-rise motion. In

Proceedings of the IEEE/RSJ International Conference on Robotics and Intelligent Systems (IROS), Las Vegas. IEEE Press, Piscataway, NJ.

Thelen, E. and Smith, L. B. (1994). *A Dnamical Systems Approach to the Development of Cognition and Action*. MIT Press, Cambridge, MA.

Theraulaz, G. and Bonabeau, E. (1995). Coordination in distributed building. *Science*, 269:686–688.

Theraulaz, G., Bonabeau, E., Nicolis, S., Solé, R. V., Fourcassié, V., Blanco, S., Fournier, R., Joly, J.-L., Fernandez, P., Grimal, A., Dalle, P., and Deneubourg, J.-L. (2002). Spatial patterns in ant colonies. *Proceedings of the National Academy of Sciences USA*, 99:9645–9649.

Thompson, A. and Layzell, P. (1999). Analysis of unconventional evolved circuits. *Communications of the ACM*, 42(4):71–79.

Thompson, A. and Layzell, P. (2000). Evolution of robustness in an electronics design. In Miller, J., Thompson, A., Thomson, P., and Fogarty, T., editors, *Proceedings of the Third International Conference on Evolvable Systems (ICES2000): From Biology to Hardware*, volume 1801 of *Lecture Notes in Computer Science*, pages 218–228. Springer-Verlag, Berlin.

Thompson, A., Layzell, P., and Zebulum, R. S. (1999). Explorations in design space: unconventional electronics design through artificial evolution. *IEEE Transactions on Evolutionary Computation*, 3(3):167–196.

Thompson, D. W. (1941). *On Growth and Form*. Cambridge University Press, Cambridge, UK.

Thompson, D. W. (1992). *On Growth and Form*. Cambridge University Press, Cambridge, UK. Abridged edition edited by John Tyler Bonner.

Thompson, E. and Varela, F. J. (2001). Radical embodiment: Neural dynamics and consciousness. *Trends in Cognitive Science*, 5:418–425.

Timmis, J. (2007). Artificial immune systems–today and tomorrow. *Natural Computing*, 6(1):1–18.

Toffoli, T. (1984). Cellular automata as an alternative to (rather than an approximation of) differential equations in modeling physics. *Physica D: Nonlinear Phenomena*, 10(1-2):117–127.

Toffoli, T. (1994). Occam, Turing, von Neumann, Jaynes: How much can you get for how little? In *Proceedings of ACRI'94*, pages 1–9. Springer-Verlag, Berlin.

Toffoli, T. (1999). Programmable matter methods. *Future Generation Computer Systems*, 16(2-3):187–201.

Toffoli, T. and Margolus, N. (1987). *Cellular Automata Machines: A New Environment for Modeling*. MIT Press, Cambridge, MA.

Tolman, E. C. (1948). Cognitive maps in rats and men. *Psychological Review*, 55:189–208.

Tonti, E. (2001). A direct discrete formulation of field laws: The cell method. *CMES - Computer Modeling in Engineering & Sciences*, 2(2):237–258.

Trianni, V., Nolfi, S., and Dorigo, M. (2006). Cooperative hole-avoidance in a Swarm-bot. *Robotics Autonomous Systems*, 54:97–103.

Tuci, E., Quinn, M., and Harvey, I. (2002). An evolutionary ecological approach to the study of learning behavior using a robot-based model. *Adaptive Behavior*, 10(3-4):201–221.

Turing, A. (1950). Computing machinery and intelligence. *Mind*, 59(236):433–460.

Turing, A. (1953). The chemical basis of morphogenesis. *Philosophical Transactions of the Royal Society B*, 237:37–72.

Turrigiano, G. G. and Nelson, S. B. (2004). Homeostatic plasticity in the developing nervous system. *Nature Reviews Neuroscience*, 5:97–107.

Tyrrell, A. M., Sanchez, E., Floreano, D., Tempesti, G., Mange, D., Moreno, J.-M., Rosenberg, J., and Villa, A. E. (2003). Poetic tissue: An integrated architecture for bio-inspired hardware. In *Proceedings of the Fifth International Conference on Evolvable Systems: From Biology to Hardware*, pages 269–294. Springer-Verlag, Berlin.

Ulam, S. M. (1976). *Adventures of a Mathematician*. Scribner, New York.

Urzelai, J. and Floreano, D. (1999). Incremental evolution with minimal resources. In Rückert, U., Mondada, F., and Löffler, A., editors, *First International Khepera Workshop*. Heinz Nixdorf Institute, Paderborn, Germany.

Urzelai, J. and Floreano, D. (2001). Evolution of adaptive synapses: Robots with fast adaptive behavior in new environments. *Evolutionary Computation*, 9:495–524.

Utida, S. (1957). Population fluctuations. In *Proceedings of the Twenty-second CSHL Symposium on Population Studies: Animal Ecology and Demography*. Cold Spring Harbor Laboratory, New York.

Vaario, J., Onitsuka, A., and Shimohara, K. (1997). Formation of neural structures. In Husbands, P. and Harvey, I., editors, *Proceedings of the Fourth European Conference on Artificial Life*, pages 214–223. MIT Press, Cambridge, MA.

van Essen, D. C. and Maunsell, J. H. R. (1983). Hierarchical organization and functional streams in the visual cortex. *Trends in Neuroscience*, 6:370–375.

van Gelder, T. J. (1998). The dynamical hypothesis in cognitive science. *Behavioral and Brain Sciences*, 21:1–14.

van Schaik, A., Fragnière, E., and Vittoz, E. (1996). Improved silicon cochlea using compatible lateral bipolar transistors. In Touretzky, D. S., Mozer, M. C., and Hasselmo, M. E., editors, *Advances in Neural Information Processing Systems*, volume 8, pages 671–677. MIT Press, Cambridge, MA.

van Valen, L. (1973). A new evolutionary law. *Evolution Theory*, 1:1–30.

Vassilev, V. K., Job, D., and Miller, J. F. (2000). Towards the automatic design of more efficient digital circuits. In Lohn, J., Stoica, A., Keymeulen, D., and Colombano, S., editors, *Proceedings of the Second NASA/DoD Workshop on Evolvable Hardware, EH-2000*, Palo Alto, CA, July 13-15, 2000, pages 151–160. IEEE Computer Society, Los Alamitos, CA.

Vaughan, E., Di Paolo, E. A., and Harvey, I. (2004a). The evolution of control and adaptation in a 3D powered passive dynamic walker. In Pollack, J., Bedau, M., Husbands, P., Ikegami, T., and Watson, R., editors, *Proceedings of the Ninth International Conference on the Simulation and Synthesis of Living Systems, Artificial Life IX*, pages 139–145. MIT Press, Cambridge, MA.

Vaughan, E., Di Paolo, E. A., and Harvey, I. (2004b). The tango of a load balancing biped. In Armada, M. and Gonzalez De Santos, P., editors, *Proceedings of the Seventh International Conference on Climbing and Walking Robots, CLAWAR*. Springer-Verlag, Berlin.

Verschure, P. F. M. J., Voegtlin, T., and Douglas, R. J. (2003). Environmentally mediated synergy between perception and behaviour in mobile robots. *Nature*, 425:620–624.

Villa, A. (2000). Empirical evidence about temporal structure in multi-unit recordings. In Miller, R., editor, *Time and the Brain*. Harwood Academic, Reading, UK.

Virasoro, M. (1989). Categorization and prosopagnosia. *Physics Report*, 24:301.

Vittoz, E. A. (1985). The design of high-performance analog circuits on digital CMOS chips. *IEEE Journal of Solid-State Circuits*, 20:657–665.

Vladimirescu, A. (1994). *The SPICE Book*. Wiley, New York.

Vogel, S. (2003). *Comparative Biomechanics. Life's Physical World*. Princeton University Press, Princeton, NJ.

Volterra, V. (1926). Variazioni e fluttuazioni del numero di individui in specie animali conviventi. *Memorie dell'Accademia dei Lincei*, 2:31–113. Translation in Chapman, R. N. (1931). *Animal Ecology*, pages 409–448. McGraw Hill, New York.

von Frisch, K. (1967). *The Dance Language and Orientation of Bees*. Harvard University Press, Cambridge, MA.

von Haller, B., Ijspeert, A., and Floreano, D. (2005). Co-evolution of structures and controllers for neubot underwater modular robots. In Capcarrere, M. S., Freitas, A. A., Bentley, P. J., Johnson, C. G., and Timmis, J., editors, *Eighth European Conference on Artificial Life (ECAL'2005)*. Springer-Verlag, Berlin.

von Neumann, J. (1958). *The Computer and the Brain*. Yale University Press, New Haven, CT. Reprint 2000.

von Neumann, J. (1961). The general and logical theory of automata. In Taub, A. H., editor, *John von Neumann: Collected Works.* Volume 5: *Design of Computers, Theory of Automata and Numerical Analysis*, chapter 9, pages 288–328. Pergamon Press, Oxford.

von Neumann, J. (1966). *Theory of Self-Reproducing Automata.* University of Illinois Press, Urbana, IL. Edited and completed by A.W. Burks.

von Toussaint, U., Gori, S., and Dose, V. (2006). Invariance priors for Bayesian feedforward neural networks. *Neural Networks*, 19(10):1550–1557.

Vose, M. D. (1991). Generalizing the notion of schema in genetic algorithms. *Artificial Intelligence*, 50:385–396.

Vukobratovic, M. and Borovac, B. (2004). Zero-moment point - Thirty five years of its life. *International Journal of Humanoid Robotics*, 1:157–173.

Wagner, A. (2005). *Robustness and Evolvability in Living Systems.* Princeton Studies in Complexity. Princeton University Press, Princeton , NJ.

Wagner, G. P. (2000). What is the promise of developmental evolution? Part 1: Why is developmental biology necessary to explain evolutionary innovations? *Journal of Experimental Zoology*, 288(2):95–98.

Wagner, G. P. (2001). What is the promise of developmental evolution? Part 2: A causal explanation of evolutionary innovations may be impossible. *Journal of Experimental Zoology*, 291(4):305–309.

Wagner, G. P. and Altenberg, L. (1996). Complex adaptations and the evolution of evolvability. *Evolution*, 50:967–976.

Wagner, G. P. and Larsson, H. C. (2003). What is the promise of developmental evolution? Part 3: The crucible of developmental evolution. *Journal of Experimental Zoology Part B: Molecular and Developmental Evolution*, 300B(1):1–4.

Waibel, M., Keller, L., and Floreano, D. (2008). Genetic team composition and level of selection in the evolution of cooperation. Technical report, Laboratory of Intelligent Systems, École Polytechnique Fédérale de Lausanne (EPFL), Lausanne, Switzerland.

Wakerly, J. F. (2001). *Digital Design: Principles and Practices*, 3rd edition. Prentice Hall, Upper Saddle River, NJ.

Wallace, A. R. (1870). *Natural Selection.* Macmillan, London.

Walter, W. G. (1950). An imitation of life. *Scientific American*, 182(5):42–45.

Walter, W. G. (1951). A machine that learns. *Scientific American*, 185(2):60–63.

Watt, R. (1991). *Understanding Vision.* Academic Press, London.

Watts, D. J. (2003). *Six Degrees: The Science of a Connected Age.* Norton, New York.

Webb, B. (2001). Can robots make good models of biological behaviour? *Behavioral and Brain Sciences*, 24:1033–1050.

Webb, B. (2002). Robots in invertebrate neuroscience. *Nature*, 417:359–363.

Webb, B. and Consi, T. R., editors (2001). *Biorobotics*. MIT Press, Cambridge, MA.

Webb, B. and Scutt, T. (2000). A simple latency-dependent spiking-neuron model of cricket phonotaxis. *Biological Cybernetics*, 82:247–269.

Weber, M. (1996). Evolutionary plasticity in prokaryotes: A Panglossian view. *Biology and Philosophy*, 11(1):67–88.

Wehner, R. (1997). The ant's celestial compass system: Spectral and polarization channels. In Lehrer, M., editor, *Orientation and Communication in Arthropods*, pages 145–185. Birkäuser, Basel, Switzerland.

Wehner, R., Michel, B., and Antonsen, P. (1996). Visual navigation in insects: Coupling of egocentric and geocentric information. *Journal of Experimental Biology*, 199(1):141–146.

Wehner, R. and Räber, F. (1979). Visual spatial memory in desert ants *Cataglyphis bicolor* (Hymenoptera: Formicidae). *Experientia*, 35:1569–1571.

Weng, J., McClelland, J., Pentland, A., Sporns, O., Stockman, I., Sur, M., and Thelen, E. (2001). Autonomous mental development by robots and animals. *Science*, 291:599–600.

Werbos, P. (1974). *Beyond Regression: New Tools for Prediction and Analysis of Behavioral Sciences*. PhD thesis, Harvard University, Cambridge, MA.

Werger, B. B. and Matarić (1996). Robotic food chains: Externalization of state and program for minimal-agent foraging. In Maes, P., Matarić, M., Meyer, J., Pollack, J., Roitblat, H., and Wilson, S., editors, *From Animals to Animats IV: Proceedings of the Fourth International Conference on Simulation of Adaptive Behavior*, pages 625–634. MIT Press, Cambridge, MA.

Werner, G. M. and Dyer, M. G. (1992). Evolution of communication in artificial organisms. In Langton, C., Taylor, C., Farmer, D., and Rasmussen, S., editors, *Artificial Life II*, pages 659–687. Addison Wesley, Redwood City, CA.

West-Eberhard, M. (2003). *Developmental Plasticity and Evolution*. Oxford University Press, Oxford.

Whiteside, T. C. D. and Samuel, G. D. (1970). Blur zone. *Nature*, 225:94–95.

Whitley, D. and Kauth, J. (1988). GENITOR: A different genetic algorithm. In *Proceedings of the Rocky Mountain Conference on Artificial Intelligence*, pages 118–130. Denver.

Whitley, D., Rana, S., and Heckendorn, R. B. (1998). The island model genetic algorithm: On separability, population size and convergence. *Journal of Computing and Information Technology*, 7(1):33–47.

Whitley, D., Starkweather, T., and Bogart, C. (1990). Genetic algorithms and neural networks: Optimizing connections and connectivity. *Parallel Computing*, 14:347–361.

Widrow, B. and Hoff, M. E. (1960). Adaptive switching circuits. In *Proceedings of the 1960 IRE WESCON Convention*, volume 4, pages 96–104. IRE, New York. Reprinted in Anderson and Rosenfeld (1988).

Wiemann, B. and Starnes, C. O. (1994). Coley's toxins, tumor necrosis factor and cancer research: A historical perspective. *Pharmacology & Therapeutics*, 64(3):529–564.

Wilke, C. O., Wang, J. L., Ofria, C., Lenski, R. E., and Adami, C. (2001). Evolution of digital organisms at high mutation rate leads to survival of the flattest. *Nature*, 412:331–333.

Williams, J. (1991). *Analog Circuit Design: Art, Science, and Personalities*. Butterworth-Heinemann, Boston.

Williams, J. (1995). *The Art and Science of Analog Circuit Design*. Butterworth-Heinemann, Boston.

Williamson, M. M. (1998). Neural control of rhythmic arm movements. *Neural Networks*, 11:1379–1394.

Wilson, E. O. (1971). *The Insect Societies*. Harvard University Press, Cambridge, MA.

Wilson, E. O. (1984). The relation between caste ratios and division of labour in the ant genus *Pheidole* (Hymenoptera: Formicidae). *Behavioural Ecology and Sociobiology*, 16:89–98.

Wilson, E. O. (2000). *Sociobiology. The New Synthesis*. Harvard University Press, Cambridge, MA. Twenty-fifth Anniversary Edition.

Wilson, S. W. (1987). Classifier systems and the animat problem. *Machine Learning*, 2:199–228.

Wilson, S. W. (1994). ZCS: A zeroth-level classifier system. *Evolutionary Computation*, 2(1):1–18.

Wilson, S. W. (1995). Classifier fitness based on accuracy. *Evolutionary Computation*, 3(2):149–176.

Wineberg, M. and Oppacher, F. (2003). The underlying similarity of diversity measures used in evolutionary computation. In Cantú-Paz, E., Foster, J. A., Deb, K., Davis, L., Roy, R., O'Reilly, U.-M., Beyer, H.-G., Standish, R. K., Kendall, G., Wilson, S. W., Harman, M., Wegener, J., Dasgupta, D., Potter, M. A., Schultz, A. C., Dowsland, K. A., Jonoska, N., and Miller, J. F., editors, *Proceedings of Genetic and Evolutionary Computation Conference - GECCO 2003*, pages 1493–1504. Springer-Verlag, Berlin.

Winfield, A. F. T., Harper, C. J., and Nembrini, J. (2004). Towards dependable swarms and a new discipline of swarm engineering. In Şahin, E. and Spears, W. M., editors, *Proceedings of the Swarm Robotics Workshop*, pages 126–142. Springer-Verlag, Heidelberg, Germany.

Wittlinger, M., Wehner, R., and Wolf, H. (2006). The ant odometer: Stepping on stilts and stumps. *Science*, 312(5782):1965–1967.

Wolf, D. E. (1999). Cellular automata for traffic simulations. *Physica A: Statistical Mechanics and Its Applications*, 263(1-4):438–451.

Wolfram, S. (1983). Statistical mechanics of cellular automata. *Reviews of Modern Physics*, 55(3):601–644.

Wolfram, S. (1984). Universality and complexity in cellular automata. *Physica D: Nonlinear Phenomena*, 10(1-2):1–35.

Wolfram, S. (2002). *A New Kind of Science*. Wolfram Media, Champaign, IL.

Wolpert, L. (1969). Positional information and the spatial pattern of cellular differentiation. *Journal of Theoretical Biology*, 25(1):1–47.

Wolpert, L. (1992). *The Unnatural Nature of Science*. Faber and Faber, London.

Wolpert, L. (2003). Relationships between development and evolution. In Kumar, S. and Bentley, P. J., editors, *On Growth, Form and Computers*, pages 47–63. Academic Press, London.

Wolpert, L., Jessell, T., Lawrence, P., Meyerowitz, E., Robertson, E., and Smith, J. (2007). *Principles of Development*, 3rd edition. Oxford University Press, Oxford.

Wolpert, L. and Szathmary, E. (2002). Multicellularity: Evolution and the egg. *Nature*, 420:745–745.

Worsch, T. (1999). Simulation of cellular automata. *Future Generation Computer Systems*, 16(2-3):157–170.

Wynne-Edwards, V. C. (1986). *Evolution through Group Selection*. Blackwell, Palo Alto, CA.

Yamauchi, B. M. and Beer, R. D. (1994). Integrating reactive, sequential, and learning behavior using dynamical neural networks. In Cliff, D., Husbands, P., Meyer, J., and Wilson, S. W., editors, *From Animals to Animats III: Proceedings of the Third International Conference on Simulation of Adaptive Behavior*, pages 382–391. MIT Press, Cambridge, MA.

Yao, X. (1993). A review of evolutionary artificial neural networks. *International Journal of Intelligent Systems*, 4:203–222.

Yao, X. and Higuchi, T. (1999). Promises and challenges of evolvable hardware. *IEEE Transactions on Systems, Man and Cybernetics, Part C*, 29(1):87–97.

Yim, M., Zhang, Y., Roufas, K., Duff, D., and Eldershaw, C. (2002). Connecting and disconnecting for chain self-reconfiguration with polybot. *IEEE/ASME Transactions on Mechatronics*, 7:442–451.

Zahavi, A. and Zahavi, A. (1997). *The Handicap Principle. A Missing Piece of Darwin's Puzzle*. Oxford University Press, New York.

Zampoglou, M., Szenher, M., and Webb, B. (2006). Adaptation of controllers for image-based homing. *Adaptive Behavior*, 14(4):381–399.

Zebulum, R. S., Pacheco, M. A. C., and Vellasco, M. M. B. R. (2002). *Evolutionary Electronics: Automatic Design of Electronic Circuits and Systems by Genetic Algorithms*. CRC Press, Boca Raton, FL.

Zeil, J., Hofmann, M. I., and Chahl, J. S. (2003). Catchment areas of panoramic snapshots in outdoor scenes. *Journal of the Optical Society of America A*, 20(3):450–469.

Zeil, J., Kelber, A., and Voss, R. (1996). Structure and function of learning flights in bees and wasps. *Journal of Experimental Biology*, 199:245–252.

Zhang, L. I., Tao, H. W., Holt, C. E., Harris, W. A., and Poo, M.-m. (1998). A critical window for cooperation and competition among developing retinotectal synapses. *Nature*, 395:37–44.

Zipser, D. and Andersen, R. A. (1988). A back-propagation programmed network that simulates response properties of a subset of posterior parietal neurons. *Nature*, 331:679–684.

Zuckerkandl, E. (1976). Gene control in eukaryotes and the C-value paradox: "Excess" DNA as an impediment to transcription of coding sequences. *Journal of Molecular Evolution*, 9:73–104.

Zufferey, J. and Floreano, D. (2006). Fly-inspired visual steering of an ultralight indoor aircraft. *IEEE Transactions on Robotics*, 22:137–146.

Zufferey, J., Floreano, D., van Leeuwen, M., and Merenda, T. (2002). Evolving vision-based flying robots. In Bülthoff, H., Lee, S.-W., Poggio, T., and Wallraven, C., editors, *Second International Workshop on Biologically Motivated Computer Vision (BMCV'2002)*, pages 592–600. Springer-Verlag, Berlin.

Zufferey, J., Guanella, A., Beyeler, A., and Floreano, D. (2006a). Flying over the reality gap: From simulated to real indoor airships. *Autonomous Robots*, 21(3):243–254.

Zufferey, J., Klaptocz, A., Beyeler, A., Nicoud, J., and Floreano, D. (2006b). A 10-gram microflyer for vision-based indoor navigation. In *IEEE/RSJ International Conference on Intelligent Robots and Systems (IROS'2006)*.

Zuse, K. (1982). The computing universe. *International Journal of Theoretical Physics*, 21(6-7):589–600.

Zykov, V., Mytilinaios, E., Adams, B., and Lipson, H. (2005). Self-reproducing machines. *Nature*, 435:163–164.

Index

abstraction, 45
    cost of, 47
action potential, 167
active vision, 473
actor, 236
adaptation, 173
    as a homeostatic process, 263
adaptive resonance theory (ART), 213
AES; *see* artificial evolutionary system
affinity maturation , 348
AGE; *see* analog genetic encoding
agent-based models, 148
AIS; *see* artificial immune system
algorithm
    clonal selection, 388
    negative selection, 384
altruism, 550
ALV; *see* average landmark vector
amino acid, 7
analog, 49
    genetic encoding (AGE), 84, 242
    immune recognition, 354
android, 435
animate vision, 406
ant system, 528
antibody, 363
antigen, 337
    -binding region, 363
    -presenting cell (APC), 346, 357
AO; *see* artificial ontogeny
APC; *see* antigen-presenting cell

apoptosis, 333, 343, 348
ART; *see* adaptive resonance theory
artificial
    developmental systems, 271
    evolutionary system (AES), 329
    immune system (AIS), 336
    immune system (ARTIS), 390
    intelligence, xi, 403, 585
    life, 141
    nose, 235
    ontogeny (AO), 330
ARTIS; *see* artificial immune system
Asimo, 431
attractor, 217
autoantigen, 338
autoassociative network, 190
autoimmune diseases, 351
automatic definition of neural subnet-
    works, 315
automaton, 107
autonomous robots, 399
average landmark vector (ALV), 443
axon, 167

backpropagation of error, 221
Baldwin effect, 165
Bayesian learning, 224
Bayes' theorem, 224
B cell, 362
    antigen receptor (BCR), 363
BCR; *see* B cell antigen receptor

behavior, 410
behavior-based robotics, 406
behaviorism, 400
bias, 182
bifurcation, 517
binary representation, 16
blind bulldozing, 537
blueprint, 9
boids, 532
building blocks, 37

CA; *see* cellular automaton
Cartesian genetic programming (CGP),
      67
catabolism, 333
cell differentiation, 321
cellular
    automaton (CA), 107
      agent-based models, 148
      asynchronous, 125
      binary, 107
      block rule, 129
      elementary, 118
      evolution of, 158
      forest fire model, 126
      Game of Life, 120
      granular media models, 150
      HPP gas model, 130
      in physics, 152
      Langton parameter, 156
      Margolus neighborhood, 129
      maze solver, 134
      mobile, 126
      multilayered, 131
      natural topology, 150
      nonhomogeneous, 124
      outer neighborhood, 109
      particle, 128
      partitioning, 129
      probabilistic, 126
      random number generation, 152
      space-time diagram, 109

    traffic model, 111
    transition rule, 107
    transition table, 107
    Wolfram classes, 154
    Wolfram's rule code, 118
  computer, 134
  encoding, 311
  evolutionary models, 146
  immunity, 354
  neural network (CNN), 131
  space, 102
  system
    analysis, 153
    boundary condition, 104
    homogeneous, 104
    initial condition, 106
    neighborhood, 102
    quiescent state, 102
    vs. rewriting system, 293
    seed, 106
    state, 102
    state set , 102
    stopping condition, 106
    synthesis, 153
    transition function, 103
central pattern generator (CPG)), 447
central tolerance, 343
chaining, 536
chromosome, 6
CIAO plot, 562
circuit
    analog, 49
    combinational, 58
    digital, 49
    sequential, 58
    sizing, 42
    topology, 42
clonal selection algorithm, 388
closure, 20
clustering, 538
CML; *see* coupled map lattice
CNN; *see* cellular neural network

coding region, 9
codon, 8
COG, 433
cognitive science, 400
competing conventions, 238
complement system, 354
complexity, 139, 146
complex system, 146
computational
    irreducibility, 137
    neuroscience, 163, 401
connection machine, 251
constraints, 87
constructal theory, 294
continuous time recurrent neural net-
        work (CTRNN), 184
copying, 457
cortical canonical circuit, 173
costimulation, 347
coupled map lattice (CML), 130
CPG; *see* central pattern generator
credit assignment problem, 236
    structural, 236
    temporal, 236
critic, 236
crossover, 26
    arithmetic, 27
    one-point, 27
    uniform, 27
CTRNN; *see* continuous time recurrent
        neural network
C-value, 12

danger
    model, 374
    signal, 343
delta rule, 220
dendrite, 167
dendritic cell, 355
density classification task, 137
deoxyribonucleic acid (DNA), 5
    genic, 9

nongenic, 12
depression
    long-term (LTD), 173
development, 269
developmental
    program, 310
    representation, 269
    robotics, 450
    system
        artificial, 271
        classification, 299
        inference problem, 296
        intrinsic, 328
digital, 49
    immune recognition, 354
diversity, 2, 31
DNA; *see* deoxyribonucleic acid
DNA computing, 36
dry adhesion, 427
dynamic encoding, 238

eater, 122
echo state network, 190
ecological approach, 401
elementary motion detectors (EMDs),
        421
elitism, 26
embodied cognitive science, 403
embodiment, 409
embryo, 271, 289
    mosaic , 289
    regulative , 289
EMD; *see* elementary motion detector
enactive perception, 402
EP; *see* evolutionary programming
epigenetic robotics, 449
ES; *see* evolutionary strategy
estimation-exploration algorithm, 510
evo-devo; *see* evolutionary developmen-
        tal biology
evolution
    artificial, 13

extrinsic, 53
intrinsic, 53
Lamarckian, 250
of learning, 511
of learning rules, 245
neutral, 4
steady-state, 34
unconstrained, 48
viability-based, 95
evolutionary
algorithm, 13
multiobjective, 87
developmental biology, 298
electronics, 42
measures, 29
programming (EP), 33
robotics, 461
strategy (ES), 33
system, 1
theory, 2
evolvability, 20, 271, 327
evolvable hardware, 58
excitable media, 146
extrinsic
developmental process, 328
evolution, 53

fabbers, 504
false negatives and positives, 390
fault
detection, 394
tolerance, 328, 394
feedback, 517
feedforward, 189
FET; *see* field-effect transistor
field-effect transistor (FET), 259
field-programmable gate array (FPGA),
61, 72, 253
finite state machine (FSM), 394
firing squad synchronization problem,
135
fitness

evaluation, 22
function, 22
graph, 30
landscape, 29
subjective, 23
FPGA; *see* field-programmable gate array
fractal, 279
Framsticks, 501
FSM; *see* finite state machine
functional genomics, 10
functions, 19

GA; *see* genetic algorithm
Game of Life, 120
gantry robot, 468
GasNet, 261
gene, 7
duplication, 12
expression, 7
expression threshold, 318
regulatory network, 10
generalized
delta rule, 221
shape, 376
generational replacement, 25
genetic
algorithm (GA), 33
encoding, 16
binary, 16
dynamic mapping, 21
implicit, 84
real-valued, 18
schematic-based, 80
tree-based, 19, 81
mutation, 10
operators, 26
programming (GP), 19, 33, 81, 307
closure, 20
functions, 19
sufficiency, 20
terminals, 19

genotype, 5
glider, 122
golem project, 503
good old-fashioned AI, 407
GP; *see* genetic programming
group selection, 553
growth, 271

Hall of Fame, 561
head-direction neurons, 481
Hebb, Donald, 173
Hebb's rule, 173, 196
heredity, 2
heterochrony, 270
hierarchical structure, 46
Hopfield network, 215
host, 335
HPP gas, 130
humoral immunity, 354
hypermutation; *see* somatic hypermu-
    tation

imitation
    active route, 459
    passive route, 459
immune
    detector, 337
    effector, 339
    gene library, 359
    memory, 348
    receptor
        affinity, 376
        permutation mask, 381
        recognition region, 376
        specificity, 376
    repertoire, 377
        hole, 377
    response
        primary, 351
        secondary, 351
    synapse, 347
    system, 336

artificial, 336
    danger model, 374
    network model, 375
    traditional model, 374
immunity
    cellular, 354
    danger model, 374
    humoral, 354
    traditional model, 374
immunological synapse, 347
immunotronics, 394
inclusive fitness, 550
independent noise, 508
indoor flying robot , 422
intelligence, 585
interneuron, 170
intrinsic
    developmental process, 328
    evolution, 53
inverse
    model, 457
    problem
        for cellular systems, 153
        for developmental systems, 296
island models, 33

Khepera, 462
kin selection, 551
Kismet, 435
knacks, 433
Koch curve, 279

Lamarckian evolution, 250
Langton's loop, 143
lattice Boltzmann model, 131
lattice-gas automaton, 128
leaky integrator, 185
learning
    plasticity-stability dilemma, 212
    supervised, 196
    unsupervised, 196
Lindenmayer system; *see* L-system

linearly separable, 181
liquid state machines, 191
logistic function, 179
Lotka-Volterra model, 547
LPD; *see* depression, long-term (LTD)
LPT; *see* potentiation, long-term (LTP)
L-system, 272
 alphabet, 272
 axiom, 272
 bracketed, 283
 context-free, 289
 context-sensitive, 289
 deterministic, 286
 evolutionary, 301
 graph interpretation, 276, 281
 parametric, 287
 production rule; *see* rewriting rule
 rewriting rule, 272
  predecessor, 272
  successor, 272
 stochastic, 286
 stopping condition, 273
 turtle graphics, 276
lymphatic system, 355

major histocompatibility complex (MHC),
  355, 358
 restriction , 360
master tournaments, 566
matrix rewriting, 305
MEA; *see* multielectrode array
meiosis, 7
memory units, 230
MHC; *see* major histocompatibility complex
minimal simulations, 508
mirror neurons, 458
mitosis, 6
modularity, 270
modular networks, 284
molecubes, 506
momentum, 229

morphogen, 320
morphogenesis, 319
morphogenetic system, 323
 robustness, 327
morphological computation, 434
motor babbling, 456
multielectrode array (MEA), 256
multiple objectives, 22
mutation, 10, 28
 deletion, 11
 insertion, 11
 inversion, 11
 recombination, 11
 substitution, 10

natural killer cell (NK), 358
NEAT; *see* neuroevolution of augmenting topologies
negative selection, 342
 algorithm, 384
neighborhood, 102
 radius, 103
NETtalk, 232
network transformation operations, 311
neural
 engineering, 163
 network (NN), 175
  architecture, 189
  feedforward, 189
  recurrent connections, 189
neuroevolution of augmenting topologies (NEAT), 241
neuromodulatory system, 451
neuromorphic engineering, 254
neuron
 activation level, 167
 axon, 167
 bias, 182
 dendrites, 167
 direct connections, 171
 excitatory, 169
 firing

rate, 170
time, 170
inhibitory, 169
model, 177
continuous time, 184
discrete-time, 183
dynamic, 183
integrate and fire, 187
McCulloch and Pitts, 178
spike response, 188
motor, 170
potential, 167
receptive field, 193
sensory, 169
spiking, 187
neurotransmitter, 167
long-range, 171
neutral
evolution, 4
path, 31
neutralist hypothesis, 12
new AI, 407
nitric oxide, 171
NK; *see* natural killer cell
NN; *see* neural network
normalization, 194
nucleotide, 6
nucleotypic hypothesis, 13
numerical methods, 133

objective, 22, 87
priority-ranked, 87
tradeoffs, 88
with targets, 88
Oja rule, 200
operational envelope, 91
optical character recognition (OCR), 234
optical systems, 251
overfitting, 227

PAMP; *see* pathogen-associated molecular pattern

parallel rewriting system, 273
Pareto dominance, 88
partial differential equation (PDE), 131, 133
passive walker, 431
pathogen, 335
-associated molecular pattern (PAMP), 339
pattern recognition receptor (PRR), 337
PBIL; *see* population-based incremental learning
PDE; *see* partial differential equation
periodic boundary condition, 111, 119, 127, 147
peripheral tolerance, 346
phagocyte, 354
phase transition, 113
phenotype, 5
plasticity, 316
pheromone robotics, 535
physical entrainment, 455
PLA; *see* programmable logic array
place cells, 481
plasticity-stability dilemma, 212
population, 2
-based incremental learning (PBIL), 35
adaptive, 35
diversity, 31
all-possible-pairs, 32
entropic, 32
initial, 21
size, 21
positional information theory, 320
positive selection, 342
potentiation
long-term (LTP), 173
primary immune response, 351
Prisoner's dilemma, 552
programmable logic array (PLA), 60, 63
progress, 3

protein, 5
  production, 8
PRR; *see* pattern recognition receptor
pseudogene, 12
pulse; *see* spike

reality gap, 507
recombination, 26
recurrent connections, 189
Red Queen effect, 560
regulatory region, 9
reinforcement learning, 235
  actor and critic, 236
replacement
  generational, 25
  steady-state, 34
reproduction, 23
response threshold mode, 523
reversibility, 123
rewriting system, 272
  vs. cellular system, 293
RHex, 424
ribonucleic acid (RNA), 6
RMS error surface, 445
RNA; *see* ribonucleic acid
robot, 399, 404
  autonomous, 399
  mobile, 404
  reconfigurable
    chain-type, 541
    lattice-type, 541
robustness, 270, 327
robust self-reproduction, 141

Sanger rule, 201
satisficing, 89
scalability, 270
schema, 37
  theory, 37
secondary immune response, 351
selection, 3, 23
  elitism, 26

natural, 3
negative, 342
positive, 342
pressure, 23
proportionate, 24
rank-based, 25
roulette wheel, 24
selective reproduction, 3
tournament, 25
truncated rank-based, 25
selectionist hypothesis, 12
selfish DNA hypothesis, 12
self/nonself discrimination, 374
self-organization, 145, 329, 334
self-organizing map, 206
  convergence stage, 210
  ordering stage, 210
self-protection paradox, 371
self-reproduction, 123
  robust, 141
self-similarity, 279, 282, 294
sensitivity level, 390
separation line, 180
SGOCE; *see* simple geometry-oriented
    cellular encoding
shape space, 376
sigmoid function, 179
signal encoding
  distributed, 192
  local, 192
  with spikes, 194
simple geometry-oriented cellular en-
    coding (SGOCE), 495
simulated annealing, 34
simultaneous localization and mapping
    (SLAM), 446
situatedness, 409
sizing; *see* circuit sizing
SLAM; *see* simultaneous localization and
    mapping
snapshot model, 441
sociable robots, 435

socially evocative, 435
somatic hypermutation, 348, 364
species adaptation genetic algorithm
    (SAGA), 95
spike, 169
    response model, 188
    time-dependent plasticity (STDP), 174,
    197
spin glass theory, 217
SRI; *see* syllable repetition interval
staged evolution, 464
standard competitive learning, 209
STDP; *see* spike time-dependent plas-
    ticity
Stickybot, 428
stigmergy, 518
subjective fitness, 23
subsumption architecture, 410
sufficiency, 20
superlinearity, 532
swarm intelligence, 523
syllable repetition interval (SRI), 438
synapse, 167

T cell, 358
    antigen receptor (TCR), 358
    cytotoxic (T), 358
    helper (T), 358
    monospecificity of , 359
TCR; *see* T cell antigen receptor
temporal difference reinforcement learn-
    ing, 237
terminals, 19
testing phase, 198
time delay neural network (TDNNs),
    230
tolerance
    central, 343
    peripheral, 346
topology; *see* circuit topology
training phase, 198
transcription, 8

transition rule
    null state quiescent, 109
    outer totalistic, 108
    symmetric, 109
    totalistic, 108
transposon, 12
transputers, 251
traveling salesman problem (TSP), 18
true imitation, 457
TSP; *see* traveling salesman problem
Turing, Alan
    morphogenetic model, 320
    pattern, 320
turtle graphics, 276

Ulam, Stanislaw, 117
uncanny valley, 437
unconstrained evolution, 48
universal
    computation, 137
    constructor, 141
unsupervised learning, 196

vaccination, 382
value system, 451
viability evolution, 95
von Neumann, John, 117, 138
    cellular model, 139
    kinematic model, 139
    self-reproducing automaton, 139, 295
        universal constructor, 141

Waalbot, 428
WABOT, 430
Walter, Grey, 404
weight sharing, 234
Whegs, 424

zero-moment point, 431
zygote, 269